Agricultural Law
in Scotland

Agricultural Law in Scotland

Sir Crispin Agnew of Lochnaw Bt, QC

Butterworths/Law Society of Scotland
Edinburgh
1996

United Kingdom	Butterworths, a Division of Reed Elsevier (UK) Ltd, 4 Hill Street, EDINBURGH EH2 3JZ and Halsbury House, 35 Chancery Lane, LONDON WC2A 1EL
Australia	Butterworths, SYDNEY, MELBOURNE, BRISBANE, ADELAIDE, PERTH, CANBERRA and HOBART
Canada	Butterworths Canada Ltd, TORONTO and VANCOUVER
Ireland	Butterworth (Ireland) Ltd, DUBLIN
Malaysia	Malayan Law Journal Sdn Bhd, KUALA LUMPUR
New Zealand	Butterworths of New Zealand Ltd, WELLINGTON and AUCKLAND
Singapore	Reed Elsevier (Singapore) Pte Ltd, SINGAPORE
South Africa	Butterworth Publishers (Ptd) Ltd, DURBAN
USA	Michie, CHARLOTTESVILLE, Virginia

Law Society of Scotland
26 Drumsheugh Gardens, EDINBURGH EH3 7YR

A CIP Catalogue record for this book is available from the British Library.

ISBN 0 406 11514 1

Typeset by Phoenix Photosetting, Chatham, Kent
Printed and bound in Great Britain by
Mackays of Chatham PLC, Chatham, Kent

Preface

In this book I have tried to deal with aspects of agricultural law, in particular as they relate to agricultural holdings, succession planning, game and deer, environmental and other constraints on agriculture, and CAP subsidies and quotas.

This book is presented in, I hope, a practical form, with 'Key points' and 'Time limits' highlighted at the start of each chapter, which I trust will provide a ready reference book for the busy practitioner.

I have also given a brief outline of the law of smallholdings, which raises its head from time to time in any agricultural context.

I would like to thank W & J Burness WS for kindly allowing me to reproduce their styles for an agricultural lease, post-lease agreement and limited partnership.

I would like to thank my publishers for their patience in waiting for a book which is now four years behind schedule, and for seeing my idea through to publication. I am also grateful to Butterworths for producing the tables of statutory materials and of cases, patiently correcting my many inconsistencies in citation, and for editing the text and appendices.

I have tried to state the law at 31 March 1996, although only brief mention could be made of developments post-31 December 1995, when the bulk of the book was submitted for typesetting.

Crispin Agnew of Lochnaw
15 April 1996

Contents

Table of statutes

Table of orders, rules and regulations

Table of cases

Abbreviations

Unless otherwise specified all references are to the Agricultural Holdings (Scotland) Act 1991 (the 1991 Act).

Statutes

1886 Act	Crofters Holdings (Scotland) Act 1886
1907 Act	Sheriff Courts (Scotland) Act 1907
1911 Act	Small Landholders (Scotland) Act 1911
1923 Act	Agricultural Holdings (Scotland) Act 1923
1931 Act	Small Landholders and Agricultural Holdings (Scotland) Act 1931
1946 Act	Hill Farming Act 1946
1948 Act	Agriculture (Scotland) Act 1948
1949 Act	Agricultural Holdings (Scotland) Act 1949
1958 Act	Agriculture Act 1958
1964 Act	Succession (Scotland) Act 1964
1973 Act	Land Compensation (Scotland) Act 1973
1976 Act	Agriculture (Miscellaneous Provisions) Act 1976
1981 Act	Wildlife and Countryside Act 1981
1983 Act	Agricultural Holdings (Amendment) (Scotland) Act 1983
1986 Act	Agriculture Act 1986
1991 Act	Agricultural Holdings (Scotland) Act 1991
DPQRs 1994	Dairy Produce Quotas Regulations 1994, SI 1994/672

Law reports

AC	Law Reports, Appeal Cases 1890–
All ER	All England Law Reports 1936–
Broun	Broun's Justiciary Reports 1842–45
CLY	Current Law Year Book 1947–
CMLR	Common Market Law Reports 1962–
Ch	Law Reports, Chancery Division 1890–
D	Dunlop's Session Cases 1838–62
Dow & L	Dowling and Lowndes' Practice Reports 1843–1849
ECJ	European Court of Justice
ECR	European Court of Justice Reports 1954–
EG	Estates Gazette 1858–
EGD	Estates Gazette Digest 1902–
EGLR	Estates Gazette Law Reports 1985–
Env LR	Environmental Law Reports
F	Fraser's Session Cases 1989–1906

GWD	Green's Weekly Digest 1986–
Hume	Hume's Decisions (Court of Session) 1781–1822
KB	Law Reports, King's Bench Division 1900–52
LGR	Knight's Local Government Reports 1902–
LJ	Law Journal newspaper 1866–1965
LJKB	Law Journal, King's Bench 1900–52
Lloyd's Rep	Lloyd's List Law Reports 1951–67; Lloyd's Law Reports 1968–
M	Macpherson's Session Cases 1862–73
Macq	Macqueen's House of Lords Reports 1851–65
Mor	Morison's Dictionary of Decisions (Court of Session) 1540–1808
P & CR	Planning and Compensation Reports 1949–67; Property and Compensation Reports 1968–
PLR	Planning Law Reports 1988–
Pat	Paton's House of Lords Appeal Cases 1726–1821
QB	Law Reports, Queen's Bench Division1891–1901, 1952–
R	Rettie's Session Cases 1873–98
S	P Shaw's Session Cases 1821–38
SC	Session Cases 1907–
SC (HL)	House of Lords Cases in Session Cases 1907–
SC (J)	Justiciary Cases in Session Cases 1907–
Sh Ct Rep	Sheriff Court Reports in Scottish Law Review 1885–1963
SLCR	Scottish Land Court Reports in Scottish Law Review (1913–63) and Scottish Land Court Reports 1982–
SLCR App	Appendix to the annual reports of the Scottish Land Court 1963-
SLR	Scottish Law Reporter 1865-1925
SLT (Land Ct)	Scottish Land Court Reports in Scots Law Times 1964–
SLT (Lands Tr)	Lands Tribunal for Scotland Reports in Scots Law Times 1971–
SLT (Sh Ct)	Sheriff Court Reports in Scots Law Times 1893–
Sol Jo	Solicitor's Journal 1856–
TLR	Times Law Reports 1884–1952
WLR	Weekly Law Reports 1953–

Textbooks

Connell	Connell on the Law of Agricultural Holdings in Scotland (T & T Clark, 6th edn, 1970 by C H Johnston and K M Campbell)
Duncan	A G M Duncan, notes to the Agricultural Holdings (Scotland) Act 1991 in Current Law Statutes
Erskine	J Erskine Institute of the Law of Scotland (2 vols, 1773)
Gill	B Gill QC The Law of Agricultural Holdings in Scotland (W Green, 2nd edn, 1990)
Halliday	J M Halliday Conveyancing Law and Practice in Scotland (W Green, 1985)
Macphail	I D Macphail Sheriff Court Practice (W Green, 1989)
Marshall	D Marshall Agricultural Outgoing Claims (1929)
Muir Watt	Agricultural Holdings (Sweet & Maxwell, 13th edn, 1987)
Paton and Cameron	G C H Paton and J G S Cameron Landlord and Tenant (W Green, 1967)
Rankine	Professor Sir J Rankine KC The Law of Leases in Scotland (W Green, 3rd edn, 1916)
Rodgers	C P Rodgers Agricultural Law (Butterworths, 1991)
Scammell and Densham	Scammell and Densham's Law of Agricultural Holdings (Butterworths, 7th edn, 1989 and First Supplement by H A C Densham)
Scott	J Scott The Law of Smallholdings in Scotland (W Green, 1933)
Williams	R G Williams Agricultural Valuations (The Estates Gazette, 2nd edn, 1991)

CHAPTER 1

Introduction

Key points

- It is not safe to assume that the 1991 Act did not change the law applicable under the 1949 Act (as amended).

- An agent, including one of the joint landlords or tenants acting as agent of the others, serving a notice should ensure that he has the particular authority to serve that notice.

- Where there is a change of landlord it is essential that the tenant is notified of the change and the name and address of the new landlord: s 84(3).

Time limits

- None. (See definitions of particular time limits below.)

THE AIM OF THE BOOK

Agricultural holdings and related legislation are a minefield for the unwary. The legislation requires precise notices and responses to particular events to be issued within tight time limits to end or to secure the protection of the Acts.

The aim of this book is to try to provide a ready guide to the actions and reactions that should be taken by the landlord or tenant on the happening of particular events. Further, it deals with some of the more general aspects of agricultural law, the Common Agricultural Policy and quotas with some consideration of related environmental legislation. There is a short chapter on landholdings and statutory small tenancies[1].

Each chapter, while trying to summarise the law, opens with a summary of 'Key points' and 'Time limits' which are applicable in the particular circumstances. The aim of these summaries is to provide the legal practitioner, factor, landlord or tenant with an *aide memoire* of what action requires to be taken and by when, the details of which can then be found in the body of the chapter.

THE SCOPE OF THE BOOK

The book aims to cover:

(1) The legal relationship of landlord and tenant of an agricultural holding which arises out of the 1991 Act. This involves a consideration of that Act and the related legislation, in particular:
a.   Ground Game Act 1880
b.   Sheriff Courts (Scotland) Act 1907
c.   Hill Farming Act 1946
d.   Agriculture (Scotland) Act 1948
e.   Opencast Coal Act 1958
f.   Succession (Scotland) Act 1964, ss 14 and 16
g.   Land Compensation (Scotland) Act 1973
h.   Agriculture Act 1986

(2) Aspects of agricultural law arising out of the Common Agricultural Policy and environmental legislation. This includes consideration of:
a.   The quota regimes
b.   Set-aside and arable area payments
c.   Conservation legislation, designation of special areas for protection, management agreements, nitrate sensitive areas, etc.

THE 1991 ACT

The Agricultural Holdings (Scotland) Act 1991 (the 1991 Act), which came into force on 25 September 1991, set out to consolidate the Agricultural

---

1   Landholders (Scotland) Acts 1886 to 1931.

Holdings (Scotland) Act 1949 (the 1949 Act) and the numerous amendments and supplements to that Act introduced over the years[1].

The Act also incorporated the provisions relating to sheep stock valuations previously found in the Sheep Stocks Valuations (Scotland) Act 1937 and the Hill Farming Act 1946[2]. It is surprising that the Act did not seek to incorporate the few remaining provisions of the Agriculture (Scotland) Act 1948[3].

While a consolidation Act is supposed to consolidate and perhaps clarify the law, rather than to amend it, the draftsman of the 1991 Act appears, perhaps unwittingly in some cases, to have made some changes to the law[4]. These clarifications and changes are noted in the text.

An omission from the 1991 Act[5] of particular note is the Agricultural Holdings (Scotland) Act 1949, s 100.

OBJECTIVES OF THE AGRICULTURAL HOLDINGS LEGISLATION

It has long been recognised that the agricultural holdings legislation was passed not only in the interests of the tenant, but also in the interests of the wider community[6].

The Agricultural Holdings (Scotland) Act 1883 was introduced to secure to the tenant the right to compensation for improvements at his waygoing as he was not entitled to compensation at common law. It was also designed to encourage the tenant to maintain the fertility of the land in the last years of his lease.

The Agriculture (Scotland) Act 1948[7] gave tenants and their successors security of tenure for the first time. The security of tenure given to tenants and their successors was for the protection of the nation itself and arose out of the important place that agriculture had played in feeding the nation during the battle of the Atlantic.

The thinking behind this policy is highlighted by the speeches in *Johnson v Moreton*[8]:

"During the 1939–45 war, the submarine menace was such that it would have been virtually impossible to import into this country any more goods vital for our survival than we in fact did. Accordingly, it is extremely doubtful whether we could have survived had it not been for the food produced by our own farms. Even in 1947 when the Agriculture Act of that year was passed, food rationing was still in existence. It must have been clear to all it was then and always would be of vital importance, both to the

---

1 Cf Succession (Scotland) Act 1964, Agriculture Act 1958, Agriculture (Miscellaneous Provisions) Act 1968 and Agricultural Holdings (Amendment) (Scotland) Act 1983.
2 As amended by the Agriculture (Miscellaneous Provisions) Act 1963 and the Hill Farming (Variation of Second Schedule) (Scotland) Order 1986, SI 1986/1823.
3 Eg the provisions regarding the rules of good estate management and good husbandry.
4 See A G M Duncan 'The Agricultural Holdings (Scotland) Act 1991' 1992 SLT (News) 1 (*Duncan*) where some of the alterations are noted. Others are noted in this book.
5 But not omitted from the Agricultural Holdings Act 1986 - see s 97.
6 *Earl of Galloway v McClelland* 1915 SC 1062 at 1099, 1100 per Lord Salveson; *Turnbull v Millar* 1942 SC 521 at 532 per LJC; *Findlay v Munro* 1917 SC 419 at 428, 429.
7 Replaced by the Agricultural Holdings (Scotland) Act 1949.
8 [1980] AC 37.

national economy and security, that the level of production and the efficiency of our farms should be maintained and improved. This could be achieved only by the skill and hard work of our farmers and the amount of their earnings, which they were prepared to plough back into the land from which those earnings had been derived. A very large proportion of those farmers were tenant farmers. They were tenants because they did not have the necessary capital to buy land or they could not find any land which they wanted that was for sale, or for sale at a price which they could afford. In spite of ss 23 and 25 of the 1923 Act which had put them in a somewhat better position than did the common law, the sword of Damocles was always hanging over their heads. If they were tenants for a term of years, they might receive an effective notice to quit on the date when the term expired – and this term was rarely for more and usually for less than ten years. If they were tenants from year to year, and very many of them were, they might in any year receive an effective notice to quit at the end of the next ensuing year. Accordingly there was no great inducement for these farmers to work as hard as they could, still less to plough money back into land which they knew they might well lose sooner or later.

The security of tenure which tenant farmers were accorded by the 1947 Act was not only for their own protection as an important section of the public, nor only for the protection of the weak against the strong; it was for the protection of the nation itself. This is why s 31(1) of the 1947 Act, reproduced by s 24(1) of the 1948 Act, gave tenant farmers the option to which I have referred and made any agreement to the contrary void. If any clause such as cl 27 was valid landlords might well insist on a similar clause being introduced into every lease; and prospective tenants, having no money with which to buy the land they wanted to farm, would, in reality, have had little choice but to agree. Accordingly, if cl 27 is enforceable the security of tenure which Parliament clearly intended to confer, and did confer on tenant farmers for the public good would have become a dead letter.[1]'

Lord Hailsham of Marylebone put the policy in its historical context[2]:

'The first is the nature of farming itself. At least since the 1880s successive Parliaments have considered the fertility of the land and soil of England and the proper farming of it as something more than a private interest. Fertility is not something built up as a result of a mere six months' activity on the part of the cultivator, which was all the period of notice given by the common law to the individual farming tenant, by whom in the main the land of England was cultivated then, as now, mainly under a yearly tenancy. It takes years (sometimes generations) of patient and self-abnegating toil and investment to put heart into soil, to develop and gain the advantage of suitable rotations of crops, and to provide proper drains, hedges and ditches. Even to build up a herd of dairy cattle, between whose conception and first lactation at least three years must elapse, takes time and planning, whilst to disperse the work of a lifetime of careful breeding is but the task of an afternoon by a qualified auctioneer. Even within the space of a single year the interval between seed time and harvest, between expenditure and return, with all the diverse dangers and chances of weather, pest or benignity of climate is sufficient to put an impecunious but honest cultivator at risk without adding to his problems any uncertainty as to his next year's tenure. At first Parliament was concerned simply with compensation for cultivation, manuring and improvement. But it never regarded these as matters simply for private contract, or something wholly unconnected with any public interest. From the first, Parliament was concerned with the management of the soil, the land of England which had grown gradually into its present fertility by the toil of centuries of husbandmen and estate owners. By the late 1920s Parliament similarly concerned itself with the length of notice to which the yearly tenant was entitled. Such provisions are now to be found in ss 3 and 23 of the 1948 Act. But they date from this time. In 1947 a new and momentous step was taken. The land-

1 At 52C per Lord Salmon.
2 At 59A.

lord's notice to quit, save in certain specified instances, was at the option of the tenant to be subject to consent, at first of the Minister, but latterly of quasi judicial tribunal, the agricultural land tribunal, whose jurisdiction cl 27 of this lease seeks by its express terms to eliminate and oust. Even the consent of the agricultural land tribunal is carefully regulated by s 25 of the 1948 Act (consolidating and amending the 1947 provisions). The circumstances in which its consent may be accorded are thus defined and limited by objective and justiciable criteria. These are not simple matters of private contracts from which the landlord can stipulate that the tenant can deprive himself as if it were a "jus pro se introductum". It is a public interest introduced for the sake of the soil and husbandry of England of which both landlord and tenant are in a moral, though not of course a legal, sense the trustees for posterity. Silence is not an argument, particularly when the words are prima facie mandatory, for excusing a term in a contract introduced for the purpose of annulling the protection given to the tenant by s 24.'

An unfortunate side-effect of the security of tenure provisions of the 1949 Act and its sister legislation in England has been the reluctance of landlords to re-let farms as they have become vacant[1].

As a result of discussions in England between landlord and tenant interests the Agricultural Tenancies Act 1995 has been passed. This Act effectively provides for the granting of agricultural tenancies for a fixed term with no rights of succession. Existing tenancies continue with their security of tenure. At present there are no proposals for similar legislation in Scotland, because the Scottish landlord and tenant interests cannot agree on appropriate legislation.

In view of the agricultural surpluses generated within the EU, the policy behind the Agricultural Holdings Acts, formulated at a time of wartime food shortages, appears to be no longer entirely apposite.

GENERAL POINTS

The 1991 Act and related legislation include a number of general points applicable in a whole range of circumstances, which are noted here for convenience.

(a) Time limits

Where a notice has to be served or money has to be paid within a particular period, in general the notice or money must be in the hands of that person within the time limit[2]. Placing the notice or money in the post within the time limit does not in general comply with the time limit.

The meanings of the time limits that appear in the 1991 Act are as follows:

from year to year – In general this phrase means from the date of the start of the tenancy to the anniversary of that date (ie 28 May to 28 May next). Where a

---

1 The amount of farmland leased to tenants has fallen from 90 to 35 per cent this century: Smiths Gore's Newsletter, December 1994.
2 *Scott v Scott* 1927 SLT (Sh Ct) 6; no extension is allowed because a day is a Sunday.

tenancy takes effect as a lease from year to year[1] the lease runs from the date of entry to the year ending the day before the anniversary (ie 1 March to 28 February next)[2].

within [one] month; later than [one] month; before the expiry of [2] months –

'Reference to a month in a statute is to be understood as a calendar month . . . in calculating the period that has elapsed after the occurrence of the specified event such as giving notice, the day on which the event occurs is excluded from the reckoning. It is equally well established . . . that when the period is a month or a specified number of months after the giving of the notice, the general rule is that the period ends on the corresponding date in the subsequent month, ie the day of that month that bears the same number as the day of the earlier month'[3].

Where there is no 'same number' date the month ends 'on the last day of the month in which the notice expires'[4] (ie service on 31 May – the one month then ends on 30 June as there is no 31 June).

The period of time expires at any time up to midnight on the final day of the period[5].

not less than [3] months – See 'within [one] month' above for meaning of 'month'. The first and the last day must be included in the calculation[5].

within [21] days after – 'after' means that the day of the event is not counted[5]. The act can be performed timeously at any time up to midnight of the final day[5].

before the [termination of the tenancy], before the expiry of [2] months – Where an act has to be done not later than a given period before the happening of an event, the day on which the event happens is excluded[6]. Where an act has to be done 'before the termination of the tenancy' the day of the termination is excluded[7].

(b) Term dates

Whitsunday and Martinmas are now defined by the Term and Quarter Days (Scotland) Act 1990 to mean 28 May and 28 November respectively.

Those dates apply to leases and other agreements entered into before the passing of the Act, where there is reference to Whitsunday and Martinmas

1 Eg under the 1991 Act, ss 2(1) or 3.
2 *Morrison's Exrs v Rendall* 1989 SLT (Land Ct) 89; *McGill v Bichan* 1970 SLCR App 122. Cf D C Coull 'Termination Date in a Notice to Quit' 1989 SLT (News) 431.
3 *Dodds v Walker* [1981] 1 WLR 1027 at 1029A per Lord Diplock.
4 *Dodds v Walker* above at 1029F per Lord Diplock.
5 22 *Stair Memorial Encyclopaedia* para 826.
6 *Craparayoti & Co Ltd v Comptoir Andre & Cie SA* [1972] 1 Lloyd's Rep 139, CA.
7 The day of the termination of the tenancy may be specified in the lease (ie Whitsunday or Martinmas) or it may be the anniversary date, the day before the date of entry in a lease running from year to year; see *Morrison's Exrs v Rendall* 1989 SLT (Land Ct) 89.

without further specification[1], unless the parties made application to the sheriff for a declaration that a specific date applied in lieu[2]. It is unlikely that this provision will have been invoked in respect of agricultural leases and agreements.

(c) Agents

Section 85(5) provides:

'(5) anything which by or under this Act is required or authorised to be done by, to or in respect of the landlord or tenant of an agricultural holding may be done by, to or in respect of any agent of the landlord or of the tenant.'

No difficulty arises where the landlord or tenant authorise their agent to carry out an act, such as the service of a particular notice.

Difficulties can arise where there are joint landlords or tenants. In those circumstances the agent serving the document must ensure that he has the authority of all the joint landlords or tenants[3].

Although the agency does not have to appear on the face of the notice, good practice suggests that the agency should be stated specifically on the notice. It is a matter of proof whether the person serving the notice had the particular authority[4]. A general agent such as a solicitor or factor probably does not have specific authority to serve a notice to quit, unless authorised to grant leases[5].

Subsequent adoption of a notice served without authority is probably not sufficient to validate the notice[6].

A landlord or tenant should be slow to serve a notice on an agent, without first ascertaining that the agent has authority to accept service of the particular notice. While a person may have been agent for certain purposes he might not have authority for the purpose of accepting the particular notice. If he does not have authority, the notice will not have been effectively served.

A landlord or tenant (or their agents) would be wise to serve particular notices on the landlord or tenant personally with a copy to the purported agent.

(d) Notices

Section 84(1) provides:

1 Term and Quarter Days (Scotland) Act 1990, s 1(4).
2 Ibid, s 1(5).
3 Cf *Graham v Stirling* 1922 SC 90 (notice at instance of one of two joint tenants); *Walker v Hendry* 1925 SC 855 (notice on the face of it by husband alone where farm owned *pro indiviso* with wife); *Combey v Gumbrill* [1990] 2 EGLR 7 (divorced wife held to have had husband's authority to serve notice requiring arbitration by virtue of the orders made in the divorce proceedings); *Divall v Harrison* [1992] 2 EGLR 64 (notice failed to identify landlord and after proof held that solicitors had purported to act on instructions of person not the landlord).
4 *Laing v Provincial Homes Investment Co* 1909 SC 812; *Walker v Hendry* 1925 SC 855 (notice borne to be at instance of husband only, where farm owned *pro indiviso* with wife; held competent to prove by parole evidence that husband had wife's authority).
5 Cf *Danish Dairy Co v Gillespie* 1922 SC 656.
6 *Walker v Hendry* 1925 SC 855 at 861 per Lord Constable. Cf Requirements of Writing (Scotland) Act 1995, s 1(5).

'(1) Any notice or other document required or authorised by or under this Act to be given to or served on any person shall be duly given or served if delivered to him, or left at his proper address, or sent to him by registered post or recorded delivery.[1]

(2) Any such document required or authorised to be given or served on an incorporated company or body shall be duly given or served if it is delivered to or sent by registered post or recorded delivery to the registered office of the company or body.[2]'

The proper address of a person is 'the last known address of the person in question'[3]. The proper address of an incorporated company or body is 'the registered or principal office of the company or body'[3].

Where the notice is left at the proper address, it matters not if the party does not receive it. A notice left at the proper address of a person:

'must be left there in a proper way; that is to say, in a manner which a reasonable person, minded to bring the document to the attention of the person to whom the notice is addressed, would adopt.'[4]

The notice is served on the date of delivery[5].

Until the tenant receives notice that the original landlord has ceased to be the landlord and has received notice of the name and address of the new landlord, any notice by the tenant may be validly served on the original landlord[6].

A legatee or an acquirer of a lease has to serve his notice[7] on the actual landlord and cannot rely on the failure of the original landlord to give notice of change of ownership[8].

Similar provisions are made in respect of claims for compensation for milk quotas[9].

Notices to quit are an exception to the 1991 Act, s 84(1) and require to be served in terms of the special provisions of the 1991 Act, s 21(5)[10].

(e) Consents and agreements

Where the Act requires any consent or agreement to be in writing the writing can be an informal deed provided it is signed by the parties or their agents[11].

---

1 Service by ordinary post is competent, but the server will have to prove the notice was received, which is difficult; *Sharpley v Manley* [1942] 1 KB 217.
2 Cf Agriculture Act 1986 (1986 Act), Sch 2, para 14 in relation to notices claiming compensation for milk quota.
3 1991 Act, s 84(3).
4 *Lord Newborough v Jones* [1974] 3 All ER 17 at 19d (a notice pushed under the back door of the farm house by the landlord and found months later by the tenant under the linoleum was held to be a good notice); see *Datnow v Jones* [1985] 2 EGLR 1 (notice put through back-door letter box of farm but not found by tenant held to be good service).
5 *Scott v Scott* 1927 SLT (Sh Ct) 6; *Lord Newborough v Jones* [1974] 3 All ER 17.
6 1991 Act, s 83(4).
7 Under ibid, ss 11 and 12.
8 Ibid, s 84(4).
9 1986 Act, Sch 2, para 14.
10 See chap 15.
11 1991 Act, s 78; cf *Connell* p 192.

(f) 1949 Act, s 100

Section 100[1] reserved the other powers, rights and remedies of a landlord, tenant or other person, not expressly otherwise provided for in the 1949 Act[2].

Section 100 preserved the landlord's right to go against the tenant during the currency of the lease for any breach of a term of the lease[3] and could be used to enable the landlord to require the tenant, after the expiry of the lease, to re-enter to carry out any unfulfilled obligations incurred by the tenant on the termination of the lease[4]. It was also relied upon to found and justify a claim for damages for dilapidation and deterioration during the currency of a lease, where it was argued that the equivalent provision (s 45) of the 1991 Act restricted such a claim to the termination of the lease[5].

This section was not carried forward into the 1991 Act[6]. It is not clear what the effect of this will be. It is arguable that the section was unnecessary, in that an Act is not presumed to take away rights unless it expressly so provides.

1 1949 Act, s 100 provided: 'Subject to the provisions of subsection (2) of section 12 and subsection (1) of section 68 of this Act in particular, and to any other provision of this Act which otherwise expressly provides, nothing in this Act shall prejudicially affect any power, right or remedy of a landlord, tenant or other person, vested in or exercisable by him by virtue of any other Act or law, or under any custom of the country, or otherwise, in respect of a lease or other contract, or of any improvements, deteriorations, away-going crops, fixtures, tax, rate, teind, rent or other thing'.
2 See *Gill* para 439.
3 *Gill* para 554.
4 *Gill* para 439; *Coventry v British Gas Corp* (15 August 1984, unreported) Outer House.
5 Cf *Kent v Conniff* [1953] 1 QB 361, which held that such a claim was competent under the Agricultural Holdings Act 1948 because of the reservation of rights in s 100. *Duncan*, note to s 45(3) in *Current Law Statutes* suggests that such a claim might no longer be competent as s 100 has not been re-enacted.
6 See *Duncan* above.

CHAPTER 2

The holding

Key points

- Check that the holding is an agricultural holding within the terms of the 1991 Act, s 1(1).

- Confirm that the holding is not (a) a croft in the crofting counties, or (b) a landholding in terms of the Small Landholders (Scotland) Act 1911 outwith the crofting counties.

- Check that even if the holding was originally an agricultural holding, that it has not ceased to be governed by the Act; alternatively if it was not subject to the Act, if it has become subject.

Time limits

- None

THE HOLDING

An 'agricultural holding'[1] is defined by the 1991 Act[2], s 1(1) as:

'The aggregate of the agricultural land comprised in a lease, not being a lease under which the land is let to the tenant during his continuance in any office, appointment or employment held under the Landlord.[3]'

'Agricultural land' for the purposes of the 1991 Act, ss 1 and 2 is defined to mean[4]:

'. . . land used for agriculture for the purposes of a trade or business, and includes any other land which, by virtue of a designation of the Secretary of State under section 86(1) of the Agriculture (Scotland) Act 1948[5], is agricultural land within the meaning of that act.'

Although the definition includes land designated by the Secretary of State, this power has not been used for many years[6].

'Agriculture' and 'livestock' are defined to mean:

'"agriculture" includes horticulture; fruit growing; seed growing; berry farming; livestock breeding and keeping; the use of land as grazing land, meadow land, osier land, market gardens and nursery grounds; and the use of land for woodlands where that use is ancillary to the farming of land for other agricultural purposes: and "agricultural" shall be construed accordingly; . . .

"livestock" includes any creature kept for the production of food, wool, skin or fur, or for the purpose of its use in the farming of land;[7]'

---

1 Except for the purposes of the 1991 Act, ss 67-71 (sheep stock valuation).
2 While the 1991 Act definition is the same as the definition in the Agriculture Act 1948, the definition in the Agricultural Holdings Act 1986 is different; see *Scammell and Densham* pp 24–26 for a consideration of the effect of the change of definition.
3 The definition of holding under the 1883 to 1923 Acts was '. . . any piece of land held by a tenant which is either wholly agricultural or wholly pastoral or in part agricultural and as to the residue pastoral'. This definition was considered in *McNeill v Duke of Hamilton's Trs* 1918 SC 221 under the Small Landholders (Scotland) Act 1911 and followed, erroneously, in *McGhie v Lang* 1953 SLCR 34 by the Land Court which failed to take into account (1) that *McNeill* related to a small holding under the Small Landholders (Scotland) Act 1911 and (2) the changed definition in the 1949 Act.
4 1991 Act, s 1(2).
5 1948 Act, s 86(1) provides:

'(1) In this Act the expression "agricultural land" means land used for agriculture which is so used for the purposes of a trade or business, or which is designated by the Secretary of State for the purpose of this subsection, and includes any land so designated as land which in the opinion of the Secretary of State ought to be used for agriculture:
    Provided that no designation under this subsection shall extend-
    (a) to land used as pleasure grounds, private gardens or allotment gardens, or
    (b) to land kept or preserved mainly or exclusively for the purpose of sport or recreation, except where the Secretary of State is satisfied that its use for agriculture would not be inconsistent with its use for the said purposes and it was so stated in the designation.'

The definitions of 'agriculture' and 'livestock' in the 1948 Act, s 86(3) are the same as those in the 1991 Act.
6 *Gill* para 14. It appears never to have been used in England - *Scammell and Densham* p 24.
7 1991 Act, s 85(1).

The definition of livestock is again inclusive and not exclusive. The definition is *habile* to include birds[1], fish[2] and horses, if kept for the production of food.

Section 85(3) of the 1991 Act provides that:

'(3) References in this Act to the farming of land includes references to the carrying on in relation to the land of any agricultural activity.'

In order to qualify as an agricultural holding for security of tenure under the 1991 Act there must be a lease of agricultural land lawfully used under the lease for agriculture[3].

Land includes buildings and can be a building by itself provided the building is used for agriculture[4].

The holding, in terms of the Act, can relate to only one tenancy. Thus if land is held separately under two tenancies but farmed together as one holding, the combined lands do not become one 'agricultural holding' to which the 1991 Act applies. There may in fact be two separate holdings in the possession of one person. The facts of the case might establish that the parties have agreed that two tenancies should be consolidated into one lease[5], thereby making the two separate holdings into one holding. This is not necessarily achieved by agreement that a single rent will apply to the two tenancies[6].

The definitions, which are continued from the 1949 Act, provide the basis upon which a determination can be made as to whether or not a holding is an agricultural holding.

No minimum size is specified for an agricultural holding[7].

In order to have qualified as a croft or a landholding in 1886 or 1912, the holding must in the first instance have qualified as an agricultural holding. It is therefore important to confirm, in the case of tenancies predating 1912 and particularly of 50 acres or less, that the holding is not a holding to which either the Crofters Acts or the Landholders Acts apply[8].

In determining whether or not a holding is an agricultural holding:

'one must look at the substance of the matter and see whether, as a matter of substance, the land comprised in the tenancy, taken as a whole, is an Agricultural Holding. If it is, then the whole of it is entitled to the protection of the Act. If it is not, then none of it is so entitled.[9]'

---

1 *Earl of Normanton v Giles* [1980] 1 WLR 28 (held that pheasants were not livestock under the Rent (Agriculture) Act 1976, since they were not kept for the production of food).
2 *Jones v Bateman* (1974) 232 EG 1392; cf *Cresswell v British Oxygen Co Ltd* [1980] 1 WLR 1556 (a rating case).
3 See chap 3.
4 *Blackmore v Butler* [1954] 2 QB 171, [1954] 2 All ER 403 (held that a cottage let for the purpose of housing agricultural workers employed on neighbouring land was 'agricultural land'); *Godfrey v Waite* (1951) 157 EG 582; *Adsett v Heath* (1951) 95 Sol Jo 620; *Hasell v McAulay* [1952] 2 All ER 825. Cf *Barr v Strang* 1935 SLT (Sh Ct) 10 (a decision under a different definition in the 1923 Act, to the contrary effect).
5 *Blackmore v Butler* above.
6 *Murray v Nisbet* 1967 SLT (Land Ct) 14.
7 *Malcolm v M'Dougall* 1916 SC 283 (1 rood of ground and a share in common grazings); *Stevens v Sedgeman* [1951] 2 KB 434 (half-acre field).
8 See chap 22.
9 *Howkins v Jardine* [1951] 1 KB 614 at 628 per Jenkins LJ; cf *Taylor v Earl of Moray* (1892) 19 R 399.

There cannot be severance or an excision between part of the land which may be used for agriculture and part that is not so used[1].

If the tenancy is principally non-agricultural in character then the whole subjects, including any part that may be in agricultural use, fall outwith the protection of the Act except in regard to limited rights of compensation[2]. Conversely where the principal use is agriculture the whole holding is an agricultural holding even if some part is used for another purpose.

An agricultural tenant may not use the holding for non-agricultural purposes, which amounts to an inversion of possession, unless the lease permits the use[3].

The Act does not apply to the lease of land which is 'let to the tenant during his continuance in any office, appointment or employment held under the Landlord'[4]. It is a question of fact whether the contract is a tenancy or a service occupancy[5]. In *Budge v Gunn*[6] the Land Court said:

'The point of the exclusion . . . is that the letting of the holding of the tenant shall have been limited to, or conditional on, his continuance in the employment of the Landlord.'

The lease must authorise the use of the land for agricultural purposes because mere use without warrant in terms of the lease does not give rise to a protection under the Act, unless the use is consented to or acquiesced in by the landlord[7].

It is a question of fact whether the land is used for agriculture. The agricultural land must be used for a 'trade or business'[8].

In *McGill v Bichan*[9] the Land Court said:

'. . . occupation of an agricultural holding involves a "trade or business" and must therefore be primarily directed towards an economic end and a profitable return taking one year with another; but not, of course, short-term exploitation of the land, which would be contrary to Rule 1 [of the Rules of Good Husbandry].'

The 'trade or business' must not be a sideline[10].

A requirement that the land should be used for the purposes 'of a trade or

---

1 *Howkins v Jardine* above not following the dicta in *Dunn v Fidoe* [1950] 2 All ER 685 which suggested the possibility of limited severance. Cf *McNeill v Duke of Hamilton's Trs* 1918 SC 221 to the contrary effect in relation to a landholding, which decision does not apply to an agricultural holding as defined by the 1991 Act - see *Gill* para 18.
2 1991 Act, s 48(3).
3 *Rankine* p 236; *Duke of Argyll v McArthur* (1861) 23 D 1236; *Sinclair v Secretary of State* 1966 SLT (Land Ct) 2; *BTC v Forsyth* 1963 SLT (Sh Ct) 32; *Cayzer v Hamilton (No 2)* 1995 SLCR 13. Cf Crofters (Scotland) Act 1993, Sch 2, para 3, which authorises a crofter to use the croft for 'subsidiary or auxiliary occupations'.
4 1991 Act, s 1(1).
5 *Dunbar's Trs v Bruce* (1900) 3 F 137; *MacGregor v Dunnett* 1949 SC 510.
6 1925 SLCR 74 (a case under the Landholders Acts).
7 *Muir Watt* p 6; *Kempe v Dillon-Trenchard* [1951] EGD 13 (letting of 9 acres of land with dwellinghouse with restriction against use 'otherwise than as a private dwellinghouse' and no agricultural land). Cf *Iredell v Brocklehurst* [1950] EGD 15 (let excluding 'trade or business' was construed to mean any trade or business except agriculture); and *R v Agricultural Land Tribunal (SE Province) ex p Palmer* (1954) 163 EG 106.
8 1991 Act, s 1(2); see *Stroud's Judicial Dictionary* sub nom 'trade or business'.
9 1982 SLCR 33 at 40.
10 *Hickson and Welch v Cann* (1977) 40 P & CR 218 (breeding and fattening of pigs as a sideline held not to be a trade or business).

business' does not require that the trade or business be an agricultural trade provided the land is used for agriculture[1].

'Agriculture' as defined in s 93(1) which, while it is wide, is not exhaustive, in that the definition provides that the activities specified are only included and are not exclusive[2].

Definitions of what amounts to agriculture have tended to relate to the type of activities undertaken or the nature of the livestock run on the holding. Some guidance can be attained from similar definitions used in valuation[3] and planning cases[4].

Land used for growing crops and weeds for testing weedkillers has been held not to be agriculture[5], as has land used for stud farms[6], or pheasant rearing for sport[7]. A fish farm may be an agricultural holding[8].

'Livestock' is defined to include any creature kept for the production of food, wools, skin or fur or for the purpose of its use in the farming of land. While the definition envisages the keeping of livestock for purposes other than the production of food, such as the production of wool, skins or fur, where the purpose for which the livestock is kept, which might normally be agricultural livestock, ceases to be a commercial use for the farming of land then that purpose might cease to be agricultural under the Act[9].

The reference to 'the use of lands for woodlands where that use is ancillary' relates to the use of the woodland as shelter belts which are 'fixed equipment'[10].

Market gardens are included in the definition of agriculture[11].

Agricultural land does not cease to be part of an agricultural holding just because it is being used for opencast coal mining[12].

CHANGE OF STATUS

An agricultural holding can lose its status as such if the agricultural use is wholly or substantially abandoned[13].

The test applied appears to be:

1 *Rutherford v Maurer* [1962] 1 QB 16 (grazing by horses used in a riding school). Criticised by *Gill* para 24 as a decision 'contrary to the plain purpose of the Act'; followed in *Crawford v Dun* 1981 SLT (Sh Ct) 66.
2 *McLinton v McFall* (1974) 232 EG 707.
3 *Armour on Valuation for Rating* (5th edn) paras 7-40 to 7-47.
4 Town and Country Planning (Scotland) Act 1972, s 275(1) - 'agriculture'.
5 Cf *Dow Agrochemicals Ltd v EA Lane (North Lynn) Ltd* [1965] 192 EG 737.
6 *Belmont Farms Ltd v Minister of Housing and Local Government* (1962) 13 P & CR 417; *Forth Stud Ltd v Assessor of East Lothian* 1969 SC 1; *Hemens (VO) v Whitsbury Farm & Stud* [1988] AC 601.
7 *Earl of Normanton v Giles* [1980] 1 WLR 28.
8 *Scammell and Densham* p 22; *Muir Watt* p 358, note to subsection (4).
9 *Earl of Normanton v Giles* [1980] 1 WLR 28 (studfarming); *Lord Glendyne v Rapley* [1978] 1 WLR 601 (breeding for sport); *National Pig Progeny Testing Board v Assessor for Stirlingshire* 1959 SC 343 (research).
10 *Sykes and Edgar* 1974 SLT (Land Ct) 4.
11 See chap 21.
12 Opencast Coal Act 1958 (as amended), s 14A(3).
13 *Wetherall v Smith* [1980] 1 WLR 1290; *Short v Greeves* [1988] 1 EGLR 1; *Hickson and Welch v Cann* (1977) 40 P & CR 218; *Russell v Brooker* (1982) 263 EG 513, CA.

'The cases show that the tenancy is not to be regarded as alternating between being within or outside the 1948 Act as minor changes of user take place, and that, when the tenancy is clearly an agricultural one to start with, strong evidence is needed to show that agricultural use has been abandoned.[1]'

It was suggested that a change of use for at least two years was required prior to the date of assessment and that the consent of the landlord was not required. While a holding might be used for agriculture, if the use ceased to be for a trade or business or the trade or business ceased to be agricultural, the holding would cease to be an agricultural holding as defined by the 1991 Act.

In *Short v Greeves* it was said that 'The court is not lightly to treat a tenancy as having ceased to be within the protection of the Agricultural Holdings Acts'[2].

Similarly a non-agricultural tenancy can come under the 1991 Act if the use has become predominantly agricultural with the consent or acquiescence of the landlord.

The change of status test would appear to be more strict than the test to be applied *ab initio*[3].

CROWN LAND

The 1991 Act applies to any agricultural holding on land belonging to Her Majesty in right of the Crown or to land belonging to a government department or held by Her Majesty for the purposes of a government department, subject to such modifications as may be prescribed[4].

Where the Secretary of State is the landlord or tenant of the holding any provision in the Act which requires reference to the Secretary of State has effect with the substitution of 'the Land Court' for 'the Secretary of State'[5].

---

1 *Wetherall v Smith* [1980] 1 WLR 1290 at 537g; *Short v Greeves* [1988] 1 EGLR 1.
2 Above at 3D per Dillon LJ.
3 *Scammell and Densham* pp 26 and 27.
4 1991 Act, s 79. Cf the 1986 Act, s 16(7) regarding milk quota rent arbitrations. The power to prescribe modifications has not been exercised.
5 1991 Act, s 80.

CHAPTER 3

The lease

Key points

- Does the 'tenant' actually have a lease of the subjects?

- Does the lease fall within the statutory definition in s 85(1) so as to attract security of tenure?

- Changes in ownership of the landlord's interest require to be intimated to the tenant as soon as possible: see s 84(4).

- Where there is divided ownership all proprietors have to join in the service of any notices or demands on the tenant. Equally a tenant should serve all notices etc on all the proprietors.

- Notices served by joint tenants have to be served by or with the authority of all joint tenants.

- A legatee or successor in a lease has to serve his notices on the actual landlord of the holding and cannot rely on the original landlord's failure to notice a change of ownership: s 84(4).

- Where parties agree to vary a lease, care must be taken not to create a new lease in place of the old lease.

Time limits

- A Record of the condition of the fixed equipment on the holding, which is deemed part of the lease, must be made forthwith after a lease is entered into: s 5(1).

- In a demand under s 4 by either party for a written lease or variation of a lease, the parties have six months to reach agreement before arbitration is required.

- Where, in a reference under s 4, an arbiter's award varies a lease to transfer the liability for maintenance of fixed equipment from the tenant to the landlord, the landlord has one month in which to require arbitration to determine what compensation should be paid to him by the tenant for past failures to discharge the liability.

THE LEASE

Section 85(1) 'unless the context otherwise requires' provides that:

'lease, means the letting of land for a term of years, or for lives, or for lives and years, or from year to year.'

In order to obtain the security of tenure provided by the 1991 Act a lease of the agricultural holding[1] has to be constituted at common law[2]. Thus the cardinal elements[3] have to be agreed, namely (1) that there is consensus between the parties agreeing to a lease, (2) the subjects of let must be heritable, (3) there must be a consideration for the let and (4) a period of time for the let must be agreed. Where the lease is granted by an agent, the agent will require to have had specific authority to grant the lease[4].

Shared agricultural occupation of the subjects does not confer a protected lease on the licensee[5], nor does a shared milking arrangement[6].

If there is no lease or if the subjects are not an agricultural holding there can be no security of tenure. Further the lease must fall within the statutory definition.

As the s 85(1) definition of lease includes a lease from year to year, a lease may be constituted verbally and proved *prout de jure*. Where there is a verbal lease or a lease in informal writings for a duration of more than one year, the normal rules that such leases could only be proved by writ or oath of the granter perfected by *rei interventus* or homologation have now been superseded[7]. An informal agreement for a lease for more than one year will be set up if one of the parties has acted or refrained from acting and been affected to a material extent or would be affected to a material extent if the other party withdrew from the lease[8].

There is no requirement to have a probative lease, although a landlord would be advised to have a written lease. A style is annexed at Style 2.

Where a verbal lease of agricultural land is granted and the tenant has entered into possession, but no duration is stipulated, such a lease is deemed to be a lease for at least one year[9]. It accordingly continues by tacit relocation in terms of s 3. In *Morrison-Low v Paterson*[10] Lord Keith of Kinkel said:

'In the ordinary case, there can be no doubt that where a proprietor admits someone into possession of an agricultural holding, or maintains him in such possession without any pre-existing right thereto, and regularly accepts rent from him, there is an inescapable inference that a tenancy has been brought into existence, and it is of no moment that no particular occasion can be pointed to upon which the parties agreed to the one granting and the other taking a tenancy.'

1 See chap 2.
2 See *Paton and Cameron* p 5.
3 See 13 *Stair Memorial Encyclopaedia* para 104.
4 *Danish Dairy Co v Gillespie* 1922 SC 656.
5 *Finbow v Air Ministry* [1963] 2 All ER 647; *Harrison-Broadley v Smith* [1964] 1 WLR 456; *Evans v Tompkins* [1993] 2 EGLR 6. Cf *Broomhall Motor Co v Assessor for Glasgow* 1927 SC 447; *Magistrates of Perth v Assessor for Perth and Kinross* 1937 SC 549.
6 *McCarthy v Bence* [1990] 1 EGLR 1.
7 Requirements of Writing (Scotland) Act 1995, s 1(5).
8 Ibid, s 1(3) and (4).
9 *Gray v Edinburgh University* 1962 SC 157 at 164.
10 1985 SC(HL) 49 at 78.

However in *Strachan v Robertson-Coupar*[1] the First Division said:

'the inquiry must always be, where there is no direct evidence of a contract of lease entered into at a particular time, whether the whole circumstances are only to be explained upon the basis that both parties must have agreed to the creation of the essential relationship of landlord and tenant in respect of the defined subjects. That relationship, in the case of agricultural land for use as such, necessarily involves the implication that the alleged tenant is to enjoy all the rights and privileges of a tenant, including security of tenure of the particular subjects.'

If a lease is proved, security of tenure follows unless the lease falls outwith the terms of the definition in s 85(1).

The statutory terms in relation to endurance all appear to contemplate a lease for a duration of more than one year. In England[2] a lease for 12 months certain has been held not to be covered by the term 'from year to year'. A lease for one year only would be converted into a lease from year to year by s 2(1)[3] and thus obtain security of tenure under s 3(1). The exception in England deriving from *Gladstone v Bower* does not apply in Scotland because of the very different provisions in the respective ss 3 of the 1986 and 1991 Acts.

The statutory definition excludes leases of land let to 'the tenant during the continuance in any office, appointment or employment under the landlord'[4]; a lease for a rotation of cropping terminating on the sale of the land[5]; a lease of fields in rotation where the crop was stipulated by the landlord[6]; and, it has been suggested, a liferent lease[7].

A let for four years for cropping or a let 'for years and crops' has been held to be leases for a term of years and accordingly within the statutory definition[8].

PARTIES TO A LEASE

(a) The landlord

In terms of s 85(1) 'unless the context otherwise requires' landlord means:

'any person for the time being entitled to receive the rents and profits or to take possession of the agricultural holding, and includes the executor, administrator, assignee, heir-at-law, legatee, disponee, next-of-kin, guardian, curator bonis or trustee in bankruptcy, of a landlord.'

The definition of landlord includes limited owners or persons whose interest is not that of absolute owner. An interposed head tenant[9] could be a landlord. Section 74 provides that limited owners may for the purposes of the Act give any consent or make any agreement etc, which they might have done if they were the absolute owner.

1 1989 SC 130 at 134.
2 *Gladstone v Bower* [1960] 1 QB 170 at 178, affd [1960] 2 QB 384; *Lower v Sorrell* [1963] 1 QB 959; *Bernays v Prosser* [1963] 2 QB 592 and *EWP Ltd v Moore* [1991] 2 EGLR 4.
3 *Bernays v Prosser* above.
4 s 1(1); *Budge v Gunn* 1925 SCLR 74.
5 *Stirrat v Whyte* 1968 SLT 157.
6 *Strachan v Robertson-Coupar* 1989 SC 130.
7 Gill paras 36 and 57; 1 *Stair Memorial Encyclopaedia* 728.
8 *McKenzie v Buchan* (1889) 5 Sh Ct Rep 40; *Stonehaven v Adam* 1970 SLT (Sh Ct) 42.
9 Under the Land Tenure Reform (Scotland) Act 1974, s 17; *Kildrummy (Jersey) Ltd v Calder* 1996 GWD 8–458.

The limited owner has recourse against the absolute owner for sums paid by him to the tenant in respect of compensation for improvements or disturbance or for the cost of any improvements proposed to be executed by the tenant of which he has defrayed the cost. He may obtain a charging order from the Secretary of State burdening the holding with an annuity to repay the sums paid[1].

The primary objective of the definition was to extend the meaning to include heritable creditors in possession, liferenters etc. The phrase 'for the time being' means that the person who is landlord in relation to any obligation has to be ascertained at the relevant date, when the obligation arose[2].

The definition is in the alternative being either the person entitled to receive the rents or the person entitled to take possession[3]. There can, therefore, be separate landlords for different purposes. The definition includes disponee and assignee[4]. A landlord need not be infeft to serve a demand for rent[5] or to raise an action of removing, although he must be infeft before decree is pronounced[6]. While a purchaser under missives might come under an obligation to pay compensation upon termination of a tenancy[7], he has no right to serve a notice to quit or pursue a removing prior to entry[8].

While the definition includes guardians and curators bonis appointed by the civil courts, the 1991 Act, s 77 makes special provision for the appointment by the sheriff of a curator or guardian to a landlord or tenant who is a pupil or of unsound mind for the purposes of the Act upon the application of any person interested. Thus a landlord or tenant is not dependent upon the next-of-kin of the incapax to seek such an appointment.

Where parts of the holding are owned by different persons all the heritable proprietors are the landlord. They are required to act in concert, because the holding is deemed to be an entity. All notices and demands under the Act, such as notices to quit, counternotices, demands for rent review or arbitration, references to the Land Court, and objections to a bequest or acquisition of a lease by succession require to be served by all the proprietors together[9].

A single proprietor cannot seek to act in relation to his own part of the holding by eg serving a notice to quit that part[10]. Nor can one proprietor seek to act in relation to the whole holding without the concurrence of the others[11].

A tenant should in safety serve any notices, demands or other legal processes at his instance on each of the proprietors of the holding as they are all his landlord.

In his dealings with the proprietors the tenant is entitled to treat with them as if the holding had not been divided. Where the proprietors have agreed an apportionment of the rent between themselves and the tenant has consented,

1 1991 Act, s 75(3).
2 *Waddell v Howat* 1925 SLT 403.
3 *Alexr Black & Sons v Paterson* 1968 SLT (Sh Ct) 64.
4 Cf *Cunningham v Fife CC* 1948 SC 439.
5 *Alexr Black & Sons v Paterson* above; *Gordon v Rankin* 1972 SLT (Land Ct) 7.
6 *Walker v Hendry* 1925 SC 855.
7 Cf *Cunningham* above.
8 Erskine *Institutes* II, 5, 52.
9 *Styles v Farrow* (1976) 241 EG 623.
10 *Bebington v Wildman* [1921] 1 Ch 559.
11 *Fforde v McKinnon* (22 February 1996, unreported) Court of Session (it was held that a beneficiary under a trust, entitled to receive the rents, was entitled to serve a notice to quit). *Stewart v Moir* 1965 SLT (Land Ct) 11; *Secretary of State v Campbell* 1959 SLCR 49.

then claims for compensation for improvements or disturbance, reorganisation payments and milk quota compensation fall to be apportioned in proportion to the rental apportionment[1].

Where an apportionment of the rent has not been agreed with the tenant or fixed by statute an arbiter is required by s 50 to apportion the amount awarded between the persons constituting the landlord. The arbiter is required to direct that any additional expense of the arbitration so caused shall be paid by those persons in the proportions he shall direct.

Similar restrictions apply to *pro indiviso* proprietors. They must all concur in the granting of a lease[2]. They all have to agree to notices to quit and to actions of removing[3].

One *pro indiviso* proprietor can be the tenant of himself and the other joint *pro indiviso* proprietors[4].

(b) Changes in ownership

In terms of s 84(4) a tenant is entitled to serve notices on the person who has been notified to him as the landlord. Until the tenant receives notice that he has a new landlord he is entitled to continue to serve notices on the old landlord at his address. Thus it is important for a new landlord to intimate his acquisition of the landlord's interest to the tenant as soon as possible.

This provision applies only to the tenant. It is not available to a legatee or a successor to a lease[5] giving notice of succession to a lease, because he is not the tenant at that stage.

A notice to quit falls upon the landlord concluding a contract for sale after giving the notice to quit and before its expiry[6]. In contrast a notice of irritancy under the lease, specifically assigned to the new landlord, would appear to be effective[7].

(c) The tenant

In terms of s 85(1) 'unless the context otherwise requires' tenant means:

'the holder of land under a lease and includes the executor, administrator, assignee, heir-at-law, legatee, disponee, next-of-kin, guardian, curator bonis or trustee in bankruptcy, of a tenant.'

While the definition includes executor, legatee etc, persons in that capacity are merely the tenant while the deceased tenant's estate is being wound up. A lega-

---

1 1991 Act, ss 50 and 55(5); 1986 Act, Sch 2, para 11(5).
2 *Bell's Exrs v IRC* 1987 SLT 625.
3 *Rankine* p 82; *Walker v Hendry* 1925 SC 855 (a notice was served purportedly in the name of the husband alone, but it was proved that his wife, the other *pro indiviso* proprietor, had agreed to the husband so acting).
4 *Barclay v Penman* 1984 SLT 376; *Clydesdale Bank plc v Davidson* 1996 SLT 437; *Trs of Halistra Grazings v Lambert* 1995 Highland RN 496.
5 1991 Act, ss 11(2) or 12(1).
6 Ibid, s 28.
7 *Life Association of Scotland v Blacks Leisure Group plc* 1989 SC 166.

tee does not become tenant in his own right until the bequest has been intimated and accepted or the Land Court declares him to be tenant[1].

Normally the tenant is an individual, partnership or company holding under an express or implied lease. Where a lease is to a partnership the lease terminates upon the death or retirement of a partner, unless the lease was to the 'house' or a continuing partnership[2]. Where parties intend that a lease to a partnership should continue beyond the death or retiral of a partner, it is important to ensure that both the partnership and the lease make provision for such a continuance.

Neither a partner nor his executor winding up a partnership become the tenant by the payment of the rent[2], because the lease will have terminated upon the dissolution of the partnership. This will not prevent a new lease being established in favour of the partner or executor if they are maintained in possession and rent is regularly accepted from them[3]. This is contrasted with the situation where rent is paid for a period of occupation while the affairs of the partnership or executory are wound up, which will not constitute a new lease[4].

The fact that the tenant pays the rent from a partnership account does not necessarily detract from the fact that he is tenant[5].

Joint tenants are treated as 'the tenant' in terms of the statutory definition. Where there are joint tenants, they are all bound jointly and severally by the terms of the lease. They are each liable for the whole rent. They are all bound to comply with provisions such as a residence clause in the lease[6]. Where one of the joint tenants dies his *pro indiviso* share requires to be transferred on death in the same way as if his interest were the whole interest under the lease[7].

This applies where the tenant is a trustee/executor or a number of trustees or executors[8]. There is no limitation on the trustees liability flowing from their character as trustees[9].

Where there is a joint tenancy, on the death of one of the tenants it is important to ascertain if there is a survivorship clause in favour of the other joint tenant or if the interest of the deceased joint tenant requires to be transferred by his executors. Failure to transfer the interest of a deceased joint tenant, where there is no survivorship clause, when the lease is running on tacit relocation means that the lease will terminate[10].

All notices, counternotices or other intimations under the statute should normally be given by or with the authority of all the joint tenants[11]. One joint tenant can terminate the lease without the consent of the other, although if the other joint tenant is the landlord or his nominee then the 'landlord tenant' probably cannot terminate the lease without the consent of the other joint tenant[12]. Further one of the joint tenants cannot adopt or ratify a notice served by

1 See chap 11.
2 *IRC v Graham's Trs* 1971 SC (HL) 1.
3 *Morrison-Low v Paterson* 1985 SC (HL) 49 at 78 per Lord Keith of Kinkel.
4 *Jardine-Paterson v Fraser* 1974 SLT 93 at 98.
5 *Morrison-Low v Paterson* above at 79 per Lord Keith of Kinkel.
6 *Morrison-Low v Howison* 1961 SLT (Sh Ct) 53; *Lloyds Bank v Jones* [1955] 2 QB 298.
7 See chaps 12 and 13.
8 *Lloyds Bank v Jones* above.
9 *Dalgety's Trs v Drummond* 1938 SC 709 at 718.
10 *Coats v Logan* 1985 SLT 221 at 225.
11 *Newman v Keedwell* (1977) 35 P & CR 395; *Featherstone v Staples* [1986] 1 WLR 861.
12 *Smith v Grayston Estates Ltd* 1960 SC 349 (dicta in *Graham v Stirling* 1922 SC 90 not followed); cf *Featherstone v Staples* above, where held landlord's nominee could not effect termination of lease at own hand.

another joint tenant after the time limit in which the notice required to be served. The authority must have been obtained before the notice was served[1].

Where a landlord or his nominee is inserted into the tenancy arrangement as a joint tenant for the purposes of being able to withhold his authority to certain notices, with a view to defeating the tenant's statutory rights, which flow from such notices timeously served, then the court might well treat the landlord as having given authority[2].

In limited cases the courts have allowed one joint tenant to serve a notice under the Act. In *Howson v Buxton*[3] where one of the joint tenants made a claim for compensation under the Agricultural Holdings Act 1923 the court held that the relevant notice was valid, in circumstances where the whole loss was that of the one joint tenant. While such a notice might be effective in respect of individual waygoing claims, it would not be prudent to rely on the exception. In *Combey v Gumbrill*[4] the court held that a divorced wife, who was joint tenant with her husband, had her husband's authority to serve a notice requiring arbitration by virtue of court orders made in the divorce proceedings.

Similarly any notices by the landlord require to be given to all the joint tenants[5].

Servicemen abroad are given special protection when a notice to quit is served[6].

(d) Sub-tenant

Most agricultural leases prohibit sub-tenancies, which in any event are prohibited at common law[7].

Where an agricultural lease permits, or a landlord agrees to a sub-tenancy being created, the sub-tenant can be the 'tenant' of the agricultural holding standing in that relationship to his landlord, the head tenant.

A sub-tenant's position does not have the security of tenure of the head tenant, because the sub-tenant has no right to serve a counternotice to a notice to quit if in consequence of a notice to quit given by the head landlord. If the notice to quit on the head tenant does not have effect, his notice to quit on the sub-tenant similarly is of no effect[8]. Notice to quit part of the holding under s 30, which is accepted by the head tenant as notice to quit the entire holding, is treated as a notice to quit the holding by the sub-tenant[9].

If the head tenant has served a counternotice on the head landlord, then the sub-tenant is entitled to be a party to the Land Court proceedings[10]. The head tenant requires to serve on the sub-tenant notice that he has served a counternotice[10].

1 *Graham v Stirling* 1922 SC 90; *Walker v Hendry* 1925 SC 855.
2 *Featherstone v Staples* [1986] 1 WLR 861.
3 (1928) 97 LJKB 749; followed in *Lloyd v Sadler* [1978] QB 774.
4 [1990] 2 EGLR 7.
5 *Jones v Lewis* (1973) 25 P & CR 375.
6 Reserve and Auxiliary Forces (Protection of Civil Interests) Act 1951, ss 22 and 24; Agricultural Holdings (Servicemen) (Scotland) Regulations 1952, SI 1952/1338.
7 *Rankine* p 192.
8 1991 Act, s 23(6).
9 Ibid, s 23(7).
10 Ibid, s 23(8).

A notice to quit by the head landlord which is given effect terminates sub-tenancies as well[1]. This is in contrast to the common law position where the termination of a tenancy does not automatically entitle a landlord to remove a sub-tenant without further notice[2].

The grant of a sub-tenancy, with the purpose of avoiding the protection which would otherwise be given to a tenant under the Acts, has been ignored by the courts[3], holding that the so called sub-tenant had security of tenure.

A tenancy having security of tenure under the 1991 Act, which is converted into a sub-tenancy by the interposition of a head tenant under the Land Tenure Reform (Scotland) Act 1974, s 17, has statutory security of tenure. The new sub-tenant cannot be removed as a sub-tenant and cannot be removed by the head landlord[4].

(e) Secretary of State

Where the Secretary of State is either the landlord or the tenant under a lease his functions under the 1991 Act are vested in the Land Court and any right of appeal to the Secretary of State from an arbiter ceases to have effect[5].

TERMS IMPLIED INTO AGRICULTURAL LEASES

Certain terms of particular significance are implied into agricultural leases either at common law or under statute. The terms of particular significance relating to leases entered into on or after 1 November 1948 are noted below:

Lease. The parties have the right to demand a written lease, where none exists, or to have terms of the lease varied, where they are inconsistent with the statutory provisions[6].

Security of tenure. The 1991 Act gives the tenant security of tenure[7].

Succession. The 1991 Act gives certain rights of succession to near relatives of the tenant and more limited rights to others entitled to succeed on intestacy[8].

Rent. Once the lease is running on tacit relocation there is a right to have rent fixed in accordance with the statutory formula[9].

Sub-tenants. See p 22 above.

Husbandry. The 1948 Act[10] imposes and defines obligations of good estate management and husbandry on the parties to an agricultural lease[11].

1 *Baron Sherwood v Moody* [1952] 1 All ER 389.
2 *Robb v Brearton* (1895) 22 R 885.
3 *Gisbourne v Burton* [1989] QB 390.
4 *Kildrummy (Jersey) Ltd v Calder* 1996 GWD 8-458.
5 s 80. This also applies to milk quota compensation arbitrations under the 1986 Act, Sch 2, paras 11 and 12.
6 See below.
7 See chap 4.
8 See chaps 11–13.
9 See chap 7.
10 Schs 5 and 6.
11 See chap 15.

Freedom of cropping. A tenant has certain freedoms in relation to cropping and disposal of produce[1]. These are modified by the Common Agricultural Policy arable area payment schemes[2].

Muirburn. The provisions of a lease of a hill sheep farm regarding muirburn are modified by statute[3] to the effect of allowing the tenant to muirburn between 30 September and 16 April in the following year[4].

Inversion of possession. At common law, unless the lease specifically permitted it, the tenant is not entitled to invert the possession by using the subjects for non-agricultural purposes[5]. Participation in a statutory scheme may not amount to inversion[6], although the landlord's consent should, if appropriate, be obtained.

Fixed equipment. The respective obligations of the parties in relation to fixed equipment under s 5 are deemed incorporated into every lease[7].

Record[7]. The parties have a right to have a record made of the condition of the fixed equipment and holding in certain circumstances.

Residence. At common law a tenant is not obliged to reside on the holding. Many leases provide for a residence clause. Whether or not a particular absence constitutes a breach of such a clause has been considered in a number of cases[8]. The clause is valid against joint tenants who will all, in the absence of an agreement to the contrary, be required to comply with it[9]. Upon the death of a tenant it is probably implied that the residence clause is suspended for a reasonable time until the next tenant can take up occupation[10].

Landlord's right to enter holding. At common law a landlord is entitled to enter a holding for the purposes of inspection. Under s 10 the landlord or anyone authorised by him may enter the holding for the purpose of viewing the holding, fulfilling his responsibilities to manage the holding in accordance with the rules of good estate management or to provide, improve, replace or renew fixed equipment. The landlord entering under the statutory provision is probably not entitled to handle the tenant's stock[11].

Penal rent etc. The landlord's right to claim penal rent or liquidated damages is limited by statute, notwithstanding any provision to the contrary in the lease[12].

Game. At common law, and usually under a written lease, the sporting rights are reserved to the landlord. The tenant has a limited right to kill ground game

---

1 See chap 8.
2 See chap 25.
3 Hill Farming Act 1946, ss 23–27.
4 See chap 10.
5 *Rankine* pp 236–238; *Cayzer v Hamilton (No 2)* 1995 SCLR 13.
6 Eg set-aside (arable area payment scheme).
7 See chap 6.
8 *Edmond v Reid* (1871) 9 M 782; *Stuart v Warnocks* (1883) 20 SLR 863; *Blair Trust Co v Gilbert* 1940 SLT 322; *Summal v Statt* (1984) P & CR 367.
9 *Morrison-Low v Howison* 1961 SLT (Sh Ct) 53.
10 *Lloyds Bank v Jones* [1955] 2 QB 298 at 324.
11 *Luss Estates Co v Firkin Farm Co* 1985 SLT (Land Ct) 17.
12 See 1991 Act, s 48; *Willson v Love* [1896] 1 QB 626.

to protect crops from injury. The tenant is given a right to claim for damage by game[1].

A tenant may be authorised by the Deer Commission to kill red or sika deer that are causing serious damage to agricultural land[2].

Tenant's right to remove certain fixtures and buildings. In particular circumstances the tenant is given a right to remove fixtures and buildings for which he is not entitled to compensation at the end of the tenancy[3]. The landlord may defeat this right by purchase under the statutory provisions.

Removal of tenant for non-payment of rent. Apart from the right to remove a tenant for non-payment following a demand for payment as a prelude to a s 22(2)(d) notice to quit, a tenant can be removed from the holding when six months' rent are overdue by action of removing in the sheriff court[4].

Termination of tenancy[5]. The landlord is given certain limited rights to terminate the tenancy in specific cicumstances.

Compensation on termination of tenancy or part of the tenancy[6]. The statute gives (1) the landlord a right to claim compensation for dilapidations and deteriorations if a record has been made[7] and (2) the tenant a right to compensation for improvements[8] and, in certain circumstances, a right to compensation for high farming[9], disturbance[10] and an additional payment[11]. The tenant has a right to claim compensation for milk quota[12]. Compensation is payable in respect of removal from part of a holding by notice to quit under s 29 or a resumption[13].

Any agreement requiring the incoming tenant to pay compensation to the outgoing tenant, except in regard to Part 3 improvements up to a specific sum, is null and void[14].

Sale of implements to landlord or incoming tenant. Section 19 implies a term into the lease or agreement that the property in the goods shall not pass until the price is paid. The price requires to be paid within one month of quitting the holding or within one month of a valuation being delivered.

Arbitration[15]. The statute confers a right to arbitration, by a single arbiter, on any question or difference between the landlord and tenant[16].

1 See chap 11.
2 Deer (Scotland) Act 1959, s 6.
3 See 1991 Act, s 18 and chap 16.
4 Ibid, s 20. Rarely used though as a s 22(2)(d) demand is more effective.
5 See chap 15.
6 See chaps 18, 19 and 20.
7 1991 Act, ss 45-47.
8 Ibid, s 34.
9 Ibid, s 44.
10 Ibid, s 43.
11 Ibid, s 54.
12 1986 Act, Sch 2.
13 1991 Act, s 49.
14 Ibid, s 35(2).
15 See chap 17.
16 s 60 and certain other specific provisions. See chap 23.

STATUTORY WRITTEN LEASE

Section 4 gives both parties to a lease, in particular circumstances, the right to obtain a written lease containing the statutory provisions set out in Sch 1 or to obtain a revisal of a written lease, which does not conform to the statutory provisions.

Under s 4(1)(a) where there is no written lease in force, either party may give notice to the other in writing requesting the other to enter into a written lease embodying the matters specified in Sch 1 or containing matters consistent with that Schedule and s 5. Where there is an informal written lease the provision does not apply[1].

Where there is a written lease in force entered into on or after 1 November 1948 or a written lease entered into before that date, which is running on tacit relocation, and that lease contains no provision for one or more of the matters specified in Sch 1 or contains a provision inconsistent with Sch 1 or s 5, then either party may give notice in writing requiring the lease to be varied to bring it into line with the statutory provisions[2].

After either party has served the foregoing notice in writing s 4(1) provides for a period of six months in which parties can reach agreement on a written lease or variation of the existing lease. If matters are not resolved within that time the matter is referred to arbitration.

On a reference to arbitration under s 4(1) the arbiter specifies the terms of the tenancy. This means that the arbiter is required to determine, after competent proof, what was verbally agreed between the parties, or where the lease is in non-probative documents, what documents constitute the lease.

In so far as the terms of the tenancy do not make provision for matters specified in Sch 1 or make provisions which are inconsistent with the Schedule or with s 5 the arbiter shall make provision for those matters 'as appears to the arbiter to be reasonable'. It is only where the lease includes a provision which is 'inconsistent with' the statutory provisions that the arbiter may then make provision in respect of the inconsistent matters[3]. He may only include matters relating to Sch 1, but he cannot include matters relating to s 5, unless the lease contains matters inconsistent with that section.

The arbiter may include in his award any further provisions relating to the tenancy which may have been agreed between the landlord and the tenant which are not inconsistent with the Act[4]. He has no power to include additional provisions which are not agreed between the parties. The purpose of this subsection is to allow parties to agree additional matters and have them included in the lease. The parties may agree after the reference on the additional provisions to be included in the lease.

The effect of the award, which may take effect from the making of the award or from such later date as the award may specify, is the same as an agreement in writing between the parties. The arbiter's discretion to specify a later date gives him the power to specify such a date, where an earlier date may cause hardship[5].

1 *Grieve v Barr* 1954 SLT 261.
2 1991 Act, s 4(1)(b).
3 *Connell* S. 4.
4 1991 Act, s 4(3).
5 *Connell* S. 4.

Where, under a reference in terms of s 4, the arbiter has included provisions in his award, which he is required to include, he may vary the rent of the holding if it appears equitable to him to do so[1]. He may not vary the rent in respect of any provisions incorporated in his award by agreement under s 4(3), although there is no reason why the parties should not agree a rent variation as part of those agreed provisions.

Where an arbiter transfers the liability for the maintenance or repair of fixed equipment from the tenant to the landlord, the landlord may, within one month, require arbitration to determine the amount of compensation that the tenant should pay to him in respect of the tenant's previous failure to discharge such liability[2].

VARIATION OF LEASE

Where entry is given before a formal lease is executed the terms of the informal agreement govern the parties obligations. Where the subsequently executed lease does not innovate on the original agreement, the lease is held binding from the date of entry, but where the executed lease imports additional conditions into the lease, the parties cannot be liable for a breach of such a condition committed before the date of the execution of the lease[3].

At common law the parties may at any time during the currency of a lease agree to modify its terms. An informal verbal lease for less than a year can be varied verbally, but the variation of a written lease for more than one year requires to be varied in writing[4]. Where there is a verbal variation of a written lease one party cannot withdraw from the variation and nor is the variation invalid if the other party has acted or refrained from acting on it and has been affected to a material extent or would be so affected by a withdrawal from the variation[5].

It may be difficult to prove a variation of a lease unless that variation is in writing[6].

The addition of another person as joint tenant with the existing tenant does not create a new lease[7].

A variation of a lease may in certain circumstances amount to a new lease if 'any variation is made from the first (lease) in relation to rent, duration, or other essential stipulation'[8]. A surrender of part of the holding back to the

1 1991 Act, s 14.
2 Ibid, s 46(2). The provision for compensation in favour of the tenant in s 46(3) is meaningless in the context of the section; cf *Duncan*, note to s 46(3) in *Current Law Statutes*.
3 *Korner v Shennan* 1950 SC 285; cf *Pahl v Trevor* [1992] 1 EGLR 22 - in England where a lease is executed after the date of entry the tenant's legal interest only starts on the date of execution.
4 Requirements of Writing (Scotland) Act 1995, s 1(2)(a)(i) and (7).
5 Ibid, s 1(3) and (4). The former rule that a written lease requires to be varied in writing otherwise proof of any agreed variation may be limited to proof by writ or oath completed by *rei interventus* or homologation has been abolished; Requirements of Writing (Scotland) Act, ss 1(5) and 11.
6 *Rankine* pp 110-114 and *Carron Co v Henderson's Trs* (1896) 23 R 1042 at 1048.
7 *Francis Perceval Saunders Dec'd, Trs of v Ralph* [1993] 2 EGLR 1.
8 Hunter *Landlord and Tenant* (4th edn, 1876) vol II, p 1050; *Erskine* II, 6, 44; *Tufnell v Nether Whitehaugh Co Ltd, Applicants* 1977 SLT (Land Ct) 14; *Mackie v Gardner* 1973 SLT (Land Ct) 11; *Jenkin R Lewis & Sons Ltd v Kerman* [1971] Ch 477.

landlord with an agreed rent variation, unless under a resumption clause, may well amount to the surrender of the previous lease and the acceptance of a new lease[1].

A variation amounting to a new lease can have important consequences as particular rights and liabilities under a lease under the Act depend on the date of the lease[2].

A tenant who remains in occupation of a holding during two or more tenancies does not lose his right to compensation for improvements by reason only that the improvements were not carried out during the tenancy on the termination of which he is quitting the holding[3]. Continuous occupation is implied.

Section 16 provides that a lease of an agricultural holding shall not be brought to an end or treated as at an end except with the consent of the other party 'by reason only that any new term has been added to the lease or any terms of the lease (including the rent payable) have been varied or revised in pursuance of this Act'. Thus any variation or revisal under ss 2(1), 4, 5(2), 9, 13, 14 and 15 of the Act do not, in terms of the statute, amount to a termination of the old lease and the substitution of a new lease. The section does not apply to a variation, other than 'in pursuance of this Act'[4].

---

1 *Jenkin R Lewis & Sons Ltd v Kerman* above. This decision should be treated with caution as the English law of leases has its own specialities.
2 Eg compensation for pre-1923 Act, 1931 Act or 1 November 1948 improvements (1991 Act, ss 33 and 34); grounds for consent to notice to quit relating to near relative successors relating to pre or post-1 January 1984 leases (1991 Act, s 25(3) and Sch 2); and sheep stock valuations depending on whether the lease predates 6 November 1946 or post-dates 1 December 1986 (1991 Act, ss 68-72).
3 1991 Act, s 34(5).
4 1949 Act, s 10 used the wording 'in pursuance of any of the foregoing provisions of this Act in that behalf'. *Gill* at para 100 said 'The language of the section is obscure'. The obscurity has been clarified by the change in wording.

CHAPTER 4

Security of tenure and contracting out

Key points

- It is not lawful to contract out of the security of tenure provisions. Schemes to avoid security of tenure can be set aside by the courts.

- Any short term or long term alternative to security of tenure has to be operated strictly in terms of the statutory provisions or the contractual agreement to try to avoid the establishment of a secure tenancy.

Time limits

- None

SECURITY OF TENURE

The security of tenure provisions for an agricultural tenant were introduced into Scotland by the Agriculture (Scotland) Act 1948, consolidated in the following year into the Agricultural Holdings (Scotland) Act 1949.

The right to security of tenure arises first in the 1991 Act from s 3 which provides:

'3. Notwithstanding any agreement or any provision in the lease to the contrary, the tenancy of an agricultural holding shall not come to an end on the termination of the stipulated lease, but shall continue in force by tacit relocation for another year and thereafter from year to year, unless notice to quit has been given by the landlord or notice of intention to quit has been given by the tenant.'

Thus all agricultural leases falling within the definition of 'lease' in s 85(1) are by statute continued by tacit relocation from year to year after the termination of the lease. This is a substantial variation to the common law which imputed constructive consent to both parties to allow a lease to continue on tacit relocation.

A lease running on tacit relocation is deemed to be the original lease and not a new lease[1].

A lease which does not fall within the definition of lease in s 85(1) does not attract security of tenure[2]. As the wording of s 3 is significantly different from the wording of s 3 of the English Agricultural Holdings Act 1986, the exclusion of leases of a duration of between 12 and 24 months from the security of tenure provisions of the English Act, under the *Gladstone v Bower*[3] principle, does not apply.

A lease which attracts security of tenure may only be terminated by notice to quit or notice of intention to quit given in terms of s 21(1) which provides that 'a tenancy of an agricultural holding shall not come to an end except by operation of a notice which complies with this subsection notwithstanding any agreement or any provision in the lease to the contrary'.

Anti-avoidance provisions

The principal anti-avoidance provision comes from the words in both ss 3 and 21(1): 'notwithstanding any agreement or any provision in the lease to the contrary'.

The word 'or' is read disjunctively to cover both agreements separate from the lease and agreements or provisions in the lease[4]. These words have been construed to mean that parties cannot contract out of the provisions of these sections.

Section 2(1) provides that a lease of an agricultural holding for a shorter period than from year to year 'shall take effect, with the necessary modifications, as if it were a lease of land from year to year'. There are two principal

---

1 *Douglas v Cassils and Culzean Estates* 1944 SC 355.
2 See eg *Stirrat v Whyte* 1967 SC 265; *Strachan v Robertson-Coupar* 1989 SC 130.
3 [1960] 1 QB 170, affd [1960] 2 QB 384 (CA).
4 *Morrison v Rendall* 1986 SC 69.

exceptions, namely such a lease entered into with the prior approval of the Secretary of State and a grazing lease falling within the provision in s 2(2)[1].

Thus leases for a shorter period than from year to year take effect as a lease from year to year and are therefore continued in terms of s 3 by tacit relocation. It has been held that the effect of s 2(1) is to continue the first shorter lease to the end of the anniversary of the year in which it started so that the ish of such a lease is the date one year later before the date upon which the lease started[2].

The 'necessary modification' must be such that the agreement remains recognisably the same agreement. If the modifications which are necessary to convert the lease into a lease from year to year are such that the agreement is no longer recognisably the same, or is radically different from the original agreement, then the provisions of s 2(1) cannot apply[3].

'It is not permissable to substitute for the original agreement a radically different agreement and make that take effect instead of the original agreement.[4]'

A lease modified under s 2(1) is not a new lease, but is the old lease continuing with such variations as the modifications require[5].

CONTRACTING OUT

The security of tenure provision for agricultural tenants was introduced into Britain in 1948 in the public interest as a matter of public policy. Security of tenure was in the national or public interest to encourage efficient farming and good husbandry by conferring security of tenure on tenant farmers, and such farms, because of their inequality of bargaining power, required statutory protection in the making of contracts for the leasing of land[6].

Johnson v Moreton makes it clear that in general the parties to an agricultural lease may not contract out of the statutory provisions, particularly in relation to security of tenure, unless the Act specifically permits contracting out[7].

In *Morrison v Rendall*[8] the Second Division held that an agreement during the currency of a lease to renounce the lease was not enforceable because the parties were not entitled to contract out of the provisions of (now) s 21(1), even during the currency of the lease. The Lord Justice Clerk said:

'In my opinion, the plain words of section 24(1) mean that an agricultural tenancy shall not come to an end unless one of the parties has given written notice to the other in terms of the subsection. Of course, parties may circumvent these provisions by one of

1 See chap 5.
2 See *Morrison's Exrs v Rendall* 1989 SLT (Land Ct) 89 holding that a lease from 1 March to 31 January was continued to an ish of 28 February. But cf D C Coull 'Termination Date in a Notice to Quit' 1989 SLT (News) 431.
3 *Bahamas International Trust Co Ltd v Threadgold* [1974] 1 WLR 1514 (CA), 1525 (HL), where the Court of Appeal's decision on the s 2(1) point was not disturbed; *Goldsack v Shore* [1950] 1 KB 708 and *Harrison-Broadley v Smith* [1964] 1 WLR 456.
4 *Harrison-Broadley v Smith* above at 467 per Pearson LJ.
5 s 16.
6 *Johnson v Moreton* [1980] AC 37; see pp 3–5.
7 Eg post-lease ageement to vary the parties repair/replacement obligations in relation to fixed equipment (s 5(3)); agreement as to substitute compensation for improvements (s 38(5)) in lieu of statutory compensation under the Act.
8 1986 SC 69.

them renouncing the lease, or by their agreeing to a new lease in substitution of the old lease. That apart, I am of opinion that the provisions of section 24(1) clearly apply to all leases, and that it is not therefore open to parties to terminate a lease in the manner suggested by the pursuers here, that is, by an agreement. ... The result is that parties are not entitled to contract out of section 24(1) either by making a provision in the lease to the contrary or by making a provision to the contrary in a separate agreement. Unless the words are construed in this way, the result would be that parties could execute a lease, and shortly thereafter could execute an agreement containing provisions contravening this section. ... In my opinion, there is no justification for concluding that the words in the final sentence of section 24(1) are confined to preventing parties contracting out of the provisions of the Act *ab ante*; in my opinion the words go further than that and prohibit parties at any time from agreeing to end a tenancy other than by giving notice to quit in terms of the subsection. I appreciate that in practice parties may agree to terminate the tenancy without notice to quit having been given in terms of the subsection, and if they both act upon such agreement it will no doubt be effective. Likewise if such an agreement is made and one party acts upon it, the other party may be personally barred from founding on the provisions of the subsection. Apart from that, I am of opinion that an agreement made in contravention of the plain provisions of the final sentence of subsection (1) of section 24 would not be enforceable.'

It is thus clear that parties cannot contract out of the security of tenure provisions of the Act either in the lease or in any agreement, whether made before or during the currency of the lease. Any such agreement is unenforceable.

Any agreement that obliges a tenant to serve a notice of intention to remove[1] or by which the tenant obliges himself not to serve a counternotice under s 22(1)[2] is void and unenforceable.

Personal bar

In *Morrison v Rendall*[3] the landlord had argued that the tenant, having entered into a verbal agreement to quit, was thereafter by his actings personally barred from, or should be held to have waived his right to, objecting to the agreement. The court held that the averments of personal bar in that instance were not relevant to sustain such an argument. However the Lord Justice Clerk did say:

'Likewise if such an agreement is made and one party acts upon it, the other party may be personally barred from founding on the provisions of the subsection.'

What was not argued in *Morrison's Exr v Rendall* was whether or not personal bar could in fact operate in the face of the Agricultural Holdings (Scotland) Acts, which were Acts introduced in the public interest and as matters of public policy.

Rankine[4] states:

'The doctrine of estoppel cannot be applied to an Act of Parliament. Estoppel only applies to a contract *inter partes* and it is not competent to parties to a contract to estop themselves or anyone else in the face of an Act of Parliament, as for example, to evade it.'

1 *Johnston v Moreton* [1980] AC 37; *Featherstone v Staples.*
2 *Featherstone v Staples* above.
3 1986 SC 69.
4 Rankine *A Treatise on the Law of Personal Bar in Scotland* (1921) p 6.

In *Kok Hoong v Leong Cheong Kweng Mines Ltd*[1] the Privy Council suggested the test that should be applied to determining if personal bar could be pled in the face of a statute. Lord Radcliffe said[2]:

'Given a "statutory obligation of unconditional character" it is not open to the court to allow the party bound by that obligation to be barred from carrying it out by the operation of an estoppel. Similarly, there is, in most cases, no estoppel against a defendant who wishes to set up the statutory invalidity of some contract or transaction on which he is being sued, despite the fact that by conduct or other means he would otherwise be bound by estoppel. ... It has been said that the question whether an estoppel is to be allowed or not depends on whether the enactment or rule of law relied on is imposed in the public interest or "on grounds of general public policy"[3]. However a principle as widely stated as this might prove to be rather an elusive guide, since there is no statute, at least public general statute, for which this claim might not be made. In their lordships' opinion a more direct test to apply in any case such as the present, where the laws of money lending or monetary security are involved, is to ask whether the law that confronts the estoppel can be seen to represent a social policy to which the court must give effect in the interests of the public generally or some section of the public, despite any rules of evidence as between themselves that the parties may have created by their conduct or otherwise. Thus the laws of gaming or usary[4] override an estoppel, so do the provisions of the Rent Restriction Acts with regard to orders for possession of controlled tenancies[5].'

Note in particular the reference 'to orders for possession of controlled tenancies'.

In the light of the test laid down in *Kok Hoong v Leong Cheong Kweng Mines Ltd* and the observations in *Johnson v Moreton* regarding the public policy of the 1991 Act, there must be doubt as to the correctness of the Lord Justice Clerk's dicta in *Morrison v Rendall,* that personal bar might operate so as to bar a person founding on his rights under s 21(1).

The Land Court has held that personal bar does not operate in the face of the Crofting Acts, which were passed for very similar public policy purposes[6].

Sham transactions

In line with the courts' refusal to allow parties to contract out of the statutory protection of the 1991 Act, the courts will also refuse to give effect to pretences, whose purpose is to evade the statutory provisions of security of tenure.

The courts have been advised to be 'astute to detect and frustrate sham devices and artificial transactions whose only object is to disguise the grant of a tenancy and to evade'[7] the security intended to be given by Parliament.

This approach has been followed in a number of cases dealing with the

1 [1964] AC 993.
2 Above at 1015 in delivering the Opinion of the Privy Council.
3 See *Re a Bankruptcy Notice* [1924] 2 Ch 76 at 97 per Atkin LJ.
4 See *Carter v James* (1844) 2 Dow & L 236.
5 See *Welch v Nagy* [1949] 2 All ER 868.
6 See eg *Guthrie v MacLean* 1990 SLCR 47 at 55.
7 *Street v Mountford* [1985] AC 809 at 825H per Lord Templeman (a case concerning security of tenure under the Rent Acts). Cf *A G Securities v Vaughan* [1990] 1 AC 417 where Lord Templeman said at 462H that 'pretence' should be substituted for 'sham devices' or 'artificial transactions'.

Agricultural Holdings Acts. In *Featherstone v Staples*[1] the parties set up a complicated arrangement whereby the landlord let a farm on a yearly tenancy to a partnership of two brothers and a limited company owned by the landlord[2], which provided that upon the service of a notice to quit the partnership was to come to an end and the brothers were to be barred from serving a counternotice without the consent of the limited company. A notice to quit was served, the partnership brought to an end, but contrary to the agreement the brothers served a counternotice. The Court of Appeal held that the agreement that the brothers could not serve a counternotice was contrary to public policy and void, and accordingly that the counternotice was good.

This case was followed by *Gisborne v Burton*[3], which decided that a sub-tenancy granted by a wife, who was the 'artificial' tenant, could not be brought to an end except by way of notice to quit, because the sub-tenant was in fact the tenant. It was held:

'Where there were a series of transactions which taken together constituted a composite scheme intended to avoid a mandatory statutory provision the court would look to the overall result sought to be achieved by the scheme rather than considering the individual transactions in isolation.'

In delivering the leading judgment Dillon LJ said[4]:

'If the present artificial scheme, avowedly adopted for the sole purpose of depriving the defendant of statutory security of tenure is effective to that end ... Security of tenure will thus be at the whim of the landlord. I cannot regard this as consistent with the policy of the 1948 Act and the succeeding Acts ... Essentially the scheme must fail because the Christophersons were trying to do by document what, for the reasons given in *Johnson v Moreton* the law does not permit, viz to grant the defendants an agricultural tenancy without the statutory protection.'

Russell LJ went on to say:

'I am firmly of the view that the lease ... was an artificial device the only object of which was to disguise the grant of a tenancy to Mr Burton and to evade the 1948 Act. In striking down the lease ... as an artificial device, I do not think the court is guilty of a procedure which it is not entitled to take; on the contrary, in one sense it is giving effect to the true intention of the parties, although in the process not permitting the legal consequences to flow to which the parties were prepared to accede.'

In considering whether or not a limited partnership arrangement was a sham arrangement the Land Court has said[5]:

'If the original arrangement in the present case had been designed entirely to avoid security of tenure under the 1949 Act, with no genuine partnership enterprise it might be open to challenge by the tenant, as being a "pretence" following *Featherstone* and *Gisborne* supra – but with an added complication of a Scottish partnership as tenant.'[6]

1   [1986] 1 WLR 861.
2   In England a partnership has no separate persona. Therefore a let by the landlord to a partnership of which he is a partner amounts to a lease to himself and another. Whether or not this is a valid lease gives rise to considerable difficulty in English law; see *Scammell and Densham* p 54. It is for this reason that one often finds English limited partnerships involving a nominee of the landowner as limited partner, whereas in Scotland it is quite common for the landowner to be the limited partner.
3   [1989] QB 390.
4   Above at 400 F.
5   *Dickson v MacGregor* 1992 SLT (Land Ct) 83 at 88F.
6   Cf *Featherstone v Staples* [1986] 1 WLR 861, where the partnership was brought to an end, but nevertheless the brothers were held entitled to serve a counternotice.

The key to the approach of the courts probably lies in the phrase 'giving effect to the true intention of the parties'. Where the true intention is to grant a lease of an agricultural holding the court may strike down any agreement, sham devices, artificial transaction or pretence[1] which seek to evade statutory security of tenure.

ALTERNATIVES TO SECURITY OF TENURE

The obvious alternative to security of tenure is for the landlord to farm the land himself. There is no reason why an arable and stock farm cannot be farmed for a long period by means of contracting out the ploughing, planting and harvesting of the crops, with the use of some of the short-term alternatives noted below, such as grazing lets of the grass parks, or potato joint ventures, or lets of some of the fields. While such a system of husbandry may not be the most efficient use of the land, it may well serve a need to keep vacant possession over a few years until, say, a son is old enough to take over from an ageing father.

While the principal objective of the 1991 Act is to provide security of tenure for agricultural tenants, there are provisions in the Act and at common law whereby security of tenure might be avoided.

Short-term alternatives

Where any short-term alternative to security of tenure is considered it is important that (1) the terms of the short-term let are specifically agreed, preferably in writing, before entry is given and (2) at the end of the agreed period of occupation, the tenant is made to remove from the holding, if necessary by court action raised immediately.

It is almost impossible to prove that a so called series of occupations, where there is no vacating of the subjects between periods, was not in fact an agreement for longer-term occupation giving rise to security of tenure. Permitted continued occupation after the contractual period is over generally amounts to relocation, which then gives rise to security of tenure.

The short-term alternatives include:

(1) *Section 2(1) leases.* The 'let for use as agricultural land for a shorter period than from year to year ... [where] ... the letting was approved by the Secretary of State before the lease was entered into ... '. See chapter 5.

(2) *Grazing lets.* The s 2(2) 'lease entered into ... in contemplation of the use of the land only for grazing or mowing during some specified period of the year'. See chapter 5.

(3) *Section 22(2)(a) lets.* The lease of permanent pasture 'which the landlord has been in the habit of letting annually for seasonal grazing or of keeping in his own occupation' does not give rise to a secure tenancy if 'let to the tenant

1 *A G Securities v Vaughan* [1990] 1 AC 417.

for a definite and limited period for cultivation as arable land on condition that he shall, along with the last or waygoing crop, sow permanent grass seed'.

(4) *Potato lets.* The practice of letting potato land on an annual let probably confers security of tenure[1], although this is seldom claimed as the potato grower requires to lease land which has been potato-free for at least five years. The safe course is to enter into a joint venture - see (5).

(5) *Joint venture for a crop.* A safe short-term arrangement is for the landowner to enter into a joint venture with the crop grower or merchant for the growing of a season's crop[2]. Such joint ventures were common in pre-World War II times[3]. A longer-term joint venture may in effect be held to amount to a lease[4].

(6) *License*[5]. A license to occupy may be an appropriate short-term expedient. If no rent were paid that would be a license and not a lease. An agreement to pay a lump sum for a period of occupation has been held to be a license and not a lease[6]. The 'arrangement' in *Strachan v Robertson-Coupar*[7], where a farmer was allowed to crop particular fields in each year and different fields in the next year, all in accordance with the landlord's rotation, was held not to be a lease but an arrangement. Because the parties had not contemplated entering into the relationship of landlord and tenant this was held to be a form of licence[8].

(7) *Shared occupation.* Shared occupation with the landowner where the agricultural possession is not exclusive (eg each runs part of the stock on the fields) does not amount to a lease attracting security of tenure[9]. A shared milking arrangement, particularly where the landowner reserves the right to vary the land used for the milk production and shares the occupation of the land, does not confer a secure tenancy[10].

Longer-term alternatives

Any long-term alternative to a lease conferring security of tenure runs the risk of being held to be a sham arrangement, which in fact confers security of tenure.

1 *Prior v J & A Henderson Ltd* 1983 SLCR 34 at 37; but cf *Strachan v Robertson-Coupar* 1989 SLT 488.
2 See *Encyclopaedia of Scottish Legal Styles* vol 7, p 192 for a suitable crop joint venture style.
3 *McKinley v Hutchison's Tr* 1935 SLT 62.
4 *Paton and Cameron* p 11.
5 It should be noted that the Agricultural Holdings Act 1986, s 2(2) confers an agricultural tenancy on a person granted 'a license to occupy land for use as agricultural land'. Therefore English cases on a licence to occupy should be treated with caution.
6 *Mann v Houston* 1957 SLT 89 (£200 was paid for a ten-year occupancy of a garage, which was held not to be a lease).
7 1989 SC 130.
8 The facts of the case are complex; the report suggests that the correct case might not have been pled on record, so a *Strachan* arrangement cannot really be recommended.
9 *Finbow v Air Ministry* [1963] 2 All ER 647; *Harrison-Broadley v Smith* [1964] 1 WLR 456; *Evans v Tompkins* [1993] 2 EGLR 6; *McCarthy v Bence* [1990] 1 EGLR 1. Cf *Broomhall Motor Co v Assessor for Glasgow* 1927 SC 447; *Magistrates of Perth v Assessor for Perth and Kinross* 1937 SC 549.
10 *McCarthy v Bence* [1990] 1 EGLR 1.

Possible alternatives include:

(1) *The limited partnership.* The lease to a limited partnership, where the landlord or his nominee is the limited partner, has been the most generally accepted method of trying to avoid security of tenure. The scheme requires that the landlord or his nominee enters into a limited partnership agreement with a general partner who is to farm the holding for an agreed duration. The landlord then lets the farm to the limited partnership on a normal agricultural lease.

The lease is automatically terminated when the partnership terminates[1] either at the expiry of a fixed duration of the partnership or upon the partnership being terminated with notice after it has run on tacit relocation after the specified period[2].

In the granting of a limited partnership lease it is essential that (a) the limited partnership agreement is entered into and executed before the lease is granted and executed and (b) the general partner is not given entry to the lands before the partnership agreement and the lease have both been executed.

If the landlord is to have the protection of a limited partnership against the general partner's debts, it is important that the limited partnership is registered[3].

In *Featherstone v Staples* Slade LJ[4] suggested that there was nothing contrary to public policy in any limited partnership agreement, where the landlord was the sleeping partner. His observations were obiter. He qualified them by saying:

'Arrangements are not infrequently made by virtue of which landowners enter into a partnership with one or more other persons, on the basis that the partnership will be granted a tenancy of an agricultural holding and that the landowner himself will be what is colloquially known as a sleeping partner. Quite apart from the considerations relating to security of tenure, there may well be good and sufficient reasons (whether of a commercial, family, fiscal or practical nature) why all interested parties regarded such an arrangement as sensible and beneficial.'

In *Dickson v MacGregor*[5] the Land Court was asked to adjudicate on whether or not a lease to a limited partnership was a sham device to avoid security of tenure and so to accept that the general partner was in fact the tenant under the lease. The issue arose out of a rent arbitration under s 13, where the arbiter stated a case for the opinion of the court on what was the lease between the parties.

The facts were that the landlord and tenant had entered into a limited partnership in 1981 with a subsequent lease to the limited partnership. The partnership was terminated in 1986 with the tenant being allowed to continue in occupation paying rent. At the rent review the tenant contended that he had an

---

1 *IRC v Graham's Trs* 1971 SC (HL) 1.
2 A landlord limited partner, who terminated the partnership for the purpose of obtaining vacant possession for himself, may be in breach of the rule that a partner owes a duty to his co-partners not to acquire a special advantage over them; cf Brough *Miller on Partnership* (2nd edn, 1994) pp 167-168.
3 Limited Partnership Act 1907, s 5 - 'Every limited partnership must be registered . . . or in default thereof it shall be deemed to be a general partnership, and every limited partnership, and every limited partner, shall be deemed to be a general partner'.
4 [1986] 1 WLR 861 at 879G.
5 1992 SLT (Land Ct) 83.

unwritten lease governed by the provisions of the 1991 Act. The landlord contended that the partnership transaction was a sham, so that the tenant was in fact still the tenant under the 1982 full repairing lease. The court held that the author of the 'pretence', if indeed it was one, could not invoke the pretence as a ground to nullify the limited partnership arrangement *ab initio*.

This case suggests that if the limited partnership arrangement is a pretence designed from the beginning to avoid security of tenure then the court may well look behind the pretence and hold that the tenant has security of tenure. Clearly if there are genuine reasons for entering into a limited partnership, where a landlord or other party puts a reasonable amount of capital into the farming enterprise but does not want the unlimited risk, a limited partnership arrangement is probably acceptable.

It is difficult to see how any limited partnership where the limited partner puts in only £100 and takes no interest in the farming thereafter can ever be said to be anything other than a 'pretence' designed to avoid security of tenure.

If there is to be a limited partnership the landlord must ensure that all the provisions of the arrangement are operated; eg the capital is contributed, the annual payment is paid, the partnership operates a separate bank account, the landlord insists on seeing and examining the annual accounts, which are produced in the name of the partnership, the annual agricultural returns are made in the name of the partnership etc. If these basics are not followed and are allowed to lapse it will be much easier for a general partner to challenge the arrangement in any subsequent proceedings.

Featherstone v Staples[1] was argued on the basis that the partnership arrangement was not a pretence but a bona fide arrangement. Nevertheless the court held that the provision in the partnership agreement, which gave the landlord the power to veto a counternotice, was of no effect. On a similar ratio, it would appear that a provision in a bona fide Scottish limited partnership, whereby the landlord could terminate the partnership for the purpose of terminating the tenancy, may equally be held to be void and of no effect.

(2) *A partnership.* While many landowners will not wish to be involved in the financial risks of a farming partnership, a genuine partnership is an alternative to security of tenure in that the partnership deed can provide for termination. A partnership may be appropriate in a family situation, where the parties want the lease to end upon a particular death, but this raises tax implications[2].

(3) *Share farming*[3]. Share farming is another alternative. In Scottish terms it is a form of contracting arrangement or joint venture[4], whereby the landowner provides the land and some of the inputs, with the contractor/manager providing machinery, labour, cultivation and day-to-day management. Both parties share part of the risks and benefit from the profits.

---

1 [1986] 1 WLR 861.
2 *IRC v Graham's Trs* 1971 SC (HL) 1.
3 For a general discussion of the topic see *Agricultural Law, Tax and Finance* (Longman, 1992) ed by A A Lennon at section D (*Lennon*); 'The New Share Farming Agreement', lecture by W Wallis in 'The 1990 Changes in Agriculture in Scotland' Seminar Documentation, October 1990 (*Wallis*); and *Scammell and Densham* p 53.
4 It is difficult to characterise the agreement in precise legal terms; see *Lennon* at D2 and *Scammell and Densham* at p 53. The agreement has to be drafted with care because it can easily be held to be either a tenancy or a partnership if the wording of the agreement does not exclude those relationships.

A share-farming agreement is not a long-term solution, but provides a temporary expedient that can last for a few years, say to see out a retiring farmer who requires to retain a business for the purpose of retirement relief in the few years before he can retire.

Share farming has been defined as:

'an arrangement under which a person entitled to the occupation of the land arranges for a farmer to carry out farming operations on that land and the remuneration of the farmer is substantially dependent on the results achieved and where the agreement is so structured that it neither amounts to a partnership, nor a tenancy, nor a contract of employment. A share farmer will typically provide for the working of the farm and a degree of working capital[1]';

or as a means whereby the landowner may:

'retain the sole right to occupy, but exchange the right to enjoy the whole income for a right to enjoy part of the income in return for assumption, by another, of responsibility for day to day management and the provisions of sundry services and inputs[2]';

or

'as an arrangement entered into between two or more persons to make available from their separate businesses assets or services for the carrying out of specified farming operations, the gross receipts from which will be divided between and paid to the separate businesses in agreed proportions.[3]'

The main advantage of a share-farming arrangement is that if it is properly structured both the landowner and the farmer are deemed to be carrying on business for the purposes of income, capital gains and other tax purposes. Thus income is earned income and roll over and retirement relief remain available to the landowner. It would be wise to confirm with the Inland Revenue that the particular agreement is accepted by it as a proper share-farming agreement capable of attracting those particular tax advantages.

The agreement is appropriate only for those areas of the farm which are under crop. A landowner should keep the grazings out of the agreement and let them on s 2(2) grazing lets.

To avoid a share-farming agreement[4] being classified either as a partnership or as a tenancy the agreement should provide that the landowner remains the sole owner of the growing crops[5]. The landowner should retain the effective management and control. Where the agreement provides for consultation the landowner should retain the right to make the final decisions[6].

Payment to the share farmer should probably take the form of a fixed payment to cover the contract works and a variable amount linked to the profits – eg fixed fee of £75 per acre and 85 per cent of the profit which exceeds £10,000 with a provision to carry forward any shortfall into the following years[7], although a sharing of the gross receipts or of the crops may suffice.

---

1 *Lennon* D2.1.
2 *Wallis* p 1.
3 *Scammell and Densham* p 53.
4 See *Lennon* and *Wallis* for checklists of what should be included in the agreement.
5 *Lennon* D2.3, D3.4 at H.
6 *Lennon* D3.4 at C and D.
7 *Lennon* D3.4 at E and F.

(4) *The liferent lease.* It has been suggested that a lease for a single life, the liferent lease, may not be a lease in terms of the 1991 Act[1]. This question is entirely untested in law, but there may be circumstances where a landowner is willing to grant a liferent lease[2] and take the risk of having to run a test case, if on the death of the tenant the executor claims an entitlement to transfer the lease.

(5) *The lease by a liferenter. Gill*[3] suggests

'it is unlikely that a liferenter could, in the absence of express power, grant an agricultural lease under the [1991] Act. At common law, a liferenter cannot grant a lease beyond the duration of his lifetime. It would seem that a liferenter cannot grant a lease even from year to year, if it is a lease to which the [1991] Act applies, because of the security of tenure and rights of succession which necessarily follow thereon.'

As a liferenter has power at common law to confer a lease which will either last from year to year or last for less than a year if he dies within the year, it may be that any lease a liferenter grants acquires security of tenure, not from the liferenter's grant, but from ss 2(1) or 3.

The situation may be different if the liferent is under an English trust, as English liferenters, called 'tenants for life', appear to have power to confer leases beyond their lifetime[4].

The question is likely to arise where a liferenter has inadvertently granted a lease, which has taken effect from year to year, where on the liferenter's death the fiar may want to challenge the continuing right of occupancy. As a prospective tenant is taken to have been bound to investigate the title of his landlord to grant a lease, he would be barred from maintaining that he did not know of the liferent[5].

(6) *Choice of tenant.* It may be possible to evade security of tenure by choosing a tenant who cannot, or is unlikely to have, near relative successors. The lease could be terminated under s 25 if the executor transfers the tenancy to a person who is not a near relative successor.

(5) *Special destination in lease.* A lease granted with a special destination is not available to be transferred by the executors[6]. A lease to the proposed tenant with a special destination on his death to the landlord's nominee should bring the tenancy back into the landlord's control, if it is not deemed to be void.

---

1 A G M Duncan 'Agricultural Tenancies – the Exclusion of Successors' [1988] November JLSS 384; 1 *Stair Memorial Encyclopaedia* 728.
2 See *Duncan* above for suggested terms.
3 Para 36.
4 Settled Land Act 1925, s 41(iv) allows a tenant for life to grant a lease for 50 years.
5 *Trade Development Bank v Crittall Windows Ltd* 1983 SLT 510.
6 Succession (Scotland) Act 1964, ss 16 and 36(2).

CHAPTER 5

Leases for less than year to year

Key points

- To avoid security of tenure under a short-term s 2(1) let, the Secretary of State's approval must predate the granting of the lease.

- The land let under an approved short-term let must not include land not subject to approval. It is safer to relate each lease to the land included in each specific approval.

- In a grazing lease, ensure that both parties 'contemplate' that the let is for grazing or mowing alone.

- The grazing let must be for a specific period of less than one year.

- The grazing tenant is not permitted to plough, crop or otherwise deal with the land.

- The short-term or grazing tenant is not allowed to remain on the land after the period of the let expires.

- If the farm is a dairy farm steps must be taken to restrict the short term or grazing let to a period of less than eight months to ensure that milk quota is not inadvertently transferred.

- Arbitration in relation to lets for less than from year to year applies to the operation of the section, but not to its applicability: s 2(3).

Time limits

- The Secretary of State's approval for a short-term let under s 2(1) has to be obtained prior to the grant of the lease.

- Any grazing lease has to be for 'some specified period of the year': s 2(2)(a).

GENERAL

Where land is let for use as agricultural land for a shorter period than from year to year the lease takes effect with the necessary modifications as if it were a lease from year to year, unless the lease is excluded by the provisions of the 1991 Act, s 2.

In considering the operation and effect of s 2 care should be taken in consulting English case law, because s 2 of the English Agricultural Holdings Act 1986 also applies to 'a license to occupy land as agricultural land' and to 'a license to occupy land' for grazing or mowing during some specified period of the year. A true license to occupy land in Scotland is not a 'let' or 'lease' of land.

SHORT-TERM LETS WITH SECRETARY OF STATE'S APPROVAL

A lease of land for use as agricultural land for a shorter period than from year to year in general takes effect as a lease of that land from year to year[1] 'unless the letting was approved by the Secretary of State before the lease was entered into[2]'.

The applicant for a consent requires to satisfy the Secretary of State that it is reasonable that the consent should be granted. The application should include a plan of the land to be let, details of the proposed dates of the let and give a summary of the reasons why the request for permission is being made.

The circumstances in which a short-term let are often approved include (1) where the land is to be developed in early course[3], (2) where a short term arrangement for the cultivation of the land is required to cover, say, an *inter regnum* before a sale or between tenancies, where a child is shortly to return to take over the farm[4], (3) to cover a crop rotation in otherwise permanent grazing; or (4) to cover a trial period on a prospective tenant[5].

The Secretary of State has to apply his mind to the letting of a particular piece of land, but is not concerned with particular terms of the agreement that have, or may subsequently be, entered into[6].

The Secretary of State's approval, which cannot be backdated, must precede

---

1 See p 30 'Anti-avoidance provisions'.
2 1991 Act, s 2(1).
3 *NCB v Drysdale* 1989 SLT 825, land shortly to be used for coal extraction.
4 *Pahl v Trevor* [1992] 1 EGLR 22.
5 Cf *Verrall v Farnes* [1966] 1 WLR 1254 (a trial rent-free licence to occupy for a year to a prospective tenant was held, under the English legislation, to have become a lease from year to year).
6 *Epsom and Ewell BC v C Bell (Tadworth) Ltd* [1983] 1 WLR 379; *Finbow v Air Ministry* [1963] 1 WLR 697 (a blanket approval of proposed short-term lets by specified named authorities was held valid). *Gill* para 64, n 42 describes this as 'a dubious decision'. *Pahl v Trevor* [1992] 1 EGLR 22 (it was held that under the Agricultural Holdings Act 1986, s 5 the minister was not concerned with the start date of the tenancy).

the granting of the lease[1]. A lease entered into after the approval has expired takes effect as a lease from year to year[2].

The approval must relate to all the land let and if land not subject to approval is included in the let then the lease takes effect as a lease from year to year[3]. It may be competent to let the land subject to approval in part only or under a number of separate leases, but this would be unwise[4].

If a tenant under an approved short-term let does not remove at the ish, it is essential that the landlord starts proceedings for removal immediately, otherwise the tenant may be able to claim that the landlord allowed the tenancy to continue by tacit relocation, thus creating a lease from year to year[5].

GRAZING LEASES

Leases of land for grazing or mowing 'during some specified period of the year' are excluded from the protection of the 1991 Act by s 2(2)(a) which provides:

'2(2) Subsection (1) above shall not apply to-

(a) a lease entered into (whether or not the lease expressly so provides) in contemplation of the use of the land only for grazing or mowing during some specified period of the year;'

To come within the exception the lease must fulfil both criteria, namely that of the purpose of the lease and that it should continue for 'some specified period of the year'. The 'contemplation' requires to be that of both parties and qualifies both the purpose and the duration of the let. It has to be what was in contemplation at the time of the agreement.

Lord Denning in *Scene Estate Ltd v Amos* said[6]:

'... the object of the word "contemplation" in the proviso is to protect a landlord who has not expressly inserted a provision that it is for grazing only, or for mowing only, or that it is for a specified part of a year; but nevertheless both parties know that is what is contemplated.'

In general a true letting of the grass parks on a farm for grazing during the growing season by annual auction presents no difficulty. It is only when parties enter into an oral arrangement or seek to invoke s 2(2)(a) to cover or cloak other activities, or allow a grazing lease for a specified period to be continued indefinitely, ostensibly under a series of agreements, that difficulties arise.

---

1 *NCB v Drysdale* 1989 SLT 825; *Bedfordshire CC v Clarke* (1974) 230 EG 1587 (the minister's consent and the lease agreement both bore the same date and the landlord was unable to prove that the consent preceded the tenant's acceptance of the lease).
2 *Secretary of State for Social Services v Beavington* (1982) 262 EG 551.
3 *NCB v Drysdale* 1989 SLT 825 (lease of about 85 acres included an additional 1.52 acres not subject to approval).
4 *NCB v Drysdale* 1989 SLT 825 per Lord Dervaird *obiter*. *Gill* para 65, citing *Epsom and Ewell BC v C Bell (Tadworth) Ltd* [1983] 1 WLR 379 (not cited in *NCB v Drysdale*) and *Finbow v Air Ministry* [1963] 1 WLR 697, suggests that this may be a misconstruction of the 1991 Act, s 2(1).
5 Cf *NCB v Drysdale* 1989 SLT 825 at 826H.
6 [1957] 2 QB 205 at 211.

Where the agreement was a sham, the lease was not entered into 'in contemplation' of grazing[1].

A lease merely for a period of less than a year is caught by s 2(1).

In England it has been held that one is entitled to look at extrinsic evidence in appropriate cases to determine what was being contemplated by the parties, because of the phrase 'whether or not the lease expressly so provides'[2]. This is generally taken to mean either when the agreement is a sham or when the agreement is silent on the question of 'contemplation'[3]. In general, extrinsic evidence may not be looked at to explain written contracts. In *Mackenzie v Laird*[4], notwithstanding that there were written missives, the court paid regard to the tenant's admission in evidence that when the lets were entered into both parties knew that the agreement was such as would not bring the let under the Act.

Agricultural returns may be relevant evidence of whether the let was a grazing lease or a let of agricultural land, because land let for grazing should be included in the landlord's return[5].

If the 'purpose' specified in a written lease was a sham and the tenant was allowed to effect full husbandry, then the court may look behind the wording of the lease[6]. Alternatively the original purpose may have been varied by the subsequent actings of the parties, thus converting a grazing let into an agricultural tenancy.

It is competent to have a series of grazing lets, provided that the parties never had it in contemplation that the arrangement would in fact continue for a year or more. A series of 21 3–month lets under written agreements fell within the proviso, where both parties had in contemplation that each of these lets successively were grazing lets under the exception[7].

A single agreement for a succession of grazing lets annually for the grazing season can fall within s 2(2)(a)[8].

Such arrangements are ill-advised, particularly where the arrangements are oral with perhaps continuous occupation, when it may be difficult to prove that (1) there was a series of lets and (2) that the parties never contemplated that the arrangement would continue for a year or more.

In *Commercial Components (Int) Ltd v Young*[9] a farmer occupied two fields for grazing continuously over a number of years, but annually paid rent for the grazing in period April to October. The Sheriff Principal held, notwithstanding the fact that the stock was never removed, that the facts did not support the irresistible inference that the farmer occupied the fields as an agricultural tenant from year to year, that being the appropriate test.

A grazing let for a period of years, where the farmer was bound not to graze cattle in the fields during the daffodil season, did not fall within the exception, because the farmer was entitled under the lease to graze horses during the daffodil period[10].

1 *Scene Estate Ltd v Amos* [1957] 2 QB 205 at 211 per Lord Denning.
2 *Scene Estate Ltd* above at 213 per Lord Parker.
3 *Lampard v Barker* (1984) 272 EG 783.
4 1955 SC 266.
5 *Watts v Yeend* [1987] 1 WLR 323.
6 See eg *Street v Mountford* [1985] AC 809; *Scene Estate Ltd v Amos* [1957] 2 QB 205.
7 *Scene Estate Ltd v Amos* above.
8 *Mackenzie v Laird* 1959 SC 266; *Watts v Yeend* above.
9 1993 SLT (Sh Ct) 15.
10 *Brown v Tiernan* [1993] 1 EGLR 11.

The tenant of a farm may let the grazings on a seasonal basis, provided this is not specifically excluded by the lease. This probably does not constitute a sub-lease which might be in breach of the terms of the lease[1].

Grazing or mowing

The exception given by s 2(2)(a) is restricted to leases for grazing or mowing. In general the allowance of any other agricultural activity will transfer the lease into a secure lease under s 2(1). An obligation to maintain fences, drains, to cut the weeds and maintain the turf does not prevent the lease remaining within s 2(2)(a)[2]. In *Duncan v Shaw*[3] the sheriff held that the grazing tenant was under an obligation to fence off the grazings from neighbouring crops.

In England it has been held that ploughing, cropping and stabling, provided they are subservient to the maintenance of a reasonable standard of efficient grass production, may not take the lease outwith the provisions of the subsection[4]. The issue was touched on but not decided in *Sansom and Chalmers*[5] because it was held that such other agricultural activities as the grazing tenant had undertaken were undertaken for the landlord and not for his own behoof under his grazing let.

If buildings are included in the let this will probably take the let outwith the exception unless the contemplated use of the buildings is incidental to the grazing or mowing[6].

Period of the year

The period of the year can be specified by reference, using any phrase which is indicative of a period of the year which is reasonably understood in farming circles; eg 'grazing season'[7], 'seasonal grazing'[8], 'growing season'[9] and 'the winter 1948–49'[10].

A grazing or mowing lease for 364 days comes within the proviso[11], although lets for such a duration always give rise to a suspicion that the intention was to try to avoid the effect of s 2(1), particularly when they are continued from year to year.

1 *Morrison v Nicolson* 1913 SLCR 90; *Little v McEwan* 1965 SLT (Land Ct) 3. These cases were decided under the Landholders (Scotland) Acts, but the principle would appear to apply to agricultural leases. In *Duke of Portland v Samson* (1843) 5 D 476 at 477 the Lord Ordinary said 'It is doubted if a landlord could prevent a tenant from giving a neighbour a servitude of pasturage . . . while he continued a joint possession and had the land fully stocked'.
2 *Mackenzie v Laird* 1959 SC 266; cf *Scene Estate Ltd v Amos* [1957] 2 QB 205 where the tenant carried out fencing works, but this did not make the let into an agricultural tenancy.
3 1944 SLT (Sh Ct) 34.
4 *Lory v Brent LBC* [1971] 1 WLR 823; *Avon CC v Clothier* (1977) 242 EG 1048 and *Boyce v Rendalls* (1983) 268 EG 26.
5 1965 SLCR App 135.
6 *Avon CC v Clothier* (1977) 75 LGR 344 (CA); eg stabling of ponies grazing the land.
7 *Mackenzie v Laird* 1959 SC 266.
8 *Watts v Yeend* [1987] 1 WLR 323.
9 *Gairneybridge Farm Ltd and King* 1974 SLT (Land Ct) 8.
10 *Goldsake v Shore* [1950] 1 KB 708.
11 *Reid v Dawson* [1955] 1 QB 214.

It should be noted that it has been held that a lease from 1 January to 31 December, or any other such period, is a lease for a year and not for a period of less than a year[1].

SUB-TENANCIES

A sub-tenancy granted for a shorter period than from year to year by a person whose interest in the land is that of a tenant under a lease for a shorter period than from year to year, which has not taken effect as a lease from year to year, does not obtain security of tenure as a sub-tenant from year to year[2].

This is consistent with the rule that a sub-tenant can have no better right than his author in title.

MILK QUOTA

If the farm is a dairy farm the approved short term or grazing let must be for less than eight months, otherwise milk quota may be inadvertently transferred[3].

ARBITRATION

Section 2(3) provides:

'(3) Any question arising as to the operation of this section in relation to any lease shall be determined by arbitration.'

This subsection has given rise to considerable difficulty in determining the respective jurisdictions of the courts and of the arbiter[4].

It would appear that the subsection refers to arbitration of all questions as to the operation of s 2(1) and (2), but not to questions as to its applicability.

It is for the courts to determine whether or not there is a lease or enforceable agreement, and if there is a lease whether or not it falls within the provisions of s 2(3)[5]. Where a party sues for vacant possession following a grazing let under

---

1 *Morrison's Exrs v Rendall* 1989 SLT (Land Ct) 89; *McGill v Bichan* 1970 SLCR App 122; *Cox v Husk* (1976) 239 EG 123. The decision in *Morrison's Exr* was criticised by D C Coull in 'Termination Date in a Notice to Quit' 1989 SLT (News) 431. He pointed out that in *Lady Bangour v Hamilton* (1681) Mor 248 the court held that 'the year was not to be counted by the number of days, but by the return of the day of the same denomination of the next year'. Cf *Dodds v Walker* [1981] 1 WLR 1027 (HL) where it was held that 'one month' means the corresponding date in the next month. If *Morrison's Exr* was wrongly decided then a lease from 1 January to 31 December might be a lease for less than a year.
2 1991 Act, s 2(2)(b).
3 Dairy Produce Quotas Regulations 1994, SI 1994/672, reg 7(6).
4 Scottish and English decisions are not always to the same effect.
5 *Goldsack v Shore* [1950] 1 KB 708.

s 2(2)(a) it is for the courts to determine whether or not the grazing provision applies to prevent the lease operating as a lease from year to year[1].

It is for an arbiter to determine whether or not the Secretary of State consented to the particular short let prior to the grant of the tenancy, because this relates to the operation of s 2(1)[2]. This would include determining whether or not the consent related to the land comprised in the lease[3]. If the arbiter holds that the lease does not have prior approval, then he has jurisdiction to determine the necessary modifications to convert the lease into a lease from year to year.

Where the courts determine that a lease for a shorter period than from year to year is not excluded from becoming a lease from year to year, under either a Secretary of State's consent or as a grazing lease under s 2(2)(a), it is for an arbiter to determine 'the necessary modifications'[4] under which the lease takes effect as a lease from year to year.

1 *Goldsack v Shore* [1950] 1 KB 708; *Love v Montgomerie* 1982 SLT (Sh Ct) 60, overruling *Maclean v Galloway* 1979 SLT (Sh Ct) 32. Cf *Mackenzie v Laird* 1959 SC 266 in which, wrongly, there was a remit to arbitration; *Sansom v Chalmers* 1965 SLCR App 135 and *Gairneybridge Farm Ltd and King* 1974 SLT (Land Ct) 8 and *Craig* 1981 SLT (Land Ct) 12, which are wrong in suggesting that the question of whether a let is a grazing let or not is a question for arbitration – see *Gill* para 81.
2 *NCB v Drysdale* 1989 SLT 825; *Exven v Lumsden* (1979, unreported) Lord Ross (briefly reported in *Gill* para 83). Cf that in England this is a question for the courts; *Epsom and Ewell BC v C Bell (Tadworth) Ltd* (1983) 266 EG 808; *Bedfordshire CC v Clarke* (1974) 230 EG 1587.
3 *NCB v Drysdale* above.
4 1991 Act, s 2(1).

CHAPTER 6

Provision and maintenance of fixed equipment and the record

Key points

- Section 5 obligations relating to the provision, repair and replacement of fixed equipment do not apply to pre-1 November 1948 leases.

- Where a pre-1 November 1948 written lease is running on tacit relocation and contains either no provision, or a provision which is inconsistent with s 5 relating to fixed equipment, either party can seek a variation of the lease under s 4.

- A record of the fixed equipment and cultivation should be made at the commencement of the lease. If there is no record, the landlord has no right to recover compensation for dilapidations: ss 5(1) and 8.

- From a landlord's point of view it is important to stipulate the purpose for which the holding is let to limit his liability for the provision of fixed equipment under s 5(2)(a)(i).

- A tenant in breach of his obligations for the maintenance of fixed equipment renders himself liable to a notice to quit under s 22(2)(d) or (e).

- A party having a right to enforce renewal, repair or maintenance obligations under a lease cannot carry out the work and sue for the cost, except in terms of a court order authorising the work.

Time limits

- Where a landlord carries out an improvement at the request of or with the agreement of the tenant or in pursuance of an undertaking given under s 39(3), or in compliance with a direction by the Secretary of State, then the landlord has six months from the completion of the improvement to serve a notice in writing on the tenant seeking a rent increase: s 15(1).

FIXED EQUIPMENT

Fixed equipment is defined by s 85(1) to include:

'any building or structure affixed to land and any works on, in, over or under land, and also includes anything grown on land for a purpose other than use after severance from the land, consumption of the thing grown or of produce thereof, or amenity, land, without prejudice to the foregoing generality, includes the following things, that is to say–

(a) all permanent buildings, including farm houses and farm cottages, necessary for the proper conduct of the agricultural holding;
(b) all permanent fences, including hedges, stone dykes, gate posts and gates;
(c) all ditches, open drains and tile drains, conduits and culverts, ponds, sluices, flood banks and main water courses;
(d) stells, fanks, folds, dippers, pens and bughts necessary for the proper conduct of the holding;
(e) farm access or service roads, bridges and fords;
(f) water and sewerage systems;
(g) electrical installations including generating plant, fixed motors, wiring systems, switches and plug sockets;
(h) shelter belts,

and reference to fixed equipment on land shall be construed accordingly.'

The rules of good husbandry include the specific obligation 'to carry out necessary work of maintenance and repair of fixed equipment'[1]. Upon whom this obligation falls will depend upon the common law, the terms of a pre-1 November 1948 lease, the 1991 Act, s 5, or any post-lease agreement.

The common law

At common law a landlord was impliedly bound to put buildings and fences into a tenantable order so that they were capable of lasting with ordinary care for the stipulated endurance of the lease[2]. The landlord's obligation does not extend to drains[3].

The tenant's obligation at common law was to maintain the fixed equipment and leave it in the condition he got it, ordinary wear and tear excepted[4]. Where the fixed equipment, during the course of the lease, got into a state of natural decay without any fault on the part of the tenant, the obligation was on the landlord to replace it[5].

A tenant was under no obligation to repair fixed equipment which was not put into repair or accepted as being in a tenantable state of repair at the beginning of the lease[6].

Where decay is caused by the tenant's neglect, the tenant is under an obligation to effect the repairs or pay damages for such neglect[7]. If the decay is

1 Agriculture (Scotland) Act 1948, Sch 6, para 2(f)(iv).
2 *Wight v Newton* 1911 SC 762; *Christie v Wilson* 1915 SC 645; *Davidson v Logan* 1908 SC 350.
3 *Wight v Newton* above at 772; but cf *Lyon v Anderson* (1886) 13 R 1020.
4 *Wight v Newton* above.
5 *Johnston v Hughan* (1894) 21 R 777.
6 *Austin v Gibson* 1979 SLT (Land Ct) 12; *Pentland v Hart* 1967 SLT (Land Ct) 2.
7 *Turner's Trs v Steel* (1900) 2 F 3673; *Johnston v Hughan* (1894) 21 R 777.

partly that of the landlord's failure and partly that of the tenant, then it is a question of fact as to what is the principal cause[1]. If the principal cause is the tenant's failure he is under the obligation to repair, but if it is the landlord's failure then the obligation is his, subject to a right to claim damages in respect of the tenant's contributory fault[2].

Prior to 1 November 1948 parties regularly contracted in the lease to seek to vary the common law obligations. To impose a higher obligation on the tenant than to leave the fixed equipment in as good a condition as he got it, fair wear and tear excepted, required clear and explicit wording[3].

Where a lease permits a landlord to carry out repairs which are the tenant's obligation, at either his or the tenant's expense, it has been held that such a provision does not impose an obligation, on the landlord to effect such repairs[4]. Further, the landlord is entitled to enforce any other remedy against the tenant for such failures as an alternative to doing the work himself[5].

It is always a question of fact as to what are the respective obligations of the parties either at common law or under the lease. The obligation remains the same even after the lease is running on tacit relocation[6]. At common law a landlord is probably not obliged to renew or replace buildings and fixed equipment which is 'done' through natural decay and fair wear and tear[7].

If a landlord has acquiesced in a past breach he may be barred from enforcing his right for the future, but in general such acquiescence must amount to variation of the lease or at least amount to personal bar or waiver. In general, mere failure to enforce a condition in the lease only operates while it continues and does not debar a landlord from enforcing the condition for the future[8].

Pre-1 November 1948 leases

If a lease was commenced before 1 November 1948, the 1991 Act, s 5 does not regulate the respective obligations of the parties as to the provision or maintenance of fixed equipment and with regard to fire insurance[9]. The parties' respective obligations will be governed by the common law and any contractual provisions in the lease.

The landlord's obligations under a written lease, including the obligation to renew fixed equipment, continue even when a lease is running on tacit relocation[10].

1 *Buchanan v Buchanan* 1983 SLT (Land Ct) 31 at 34.
2 *Macnab v Willison* (1 July 1955, unreported) First Division, cited in *Gill*.
3 Cf *Davidson v Logan* 1908 SC 350; *Cowe v Millar* (1921, unreported) First Division (reported in *Connell* p 349); *Johnston v Hughan* (1894) 21 R 777; *Macnab v Willison* above; *Nicholl's Trs v MacCarty* 1971 SLCR App 85.
4 *Allan's Trs v Allan and Son* (1891) 19 R 215; *Halliday v Wm Fergusson & Sons* 1961 SC 24, overruling *Forbes-Sempill's Trs v Brown* 1954 SLCR 36.
5 Eg to demand remedy as a prelude to a notice to quit based on s 22(2)(d); see 1991 Act, ss 32 and 66.
6 *Macnab v Willison* 1960 SLT (Notes) 25.
7 *Gill* para 119. Cf *Macnab v Willison* 1960 SLT (Notes) 25 which appears to relate to renewal obligations under a pre-1948 written lease rather than to any common law obligation.
8 *Rankine* pp 111-113; *Morrison-Low v Howison* 1961 SLT (Sh Ct) 53.
9 1991 Act, s 5(6).
10 *Macnab v Willison* 1960 SLT (Notes) 25; cf *Gill* para 119. Where there is no renewal obligation in the lease, the landlord would appear at common law to have no continuing obligation to replace buildings and fixed equipment worn out through natural decay and fair wear and tear.

Where a pre-1 November 1948 written lease, which is running on tacit relocation, contains no provision relating to fixed equipment or contains provisions inconsistent with s 5, either party may seek a variation of the lease under s 4[1] so as to impose on the landlord obligations which are consistent with s 5 obligations.

The effect of this is that a tenant burdened by repair and renewal obligations under a pre-1 November 1948 lease can have the burden shifted to the landlord upon payment of appropriate compensation in respect of any previous failure by the tenant to discharge this liability[2]. An arbiter can also vary the rent on shifting such a burden[3].

Post-1 November 1948 leases

(a) Application

Section 5 applies to all leases entered into on or after 1 November 1948[4]. Its provisions are deemed to be incorporated into every lease, whatever the lease may say.

(b) Landlord's obligations under s 5(2)(a)

Section 5(2)(a) imposes two obligations on the landlord:

(1) An obligation at the commencement of the tenancy or as soon as reasonably practical thereafter (a) to put the fixed equipment on the holding 'into a thorough state of repair' (this is arguably a higher standard than the 'tenantable order' of common law[5]) and (b) to

'provide such buildings and other fixed equipment as will enable an occupier reasonably skilled in husbandry to maintain efficient production as respects both–

(i) the kind of produce specified in the lease, or (failing such specification) in use to be produced on the holding, and
(ii) the quality and quantity thereof ....[6]'

This obligation can impose a heavy burden on the landlord to provide buildings and fixed equipment, which are not already on the holding, where the existing fixed equipment is inadequate for the use specified in the lease or for what is 'used to be produced on the holding'. It should be noted that a specific provision in the lease as to the use to which the holding is to be put does not override an existing use and so make the landlord liable to provide fixed equipment and buildings for the existing use[7]. It is important for a landlord that the

---

1 See chap 3.
2 1991 Act, s 46(2).
3 Ibid, s 14.
4 Ibid, s 5(6).
5 *Gill* para 121.
6 See *Spencer-Nairn v IRC* 1985 SLT (Land Tr) 46 at 51 for an example of the courts' approach to this obligation.
7 *Taylor v Burnett's Trs* 1966 SLCR App 139 (the lease provided that, notwithstanding the existing use as a dairy farm, the holding was let as an 'ordinary holding or general purpose farm' and the landlord was not to be liable for improvements required for other types of farming).

lease should stipulate the type of use for which the holding is let in order to limit the landlord's potential liability under this provision.

The reference to 'an occupier' in s 5(2)(a) makes it clear that the adequacy of the buildings and fixed equipment has to be assessed in relation to a hypothetical tenant as a reasonably competent farmer, rather than the actual tenant, who may in fact not require additional fixed equipment if he is farming the holding in conjunction with another farm. In circumstances such as this, if a landlord does not provide the fixed equipment at the commencement of the lease, he may find himself liable at a later stage; if, for example, the tenancy of the particular farm goes on succession to one son, while the other farm goes to another.

(2) A continuing obligation during the currency of the tenancy, which includes the period when the tenancy is running on tacit relocation, to 'effect such replacement and renewal of buildings or other fixed equipment as may be rendered necessary by natural decay or by fair wear and tear'.

The common law rules govern the extent of the obligation, what amounts to fair wear and tear and any liability on the tenant for his failures under s 5(2)(b)[1].

These are contractual obligations *ad factum praestandum* rendering the landlord liable in damages for any breach[2]. The tenant can probably insist upon any of the obligations being carried out at any time until the obligation prescribes under the long negative prescription[3].

(c) Tenant's obligation under s 5(2)(b)

This subsection is incorporated into every lease. It provides that the tenant's liability in relation to the maintenance of the fixed equipment

'shall extend only to the liability to maintain the fixed equipment on the holding in as good a state of repair (natural decay and fair wear and tear excepted) as it was in—

(i) immediately after it was put into repair as aforesaid, or
(ii) in the case of equipment provided, improved, replaced or renewed during the currency of the lease, immediately after it was so provided, improved, replaced or renewed.'

This implied obligation does not alter the common law. If the landlord does not in fact put the fixed equipment into a thorough state of repair, the tenant's obligation can be no higher than to maintain the fixed equipment (fair wear and tear excepted) in the same condition as he received it[4].

In England a condition in a lease requiring the tenant 'to repair, maintain and keep in good and substantial repair' the electrical wiring included an obligation to replace worn out electrical wiring[5].

1 *Buchanan v Buchanan* 1983 SLT (Land Ct) 31; *Haggart and Brown* 1983 SLCR 13.
2 *Christie v Wilson* 1915 SC 645; *Hamilton v Duke of Montrose* (1906) 8 F 1026.
3 *Secretary of State for Scotland v Sinclair* 1960 SLCR 10.
4 *Austin v Gibson* 1979 SLT (Land Ct) 12.
5 *Roper v Prudential Assurance Co Ltd* [1992] 1 EGLR 5.

A tenant who fails to fulfil his obligations under s 5(2)(b) renders himself liable to a demand to remedy and possible notices to quit[1].

The measure of damages for a tenant's breach will in general be the cost of putting the fixed equipment into the condition in which it ought to have been left[2].

Contracting out and post-lease agreements

It is incompetent to contract out of the provisions of s 5 in the lease[3], although parties may by agreement made after the lease has been entered into provide that one party will undertake works on behalf of the other party and provide for who shall pay for such works[4].

It is not clear how the courts or an arbiter would approach the common statement in a lease that the tenant 'accepts the buildings and fixed equipment on the holding as being in a thorough state of repair and as sufficient to fulfil the landlord's obligations under s 5(2)', where that statement was untrue. Such a statement would effectively be an attempt to contract out of the statutory provision, but is countered by the well-known rule that a party cannot lead extrinsic evidence to contradict a statement in a probative document. In view of the many dicta[5] that the courts will be 'astute to detect and frustrate sham devices and artificial transactions' it may well be that a court or arbiter would hold that they were entitled to look behind the probative document if it were averred that the statement was a sham device to avoid the landlord's statutory liability.

It is now common for a landlord to provide that before granting a tenancy the prospective tenant must agree to enter into a post-lease agreement. Commonly the parties execute the lease and post-lease agreement contemporaneously. Such provisions and procedure might render the arrangement liable to challenge on the ground that the parties have attempted to circumvent the provisions and purpose of s 5[6].

Clearly the purpose of s 5(3) is to provide that the tenant and landlord negotiate any post-lease agreement from positions of equal bargaining power, in that the tenant will by then have a secure lease incorporating s 5(2).

If there is any challenge to a post-lease agreement it may be necessary to investigate when in fact there was an enforceable tenancy agreement between the parties and if that agreement predates an enforceable post-lease agreement. The date of execution of the probative lease and post-lease agreement is not necessarily the relevant date if in fact there was an enforceable agreement at an earlier date, for example in missives incorporating revised drafts of the deeds to be executed[7].

---

1 1991 Act, s 22(2)(d) (remedial breach) or s 22(2)(e) (non-remedial breach).
2 *Fraser v Macdonald* (1834) 12 S 684; *Duke of Portland v Wood's Trs* 1926 SC 640, affd 1927 SC (HL) 1.
3 *Secretary of State for Scotland v Sinclair* 1960 SLCR 10.
4 1991 Act, s 5(3).
5 See chap 4.
6 Agreement executed in such circumstances held void; *Murray v Fane* (22 April 1996, unreported) Perth Sh Ct.
7 *Comex Houlder Diving Ltd v Colne Fishing Co Ltd* 1986 SLT 250 at 258E.

There must be a question mark over whether a tenant who has agreed to take a lease subject to a post-lease agreement can in fact be forced to execute the post-lease agreement, relying on a pre-lease agreement that he will enter into such an agreement.

Arbitration on liability under s 5

Any question as to the landlord or tenant's liability under s 5 requires to be determined by arbitration[1]. A s 5(3) arbitration does not apply to pre-1 November 1948 leases, although any question or difference relating to liability for fixed equipment under such a lease would also fall to be determined by arbitration under s 60.

The scope of an arbiter's remit under the English model clauses[2] is very similar in effect to s 5. It has been said in *Tustian v Johnston*[3]:

'... it refers to compulsory arbitration (a) the extent of a landlord's and tenant's repairing obligations under Pts I and II of the schedule to the 1973 regulations respectively, (b) the extent to which either the landlord or the tenant is in breach of those obligations. It seems to me that these are questions which have to be answered before a claim either to damages or to specific implement can succeed. That in my view drives me to the conclusion that the compulsory arbitrations provisions apply in relation to those proceedings for either damages or specific performance at any rate up to the stage of establishing (a) obligation and (b) breach of obligation.[4]'

Enforcement of repair obligations and damages

A repair or maintenance obligation should be enforced by specific implement after the extent of the obligation has been ascertained by arbitration, unless there is provision for the landlord to carry out the work at the tenant's expense.

It should be noted that a party who has a right to enforce renewal, repair and maintenance obligations on the other party to the lease cannot carry out the work and sue for the cost thereof.

The proper procedure is to seek a decree of specific implement, after having the extent of the obligation determined by arbitration. If the party in breach does not comply with the court order, then the work is carried out at the sight of the court and the party can recover the costs thereby authorised[5].

It is doubtful, under the 1991 Act, that a landlord can sue for damages for dilapidations and deteriorations during the currency of the tenancy. In *Kent v Conniff*[6] such a claim during the currency of the lease was sustained in respect

---

1 1991 Act, s 5(5).
2 Agriculture (Maintenance, Repair and Insurance of Fixed Equipment) Regulations 1973, SI 1973/1473 (amended by SI 1988/291), para 15 remits 'any claim, question or difference' to arbitration. Cf 1991 Act, s 60(1) which refers to 'any question or difference'.
3 [1993] 2 All ER 673 at 681D per Knox J; followed in *Hammond v Allen* [1994] 1 All ER 307.
4 In *Hill v Wildfowl Trust (Holdings) Ltd* 1995 SCLR 778 the sheriff refused to follow *Tustian* and held that an arbiter acting under the 1991 Act, s 60 had, in addition, a power to assess damages.
5 *Commissioner of Northern Lighthouses v Edmonston* (1908) 16 SLT 439; *Davidson v Macpherson* (1899) 30 SLR 2.
6 [1953] 1 QB 361.

of the saving provisions in the Agricultural Holdings Act 1948, s 100[1]. This section was not continued in the 1991 Act and the effect of this might be to bar a claim during the currency of the tenancy[2].

Rent variation

Where parties make a reference to an arbiter under s 5, the arbiter may, by reason of any provision in his award, if it is equitable, vary the rent of the holding[3]. Note the power to vary only arises under a s 5 reference. An arbiter does not have power to vary the rent where a dispute under a pre-1 November 1948 lease is referred to arbitration under the 1991 Act, s 60.

Increase of rent for landlord's improvements

Where a landlord has carried out an improvement on the holding (1) at the request of or with the agreement of the tenant, (2) in pursuance of an undertaking given by the landlord under s 39(3) or (3) in compliance with a direction given by the Secretary of State under powers conferred by any enactment[4], then the landlord may seek a rent increase[5].

The landlord requires to serve a notice in writing on the tenant within six months from the completion of the improvement requiring a rent increase[5].

The rent is increased from the completion of the improvement 'by an amount equal to the increase in the rental value of the holding attributable to the carrying out of the improvement'[5]. Where the landlord has received a grant out of money provided by parliament, the increase in rent is reduced proportionately[6].

Any question between landlord and tenant arising under s 15 falls to be determined by arbitration[7].

Fire insurance

In pre-1 November 1948 leases the landlord could require the tenant to effect and pay the premium for fire insurance on any or all of the fixed equipment. Where the tenant pays the premium and the landlord recovers any sum in respect of the destruction or damage of buildings by fire, unless the tenant agrees otherwise, the landlord is required to expend such sum on the rebuilding, repair or restoration of the buildings or subjects destroyed or damaged. The expenditure is to be in such manner as is agreed or, failing agreement, as may be determined by the Secretary of State[8].

---

1 Which is the same as the Agricultural Holdings (Scotland) Act 1949, s 100.
2 See p 9.
3 1991 Act, s 14(b).
4 Eg Agriculture (Safety, Health and Welfare Provisions) Act 1956, s 25(5).
5 1991 Act, s 15(1).
6 Ibid, s 15(2). NB it is money provided by Parliament, therefore an EC grant would not be discounted.
7 1991 Act, s 15(3).
8 Ibid, s 6.

The tenant under such a lease can have the lease varied to impose the fire insurance liability on the landlord[1].

In leases entered into on or after 1 November 1948 any provision that the tenant shall pay the whole or any part of the premium under a fire insurance policy is null and void[2].

THE RECORD

The parties are obliged to make a record of the condition of the whole fixed equipment for every lease entered into after 1 November 1948[3]. The record is deemed to form part of the lease. There is no penalty for failing to carry out the requirement.

The purpose of a record is 'to give such a description of the condition of the holding and its equipment as will convey to the mind of a person reading it ten, fifteen or twenty years later a clear idea of the state in which the farm was when the record was made'[4]. It is recommended that a photographic survey is always included in any record. The recorder should take soil samples and include the analysis in the record. The reporter is there to record fact not opinion[5].

At any time during the course of a tenancy the landlord or the tenant may require the making of a record of the condition of the fixed equipment on, and the cultivation of, the holding[6].

The tenant may at any time during the course of the tenancy require the making of a record of the existing improvements carried out by him for which he has, with the consent of the landlord, paid compensation to the outgoing tenant, and any fixtures or buildings that he is entitled to remove under s 18[7].

The record requires to be made in the prescribed form by a person appointed by the Secretary of State[8]. He need not be an arbiter or on the panel of arbiters. The record requires to show any consideration or allowance given by either party to the other[9]. The record may, if the landlord or tenant requires, relate only to part of the holding or fixed equipment[10]. Any question or difference between the landlord and tenant arising out of the record requires to be referred to the Land Court[11]. The cost of making the record[12], in default of any agreement, falls to be shared equally by the parties[13]. Any payment in excess of

---

1 1991 Act, s 4(1) and Sch 1.
2 Ibid, s 5(4); *Dunbar and Anderson* 1985 SLCR 1 at 14. On a proper construction of s 5(3) and (4) it would appear not to be competent to require the tenant to pay the premiums under a post-lease agreement.
3 1991 Act, ss 5(1) and 8.
4 *Marshall* p 219.
5 *Sheriff v Christie* (1953) 69 Sh Ct Rep 88.
6 1991 Act, s 8(1).
7 Ibid, s 8(2).
8 Ibid, s 8(3); Agricultural Records (Scotland) Regulations 1948, SI 1948/2817, amended by Agricultural Forms (Scotland) Amendment Regulations 1979, SI 1979/799.
9 1991 Act, s 8(4).
10 Ibid, s 8(5).
11 Ibid, s 8(6). NB arbitration is not available.
12 Fixed by the Secretary of State and subject to taxation by the auditor of the sheriff court; 1991 Act, s 8(8).
13 Ibid, s 8(7).

the share due by one party paid by him can be recovered from the other party[1].

A record made under s 8, in contrast to one made under s 5(1), is not deemed to be part of the lease, unless the parties so provide in the lease.

Apart from its importance as evidence of the condition of the fixed equipment at entry, and of the cultivation at the date made, the record may be relevant to (1) rent reviews under s 13, (2) certificates of bad husbandry, demands to remedy and subsequent notices to quit under s 21(2)(c), (d) and (e), and (3) waygoing claims under sections.

A landlord, under any lease entered into after 31 July 1931, only has a right to claim compensation for dilapidations or deterioration under s 45 upon termination of the tenancy if a record has been made[2]. In respect of leases entered into after 1 November 1948 the landlord can only claim compensation on the termination of the tenancy for dilapidations and deteriorations under the lease, if there is a record[3].

It is not clear whether or not a landlord retains a right to claim for dilapidations and deteriorations during the currency of a tenancy whether or not a record exists[4].

A tenant only has a claim for high farming if a record has been made[5].

For the purposes of claiming for dilapidations or deterioration the parties may agree in writing that a record made during a previous tenancy, subject to such modifications, if any, as may be specified, is deemed to have been made during the current tenancy[6].

1 1991 Act, s 8(9).
2 Ibid, s 47(2)(a), (b) and (3).
3 Ibid, s 47(2)(b) and (3).
4 See p 9.
5 1991 Act, s 44(2)(b).
6 Ibid, s 47(4).

CHAPTER 7

The rent and rent reviews

Key points

- The landlord should avoid granting a long lease without breaks for fear of having no right to seek a rent review during the currency of the lease.

- The significant differences in regard to procedure and appeal rights between a 'private' arbiter agreed between the parties and a 'statutory' arbiter appointed by the Secretary of State or the Land Court must be borne in mind.

- Careful preparation of relevant documentation and expert evidence is required for a rent arbitration.

Time limits

- A notice demanding a rent review requires to be served more than one year and less than two years before the next date at which the tenancy could be terminated: s 13(1).

- The arbiter must be appointed before the date from which the rent review is to take effect, although his decision can be given after the review date.

- An appeal against a rental award by a statutory arbiter to the Land Court has to be made within two months of the date of the issue of the award: s 61(3).

- A requisition for a special case from the Land Court for the opinion of the Court of Session has to be requested within one month of the intimation of the Land Court's decision: Scottish Land Court Rules, r 88.

RENT REVIEWS DURING CURRENCY OF THE TENANCY

The right to obtain a variation of rent under the 1991 Act, s 13 only arises from a date 'the tenancy could have been terminated by notice to quit'[1]. Thus a rent review can only be sought under the statutory provisions either at the date of the termination of the lease, before it begins to run on tacit relocation or at a break in the lease, or at the date when the lease could have been terminated on tacit relocation. At common law a rent review was only available at a break in the lease[2].

The case of *Moll v MacGregor*[3] has raised a question of whether or not it is competent to stipulate in the lease for rent reviews at dates prior to the termination of, or a break in, the lease. It was quite common in the 1970s when institutional investors were buying agricultural estates for their leases to be for 20 odd years with a provision that the rent would be reviewed every 5 years in line with the retail price index. In *Dunbar and Anderson*[4] the Land Court in commenting on a similar provision in a lease said:

'In our opinion, the condition in the lease regarding rent revision and its relation to the purchasing value of the pound, is binding on the parties for the duration of the lease.'

Moll v MacGregor related to a lease running on tacit relocation, where the parties had agreed that the rent would be reviewed in line with the retail price index. The Land Court said[5]:

'Having now considered the imperative tenor not only of the 1949 Act (as amended), but also of the subsidiary order governing rental arbitrations, the court conclude that it is not open to the parties, whether under the original lease or in a subsequent agreement, to contract out of the statutory rental provisions laid down in the public interest for arbiters to follow.'

Clearly parties cannot contract out of s 13 at the first break in the lease or at the termination of the lease, when it is running on tacit relocation. In so far as *Moll v MacGregor* suggests that parties cannot contract out of the statutory provisions during the currency of the lease or at least to the first break in the lease, it is *obiter*.

The object of s 13 is to provide a mechanism whereby rent is to be reviewed from the date at which the statutory security of tenure provisions take over, because without them the lease could be terminated at the particular date if the parties could not agree on future rent. It is unlikely that the purpose of s 13 was to prevent any rent reviews during the running of a long lease, where the parties had agreed such reviews, because during that period the tenant has security of tenure.

Further the parties are free to fix the first rent, which can be, and often is, well in excess of the statutory formula. In view of that objective, it is suggested that *Dunbar v Anderson* was correctly decided and *obiter* to the contrary in *Moll*

---

1 1991 Act, s 13(1).
2 *Strachan v Hunter* 1916 SC 901; *Edell v Dulieu* [1924] AC 38.
3 1990 SLT (Land Ct) 59.
4 1985 SLCR 1 at 14. This case was not cited or referred to in *Moll v MacGregor*.
5 1990 SLT (Land Ct) 59 at 66C.

v MacGregor is wrong[1]. Support for this view also comes from s 13(8)(b), which refers to variations of rent '(under this section or otherwise)', suggesting that otherwise variations are not contrary to the Act.

In view of the uncertainty a landlord would be wise to grant short leases, or at least leases with three-yearly breaks, so that the rent can be reviewed in terms of the statutory provision.

If parties can contract out of s 13 during the currency of the lease it should be noted that it is essential that the lease provides not only for the review dates, but also for the criteria upon which the rent is to be determined and the person by whom that determination is to be made, failing agreement. If there are no criteria upon which the rent can be determined by the appointed person, the provision in the lease will be void[2].

RENT REVIEWS UNDER S 13

Rent reviews are available in terms of s 13 'as from the next day after the date of the notice on which the tenancy could have been terminated by notice to quit ... given on that date'.

Parties are free to agree on a new rent at any time and probably on any basis. Such an agreement may well be binding if the parties act upon it. Section 13(1) provides that parties 'may' demand a reference to arbitration, which suggests they may otherwise agree the rent, particularly as s 13(8)(b) refers to variations '(under this section or otherwise)'. Parties cannot contract out of the statutory right to a reference of the rent which is to be paid at the next appropriate date[3].

Rent review is available under the statutory provisions at any break in the lease or at the termination of the lease provided that the rent increase or reduction made in consequence of a reference would not take effect earlier than the expiry of three years from (1) the commencement of the tenancy, (2) the date at which there took effect a previous rent variation under the section or otherwise or (3) the date as from which a previous direction took effect under s 13 that the rent should continue unchanged[4]. Consequential rent reviews under the Act[5] are disregarded for the purposes of calculating the three years[6].

If it is competent to contract out of the provisions of s 13 during the currency of the lease, if the lease provided for a rent adjustment during the last three years of the lease (eg an annual adjustment in line with the retail price index), then a review could not be sought until three years after that final adjustment, which might be a date after the ish of the lease[7].

The latter provision is quite important in that it relates to a direction under s 13 that there will be no change in the rent. If parties agree that there will be

---

1 See *Scammell and Densham* p 116 where the author suggests that there is no reason why parties should not contract out of the equivalent English provisions during the currency of the lease, although it might be contrary to public policy to contract out of the review formula.
2 Cf McBryde *The Law of Contract in Scotland* (W Green, 1987) para 4-27.
3 *Moll v MacGregor* 1990 SLT (Land Ct) 59.
4 1991 Act, s 13(8).
5 Eg under ibid, ss 14, 15(1) and 31.
6 Ibid, s 13(9).
7 See *Scammell and Densham* p 114.

no change in the rent they are not required under the statute to wait three years before demanding a reference to arbitration. Whether or not an agreement not to vary coupled with an agreement not to seek a further variation for three years amounts to a prohibited contracting out of the section is unclear.

THE NOTICE AND APPOINTMENT OF AN ARBITER

Section 13(1) requires a notice in writing demanding a reference to arbitration of the question of what rent should be payable in respect of the holding from the next date on which the tenancy could be terminated, to be served on the other party. The notice has to be served more than one year and less than two years before planned review date. This is the same timetable as for a notice to quit under s 21(3), because the review can only be at the 'next' date at which the tenancy could be terminated.

A notice should refer to s 13, specify the holding and lease to which it refers, and in particular the date at which the tenancy could next be terminated. It is essential to give the correct ish date, because a notice against the wrong ish is void. Section 84 applies to such notices.

A notice may be served by an uninfeft proprietor[1]. A notice requiring arbitration cannot be withdrawn without the agreement of the other party[2].

The parties may agree upon an arbiter, failing which the arbiter requires to be appointed by the Secretary of State[3]. It is probably essential that the arbiter should be appointed, even if he does not reach his final determination, before the date from which the revised rent is to run[4].

The parties can agree a joint reference to the Land Court[5], but such a course would preclude the parties' right of appeal to the Land Court[6], leaving them only with an appeal by special case to the Court of Session on a point of law.

Where the parties take no action on the notice, or after action has been taken allow the proceedings to lapse, they can be held to have departed from the notice and therefore the notice is spent[7].

The arbiter[8]

The parties may agree on an arbiter. Such an arbiter is a private one. The common law applies to such an arbitration, although the arbiter is required to

---

1 *Alexr Black and Sons v Paterson* 1986 SLT (Sh Ct) 64.
2 *Buckinghamshire CC v Gordon* (1986) 279 EG 853.
3 Application should be made in terms of the Agricultural Holdings (Specification of Forms) (Scotland) Order 1991, SI 1991/2154, Sch 2, Form B.
4 *Sclater v Horton* [1954] 1 All ER 712, CA; *Graham v Gardiner* 1966 SLT (Land Ct) 12; *University College Oxford v Durdy* [1982] Ch 413. Cf *Dundee Corporation v Guthrie* 1969 SLT 93. In so far as this case accepts that an arbiter can be appointed after the date from which the revised rent is to run, it should be noted that this was a second arbiter appointed to replace the first one, who had been appointed before the date from which the rent was to run.
5 1991 Act, s 60(2).
6 See below.
7 *Graham v Gardiner* 1966 SLT (Land Ct) 12; *Dundee Corporation v Guthrie* 1969 SLT 93 at 98.
8 The procedure in arbitration is dealt with in chap 23.

follow the statutory procedures set out in the 1991 Act, Sch 7. The arbiter should produce his award in the statutory form, modified as necessary[1].

Where an arbiter is appointed by the Secretary of State or the Land Court he is required to state in writing his finding of fact and the reasons for his decision and to make these available to the Secretary of State and the parties[2]. These are to be given in the statutory form[3]. Where an arbiter fails to follow the statutory provisions or give adequate reasons his award can be appealed[4].

The arbiter should conduct a formal inspection of the holding. Further evidence should not be taken at an inspection, but any material evidence which comes out during the course of the inspection upon which the arbiter may wish to rely should be revealed to both parties for their comment[5].

The arbiter is entitled to take into account his own general knowledge and expertise. He should not make use of specific comparables or other specific cases within his own knowledge of which he has not given notice to the parties and allowed them to comment thereon[6].

THE CRITERIA FOR DETERMINING THE RENT

The rent normally falls to be determined in terms of s 13(3) which provides:

'(3) ... the rent normally payable in respect of a holding shall normally be the rent at which, having regard to the terms of the tenancy (other than those relating to rent), the holding might reasonably be expected to be let in the open market by a willing landlord to a willing tenant, there being disregarded (in addition to matters referred to in subsection (5) below) any effect on the rent of the fact that the tenant is in occupation of the holding.'

This subsection provides (1) that the terms of the whole tenancy other than those relating to rent, which would include any post-lease agreement, have to be taken into account, (2) for the open market test of a willing landlord and willing tenant[7], (3) a disregard of certain tenant's improvements and (4) for a disregard of the tenant in occupation. Section 13(4) requires the arbiter to disregard any distortion of open market rents caused by scarcity.

In considering the case law in relation to s 13 it is important to bear in mind its legislative history. The 1949 Act, s 7 first provided for an arbiter to determine rent on the basis of 'what rent should be payable in respect of the hold-

---

1   *Gill* para 215. Agricultural Holdings (Specification of Forms) (Scotland) Order 1991, SI 1991/2154, Sch 1.
2   1991 Act, Sch 7, para 10.
3   SI 1991/2154. The form need not be followed slavishly; *Earl of Seafield v Stewart* 1985 SLT (Land Ct) 35. 'Seeking' in heading (vii) means considering such evidence as is presented; *Aberdeen Endowments Trust v Will* 1985 SLT (Land Ct) 23.
4   See eg *Aberdeen Endowments Trust* above; *Earl of Seafield* above; *Shand v Christie's Trs* 1987 SLCR 29 and *Strathclyde RC v Arneil* 1987 SLCR 44.
5   *Earl of Seafield* above.
6   *Dunbar and Anderson* 1985 SLCR 1 at 17; *Towns v Anderson* 1989 SLT (Land Ct) 17; *Fountain Forestry Holdings Ltd v Sparkes* 1989 SLT 853.
7   For a consideration of what is meant by the 'willing landlord, willing tenant' test see *Kilmarnock Estates Ltd v Barr* 1969 SLT (Land Ct) 10 and *F R Evans (Leeds) Ltd v English Electric Co Ltd* (1977) 36 P & CR 185.

ing'. This led to different arbiters taking into account many different factors[1]. Section 2 of the 1958 Act introduced the open market test including the disregards relating to improvements and the actual tenant, but made no provision for the distortion of the open market by scarcity. After 1958 the market became increasingly distorted by scarcity, because very few landlords were willing to re-let holding which had become vacant, preferring to take the land back in hand. The result of this was that tenants were being charged the open market rental, where such evidence as there was of open market rents generally included a substantial premium to reflect the fact that tenancies were a scarce commodity on the open market[2]. In order to remedy this imbalance, the 1983 Act, s 2 introduced the concept of reference to the open market undistorted by scarcity. The Act also amended the rent review period from five to three years[3].

It should be noted that the open market rent is the 'normal' basis for rent assessment. It is only where the evidence of open market rent is insufficient for the arbiter to determine the rent properly, or if he is of the view that the open market rents for comparable subjects 'in the surrounding area'[4] is distorted by scarcity or other factors, that he is to have regard to the secondary evidence specified in s 13(4)(a) to (d). Before an arbiter can invoke the provisions of s 13(4) he requires to be of the opinion that the open market evidence is insufficient for him to determine the rent properly payable or that the open market rents are distorted by scarcity, special factors such as marriage value[5] or other factors which prevent him relying on that evidence[6].

Having determined that s 13(4) comes into play the arbiter is required to have regard to the specific provisions in that subsection. What is the 'surrounding area' is a matter for the arbiter[7]. In considering sitting tenant rents the arbiter should take into account such factors as goodwill, lenient rental policies of a particular estate, or any particular relationship or other agreement between the landlord and tenant. 'Current economic conditions' does not include an unforeseen event occurring between valuation date and the date of the hearing[8]. The arbiter should also take account of the fact that comparable rents fixed at about the same time may well have been influenced by the current economic conditions.

---

1 Eg *Guthe v Broatch* 1956 SC 132 (the sitting tenant factor); *Secretary of State v Young* 1960 SLCR 31 and *Crown Estates Commissioners v Gunn* 1961 SLCR App 173 (tenant's personal circumstances).
2 See eg *Kilmarnock Estates and Tenants* 1977 SLCR App 141; *Witham and Stockdale* 1981 SLT (Land Ct) 27.
3 Agricultural Holdings (Amendment) (Scotland) Act 1983, s 2(d).
4 See *Moll v MacGregor* 1990 SLT (Land Ct) 59.
5 *Scammell and Densham* p 101 considers the question of whether or not the marriage value inherent in the holding itself should not be taken into account as an attribute of the holding in determining its value. A particular holding, because it is a pocket of bare land within an area of well-equipped holdings, might have an intrinsic marriage value to any one of those farms, which would enhance its open market value, as it would be unlikely to be let as bare land. *Scammell and Densham* reaches the view that it perhaps should not be taken into account, but does present a contrary argument.
6 *Dunbar and Anderson* 1985 SLCR 1 and *Kinnaird Trust and Boyne* 1985 SLCR 19 (no evidence of open market rents); *Aberdeen Endowments Trust v Will* 1985 SLT (Land Ct) 23, *Shand v Christie's Trs* 1987 SLCR 29 and *Kinnaird Trust* are all cases relating to scarcity, distortion and local considerations.
7 *Mackenzie v Bocardo SA* 1986 SLCR 53.
8 *Buccleuch Estates Ltd v Kennedy* 1986 SLCR 1.

Rents of crofts or small landholdings are not relevant as they are assessed under different statutory provisions on a 'fair rent' basis[1].

The arbiter may have regard to rent fixed in statutory arbitrations provided the full award and reasons are produced[2]. In general it is unlikely that there will be pure open market rental evidence available and most arbitrations are likely to be conducted on the basis of secondary evidence.

Of particular significance is the disregard of the actual tenant[3] and of any continuous standard of farming or system of farming more beneficial to the holding than the standard or system required by the lease, or normally practised on comparable holdings[4], practised by him. Linked to this is the disregard of dilapidations or deterioration of or damage to the fixed equipment or land of the holding caused or permitted by the tenant[5].

This means that all rent assessments have to be made on the basis of what a hypothetical tenant, taking into account the holding, its potential and the terms of the lease, would offer for the holding, disregarding scarcity.

THE VALUATION METHOD

In respect of arable and livestock farms the Land Court has suggested that the holding should be rented on an acreage basis, with the rental for each grade of land being assessed separately. It has disapproved the adjusted acreage method[6]. On hill farms the rent is assessed on a headage basis with each type of beast being rented separately[7]. On mixed farms a combination of methods may be appropriate.

PARTICULAR POINTS FOR CONSIDERATION IN RENT REVIEWS

(1) *Physical attributes.* The physical attributes of the farm, such as location (nearness to markets, local roads and transport etc), the layout of the farm, acreages, soil quality, climatic conditions, stock-carrying capacity and likely yields require to be ascertained.

The arbiter requires to take into account the rental value of fixed equipment on the holding provided by the landlord. Where the landlord has effected an improvement with grant aid the grant-aided element falls to be disregarded[8].

It is the 'holding' which is to be rented and not just the agricultural element

---

1 Crofters (Scotland) Act 1993, s 6(3).
2 *Towns v Anderson* 1989 SLT (Land Ct) 17 at 18D/E.
3 1991 Act, s 13(3).
4 Ibid, s 13(6).
5 Ibid, s 13(7).
6 *Mackenzie v Bocardo SA* 1986 SLCR 53.
7 See *Buccleuch Estates Ltd v Kennedy* 1986 SLCR 1.
8 1991 Act, s 13(5)(b); *Aberdeen Endowments Trust v Will* 1985 SLCR 38 at 45: 'If . . . there is a landlord's improvement attracting a 50 per cent grant that improvement will only attract half the rent which it would otherwise have done'. *Broadlands Properties Estates Ltd v Mann* 1994 SLT (Land Ct) 7. 'Moneys provided by Parliament' so EC grant aid would not be disregarded.

of it. The arbiter therefore requires to take into account income which the holding might produce from permitted ancillary activities, such as lets of farm cottages or caravans[1] or farm shops[2] etc.

In respect that the arbiter is required to disregard the existing tenant[3], he should take account of the potential of the holding which might be realised by the hypothetical offerer[4]. This would include ancillary non-agricultural potential not realised by the tenant.

(2) *The lease.* Section 13(3) requires the arbiter to have regard to the terms of the tenancy. If there is no formal lease and the parties cannot agree on the verbal terms of the tenancy or on what informal documents constitute the tenancy then the arbiter will have to determine the terms of the tenancy. The arbiter will have to take into account any statutorily implied terms or prohibition of terms.

The terms of the lease, including any post-lease agreement will have a material bearing on the rental which an offerer would offer on the open market. The respective repair obligations[5], a residence clause[6], an irritancy clause[7], a requirement to make the holding the tenant's full-time or main occupation[8], prohibition against letting grazings[9], or a bound sheep stock[10] are examples of factors which could have a bearing on rental.

A provision in the lease regarding the use to which the farm may be put, or restricting the right to compensation for improvements at the termination of the lease, which might arise out of a use restriction, fall to be taken into account as does any provision in the lease which varies (if such a variation is lawful[11]) statutory rights in relation to quotas.

Quite often contractual provisions in a lease are not enforced by either party. In such circumstances the arbiter has to determine to what extent such non-observance might affect the hypothetical offerer. In *Secretary of State v Davidson*[12] the Land Court held that even though the landlord was not enforcing a full-time working obligation, that nevertheless the rent should be assessed on the basis of the condition. This is consistent with the view that acquiescence in a breach of condition only operates while it continues[13].

In general, where a lease of a hill farm stipulates for a stocking rate for sheep the court will base the rental on the lease stocking rate rather than on the stock permitted to be carried in breach of the condition[14], because the lease condi-

---

1 *Kinnaird Trust and Boyne* 1985 SLCR 19 at 34, if such activities are not prohibited under the lease; see below.
2 Eg established under the Farm Diversification Grant Scheme 1987, SI 1987/1949 (as amended by SIs 1991/2 and 1991/1339); *Enfield London BC v Pott* [1990] 2 EGLR 7.
3 1991 Act, s 13(3).
4 *McGill v Bury Management Ltd* 1986 SLCR 32.
5 *Kilmarnock Estates and Tenants* 1977 SLCR App 141.
6 *Witham and Stockdale* 1981 SLT (Land Ct) 27; *Mackenzie v Bocardo* 1986 SLCR 53; *British Alcan Aluminium Co Ltd and Shaw* 1987 SLCR 1.
7 *Strathclyde RC v Arneil* 1987 SLCR 44.
8 *Secretary of State v Davidson* 1969 SLT (Land Ct) 7.
9 *Pentland v Hart* 1978 Tayside RN 13.
10 *Strathclyde RC v Arneil* above.
11 The arbiter may have to determine if such a provision is lawful, permitting the parties to contract out of the statutory provisions.
12 1969 SLT (Land Ct) 7.
13 *Rankine* p 112.
14 *Strathclyde RC v Arneil* 1987 SLCR 44.

tion is a potentially restrictive obligation. With the introduction of sheep premium quotas this will be of less importance, although the quota implications for stocking will be relevant.

Where a landlord is in breach of his repair and maintenance obligations, while the tenant does have a statutory right to demand a remedy, it has been held that a potential offerer on seeing such dilapidations would take account of the potential trouble he might have in trying to enforce such rights[1].

(3) *Milk quota.* The 1986 Act, s 16 introduced provisions for dealing with rent reviews where milk quota was registered in the name of the tenant of the holding. Section 16 applies where the tenant has milk quota, including transferred quota by virtue of a transaction, the cost of which was borne wholly or partly by the tenant, registered in the name of the tenant[2].

The arbiter or the Land Court is required to disregard any increase in rental value of the tenancy due to transferred quota in proportion to the cost of the transaction borne by the tenant[3]. Where the transferred quota affects only part of the tenancy that proportion of so much of the transferred quota as would fall to be apportioned to the tenancy[3] falls to be disregarded[4]. Any payment by the tenant for milk quota when he obtained the lease is disregarded[5]. Provision is made for successors to the original tenant to be treated as if the transferred quota had been paid for by the successors[6].

The tenant falls to be rented on allocated quota. In *Broadland Properties Estates Ltd v Mann*[7] the arbiter decided that as the allocated quota would not have been awarded to the farm but for improvements carried out by the tenant, any value attributable to the quota fell to be disregarded. The Land Court held that allocated quota could not be regarded as a tenant's improvement and that the effect of the 1986 Act, s 16 was that in determining the rent properly payable for the holding an allocated quota required to be taken into account[8].

It was said in *Broadlands Properties Estates Ltd v Mann*[9] that for the purposes of rent reviews 'where allocated quota attaches to the subject holding, an arbiter ... must assume that a successful hypothetical offerer, in making his offer, will have taken into account the extent to which the subjects on offer include allocated quota ...'. The present tenant's right to compensation at

---

1 *Strathclyde RC v Arneil* 1987 SLCR 44; *Towns v Anderson* 1989 SLT (Land Ct) 17.
2 1986 Act, s 16 relates to the situation where 'the tenant has milk quota ... registered as his'. In many cases the milk quota will be registered in the name of a partnership and not in the name of the tenant. It is not clear if s 16 operates in those circumstances. Cf p 226 'Entitlement' regarding the entitlement of a tenant 'who has milk quota as his' to compensation for milk quota on the termination of the lease. It appears that a tenant, where the quota is registered in the name of a partnership, has no right to compensation on the termination of the lease.
3 1986 Act, s 16(3).
4 Dairy Produce Quotas Regulations 1994, reg 10 and Sch 3, para 2.
5 1986 Act, s 16(4).
6 Ibid, s 16(5).
7 1994 SLT (Land Ct) 7.
8 While the Land Court's decision accords with the terms of the 1986 Act, s 16, it is open to question whether this section is contrary to EU law. In *Re the Kuchenhof Farm* [1990] 2 CMLR 289 the court held that milk quota was not an asset of the landlord. This view was reinforced in *Wachauf v Bundesamt fur Ernahrung and Fortwirtschaft* Case 5/88 1989 ECR 2609, [1991] 1 CMLR 328. If milk quota is not to be regarded as an asset of the landlord it is difficult to see why the landlord should be entitled to rental on the quota, particularly if the tenant was wholly responsible for bringing dairy production to the holding.
9 5 May 1994, Highland RN 449. The sequel to 1994 SLT (Land Ct) 7.

outgo fell to be disregarded. As quota was a restriction on the amount of milk that could be produced without superlevy 'the limitation imposed on their potential for dairying by the presence of allocated quota must be taken into account'.

Where the tenant in occupation in 1984 was dairy farming at below the level to be expected of a hypothetical tenant less quota may have been allocated to him than would have been allocated to the reasonably efficient hypothetical tenant. The effect of this is that in future rent arbitrations the arbiter has to assess the open market rent that would be paid by a hypothetical tenant. Such a hypothetical tenant would offer for the tenancy with its actual allocated quota, bearing in mind that he would have to buy or rent in additional quota in order to farm it to its reasonable productive capacity[1].

The converse, where a tenant was practising a system of farming which resulted in more quota being allocated than would have been allocated to a hypothetical tenant practising a reasonable standard of husbandry, may not however have the converse effect. It has been suggested that the arbiter will have to consider whether or not a hypothetical tenant who can only be expected to farm to a normal and reasonable system of husbandry could in fact utilise the whole quota, and if he could not, then the rent has to be based on that quota which he could use[2].

(4) *CAP and other quotas.* The arbiter may have to take account of CAP provisions (such as the arable area payment scheme) and other quota provisions, (such as sheep or suckler cow premium) as they relate to the holding under statute or in terms of the lease[3].

Where the quota belongs to the tenant[4], in considering the hypothetical offerer for a lease, it is not clear if the prospective tenant is to be viewed as already owning quota or whether in pitching his offer the prospective tenant is deemed to take into consideration the fact that he will have to buy in quota if the offer is accepted.

The impact of the Arable Area Payment Scheme will have to be considered.

(5) *Statutory restrictions*[5]. The arbiter may have to take account of any statutory restrictions on the holding such as SSSIs, nature conservation orders, environmentally sensitive areas, nitrate sensitive areas or other statutory schemes which have been imposed by statute or consented to by the landlord/tenant, which would be binding on the hypothetical tenant offering for a lease of the holding.

Payments due under environmental management agreements are also a relevant consideration.

1 See *Scammell and Densham* p 98.
2 *Scammell and Densham* p 99.
3 See chap 24.
4 Eg sheep and suckler cow premium quotas.
5 See chap 8.

PARTICULAR POINTS TO BE DISREGARDED IN A RENT ARBITRATION

(1) *The terms of the tenancy relating to rent.* The arbiter is required to have regard to the terms of the tenancy 'other than those relating to rent'. The Agricultural Holdings Act 1986[1] changed the law in England so that an arbiter now has to have regard to the terms of the tenancy, including those relating to rent. This is quite an important disregard in that the arbiter does not take into account such criteria as whether the rent is forehand or backhand or provisions as to interest and irritancy for late payment of rent.

(2) *The tenant.* The actual tenant is to be disregarded and the farm rented on the basis of the hypothetical offerer[2]. Thus any extra rent which a particular tenant might be prepared to pay for the holding (eg its marriage value to him[3]) falls to be disregarded.

(3) *Tenant's improvements and dilapidations.* The arbiter is not to take into account any increase in the rental value of the holding due to improvements (1) executed wholly or partly at the expense of the tenant, whether or not he is to be reimbursed by grants, where no equivalent allowance or benefit[4] has been given by the landlord in consideration for their execution and (2) which have not been executed under an obligation imposed by terms of the lease[5]. Where the landlord pays part of the costs of an improvement the tenant is rented on that contribution providing that the landlord did not receive grant money for that part[6].

This applies to improvements, whether or not the tenant is entitled to compensation for them at the termination of the lease[7]. This does not apply to improvements made by a predecessor in the holding or the tenant under a previous lease[8].

The Land Court suggests that the rent for the holding should be on the basis that the improvements do not exist, rather than on the basis of a calculation of the rent with the improvements and then taking out the value of the improvements[9]. Any increased carrying capacity of the holding derived from the improvement falls to be disregarded[10].

The arbiter is required to disregard any decrease in the rental value of the farm by reason of any dilapidation or deterioration of, or damage to, fixed

---

1   s 12 and Sch 2, para 1.
2   1991 Act, s 13(3).
3   See *Scammell and Densham* p 101.
4   A mere agreement to pay the statutory compensation is not an equivalent allowance. The allowance or benefit must be of near equivalent value to the improvement. It will be for the arbiter to determine if the allowance or benefit is of such value.
5   1991 Act, s 13(5)(a). *Scammell and Densham* p 105 suggests that 'terms of the lease' means the terms of the original lease and does not relate to subsequent agreement anent the construction of improvements entered into when the landlord's consent to the improvement is obtained. If the obligation imposed by the lease to provide an 'improvement' is contrary to the landlord's obligations under s 5(2), then the improvement is disregarded notwithstanding the obligation in the lease; *Grant v Broadland Properties Ltd* 1995 SLCR 39.
6   1991 Act, s 13(5)(b). This would not include EC grants.
7   *Towns v Anderson* 1989 SLT (Land Ct) 17 at 19F.
8   *Kilmarnock Estates v Barr* 1969 SLT (Land Ct) 10.
9   See *Mackenzie v Bocardo SA* 1986 SLCR 53 at 64 and *NCB v Wilson* 1987 SLCR 15 at 20.
10 *British Alcan Aluminium Co Ltd and Shaw* 1987 SLCR 1.

equipment or land caused or permitted by the tenant[1]. The holding should be rented as if the dilapidation did not exist. The disregard relates only to dilapidations caused or permitted during the current lease[2].

(4) *High farming.* The continuous adoption by the tenant of a standard of farming or a system of farming more beneficial to the holding than the standard or system required by the lease, or failing such a requirement the system normally practised on comparable holdings in the district, fall to be disregarded[3]. The word 'standard' would appear to include the skill of the farmer[4], which results in a higher standard of farming being disregarded.

(5) *Quotas.* The disregard of transferred milk quota where the tenant has borne all or part of the cost is referred to above. There may be other statutory disregards in respect of other quotas.

In respect that sheep and suckler cow premium quotas belong to the tenant, the arbiter will have to disregard these quota allocated to the tenant. However, the effect of these quotas may have to be taken into account in the consideration of open market rental values. It is not clear whether an arbiter will have to regard the hypothetical offerer for a tenancy as already owning quota or whether he will have to take into account the fact that the offerer may have to purchase quota if the offer for the tenancy is accepted.

The arbiter should probably assume that the hypothetical offerer is a newcomer to farming and then consider what quota might be available free from the national reserve and what quota will have to be bought or rented in.

(6) *Housing grants.* Any increase in rental value attributable to an improvement or repair grant paid under the Housing (Scotland) Act 1987, s 246 falls to be disregarded for the period during which the grant conditions require to be observed[5].

(7) *Opencast mining.* Any increase of or diminution in rental value of the holding attributable to the occupation of the holding or any part of it by British Coal falls to be disregarded[6].

(8) *Post-rental date occurrences.* An arbiter is required to disregard any post-rental valuation date events, which could not have been foreseen at the valuation date, even if those events might have an effect on future rental valuations[7]. He is to have regard to trends known at, or which a hypothetical tenant would take into account at, the valuation date[8].

1 1991 Act, s 13(7). See *Metropolitan Properties Co Ltd v Woolridge* (1986) 20 P & CR 64 for a commentary on a comparable (non-agricultural) provision in England.
2 *Findlay v Munro* 1917 SC 419; *Kilmarnock Estates v Barr* 1969 SLT (Land Ct) 10.
3 1991 Act, s 13(6).
4 The English provision in the Agricultural Holdings Act 1986, Sch 2, para 4 refers only to 'a system of farming'. In England the farmer's skill does not amount to high farming: *Scammell and Densham* p 109.
5 Housing (Scotland) Act 1987, s 256(4).
6 Opencast Coal Act 1958, s 14A(8) (as amended).
7 In *Buccleuch Estates Ltd v Kennedy* 1986 SLCR 1 the court held that a disastrous summer after the rental date and before the hearing was not 'a current economic condition' because it was not current at the valuation date.
8 Prospective (or rumoured, where these are affecting rental offers) Common Agricultural Policy reforms are a common factor that may require to be taken into account.

PREPARATION FOR A RENT REVIEW

The following comments are not intended to be exhaustive, but are intended to give guidance on what is required for a rent arbitration[1].

(1) *Documentation relating to holding.* The parties should produce the lease and post-lease documentation, any partnership agreement if the lease is to a limited partnership, a schedule of improvements including the cost thereof and any grant money received. The parties should specify whether or not they contend that all or part of any particular improvement falls to be rented or disregarded, and any substitute compensation agreements.

(2) *Expert witness.* It is essential that an expert witness is engaged to give evidence based upon a thorough examination of the holding and the comparables, including, where applicable, discussions with the respective tenants[2].

(3) *Productive capacity.* A budget should be prepared taking into account the productive capacity of the holding and its related earning capacity from the various enterprises that can be operated on the holding.

This budget should relate to the ordinarily competent tenant and may have to be compared to the actual budget of the holding, if the tenant is more competent or less competent than the ordinarily competent tenant.

(4) *Comparables.* Any comparables to be relied upon have to be specified in the pleadings so as to give notice to the other party with a view to allowing them to view the holding. As the English rent review provisions[3] are different from those in Scotland, in the Borders English comparables would have to be treated with care.

Full documentation[4] in relation to the comparables should be produced. The Land Court will order a landlord to produce evidence relating to comparables on the landlord's estate[5].

Comparables will have to be adjusted by the arbiter and any expert witness giving evidence, to take account of any differences between the subjects so that the rent can be interpreted and compared[5]. A letting to a limited partnership will have to be adjusted to take account of any variation in rent that the arbiter considers might apply to such a restriction on succession. Rent has to be adjusted for any difference in dates. The weight to be given to any particular comparable will depend on the closeness of the comparability. Comparables

1 See 'Guidance Notes for Valuers Acting in Reviews of Rent at Arbitration Under the Agricultural Holdings Act 1986' (RICS) which gives valuable guidance for expert witnesses in agricultural arbitrations, provided the differences between the English and Scottish Acts are borne in mind.
2 See the Land Court's comments re expert witnesses in *Kilmarnock Estates v Barr* 1969 SLT (Land Ct) 10.
3 1986 Act, s 12 and Sch 2.
4 Eg the lease, post-lease agreement, limited partnership agreement, list of improvements etc, substitute compensation agreements etc. If the comparable is an arbiter's award the award must be produced; *Towns v Anderson* 1989 SLT (Land Ct) 17. If the comparable is an open market renting then a list of offers and details of the offerors need to be produced.
5 Eg *Buccleuch Estates v Kennedy* 1986 SLCR 1.

found not to be closely relevant may nevertheless give evidence of the general market level[1]. It is for the arbiter, using his expertise, to make the appropriate comparisons between the holding and the suggested comparables or to reject them.

STATED CASES AND APPEALS TO THE LAND COURT IN RENT ARBITRATIONS

At any stage of the proceedings[2] a private arbiter may state a case on any question of law for the opinion of the sheriff and that decision may be appealed to the Court of Session[3]. There is no appeal to the Land Court against a decision of a private arbiter. Unless there are good reasons for the appointment of a private arbiter, parties would be well advised to have a statutory arbiter appointed, in order to preserve the right of appeal to the Land Court.

At any stage of the proceedings an arbiter appointed by the Secretary of State or by the Land Court may state a case on a question of law for the opinion of the Land Court, whose decision is final[4].

There is no right to state a case for the opinion of the sheriff or appeal to the Court of Session in respect of a statutory arbiter.

Where the arbiter appointed by the Secretary of State or the Land Court has issued his award the parties may appeal to the Land Court within two months of the date of issue of the award on any question of fact or law including the amount of the award[5].

The appellant to the Land Court requires to specify in his grounds of appeal why the arbiter's award is wrong, either in fact or in law. It is not sufficient just to state that the award is too low or excessive. In *Maciver v Broadland Properties Estates Ltd* the Land Court said[6]:

'... it is for the appellant to identify for the court in the arbiter's award and schedule any errors into which the arbiter had fallen in arriving at his determination, and to explain how and to what extent the award should be altered in order to rectify those errors. We consider that the court is entitled to be informed of the level of rent for which the appellant contends. It is not enough simply to say the award is "too high" or "too low".'

The Land Court can be asked to state a special case for the opinion of the Court of Session on any question of law during the course of the appeal or within one month of the date of the intimation of the decision[7].

1 *Witham and Stockdale* 1981 SLT (Land Ct) 27.
2 This means before the final award is issued.
3 1991 Act, Sch 7, paras 20 and 21.
4 Ibid, Sch 7, para 22. It may be competent to ask the Land Court to state a special case on a question of law for the opinion of the Court of Session while the proceedings are pending before it: Scottish Land Court Act 1993, s 1(7). Cf Scottish Land Court Rules, r 88 which does not appear to make provision for the statement of a special case before the issue of a decision. As the Land Court's decision is final a request for a special case after the decision is issued will not be competent.
5 1991 Act, s 61(2) and (3).
6 1995 SLT (Land Ct) 9 at 11J.
7 Scottish Land Court Act 1993, s 1(7); Scottish Land Court Rules, r 88. They do not appear to make provision for the stating of a special case while proceedings are pending in the court.

Joint reference of rent review to the Land Court

Parties who agree a joint reference to the Land Court of rent review[1] lose their right of appeal to the Land Court on any question of law or fact including the amount of the award. The loss of this right should be seriously considered before a joint reference is agreed.

The only right of review is to request a special case on any question of law for consideration by the Court of Session at any stage of the proceedings or within one month of the intimation of the Land Court's decision[2].

---

1 1991 Act, s 60(2).
2 Scottish Land Court Act 1993, s 1(7); Scottish Land Court Rules, r 88. They do not appear to make provision for the stating of a special case while proceedings are pending in the court.

CHAPTER 8

Restrictions on land use

Key points

- Where the Secretary of State or an arbiter has given a direction regarding the ploughing up of permanent pasture and has provided that the tenant will sow out an area in grass at the termination of the lease, the tenant is only entitled, by way of compensation, to the average value of the whole pasture laid down: 1948 Act, Sch 3, para 3(1)(b) and the 1991 Act, s 51(1)(b).

- A landlord should insist upon a provision in the lease as to disposal of produce and cropping to safeguard his position under s 7 at the termination of the tenancy: s 7(5).

- A tenant's freedom of cropping rights ceases in the year before the expiry of the lease and in the year before the tenant quits after receiving notice to quit or giving a notice of intention to quit. A tenant who receives a notice to quit would be well advised to cease freedom of cropping even if he intends to challenge the notice to quit: s 7(5).

- On notifying a site of special scientific interest (SSSI) to the local planning authority, the owner/occupier and the Secretary of State, Scottish Natural Heritage (SNH) requires to give parties an opportunity to make representations: Wildlife and Countryside Act 1981, s 28(2).

- Where SNH has notified an SSSI, the owner/occupier is prohibited from carrying out, causing or permitting to be carried out a prohibited operation without giving written notification to SNH and meeting the prescribed conditions: Wildlife and Countryside Act 1981, s 28(5),(6),(6A) and (6B).

- Where a nature conservation order has been made any person is prohibited from carrying out, causing or permitting to be carried out a prohibited operation without giving written notification to SNH and meeting the prescribed conditions: Wildlife and Countryside Act 1981, s 29(3),(4),(5),(6) and (7).

- Farmers and landowners require to be particularly alert to their obligations not to cause or knowingly permit pollution of controlled waters: Control of Pollution Act 1974, ss 31 and 32.

Time limits

- On notifying an SSSI, SNH is required to give three months in which representations and objections can be intimated: Wildlife and Countryside Act 1981, s 28(2).

- SNH has nine months from giving notification of an SSSI to the Secretary of State in which to withdraw or confirm the notification. If no withdrawal or confirmation is notified within nine months the notification ceases to have effect: Wildlife and Countryside Act 1981, s 28(4A).

- An owner/occupier who has given notice of a proposal to carry out a prohibited operation on an SSSI has to wait four months before such an operation is permitted, during which time SNH may negotiate regarding the proposals: Wildlife and Countryside Act 1981, s 28(6)(c) and (6A).

- Where SNH makes an offer of payment for a management agreement, the offeree has one month in which to refer the amounts offered to arbitration for a determination of the proper amount to be offered: Wildlife and Countryside Act 1981, s 50(3).

- An owner/occupier who has given notice of a proposal to carry out a prohibited operation on an SSSI and subsequently reached agreement with SNH can withdraw from the agreement on giving written notice, but thereafter is prohibited for one month from carrying out the operation: Wildlife and Countryside Act 1981, s 28(6B).

- Where a nature conservation order has been made no prohibited operation can be carried out until the expiry of either (1) the three months' notice or (2) where SNH has offered to enter into a management agreement or to purchase the interest, the date of reaching agreement or twelve months from the giving of the notice or three months from the rejection or withdrawal of the offer or (3) where a CPO is applied for, the date of confirmation or withdrawal of the order or refusal by the Secretary of State to confirm the order: Wildlife and Countryside Act 1981, s 29(5)(c),(6) and (7).

- A person aggrieved by the confirmation of a nature conservation order has six weeks in which to apply to the Court of Session to challenge the order: Wildlife and Countryside Act 1981, Sch 11, para 7.

PERMANENT PASTURE

Permanent pasture is not defined by the 1991 Act. The definition of pasture to include 'meadow' in the 1949 Act[1] has not been continued in the 1991 Act. In terms of s 7(6)(a) '"arable land" does not include land in grass which, by the terms of a lease, is to be retained in the same condition throughout the tenancy'[2]. Therefore permanent pasture can only be land which is required by the terms, actual or implied, of the lease to be retained as permanent grass throughout the lease.

Whether or not a particular pasture is to be held permanent pasture, where there is no definition in the lease, is probably a question of fact, taking into account the age and nature of the herbage and the actual and implied terms of the tenancy[3]. Generally the term will only apply to those lands which were pasture at the date of the lease[4].

A tenant is under an obligation of good husbandry to maintain 'permanent grassland (whether meadow or pasture) properly mown or grazed and in a good state of cultivation and fertility'. He is also required 'to exercise a systematic control of bracken and whins, broom and injurious weeds'[5].

The Secretary of State has a power *inter alia* to order permanent pasture to be ploughed up[6]. Compliance with such a direction, notwithstanding any provision of the lease or instrument affecting the land or any custom, does not render the person liable to sow out the land again at his expense[7]. Where the person is a tenant, the Secretary of State may provide by the relevant order that, on quitting the holding on the termination of the tenancy, the tenant should leave as permanent or temporary pasture, sown with a seeds mixture as may be specified in the order, such area of land (in addition to land required by the lease, as modified by the direction, to be maintained as permanent pasture) as may be specified in the order[8]. The area required to be left as permanent pasture or sown out per the order, should not exceed the area by which the lands required by the lease to be maintained as permanent pasture have been reduced by the order[8].

No compensation is payable to a tenant in respect of anything done in pursuance of an order under para 2[9]. In assessing compensation to an outgoing tenant, where land has been ploughed in pursuance of an order, the value per hectare of the pasture comprised in the holding is taken not to exceed the average value per hectare of the whole of the tenant's pasture comprised in the holding on the termination of the lease[10]. This is to ensure that the tenant cannot claim the good pasture as his and indifferent pasture as that laid down under the order.

---

1 1949 Act, s 93(1).
2 This definition is wider than permanent grass; *Rankine* p 417.
3 *Scammell and Densham* p 122, n 3; cf *McKenzie's Trs v Somerville* (1900) 2 F 1278.
4 *Rush v Lucas* [1910] 1 Ch 437; *Clarke-Jervoise v Scutt* [1920] 1 Ch 382.
5 1948 Act, Sch 6, para 2(a) and (f)(ii); see *Clarke v Smith* 1981 SLCR App 84.
6 1948 Act, s 35 (exercisable by statutory instrument) and Sch 3, para 1.
7 Ibid, Sch 3, para 1.
8 Ibid, Sch 3, para 2.
9 Ibid, Sch 3.
10 Ibid, Sch 3, para 3(1). 'Tenant's pasture' means pasture laid down at the expense of the tenant or paid for by the tenant on entering the holding; para 3(2).

Where a lease, whether entered into before or after the commencement of the 1991 Act, makes provision for the maintenance of specific land, or a specific proportion of the holding as permanent pasture, the tenant by notice in writing may require arbitration of the question 'whether it is expedient in order to secure the full and efficient farming of the holding that the amount of land required to be maintained as permanent pasture should be reduced'[1].

There has to be an actual (as opposed to implied) provision in the lease relating to specified land before a right to arbitration arises under s 9.

'Full and efficient use for agriculture' has been defined in a compulsory acquisition situation as meaning 'use of land as a commercial unit, ie use of land in the normal course of good farming by an active intelligent farmer'[2].

In an arbitration, the arbiter may by his award direct that the lease be modified as to the land which is to be maintained as permanent pasture or is to be treated as arable land and as to the cropping of that land[3]. Where the arbiter gives such a direction he may give a further direction that on the termination of the lease the tenant should leave, either as permanent or temporary pasture and with seeds mixture of such kind as may be specified, a specified area of land not exceeding the area by which the permanent pasture was reduced[4].

A tenant is not entitled to compensation for anything done in pursuance of a s 9(2) direction[5]. To avoid a tenant claiming compensation for his better pasture, while maintaining that the poorer pasture is the s 9(2) pasture, the average value of all the pasture is to be taken and applied to the area of pasture for which the tenant is entitled to compensation[6]. Pasture laid down under such a s 9(2) direction is not compensatable at the end of the tenancy as an improvement[7].

The landlord and the tenant are entitled to enter into an agreement in writing for a variation of the terms of a lease as could be made under s 9 and the agreement may provide for the exclusion of compensation as under s 51(1)[8].

FREEDOM OF CROPPING AND THE DISPOSAL OF PRODUCE

Many leases contain elaborate provisions regarding the rotation which is to be followed and prohibit the disposal of hay, straw and root crops from the holding. At common law the tenant was obliged to follow the rotation customary in the district or at least to leave the land in the customary rotation[9]. It is not the

---

1 1991 Act, s 9(1). This jurisdiction was formerly exercised by the Secretary of State but was transferred to arbitration under the 1958 Act.
2 *Secretary of State v MacLean* 1950 SLCR 33.
3 1991 Act, s 9(2).
4 Ibid, s 9(3).
5 Ibid, s 51(1)(a).
6 Ibid, s 51(1)(b).
7 Ibid, s 51(3)(c).
8 Ibid, s 53(2).
9 *Connell*, s 12, n 1; *Thomson's Reps v Oliphant* (1824) 3 S 194; *Hunter v Miller* (1862) 24 D 1011, (1863) 4 Macq 560.

custom of the estate[1], but the custom of the country or district, provided it can be shown that it was known to and relied upon by both parties[2].

Section 7 confers on the tenant of a holding, notwithstanding any custom of the country, any provision in the lease or any other agreement, the right to dispose of the produce[3] (other than manure[4]) of the holding and the freedom to practise any system of cropping of the arable land[5], without incurring any penalty, forfeiture or liability[6].

This general right is subject to particular restrictions. Where there is no custom of the country and no provision in the lease or agreement, which would be unusual, the tenant would be free to dispose of all the produce including manure and crop as he wished. He would not be subject to any of the restrictions in s 7 and nor would the landlord have any of the benefits of the section.

The tenant's freedom of cropping relates only to the arable land and not to any land which must be kept under grass throughout the tenancy or market garden ground[7]. Where the lease provides for a particular type of farming, the freedom to crop does not supersede that contractual restriction[8].

The tenant's freedom is subject to an obligation not to injure or deteriorate the holding, although it is implicit that he may temporarily do this. The tenant is under an obligation 'as soon as practicable' after exercising the rights to make suitable and adequate provision to return to the holding the full equivalent manurial value of any crops sold or removed off the holding in contravention of any custom or provision in the lease or agreement[9]. This obligation does not extend to crops lost through accident or fire, unless the lease so provides[10].

Further, where the tenant exercises a right to practise any system of cropping he must take steps to protect the holding from injury or deterioration[11]. Note it is the practice of 'any system of cropping', so the tenant has to have a planned system before he can avail himself of this freedom. System 'points to

---

1 *Allan v Thomson* (1829) 7 S 784; *Officer v Nicolson* (1807) Hume 827; *Anderson v Todd* (1809) Hume 842.

2 *Armstrong & Co v McGregor & Co* (1875) 2 R 339; *Anderson v M'Call* (1866) 4 M 765; *Holman v Peruvian Nitrate Co* (1878) 5 R 657.

3 'Produce' includes anything (whether live or dead) produced in the course of agriculture; 1991 Act, s 85(1).

4 This is an important exception in that manure would otherwise be included in the definition of produce in the 1991 Act, s 85(1).

5 1991 Act, s 7(6)(a) - arable land does not include permanent grazings. Market garden ground is not arable; *Taylor v Steel-Maitland* 1913 SC 562.

6 *Gore-Browne-Henderson's Trs v Grenfell* 1968 SLT 237 at 239 per Lord President: 'But s [7] does not make a lease containing such provisions illegal. It merely absolves a tenant in certain defined circumstances and subject to certain safeguards from complying with provisions in a lease which are inconsistent with his freedom of cropping'.

7 1991 Act, s 7(1) and (6)(a). The definition 'grass . . . which is to be retained in the same condition throughout the tenancy' is wider than permanent pasture; *Rankine* p 417. Market garden ground is not arable land; *Taylor v Steel-Maitland* 1913 SC 562.

8 *Muir Watt* p 68.

9 1991 Act, s 7(2)(a); see *Connell*, s 12, n 9.

10 *Hull and Meux* [1905] 1 KB 588. In terms of the 1991 Act, s 4 and Sch 1, para 6 a term can be incorporated in any lease requiring the tenant to return to the holding the full equivalent manurial value of harvested crops grown on the holding for consumption thereon in so far as required to fulfil the responsibilities of farming in accordance with the rules of good husbandry.

11 1991 Act, s 7(2)(b).

a deliberate, well-thought-out and husbandlike innovation, and bars any capricious, eccentric, exhausting, experimental, or sporadic cultivation'[1].

It is wise for a landlord to include provision in the lease as to the disposal of crops and cropping rotation, in order that he may secure the safeguards of s 7. Further, the provision in the lease will provide the norm of management agreed by the parties at the commencement of the lease against which to measure any claim that the tenant, by exercise of his rights, has injured or deteriorated the holding[2].

The landlord is protected against an abuse by the tenant of these freedoms. Where the tenant has exercised his rights so as to injure or deteriorate the holding or to be likely to injure or deteriorate the holding, the landlord has either a right to interdict at any time or to claim damages at the termination of the tenancy[3].

As previously mentioned the tenant is entitled to temporarily injure or deteriorate the holding so long as remedial steps are taken as soon as is practical thereafter. The landlord's right to interdict therefore only arises where the tenant is likely to cause gross or irreparable harm to the holding. If the tenant seems unlikely to be able to return the land to the appropriate rotation at the end of the lease this would be grounds for interdict[4].

In any action for interdict the question of whether or not the tenant has injured or deteriorated or is likely to injure or deteriorate the holding is a matter for an arbiter to determine. The arbiter's certificate is conclusive of the facts stated in the certificate for the purposes of any action or arbitration brought under s 7[5].

The landlord's right to damages only arises at the termination of the tenancy. Damages under this section are only due by the tenant in respect of any injury or deterioration to the holding caused by the exercise of the s 7 rights. An arbiter may have power to assess damages under this section[6].

The tenant's freedoms under s 7 cease when a notice to quit or notice of intention to quit is served, giving more than one and less than two years' notice to the termination of the lease, and which results in the tenant quitting the holding[7]. A tenant would be wise not to exercise these freedoms after a notice to quit is served, even if he serves a counternotice or intends to dispute the validity of the notice.

After a notice to quit or notice of intention to quit is served the tenant is prohibited from selling or removing any manure, compost, hay, straw or roots grown in the last year of the tenancy, unless he has given the landlord or incoming tenant a reasonable opportunity to purchase them on the termination of the tenancy at a fair market value[8].

1 *Rankine* p 418; see *Marshall* pp 110-111.
2 *Rankine* p 418.
3 1991 Act, s 7(3).
4 *Rankine* p 418.
5 1991 Act, s 7(4).
6 *Hill v Wildfowl Trust (Holdings) Ltd* 1995 SCLR 778; cf *McDiarmid v Secretary of State for Scotland* 1970 SLT (Land Ct) 17. See p 274 'Any question or difference' and *Gill* para 161.
7 1991 Act, s 7(5). The convoluted wording of this subsection has not altered since the 1906 Act where the distinctions were appropriate until the 1949 Act came into force. Since that date the only relevant period is the period after the notice is served.
8 1991 Act, s 17; see chap 16.

STATUTORY RESTRICTIONS

Recent years have seen the imposition of a number of statutory restrictions on land use introduced for environmental reasons or to comply with the Common Agricultural Policy. These schemes often alter year by year. Restrictions include:

(a) Quotas

Quota regimes include milk sheep annual premium, suckler cow annual premium[1].

(b) Environmentally sensitive areas (ESA)[2]

If it appears to the Secretary of State that it is particularly desirable

'(a) to conserve and enhance the natural beauty of an area;
(b) to conserve the flora or fauna or geological or physiographical features of an area; or
(c) to protect buildings or other objects of archeological, architectural or historic interest in an area,
and that the maintenance or adoption of particular agricultural method is likely to facilitate such conservation, enhancement or protection'

he may, after consultation with the Treasury and SNH, by order designate the area as an ESA[3].

The order may specify (1) the requirements as to agricultural practices, methods and operations and the installation or use of equipment which must be included in management agreements, (2) the period or minimum period for which the agreement must impose the requirements, (3) provisions regarding breach of such requirements, (4) the rates of payment and (5) requirements as to public access[4].

The Secretary of State may enter into management agreements, with any person having an interest in agricultural land which is in or partly in the ESA area, to manage the land in accordance with the agreement, for a payment[5].

---

1 See chaps 24 and 25.
2 The power to make an ESA implements in the United Kingdom the requirements of Title V of EC Council Regulation No 797/85 on improving the efficiency of agricultural structures.
3 1986 Act, s 18(1). Designation is by statutory instrument subject to annulment when laid before Parliament; ibid, s 18(12). The following ESA designation orders for Scotland are current: Whitlaw and Eildon (SI 1988/494); Breadalbane (SI 1992/1920, amended by SI 1992/2063); Loch Lomond (SI 1992/1919, amended by SI 1992/2062); Central Southern Uplands (SI 1993/996); Western Southern Uplands (SI 1993/997); Cairngorm Straths (SI 1993/2345); Central Borders (SI 1993/2767); Stewartry (SI 1993/2768); Argyll Islands (SI 1993/ 3136); Machair of Uists and Benbecula, Barra and Vatersay (SI 1993/3149); Shetland Islands (SI 1993/3150). Cf Environmentally Sensitive Areas (Scotland) Orders (Amendment) Order 1994, SI 1994/3067 which amends various designation orders allowing management agreements to include provision for management of land for public access.
4 1986 Act, s 18(4) (amended by SI 1994/249).
5 1986 Act, s 18(3). The agreement, or an agreement to terminate the agreement, can be registered in the Land Register or the Register of Sasines for enforcement against singular successors; ibid, s 19(1).

The agreement may contain such provisions as the Secretary of State thinks fit 'as are likely to facilitate such conservation, enhancement or protection' as mentioned in s 18(1)[1]. Agreements are voluntary and there are no compulsions unless the area is within an SSSI.

A tenant may enter into such an agreement provided he gives the absolute owner[2] notice and so certifies to the Secretary of State[3].

(c) Sites of Special Scientific Interest (SSSI)

In terms of the Wildlife and Countryside Act 1981, s 28[4] SNH[5] has a duty and wide powers to designate SSSIs and to notify the fact to the planning authority, the owner or occupier and to the Secretary of State. On notifying an SSSI, SNH requires to specify a time not less than three months from the date of notification in which representations and objections may be made[6]. The notification requires to specify:

'(a) the flora, fauna, or geological or physiographical features by reason of which the land is of special interest; and
(b) any operations appearing to [Scottish National Heritage] to be likely to damage that flora or fauna or other features.[7]'

Following notification SNH has nine months from the date it was served on the Secretary of State in which to give notice to persons affected, either withdrawing or confirming the notification with or without modifications[8]. The notification ceases to have effect if no notice of withdrawal or confirmation is given within nine months[9].

SNH is required to keep a register of notifications in respect of each local planning authority. The register must include copies of all notifications, plans referred to in the notifications and copies of all notices served under s 28(4C). Each local planning authority is required to keep a copy of the register relating to its area[10].

The owner or occupier of land which has been notified as an SSSI 'shall not while the notification remains in force[11] carry out, or cause or permit to be carried out' any operation specified in the notification unless SNH has been given written notice of the proposal, specifying its nature and the land on which it is to be carried out, and either (1) SNH has given written consent, (2) the oper-

---

1 1986 Act, s 18(5).
2 'Absolute owner' is no longer defined by the 1991 Act, s 85(1). He may be different from the landlord; see s 85(1).
3 1986 Act, s 18(6)(b).
4 As amended; cf Wildlife and Countryside (Amendment) Act 1985.
5 Natural Heritage (Scotland) Act 1991.
6 Wildlife and Countryside Act 1981, s 28(2).
7 Ibid, s 28(4). Notified 'operations' commonly include ploughing, re-seeding or drainage works.
8 Ibid, s 28(4A). The power to modify cannot be used to add specified operations or to extend the area; s 28(4B).
9 Ibid, s 28(4A).
10 Ibid, s 28(12), (12A) and (12B).
11 The prohibition starts with the notification, even though following representations and objections the notification may not be confirmed.

ation is carried out in terms of a management agreement[1] or (3) four months have expired from the giving of the notice[2].

During the four months SNH is required to respond to the notice. If agreement is reached that the four months is not to apply, then the owner/occupier may only carry out the operation with the consent of SNH or under a management agreement[3].

Where SNH offers to make payments in respect of a management agreement, the offer requires to be in accordance with the guidance given by the Secretary of State[4]. Within one month of receiving the offer, the offeree may require that the determination of the amount be referred to an arbiter[5]. Where the amount determined by the arbiter exceeds the amount offered, SNH may amend the offer to give effect to the determination, or, except in the case where an application for a farm capital grant has been refused in consequence of an objection by SNH, withdraw the offer[6].

The owner/occupier may give written notice terminating the agreement, in which case the operations remain prohibited for one month or such longer period as the owner/occupier has specified in the notice[7].

Further operations can only be prohibited by a nature conservation order, which SNH may ask the Secretary of State to make[8].

It is a summary offence to carry out a prohibited operation without complying with s 28(5) without reasonable excuse[9]. It is a reasonable excuse to carry out an operation (1) if the operation was authorised by planning permission[10] or (2) the operation was an emergency operation particulars of which, including the emergency, were notified to SNH as soon as practical after the commencement of the operation[11].

---

1 Under the National Parks and Access to Countryside Act 1949, s 16 or Countryside Act 1968, s 15.
2 1981 Act, s 28(5) and (6).
3 Ibid, s 18(6A). '. . . this regime is toothless, for it demands no more from the owner or occupier of an SSSI than a little patience. Unless the council can convince the Secretary of State that the site is of sufficient national importance to justify [a nature conservation order] . . . a task rarely accomplished': *Southern Water Authority v NCC* [1992] 1 WLR 775 (HL) per Lord Mustill.
4 1981 Act, s 50(2).
5 Ibid, s 50(3). Alternatively the parties could agree a joint reference to the Lands Tribunal for Scotland to act as arbiter under the Lands Tribunal Act 1949, s 1(5); see *Cameron v NCC* 1991 SLT (Lands Tr) 85.
6 1981 Act, s 50(3)(a) and (b).
7 Ibid, s 28(6B).
8 See *Southern Water Authority* above. See also 1981 Act, s 29 and the making of an NCO below.
9 Ibid, s 28(7). *Southern Water Authority* [1992] 1 WLR 775 (HL) where it was held that only the owner or occupier was liable. The water authority, which had come onto the land at the behest of the owners to carry out the operations, was held not to be liable. The expression 'occupier' referred to someone who, although lacking the title of owner, nevertheless stood in such a comprehensive and stable relationship with the land as to be in company with the actual owner, a person to whom the elaborate process of notices, waiting periods, agreements etc could sensibly apply. The section does not apply to a person whose occupation was transient, so a person who entered the land for a few weeks solely to do some work on it did not fall within the category of occupier.
10 The local planning authority has a duty to inform SNH of a planning application which affects an SSSI.
11 1981 Act, s 28(8).

(d) Nature conservation orders (NCO)

In terms of the Wildlife and Countryside Act 1981, s 29 the Secretary of State after consultation may make a nature conservation order which further restricts activities.

An NCO may be made where it appears expedient to the Secretary of State, after consultation with SNH, to do so:

'(a) ... for the purpose of securing the survival in Great Britain of any kind of animal or plant or of complying with an international obligation; or

(b) ... for the purpose of conserving any of its flora, fauna, or geological or physiographical features.'

The land in respect of which an NCO may be made must be of special interest and in respect of (b) above of national importance, by reason of its flora, fauna, or geological or physiographical features[1].

The effect of an NCO is to apply the Wildlife and Countryside Act 1981, s 29(3) to the land subject to the order[2]. Section 29(3) provides that no person[3] is to carry out on the land any operation which appears likely[4] to destroy or damage the flora, fauna, or geological or physiographical features of the land and which is specified in the order.

Section 29(3) does not apply if (1) the owner/occupier has given SNH written notice of a proposal to carry out the operation, specifying its nature and the land on which it is proposed to carry it out, and (2) the operation is carried out with the SNH's written consent or under a management agreement and three months have expired from giving the notice[5].

Schedule 11 of the 1981 Act provides a detailed procedure for advertising an order, making representations and objections, for local inquiries, and confirmation of the NCO.

A person aggrieved by an NCO has six weeks in which to apply to the Court of Session to challenge the order on the grounds that it was not within the powers conferred by ss 29 or 34 or that the requirements of Sch 11 had not been complied with[6].

Where the owner/occupier has given notice under s 29(4), SNH can give notice within the three months either offering to acquire the interest of that person or enter into a management agreement[7]. Then the three months is extended to the date where the agreement is entered into, if that is before the expiration of 12 months, or 12 months from the date of giving the notice, or three months from the date of rejection or withdrawal of the offer to enter into

---

1 1981 Act, s 29(2). *Sweet v Secretary of State for the Environment* [1989] 2 PLR 14 (it was held that the land could include not only the land of special interest, but the surrounding land whose management needed to be compatible with land of special interest to avoid its destruction).
2 1981 Act, s 29(1).
3 Cf 'no person' which is wider than the prohibition on owner/occupier in the 1981 Act, s 28(5).
4 '"Likely" falls to be understood as referring to what is probable': *North Uist Fisheries Ltd v Secretary of State for Scotland* 1992 SLT 333 at 336J.
5 1981 Act, s 29(4) and (5).
6 Ibid, Sch 11, para 7. *North Uist Fisheries* above.
7 Under the National Parks and Access to the Countryside Act 1949, s 16 or Countryside Act 1968, s 15.

the agreement[1]. If, before the expiry of the three months mentioned in s 29(5)(c) or the extension of that period by s 29(5), an order is made for the compulsory acquisition by SNH of the interest of the person who gave the notice, then the period is extended either to the day on which SNH enters the land or, in any other case, the day the order is withdrawn or the Secretary of State decides not to confirm it[2].

Where an NCO is made a person with an interest in an agricultural unit may claim compensation if he can show 'that the value of his interest is less than what it would have been if the order had not been made' equal to the difference between the two values. Where SNH has given notice of an NCO then any person with an interest in the land can claim compensation for any expenditure reasonably incurred in carrying out work rendered abortive, or for any loss and damage incurred by the extension of time limits[3].

A person who without reasonable excuse[4] contravenes s 29(3) is liable to the maximum fine on summary conviction or to a fine on conviction of indictment[5]. Where a person is convicted of an offence, then the court by which he was convicted, in addition to any other penalty, may make a restoration order[6].

(e) Nature reserves[7]

SNH or a planning authority has power to enter into agreements with owners, limited owners, occupiers and lessees of land, if it appears to SNH that the land should be managed as a nature reserve in the national interest[8].

Such an agreement may impose restrictions on the rights over the land and also impose management practices on the persons bound by the agreement.

Compulsory purchase provisions apply where agreement cannot be reached on reasonable terms[9].

Where land is being managed as a nature reserve under an agreement, or is being held and managed by SNH or an approved body as a nature reserve, and the land is of national importance, SNH may declare the land to be a national nature reserve and make byelaws for the protection of the reserve[10].

---

1 1981 Act, s 29(6).
2 Ibid, s 29(7).
3 Ibid, s 30(2) and (3). The claim must be made in the prescribed form - Wildlife and Countryside (Claims for Compensation under section 30) Regulations 1982, SI 1982/1346. Land Compensation (Scotland) Acts 1963 and 1973 apply to a claim; 1981 Act, s 30(5). Claims for disputed compensation are referred to the Lands Tribunal for Scotland; ibid, s 30(8).
4 Ibid, s 29(9); ie the operation is authorised by planning permission or the operation was an emergency notified to SNH as soon as practical after the commencement of the operation.
5 Ibid, s 29(8).
6 Ibid, s 31. The order may be discharged or varied on a change of circumstances; s 31(4). If the person fails without reasonable excuse to comply with the order he may be fined on a summary conviction; s 31(5). If the operations are not carried out within the time ordered, SNH may enter the land and carry out the operations and recover the costs; s 31(6).
7 9 *Stair Memorial Encyclopaedia* para 1237.
8 National Parks and Countryside Act 1949, s 16(1) and (5).
9 Ibid, s 17.
10 1981 Act, s 35.

(f) Country and regional parks, public paths and long-distance routes

SNH and local authorities have power to make country or regional parks, public paths and long-distance routes, which may impinge on agricultural land use[1].

(g) Nitrate sensitive areas

Under the Control of Pollution Act 1974[2] the Secretary of State, with a view to preventing or controlling nitrate from entering controlled waters as a result of, or of anything done in connection with, the use of any land for agricultural purposes, may designate areas of agricultural land as nitrate sensitive areas[3].

By agreement with the absolute owner, or with another person having an interest in the land where the owner has given his written consent, the Secretary of State may enter into an agreement under which, in consideration of payment to be made, the person accepts such obligations with respect to the management of the land or otherwise as are imposed by the agreement[4].

The agreement can contain 'requirements, prohibitions or restrictions'[5]. It can be registered in the Register of Sasines or Land Register for enforcement by the Secretary of State against singular successors[6]. A termination of the agreement may also be registered.

Failing agreement, the Secretary of State may impose by order, requirements, prohibitions or restrictions on activities in the designated area[7].

The Secretary of State and the river purification board are given powers of entry to ensure that agreements and orders are being observed and to carry out borings or other works or to install and keep monitoring apparatus on the land[8].

Any practice adopted by a tenant in compliance with any obligation under the Control of Pollution Act, s 31B is disregarded by the Land Court when considering any question of bad husbandry[9].

(h) Control of pollution of rivers and coastal waters

The provisions of the Control of Pollution Act 1974[10] for the control of pollution of, or the discharge of trade and sewage effluent into, controlled or coastal waters has a substantial impact on agricultural land usage to prevent such pollution or discharge.

Under the 1974 Act, s 31 a person is guilty of an offence if he:

1 For a more detailed treatment see 9 *Stair Memorial Encyclopaedia* paras 1239 and 1240.
2 As amended by the Water Act 1989.
3 Control of Pollution Act 1974, s 31B(1).
4 Ibid, s 31B(2).
5 Eg construction of slurry tanks.
6 1974 Act, s 31C.
7 Ibid, s 31B(3) and (4).
8 Ibid, s 31D.
9 1991 Act, s 26(2).
10 ss 31 and 32 (substituted by the Water Act 1989, Sch 23).

'causes or knowingly permits-

(a) any poisonous, noxious or polluting matter to enter controlled waters[1]; or

(b) any matter to enter any inland waters so as to tend ... to impede the proper flow of the waters in a manner leading or likely to lead to substantial aggravation of pollution due to other causes or consequences; or

(c) any solid waste matter to enter controlled waters.[2]'

'Causes' is given its common sense meaning and does not require any mens rea, knowledge or fault, although there must be some active participation in the operation or chain of operations resulting in the pollution. The effect is that the offence is almost absolute[3]. If a person erects a construction on his land, such as a slurry tank which overflows through no fault of his own[4], or has pipes with a latent defect, which result in polluting material reaching the river, or has pipes through which a third party, who has a right of discharge, discharges polluting matter[5], then he can be convicted.

A company[6] is liable for the actions of employees causing pollution[7].

'Knowingly permits' requires knowledge of the incident and thereafter failure to take the appropriate steps to prevent the discharge or continuing discharge[8].

It is a defence if the discharge was authorised, or permitted in an emergency in order to avoid danger to life and health, provided steps are taken to minimise the entry and as soon as practical thereafter the river purification board is informed[9].

The Secretary of State may designate areas in which the carrying on or restriction of activities is prohibited, if he considers that the activities are likely to result in pollution of waters[10]. The river purification board may make byelaws prohibiting or regulating the washing of or cleaning of specified things[11].

Under the 1974 Act, s 32 a person is guilty of an offence 'if he causes or knowingly permits – any trade or sewage effluent to be discharged' into controlled waters, from land into the sea outside controlled waters, from a building or plant onto or into any land or into any waters[12], unless the discharge is made in accordance with a consent obtained under s 34.

---

1 Defined by the 1974 Act, s 30A(1) to include relevant territorial and coastal waters, inland waters and ground waters.

2 Ibid, s 31(1).

3 *Alphacell v Woodward* [1972] AC 824; *Lockhart v NCB* 1981 SLT 161; *Price v Cormack* [1975] 1 WLR 988; *Wychavon DC v NRA* [1993] 1 Env LR 330; *NRA v Yorkshire Water Services Ltd* [1995] Env LR 119; *CPC (UK) Ltd v NRA* [1994] Env LR 131; *AG Reference (No 1 of 1994)* [1995] 2 All ER 1007, CA.

4 *Alphacell v Woodward* above.

5 *NRA v Yorkshire Water Services Ltd* [1995] Env LR 119; *CPC (UK) Ltd v NRA* [1994] Env LR 131.

6 And probably a Scottish partnership; Partnership Act 1890, s 4(2).

7 *NRA (Southern Region) v Alfred McAlpine Homes East Ltd* [1994] Env LR 198.

8 *Noble v Heatly* 1967 JC 5; *Wychavon DC v NRA* [1993] 1 Env LR 330 at 338 where it was observed that failure to take immediate steps to discover the cause of a discharge and to stop it might amount to knowingly permitting the discharge.

9 1974 Act, ss 31(2) and 32.

10 Ibid, s 31(4).

11 Ibid, s 31(6). Now SEPA.

12 See above regarding 'causes' and 'knowingly permits'.

It is an offence to discharge any matter other than trade or sewage effluent into controlled waters from a sewer or from a drain which the road authority is obliged to keep open[1] without a consent obtained under s 34.

The effect of these provisions is that farmers have to be particularly careful in circumstances where they generate or keep polluting matter on the farm[2] or where they spread slurry or other matter on the fields. Where there is a discharge of trade or sewage effluent from the farm it is essential that the appropriate consents are obtained and complied with.

(i) Set-aside and arable areas payment scheme

The Set-Aside Regulations 1988[3] provided under EC regulations for a system of restriction of production from arable land. This has now been superseded by the Arable Areas Payment Scheme[4].

1 1974 Act, s 32(1)(b) and (c).
2 Eg slurry and slurry tanks, and feed tanks containing liquid feed.
3 SI 1988/1352 (as amended by SIs 1989/1042, 1990/1716 and 1991/1933).
4 See chap 25.

CHAPTER 9

Improvements

Key points

- A record of the holding, made preferably at the commencement of a tenancy or at least at some stage during the tenancy, will be of inestimable evidential value in determining (1) what are the improvements and (2) the value thereof.

- Agreements that the incoming tenant will pay the compensation due to the outgoing tenant, except in so far as they relate to Part 3 improvements under an agreement in writing stating a maximum sum, are null and void: s 35(2) and (3).

- Part 1 improvements: compensation is not payable unless the landlord's consent in writing was obtained prior to the improvements being carried out. The consent can be conditional or provide for no compensation.

- Part 2 improvements: compensation is not payable unless notice in writing was given to the landlord, or dispensed with in the lease or a prior agreement, three months before the improvement was carried out.

- Part 2 improvements: the landlord's response to a notice intimating an intention to carry out an improvement is either to enter into an agreement relating to the carrying out of the improvement and substitute compensation or to object to the notice within one month of receipt.

- Part 2 improvements: where a landlord objects to an improvement the tenant's response, if he wishes the improvement carried out, is to apply to the Land Court for approval: s 39(2).

Time limits

- Part 2 improvements: the tenant requires to give the landlord notice in writing three months before commencing the improvement: s 38(3)(c).

- Part 2 improvements: the landlord has one month from receipt of the notice from the tenant intimating an intention to execute a Part 2 improvement to object to the improvement or the manner in which it is to be carried out: s 39(1).

- Part 2 improvements: where a landlord has objected to an improvement, but the Land Court granted approval, the landlord may within one month of receiving notice of the decision serve notice in writing on the tenant undertaking to carry out the improvement himself: s 39(3).

- Part 2 improvements: where a landlord effects a Part 2 improvement himself he may seek a rent increase provided he serves notice in writing on the tenant within six months of the completion of the improvement: s 15.

GENERAL

The Agricultural Holdings (Scotland) Act 1883 first gave the tenant of an agricultural holding a statutory right to compensation for improvements, where none was provided for in the lease, or any agreement or custom of the country. At common law a tenant had no right to compensation for improvements in the absence of an agreement[1].

The law relating to improvements is directed towards the objective of allowing a tenant to claim compensation for the value of the improvement at waygoing and to encourage him to keep the land in good heart during the last few years of the tenancy. It is subject to the statutory conditions with limited provisions for opting out.

Improvements are characterised as landlord's improvements or tenant's improvements. Tenant's improvements are defined as '1923 Act'[2] and '1931 Act'[3] improvements, which are called 'old' improvements. Improvements executed on an agricultural holding on or after 1 November 1948 are called 'new'[4] improvements, whether or not the tenant entered into occupation before or after that date[5].

The definition of old improvements to include 1923 and 1931 Act improvements arises out of the provisions of those Acts in regard to the conditions upon which certain improvements could be effected and the compensation payable therefor. In the 1923 Act many improvements required the landlord's consent, whereas in the 1931 Act many of the improvements which required consent under the 1923 Act were reclassified as Part 2 improvements requiring only notice[6].

The landlord was given a right by the 1931 Act to object to improvements, where notice alone was required. These differences are now only of importance in regard to the compensation which might be payable on waygoing[7].

Where a lease was entered into before 1 January 1921, a tenant is not entitled to compensation for an old or a new improvement which he was required to carry out in terms of his lease[8].

Where a tenant has remained[9] in an agricultural holding[10] during two or more tenancies he is not deprived of his right to compensation for the improvements by reason only that they were not carried out during the tenancy on the termination of which he quits the holding[11]. This covers situations where there

---

1 *Earl of Galloway v M'Lelland* 1915 SC 1062 at 1099.
2 1991 Act, Sch 3.
3 Ibid, Sch 4.
4 Ibid, Sch 5.
5 Ibid, s 33.
6 Cf ibid, Schs 3 and 4.
7 See chap 19.
8 1991 Act, s 34(2).
9 This implies that the tenancies must have been continuous.
10 The corresponding English provision is in the Agricultural Holdings Act 1986, Sch 9, para 5(1) which refers to 'has remained in *the holding*'. In England it has been suggested that this means that 'there must not be a material change in the identity of the holding for the provision to apply'; *Scammel and Densham* p 300, n 5. As the 1991 Act refers to 'a holding' it may be sufficient for the original holding to be part of a new enlarged holding held on a succeeding tenancy for the provision to apply.
11 1991 Act, s 34(5).

is a technical surrender and grant of a new lease[1] as well as situations where the parties agree a new lease. A new lease would generally make provision for liability for improvements.

In general, a tenant has a right to carry out the statutory types of improvements, some requiring the consent of the landlord ('Part 1 improvements'), others requiring written notice to the landlord ('Part 2 improvements'), and some that could be effected without notice ('Part 3 improvements')[2]. The tenant may be entitled to compensation at waygoing for some of the improvements effected by him or his predecessors in the tenancy[3], although this may depend on the terms of the lease or any agreements[4].

Where a tenant is market gardening different provisions apply[5].

The rental value of the holding depends on who provided the improvements[6].

A record of the holding made, preferably, at the commencement of the tenancy or during the course of the tenancy will be of inestimable value in determining (1) what the improvements were and (2) the value of such improvements[7].

AGREEMENT THAT INCOMING TENANT SHALL PAY OUTGOING TENANT FOR IMPROVEMENTS

Any agreement made after 1 November 1948 whereby the incoming tenant pays the outgoing tenant or refunds the landlord any compensation paid for improvements is null and void[8]. This does not apply to Part 3 improvements if the agreement is in writing and stipulates for a maximum amount[9].

The prohibition applies only to incoming tenants and the landlord. The outgoing and incoming tenants may reach an agreement with the landlord's consent.

Where an incoming tenant, with the landlord's written consent pays the outgoing tenant compensation in respect of an old improvement, in pursuance of an agreement in writing made before 1 November 1948, or where the 1991 Act, s 35(3) applies, the incoming tenant is entitled on quitting the holding to claim compensation for the improvement, if any, as the outgoing tenant would have been entitled to claim if he had remained on the holding[10].

Where an incoming tenant, in circumstances other than those in s 35(2) and (3), pays to the landlord any amount in respect of the whole or part of a new improvement, then he is, subject to any written agreement, entitled to such

---

1 Cf *Tufnell and Nether Whitehaugh Co, Applicants* 1977 SLT (Land Ct) 14; *Mackie v Gardner* 1973 SLT (Land Ct) 11; *Jenkin R Lewis & Son Ltd v Kerman* [1971] Ch 477.
2 1991 Act, Sch 5.
3 Ibid, Schs 3, 4 and 5.
4 Compensation for improvements at waygoing is dealt with in chaps 18 and 19.
5 1991 Act, Sch 6 and ss 40, 41, 73 and 79; see chap 21.
6 Cf ibid, s 13.
7 Ibid, ss 5(1) and 8.
8 Ibid, s 35(1).
9 Ibid, s 35(2) and (3).
10 Ibid, s 35(4).

compensation, if any, at outgo as he would have been entitled to if he had been tenant at the time the improvement was carried out[1].

OLD IMPROVEMENTS

(a) General

Old improvements such as buildings can still have a substantial value at waygoing, although the majority of old improvements are likely to be of little present value or may well have been written off under the agreement under which they were effected.

(b) Particular provisions

(1) A tenant whose lease was entered into before 1 January 1921 is not entitled to compensation for improvements that he was required to carry out in terms of his lease[2].

(2) A tenant is not entitled to compensation for an old improvement if the improvement was begun at a time when the land was not a holding within the meaning of the 1923 Act as originally enacted[3] or was land to which provisions of that Act relating to improvements and disturbance were applied by s 33 of the 1923 Act[4].

(3) Nothing in the 1991 Act, s 34 prejudices the right of a tenant to any compensation to which he is entitled 'under custom, agreement or otherwise'[5]. This may cover a verbal agreement. Compensation claimed under s 34(4)(a) would have to be alternative to any compensation claimed under the Act, although the claims could perhaps be put forward in the alternative[6].

(4) Where a tenant lays down temporary pasture as a 1923 Act improvement[7] in contravention of the lease or any agreement respecting the method of cropping, a right derived from the freedom of cropping provisions in s 7, then he is entitled to compensation under deduction in respect of any injury or deterioration to the holding caused by the contravention, except in so far as the landlord has recovered damages therefor[8].

1 1991 Act, s 35(5).
2 Ibid, s 34(2).
3 1923 Act, s 49(1) - '"Holding" means any piece of land held by a tenant which is either wholly agricultural or wholly pastoral, or in part agricultural and as to the residue pastoral, or in whole or in part cultivated as a market garden, and which is not let to the tenant during his continuance in any office, appointment, or employment held under the landlord'. Up until the 1920s it was quite common for a holding to be let during the continuation of the estate employee's office, appointment or employment. It is therefore worth checking that this was not the situation at the time. Valuation rolls at the time often show whether or not this was a landlord's holding let to an employee.
4 1991 Act, s 34(3).
5 Ibid, s 34(4)(a).
6 *Scammell and Densham* p 298, n 5.
7 1991 Act, Sch 4, para 28.
8 Ibid, s 34(6).

(5) Where a tenant under a written agreement made before 1 January 1921 is entitled to substitute compensation, which is fair and reasonable with regard to the existing circumstances in respect of 1923 and 1931 Part 3 improvements, such compensation shall be a substitute for the statutory compensation[1].

Whether or not the agreement was fair and reasonable at the time it was entered into will be a matter of fact for the arbiter to determine. The onus will be on the party contending that the agreement was not fair and reasonable to rebut the presumption that it is, and prove his case upon sufficient averments[2].

The provision is probably of little practical effect as Part 3 improvements are by their nature temporary and are by now likely to be spent.

(6) 'Repairs to buildings, being buildings necessary for the proper cultivation or working of the holding, other than repairs that the tenant is himself under an obligation to execute'[3] cannot be claimed as old improvements unless prior notice in writing was given to the landlord under the 1923 or the 1931 Acts, of the tenant's intention to execute the repairs and particulars thereof, and the landlord failed to exercise his right to exercise the repairs[4].

The parties can agree on compensation or otherwise[5] on which the improvement is to be carried out. The tenant is only entitled to statutory compensation where no agreement was made, the tenant did not withdraw his notice and the landlord failed to carry out the improvement or where the landlord objected, and authority was given to the improvement being carried out[6].

NEW IMPROVEMENTS

The types of new improvement which require the landlord's consent or notice to the landlord or which can be effected without notice or consent and for which compensation may be payable are set out in Sch 5. They include:

PART I – *Improvements for which consent is required*
1. Laying down of permanent pasture.
2. Making of water-meadows or works of irrigation.
3. Making of gardens.
4. Planting of orchards or fruit bushes.
5. Warping or weiring of land.
6. Making of embankments and sluices against floods.
7. Making or planting of osier beds.
8. Haulage or other work done by the tenant in aid of the carrying out of any improvement made by the landlord for which the tenant is liable to pay increased rent.

---

1 1991 Act, s 34(7).
2 *Bell v Graham* 1908 SC 1060.
3 1991 Act, Schs 3 and 4, paras 29.
4 Ibid, s 34(8).
5 This includes an agreement to pay no compensation; *Turnbull v Millar* 1942 SC 521.
6 1991 Act, s 38(4). Approval for old improvements was given by a government department or the Secretary of State.

PART II – *Improvements for which notice is required*

9. Land drainage.
10. Construction of silos.
11. Making or improvement of farm access or service roads, bridges and fords.
12. Making or improvement of watercourses, ponds or wells, or of works for the application of water power for agricultural or domestic purposes or for the supply of water for such purposes.
13. Making or removal of permanent fences, including hedges, stone dykes and gates.
14. Reclaiming of waste land.
15. Renewal of embankments and sluices against floods.
16. Provision of stells, fanks, folds, dippers, pens and bughts necessary for the proper conduct of the holding.
17. Provision or laying on of electric light or power, including the provision of generating plant, fixed motors, wiring systems, switches and plug sockets.
18. Erection, alteration or enlargement of buildings, making or improvement of permanent yards, loading banks and stocks and works of a kind referred to in the Housing (Scotland) Act 1987, Sch 8, para 13(2) (subject to the restrictions mentioned in that subsection).
19. Erection of hay or sheaf sheds, sheaf or grain drying racks and implement sheds.
20. Provision of fixed threshing mills, barn machinery and fixed dairying plant.
21. Improvement of permanent pasture by cultivation and re-seeding.
22. Provision of means of sewage disposal.
23. Repairs of fixed equipment, being equipment reasonably required for the efficient farming of the holding, other than repairs which the tenant is under an obligation to carry out.

PART III – *Improvements for which no consent or notice required*

24. Protecting fruit trees against animals.
25. Clay burning.
26. Claying of land.
27. Liming (including chalking) of land.
28. Marling of land.
29. Eradication of bracken, whins or broom growing on the holding at the commencement of the tenancy and, in the case of arable land, removal of tree roots, boulders, stones or other like obstacles to cultivation.
30. Application to land of purchased manure and fertiliser, whether organic or inorganic.
31. Consumption of the holding of corn (whether produced on the holding or not) or of cake or other feeding stuff not produced on the holding by horses, cattle, sheep, pigs or poultry.
32. Laying down of temporary pasture with clover, grass, lucerne sainfoin, or other seeds, sown more than two years prior to the termination of the tenancy, in so far as the value of the temporary pasture on the holding at the time of quitting exceeds the value at the commencement of the tenancy for which the tenant did not pay compensation.

Where a lease provides that the farm is let as a particular type of farm or that a particular type of farming is not to be carried out upon the holding, then that might restrict the type of improvement which the tenant would be entitled to effect or in any event would give grounds to the landlord to object to a particular improvement associated with a type of farming not permitted by the lease[1].

(a) Part 1 improvements requiring landlord's consent

Compensation is not payable in respect of Part 1 old or new improvements unless 'before the improvement was carried out the landlord consented to it (whether conditionally or upon terms as to compensation or otherwise agreed on between the parties)'[2]. It is essential that the agreement is in writing. The agreement could be in the lease[3].

There is no obligation on the landlord to consent. There is no provision for the tenant to obtain consent, where the consent is being capriciously or unreasonably withheld. This is understandable in the context of the s 5 obligation on the landlord to put the fixed equipment into a thorough state of repair and to provide such buildings and other fixed equipment as is required to maintain efficient production on the holding. The improvements which now require consent are, in the main, improvements which might tend to change the character of the holding.

Whether or not consent has been given is a question of fact for arbitration under s 60.

The terms on which consent are given can vary widely. Commonly they relate either to a landlord's contribution or to an agreed write-down period. As the section provides 'upon terms as to compensation or otherwise' the agreement need not relate to compensation. It could, for example, relate to a consent to the improvement in exchange for a resumption of an area of the holding. The agreement could provide for no compensation[4].

An improvement erected without consent might amount to a fixture or building which the tenant is entitled to remove unless the landlord gives a counternotice electing to purchase the building[5].

(b) Part 2 improvements requiring notice

Under the 1923 Act only drainage was a Part 2 improvement. The category of Part 2 improvements was extended by the 1923 Act.

To qualify for compensation for a 1923 Act Part 2 improvement notice in writing of not more than three months and not less than two months had to be given[6].

---

1 *Taylor v Burnett* 1966 SLCR App 139 (the court held that its approval had to be restricted to such improvements as were reasonably required to enable the tenant to carry on the type of farming specified in the lease).
2 1991 Act, s 37(1).
3 Cf *Turnbull v Millar* 1942 SC 521 at 531.
4 *Turnbull v Millar* above at 535.
5 1991 Act, s 18.
6 Ibid, s 38(3)(a).

In the case of a 1931 Act Part 2 improvement notice in writing of not less than three months and not more than six months had to be given[1].

In both cases compensation is not payable unless (1) the parties agreed on the terms as to compensation or otherwise, or (2) no such agreement was made, the tenant did not withdraw the notice and the landlord failed to exercise his rights to carry out the improvement himself, or (3) where the landlord gave notice of objection and the matter was referred for determination by the appropriate authority, which was satisfied that the improvement should be carried out and it was so carried out, in accordance with any directions given by the authority[2].

In respect of Part 2 old improvements the parties could agree by lease or otherwise to dispense with notice[3].

Whether or not the statutory requirements were complied with is a question of fact for an arbiter.

Part 2 new improvements require not less than three months' notice in writing to the landlord of the tenant's intention to carry out the improvement, and the manner in which it is to be carried out. If notice in writing is not given, there is no right to compensation[4]. The parties may by the lease or in a prior written agreement agree to dispense with the notice and regulate the terms on which the improvement may be carried out[5].

In regard to new improvements, the 1991 Act appears to have introduced an important change in the legal position regarding agreement over substitute compensation from that under the 1949 Act. Under the 1949 Act, s 51 compensation was only payable if three months' written notice was given of the intention to carry out the improvement and the manner in which it was to be carried out. On receipt of such notice the parties could enter into an agreement as to 'compensation or otherwise'[6].

In contrast, the 1991 Act, s 38(5) provides that if the parties agreed either after receipt of the notice or in an agreement to dispense with notice 'on terms as to compensation, the compensation ... shall be substitute compensation'. The import of this, leaving out the words 'or otherwise', would appear to have the effect that parties require to agree upon some compensation that is fair and reasonable even if this is ultimately nominal. Otherwise their agreement on no compensation may be held to be an attempt to contract out of the Act.

Upon receipt of a notice intimating that the tenant intends to effect an improvement the landlord's options are[7]:

(1) To enter into an agreement as to the terms upon which the improvement is to be carried out and in respect of substitute compensation.

(2) To give notice in writing within one month after receiving the notice either that he objects to the improvement or objects to the manner in which it is to be carried out[8].

The purpose of the notice is to give the landlord an opportunity to object to an improvement which is unnecessary or unproductive or inconsistent with the

1 1991 Act, s 38(3)(b).
2 Ibid, s 38(4).
3 Ibid, s 38(2).
4 Cf *Barbour v M'Douall* 1914 SC 844 for a consideration of what constitutes notice in writing.
5 1991 Act, s 38(5).
6 Which includes no compensation: *Turnbull v Millar* 1942 SC 521.
7 1991 Act, s 39.
8 Ibid, s 39(1).

type of farming specified in the lease, for which he would otherwise be liable in compensation[1]. The landlord is also protected from a reasonable improvement being executed in a manner inconsistent with the holding or which might be injurious to other parts of the holding.

When the landlord serves notice of objection the tenant may apply to the Land Court for approval[2] if he wishes to have the improvement authorised. While there is no time limit for such applications, any long delay might preclude the tenant from founding on the particular notice.

The Land Court may approve the carrying out of the improvement unconditionally, or 'upon such terms as to reduction of the compensation which would otherwise be payable or as to other matters, as appears to them to be just' or may withhold approval[2].

The test is whether the proposed improvement is reasonable and desirable on agricultural grounds for the efficient management of the holding[3]. The court has held that its jurisdiction regarding terms can only be connected with the proposed improvement; for example, the court could not authorise a new access road on condition that an existing access is given up[4].

Where the tenant sought approval for fencing, which the landlord opposed in respect of his stalking interests, the court granted approval subject to a condition that the landlord be given the option of paying for the extra cost of erecting and maintaining a deer fence on top of the proposed fence[5]. It is doubtful that the Land Court can grant retrospective approval to an improvement effected after the landlord has objected[6].

If the Land Court approves the proposed improvement the landlord is entitled to serve notice in writing on the tenant within one month after receiving notice of the decision, undertaking to carry out the improvement himself[7]. It may well be to the landlord's advantage to carry out the improvement as he might be able to employ estate labour and materials. Further, he will be under no liability for compensation at waygoing and will be entitled to a rent increase from the date of completion of the improvement[8] and for the rest of the lease. The landlord is required to carry out the improvement within a reasonable time. If he fails so to do the tenant may apply to the Land Court for a determination that the landlord has not carried out the improvement within a reasonable time[9]. The tenant may then carry out the improvement himself.

If the landlord does not serve a notice undertaking to carry out the improvement himself or is determined not to have it carried out within a reasonable time then the tenant may effect the improvement himself. Any terms, under which the Land Court gave approval, have effect as if they were contained in an agreement in writing between the landlord and tenant[9].

1 *Taylor v Burnett's Trs* 1966 SLCR App 139.
2 1991 Act, s 39(2).
3 *Fothringham v Fotheringham* 1978 SLCR App 144; *Hutcheson v Wolfe Murray* 1980 SLCR App 112; *Renwick v Rodger* 1988 SLT (Land Ct) 23.
4 *Hutcheson v Wolfe Murray* above.
5 *MacKinnon v Arran Estate Trust* 1988 SLCR 32.
6 Cf *Renwick v Rodger* 1988 SLT (Land Ct) 23 (retrospective approval was given to an improvement effected after the application was made to the court in order to beat a grant deadline, but before a decision was given, which *Gill* at para 486 describes as 'surprising').
7 1991 Act, s 39(3).
8 Ibid, s 15.
9 Ibid, s 39(4).

(c) Part 3 improvements requiring neither consent nor notice

Particular provisions relating to Part 3 old improvements have been noticed[1]. In general there is now no or very little value in such improvements.

Part III of Schedule 5 provides for the Part 3 improvements which may be carried out without notice. Particular regard should be had to para 32 (laying down of temporary pasture) where the compensation relates only to the value in so far as it exceeds the value of the temporary pasture at the commencement of the tenancy. A record of the holding would have defined the temporary pasture.

A tenant is not entitled to compensation for a Part 3 improvement if that improvement was carried out under his duty to restore fertility to the land after exercising his rights of freedom of cropping and disposal of crops under s 35(1) or the proviso to the 1949 Act, s 12(1)[2].

RENTAL PROVISIONS IN REGARD TO IMPROVEMENTS[3]

Any increase in rental value of a holding by reason of an improvement effected wholly or partly at the expense of the tenant falls to be disregarded on a rent review[4]. Such disregards apply even if the tenant has not obtained consent or given the appropriate notice which would entitle him to compensation.

High farming, for which the tenant will be entitled to compensation at way-going[5], is treated as a tenant's improvement for the purposes of a rent review[6] and falls to be disregarded.

Milk quota brought to the holding solely through the tenant's improvements which have converted a non-dairy farm to a dairy farm is not a tenant's 'improvement', and is disregarded on a rental arbitration[7].

If a landlord effects an improvement at his own hand at the request of or in agreement with the tenant, or in pursuance of an undertaking given after approval by the Land Court, or in compliance with a direction by the Secretary of State[8] the rent may be increased, providing the landlord serves notice in writing on the tenant within six months of the completion of the improvement[9]. Failing agreement the matter is one for arbitration[10].

---

1 See 'Old improvements' above.
2 1991 Act, s 51(3). No provision is made for restoring fertility after excercising s 7 rights; *quaere* whether the reference to s 9 in s 51(3)(c) is an error for s 7.
3 See chap 7.
4 1991 Act, s 13(5).
5 Ibid, s 44.
6 Ibid, s 13(6).
7 1986 Act, s 16; *Broadland Properties Estates Ltd v Mann* 1994 SLT (Land Ct) 7.
8 Eg in terms of the Agriculture (Safety, Health and Welfare Provisions) Act 1956, s 25(5).
9 1991 Act, s 15.
10 Ibid, s 15(3).

CHAPTER 10

Muirburn

Key points

- The tenant has a right to make muirburn notwithstanding any provision in the lease to the contrary, provided notice is given to the proprietor. Where muirburn is part of an approved hill farming land improvement scheme it may be carried out without notice: Hill Farming Act 1946, s 24.

- Notice has to be given to the proprietor, who may well be different from the landlord or deemed landlord: ss 84(4) and 85(1).

Time limits

- Muirburn may only be carried out on land below 450 m in the period 30 September through to 16 April in the following year, unless either an extension is approved by the landlord to 30 April or a direction has been given by the Secretary of State authorising an extension up to 1 May: 1946 Act, s 23(1) and (3).

- For land above 450 m the period for muirburn is 30 September through to 30 April in the following year, with extensions permitted by the landlord to 15 May and by a direction of the Secretary of State up to 16 May: 1946 Act, s 23(2) and (3).

- The tenant has to give 28 days' notice to the proprietor of his intention to make muirburn: 1946 Act, s 24(2).

- The proprietor has seven days following the tenant's intimation to give the tenant notice of his dissatisfaction with the muirburn proposals and to refer the matter to the Secretary of State for a decision: 1946 Act, s 24(2).

- Any person making muirburn has to give 24 hours' notice of his intention to make muirburn and the place and extent of the proposed muirburn to the proprietors of adjoining lands and woodlands, and if the tenant is muirburning to the proprietor of the land: 1946 Act, s 25(c).

GENERAL

'The expression muirburn is descriptive of the act of setting fire to ground which is muir, whatever may happen to grow upon it.[1]'

'The ground meant to be protected is where grouse and other moor game frequent and breed or may be shot.[2]'

'"Muir" is descriptive of the ground and not of its peculiar vegetable products, and that setting fire to any growth on muir ground, as grass, whin, bent or broom (and not only heather), is struck at by the Act, the only practical test being, that it must be a place which moor game frequent for breeding.[3]'

There are two separate interests in muirburn. First there is that of the tenant and landlord in connection with good estate management on a hill sheep farm[4]. In determining whether the landlord is managing the land in accordance with the rules of good estate management regard is had 'to the extent to which the owner is making regular muirburn in the interests of sheep stock'[5]. The qualification 'in the interest of sheep stock' means that muirburn undertaken by the landlord for sporting purposes will fall to be disregarded. In determining whether a tenant is farming a hill sheep farm in accordance with the rules of good husbandry regard is had to 'the extent to which regular muirburn is made'[6].

The second interest relating to muirburn is that of the proprietor in connection with the game nesting on the ground and in adjacent woodlands. This interest is given some protection by the statutory dates within which muirburn may be made. Older leases usually limited the tenant's rights to make muirburn or reserved the right to the landlord[7]. While the tenant now has a statutory right to muirburn, it may still be in the landlord's interest to provide in a lease that he is to carry out the muirburn.

THE TENANT'S STATUTORY RIGHT TO MAKE MUIRBURN

The tenant has a statutory right to make muirburn, notwithstanding any provision in the lease prohibiting absolutely or subject to conditions or restricting his right to muirburn, provided the tenant 'is of the opinion that it is necessary or expedient for the purposes of conserving or improving the land'[8].

The right can only be exercised if the tenant gives the proprietor of the land not less than 28 days' notice of the places at which and the extent to which he proposes to make muirburn[9].

If the proprietor is dissatisfied as to the places at which or the extent to which the muirburn is proposed, he may give the tenant notice within seven days of the receipt of the intimation stating the grounds of his dissatisfaction.

1 *Rodger v Gibson* (1842) 1 Broun 78 at 111 per Hope LJC; *Rankine* p 161.
2 *Rodger v Gibson* above at 113 per Lord Mackenzie.
3 *Rankine* pp 487, 488 citing *Rodger v Gibson* .
4 Agriculture (Scotland) Act 1948, s 26, Sch 5, para 2 and Sch 6, para 2(c)(v).
5 1948 Act, Sch 5, para 2.
6 Ibid, Sch 6, para 2(e)(v).
7 *Rankine* p 488.
8 Hill Farming Act 1946, s 24(1).
9 Ibid, s 24(2). NB the proprietor may be different from the landlord; see 1991 Act, s 85(1).

The proprietor may then refer the matter to the Secretary of State for his decision[1]. Upon reference the Secretary of State, after such inquiry as he thinks fit and considering any representations made by the parties, gives such directions as he may deem proper to regulate the muirburn. The decision is final[2]. It is not lawful for the tenant to engage in muirburn after such a reference until the Secretary of State has issued his decision.

A tenant has an absolute right to make muirburn, notwithstanding any provision in the lease to the contrary, without giving the foregoing notice if the muirburn is done in accordance with an approved hill farming land improvement scheme. The proprietor has no right of reference to the Secretary of State in such circumstances[3].

DUTIES IN RELATION TO MUIRBURN

Muirburn has to be carried out with reasonable care for the interests of the landlord and other neighbouring proprietors. The reasonable care incumbent upon the party making muirburn is 'that which a prudent man will observe in his own affairs and which a prudent and conscientious man will observe as to the interests of his neighbours'[4]. Negligent muirburn is actionable at common law[5]. Where the muirburn can be said not to be for the benefit of the pasture, the person burning can be found liable in damages to the proprietor whose sporting rights have been interefered with[6].

A neighbouring proprietor may go onto the land being muirburnt to prevent the spread of the fire onto his land. He may reclaim the costs incurred in fighting the fire on the muirburner's land even if no damage is caused to his own land[7].

It is also a criminal offence to make muirburn without due care so as to cause damage to any woodlands on or adjoining the land or any adjoining lands, woodlands, march fences or other subjects[8].

TIME FOR MUIRBURN

The statutory regulation of muirburn goes back to the time of Robert II[9]. It is now regulated by the Hill Farming Act 1946. Muirburn of land below 450 m above sea level may take place from 30 September in any year to 16 April in the following year[10]. It is lawful for the proprietor of the lands or the tenant with the written authority of the proprietor or his factor or commissioner to make muirburn during the period from 16 April to 30 April, both dates inclusive[10]. Where

1 1946 Act, s 24(2).
2 Ibid, s 24(3).
3 Ibid, s 24(4).
4 *Mackintosh v Mackintosh* (1864) 2 M 1357 at 1362 per Lord Neaves.
5 *Mackintosh v Mackintosh* above; *Grant v Gentle* (1857) 19 D 992.
6 *Robertson v Duke of Atholl* (1814) 6 Pat 135; cf *Grant v Gentle* above.
7 *Lord Advocate v Rodgers* 1978 SLT (Sh Ct) 31; *Cope v Sharpe (No 2)* [1912] 1 KB 496.
8 1946 Act, s 25(d).
9 *Rankine* p 487.
10 1946 Act, s 23(1).

the land is over 450 m above sea level the respective dates are 30 April and 15 May with the proprietor's written permission[1].

If it appears to be necessary or expedient for the purpose of facilitating muirburn, the Secretary of State may direct that for the purpose of the 1946 Act, s 23(1) a different day, being a date not later than 1 May for land below 450 m and 16 May for land above 450 m, may be substituted. The direction may apply to the whole of Scotland or as respects any particular lands or classes of lands[2]. Such a notice requires to be published in one or more newspapers circulating in the locality in which the lands to which the direction relates are situated.

It is unlawful to commence muirburn between one hour after sunset and one hour before sunrise[3].

NOTICES REGARDING MUIRBURN

A tenant intending to exercise his statutory right to muirburn has to give the proprietor at least 28 days' notice of his intention to muirburn and the places and the extent of the proposed muirburn. If the muirburn is taking place in accordance with an approved hill farming improvement scheme notice does not require to be given.

Any person making muirburn is required to give not less than 24 hours' notice of his intention to make muirburn to the proprietor of lands or woodlands adjoining the land and if he is a tenant to inform the proprietor of the lands of the day on which, the places at which, and the approximate extent to which he intends to make muirburn.

Any notice given under s 24(2) has to be given in writing. A notice to the proprietor may be given to the factor, commissioner or other local representative[4]. Note that notice has to be given to the proprietor and not the landlord, who may be a different individual[5]. Notice will have to be served on the actual proprietor, his factor or other local representative and the tenant will not be able to rely on the deemed landlord provisions of the 1991 Act, s 84(4).

1 1946 Act, s 23(2).
2 Ibid, s 23(3).
3 Ibid, s 25(a).
4 Ibid, s 26(1) and (2).
5 See 'landlord' as defined by 1991 Act, s 85(1).

CHAPTER 11

Game and deer

Key points

- The right to take game is reserved to the landlord or his nominee, unless specifically granted to the tenant. The landlord's sporting rights are a burden on the tenant.

- There is an implied obligation on the tenant, subject to his obligations to farm in accordance with the rules of good husbandry and his statutory rights, to do nothing to damage, decrease or destroy the game on the holding.

- Unless reserved by stipulation in the lease to the landlord, the tenant has a right to shoot rabbits himself or by an authorised person as part of his ordinary agricultural operations to protect his crops.

- The tenant has restricted rights to shoot ground game (rabbits and hares) under the Ground Game Act 1880, which cannot be contracted out of.

- The tenant has a right to claim compensation from his landlord for damage by 'game' in excess of 12 pence per hectare where the right to take and kill the game is not vested in the tenant and he does not have permission in writing to shoot the game: 1991 Act, s 52(1) and (4).

- The landlord is personally liable to pay the compensation with a right of relief against his sporting tenant: 1991 Act, s 52(1) and (4).

- The Red Deer Commission, where it is satisfied that red or sika deer are marauding, may authorise in writing any person to kill them: Deer (Scotland) Act 1959, s 6.

- The Red Deer Commission may, where it is satisfied that red or sika deer have damaged agriculture or forestry in any locality, introduce a deer control scheme, which requires to be confirmed by the Secretary of State: Deer (Scotland) Act 1959, s 7 and Second Sch.

- The occupier of agricultural land or enclosed woodland has the right, notwithstanding any agreement to the contrary, to take or kill deer on arable land, garden ground or land laid down in permanent pasture (other than moorland or unenclosed land) if he has reasonable grounds for believing that serious damage to crops, pasture etc will be caused if the deer are not killed: Deer (Scotland) Act 1959, s 33(3).

Time limits

- To claim statutory compensation for game damage (1) notice in writing must be given to the landlord as soon as is practical after the damage was first observed by the tenant and the landlord given a reasonable opportunity to inspect the damage and (2) notice in writing of the claim and the particulars thereof must be given to the landlord within one month after the expiry of the calender year, or such other 12-month period as is agreed in substitution thereof, in respect of which the claim is made: 1991 Act, s 52(2).

- To preserve a common law claim for game damage either a specific reservation of the claim has to be made annually when the rent is paid or a claim has to be instituted annually.

- Any owner or occupier upon whom the Red Deer Commission has served notice of a proposed deer control scheme has 28 days within which to object to the Secretary of State: Deer (Scotland) Act 1959, Second Sch, para 1(a).

- Any person aggrieved by a deer control scheme or by any variation or revocation thereof who wishes to question its validity or that the proper procedures have been followed has six weeks from the date of publication of the notice in the Edinburgh Gazette stating that the scheme has been confirmed, to appeal to the Court of Session: Deer (Scotland) Act 1959, Second Sch, para 13.

- Where the Red Deer Commission has carried out a requirement of a control scheme, where the owner or occupier has failed and they notify a statement of expenses and income received that they intend to recover, the owner or occupier has one month from the notification to appeal to the Land Court: Deer (Scotland) Act 1959, s 11(2).

GENERAL[1]

Game is generally taken to mean all wild animals pursued for sport and which are normally used as food for human consumption[2]. It should be noted that many statutes have their own particular definition of 'game'.

The right to take and kill game is reserved at common law to the landlord or to someone deriving title from him. The proprietor's sporting rights over the land do not pass to an agricultural tenant without an express grant of the sporting rights in the lease[3].

The landlord may lease the sporting rights to a sporting tenant. The landlord's sporting rights and those of his sporting tenant are a burden upon the agricultural tenant, who has no right to take or kill the reserved game.

Rabbits are not game and unless reserved to the landlord by a condition in the lease the tenant has a right to kill them himself or by an authorised person as an ordinary agricultural operation to protect his crops, including when the crops are not in the ground[4].

GENERAL PROHIBITION AGAINST KILLING WILD BIRDS

The Wildlife and Countryside Act 1981 implements EC Council Directive 79/409 (the Birds Directive) for the protection of wild birds. It provides for a number of offences in relation to the taking or killing of birds, unless they are on an exempt schedule and that during specified times. Game birds are, in general, exempt from the Act except during the close season, although the number of wildfowl available for shooting is more limited[5].

THE COMMON LAW

At common law the principle is that in entering into a lease of agricultural land, the parties are taken to have accepted that game in particular numbers exists on the land and it is contemplated that there will be no material change in the amount of game during the lease.

The Lord President has said[6]:

'The parties making no particular stipulation about game must be held to have looked at the land as it then stood, seeing it largely or reasonably stocked with game, as the case might be, and proceeding on the reasonable supposition that there was to be no material alterations on the land during the currency of the lease.'

---

1 See 11 *Stair Memorial Encyclopaedia* paras 926-1001 ('Game') for a fuller consideration of the game laws and the statutory restrictions on taking or killing game.
2 11 *Stair Memorial Encyclopaedia* para 802.
3 *Wemyss v Gulland* (1847) 10 D 204; *Copeland v Maxwell* (1871) 9 M (HL) 1.
4 *Crawshay v Duncan* 1915 SC (J) 64.
5 See Wildlife and Countryside Act 1981; 2 *Stair Memorial Encyclopaedia* paras 282–295 ('Animals').
6 *Morton v Graham* (1867) 6 M 71 at 73 per Lord Principal.

He cited with approval the speech of Lord Ordinary Fullerton[1]:

'It appears, then, that a tenant may have a claim for damages for injury done by game, but that, in order to support such a claim, it is necessary to prove not merely a certain visible damage arising from the game, but a certain and visible increase of game, and a consequent alteration of the circumstances contemplated in the contract, imputable to the act of the landlord.'

Lord Jeffrey said[2]:

'The common law declares, that though the landlord gives up, in terms, the entire occupation of the farm to the tenant, there is a tacit reservation of a right to keep and feed on it a certain amount of wild animals that naturally resort to it. That being the ordinary condition, the tenant agrees to keep up the existing stock of game, and engages to feed, beside his own cattle, the game on the farm when he takes it. But on the other hand the landlord cannot increase the obligation by multiplying the stock. The rent is reckoned with reference to the existing quantity, and the contract thus settled cannot be altered at the will of one of the parties.'

The tenant's obligations regarding game

As the common law position is that the parties accept as a burden on the tenancy the number of game existing at the commencement of the lease, there is an implied obligation on the tenant to do nothing which will decrease, damage or destroy that level of stock of game on the holding[3].

Reversing the dicta of Lord Jeffrey[4], the principle appears to be that (1) the tenant is bound to feed the stock of game on the farm as it existed at the time of the lease, and (2) that the tenant cannot decrease the stock by reducing the feed to be provided to the game.

The landlord would be unable to found on a breach of this implied obligation to found a s 22(2)(d) or (e) notice to quit if the tenant can say that the obligation was not consistent with the fulfilment of his responsibilities to farm in accordance with the rules of good husbandry.

In *Wemyss v Gulland*[5] the court considered what measures a tenant might take to protect his crop against game damage and inferentially the tenant's rights in respect of game.

In that case the tenant sought to scare off game by employing muzzled dogs to roam the farm and men to fire off blanks and set snares. It was held that such actings were illegal. In determining the tenant's rights to scare game the same principles had to be applied by the same measure as those of the landlord[6].

The Lord President said[7]:

---

1 *Drysdale v Jameson* (1832) 11 S 147, repeated in *Wemyss v Wilson* (1847) 10 D 194 at 202.
2 *Wemyss v Wilson* above at 203.
3 *Rankine* p 490.
4 *Wemyss v Wilson* above at 203.
5 (1847) 10 D 204.
6 *Wemyss v Wilson* (1847) 10 D 194.
7 *Wemyss v Gulland* (1847) 10 D 204 at 207.

'It has been ... laid down by Erskine [II, 6,6] that no person whatever – be he tenant or a stranger – can resort to any device that has a manifest tendency to destroy game.'

Later he said:

'If a tenant has rights, so has the landlord. While the latter is liable for injuries done by game, he is entitled to have it protected.'

In the same case Lord MacKenzie said[1]:

'Now, I think the obligation of the tenant extends farther than not to kill the game. He is not to extirpate it by ways and means never thought of, and that could not have been contemplated by the parties when the lease was entered into.'

He went on to say that a tenant might fence fields, put up scarecrows or employ boys with rattles to protect his fields, but that range or type of activities was the limit of his rights.

Lord Fullerton said[1] that a tenant could not by his own acts extinguish the rights of the landlord in the game.

Lord Jeffrey[2]:

'I have difficulty in holding that he (tenant) can encroach on the "genial couch and procreant cradle" – that is expiration.'

The principle of the case is that a tenant cannot scare game so as to expirate them.

The tenant's claim for damages at common law

While the 1991 Act, s 52 gives the tenant a statutory right to claim for damage by game as defined by that section, the common law right to claim damages applies to damage by other game reserved to the landlord, which fall outwith the statutory definition[3].

If the landlord increases the game stock during the course of the tenancy the tenant will have a claim for damages attributable to that additional stock[4]. As Lord Fullerton said[5]:

'A tenant may have a claim for damage for the injury done by game, but in order to support such a claim it is necessary to prove not merely a certain visible damage arising from game, but a certain and visible increase of the game and a consequent alteration of the circumstances contemplated in the contract, imputable to the act of the landlord. The true ground of damage seems to be, not that the game is abundant, but that its abundance has been materially increased since the date of the lease, in consequence either of active measures of the landlord or his failure to keep down the burden - which last circumstance must be held as equivalent to his act, as the right so to keep it down is one expressly withheld from the tenant.'

The tenant is also entitled to claim damages for any act of the landlord on neighbouring land which increases game or rabbits on the holding[6].

1 At 210.
2 At 211.
3 1991 Act, s 52(5) - deer, pheasants, partridges, grouse and black game.
4 *Wemyss v Wilson* (1847) 10 D 194; *Wemyss v Gulland* (1847) 10 D 204; *Morton v Graham* (1867) 6 M 71; *Kidd v Bryne* (1875) 3 R 225; *Cadzow v Lockhart* (1875) 3 R 666.
5 *Drysdale v Jameson* (1832) 11 S 147 at 149.
6 *Cameron v Drummond* (1888) 15 R 489; *Inglis v Moir's Tutor* (1871) 10 M 204; *Ormiston v Hope* (1917) 33 Sh Ct Rep 128.

A tenant is not entitled at the termination of the lease, or probably during its currency, after paying rent regularly in full without reservation, to set up a claim of damages for injury through game or rabbits or otherwise in the years past, if he can only found on grumblings at his losses or on ineffectual proceedings, but not on any serious complaint or claim for compensation[1]. The claim is restricted to the year preceding the claim or the first year of the effective complaint. This common law position is recognised in s 52 which requires a tenant to make an annual claim for statutory compensation.

THE TENANT'S STATUTORY RIGHTS

(a) Ground Game Act 1880[2]

The tenant as occupier of the land has the right to kill and take ground game, defined to mean 'hares and rabbits'[3], thereon concurrently with any other person entitled to kill and take ground game[4].

It is not competent to contract out of this right[5]. If the landlord does not reserve to himself the right to take all ground game, subject to the tenant's rights under the 1880 Act, at common law the tenant will have the right to shoot rabbits and to invite as many persons as he likes onto the holding to shoot them.

The right to kill and take ground game is subject to limitations[6]. The occupier shall kill and take ground game only by himself or persons duly authorised by him in writing. Only the occupier and one other person authorised by himself may kill ground game with firearms. The occupier may only authorise members of his household (including guests) resident on the land in his occupation, persons in his ordinary service on the land or other persons bona fide employed by him for reward in the taking or destruction of game[7].

A person authorised by the occupier on demand from any person having a concurrent right to take and kill ground game shall produce to that person the document by which he is authorised to kill and take game and in default he shall be deemed not to be an authorised person[8].

A person having a right of common grazing or a grazing tenant for not more than nine months are not deemed to be occupiers of land for the purposes of the Act[9].

---

1 *Rankine* p 493.
2 43 & 44 Vict as amended or varied by Agriculture (Scotland) Act 1948 and Wildlife and Countryside Act 1981. For a fuller consideration of this Act see 11 *Stair Memorial Encyclopaedia* para 849.
3 Ground Game Act 1880, s 8
4 Ibid, s 1.
5 Ibid, s 3.
6 Ibid, s 1(1).
7 By 1948 Act, s 48(2) on an application by the occupier of the land, and having given the landlord an opportunity to make representations or be heard, the Secretary of State may sanction the authorisation by the occupier of additional persons to kill and take ground game.
8 Ground Game Act 1880, s 1(1)(c).
9 Ibid, s 1(2).

(b) Compensation for game damage

The 1991 Act, s 52 confers on the tenant a statutory right to claim compensation from his landlord[1] for damage to his crops from game where 'the right to kill and take which is vested neither in him nor in anyone claiming under him other than the landlord[2] and which the tenant has not permission in writing to kill'[3].

The right to compensation only arises if the damage amounts to more than 12 pence per hectare[3], but the claim is for the whole amount and not the excess[4]. This alters the common law position where a tenant only had the right to claim for game damage if the landlord had increased the stock of game on the holding from the existing or contemplated stock, when the lease was entered into.

Game is defined to mean 'deer, pheasants, partridges, grouse and black game'[5]. Ground game is not included because of the tenant's rights at common law and under the Ground Game Act 1880. The tenant's rights to claim compensation in respect of other species which might be reserved to the landlord, such as wildfowl, will depend on the common law.

The statutory right to claim compensation extends to damage done by game coming from adjacent land during the close season[6].

In order to exclude a claim for game damage the tenant must have express written permission to kill the type of game causing the damage. A limited statutory right alone may not be sufficient to exclude the landlord's liability for game damage[7].

In order to make a valid statutory claim for compensation the tenant has to give the landlord two notices:

(1) *Notice of damage.* This notice must be given in writing 'as soon as is practicable after the damage was first observed by the tenant'[8]. The tenant must give the landlord a reasonable opportunity to inspect the damage[9] in the case of growing crop before the crop is reaped, raised or consumed[10] or in the case of crop that is reaped or raised before it is removed from the land[11] on which it was grown. The landlord should probably be given an opportunity to inspect

---

1 The obligation is upon the landlord, even if there is a sporting tenant who may be liable to indemnify the landlord under s 52(4).
2 The provision 'other than the landlord' is designed to defeat the device whereby the landlord let the farm and sporting rights to the agricultural tenant subject to an obligation on the tenant to sub-let the sporting rights back to the landlord.
3 1991 Act, s 52(1).
4 Cf *Rodden v McCowan* (1890) 17 R 1056.
5 1991 Act, s 52(5).
6 *Thomson v Earl of Galloway* 1919 SC 611.
7 *Lady Auckland v Dowie* 1965 SC 37. This case was decided on the specific statutory provisions of the Agriculture (Scotland) Act 1948, which by s 52 reserved a tenant's right to claim compensation which 'he would have been entitled to recover if this Act had not been passed'. Different statutory provisions authorising a tenant to shoot game might exclude the tenant's right to compensation.
8 1991 Act, s 52(2)(a).
9 *Barbour v M'Douall* 1914 SC 844; *Dale v Hatfield Chase Corp* [1922] 2 KB 282.
10 1991 Act, s 52(2)(a)(i).
11 Ibid, s 52(2)(a)(ii).

the whole crop even if damage relates only to part of it[1]. It is suggested that if time is limited notice of this fact should be given to the landlord in the written notice[1].

(2) *Notice of claim.* The purpose of the section is to fortify the common law position that a claim for game damage should be made annually. The relevant period for assessing the claim is the calendar year or such other 12-month period as the parties may have agreed. The notice of claim with the particulars thereof has to be given to the landlord in writing within one month after the expiry of the calendar year or substitute period. The claim may be intimated at anytime during the currency of the year or within the additional month. The notice should detail (a) the approximate date of the damage being caused, (b) the crop damaged, (c) the field in which it occurred, (d) the approximate area affected and (e) the sum claimed[1].

Failing agreement on the amount of compensation payable the question falls to be determined by arbitration[2].

The landlord is entitled to be indemnified by his sporting tenant against all claims for compensation under s 52(1)[3]. Any question arising out of the right of indemnification falls to be determined by arbitration[4]. It is probably competent for the landlord and his sporting tenant to contract out of this subsection[5].

GAME (EXCEPT DEER) CONTROL MEASURES

Certain statutory provisions provide for the control of game on agricultural land.

(1) If the Secretary of State considers that it 'is expedient ... for the purposes of preventing damage to crops, pasture, animal or human feedstuffs, livestock, trees, hedges, banks or any other works on land' he may serve notice on a person having right to kill or take the animals or birds[6] causing the damage, specifying the steps that have to be taken to kill, take or destroy the animals or birds[7]. Such a requirement cannot be imposed if the killing, taking or destruction would be unlawful[8].

(2) The Secretary of State has particular powers in relation to rabbits to impose a rabbit clearance order requiring an occupier of land to destroy or reduce breeding places or cover for rabbits or to exclude rabbits from land or to prevent them living in any place on the land spreading to or doing damage in any other place. An occupier has to be given time to submit written objections to the scheme, which can then be confirmed with or without modification. Where the occupier is a tenant any person to whom he pays rent has also

---

1 *Connell* p 126.
2 1991 Act, s 52(3).
3 Cf s 52(4).
4 1991 Act, s 52(4).
5 *Connell* p 126.
6 The section applies to rabbits, hares and other rodents, foxes, moles and wild birds other than those included in the Wildlife and Countryside Act 1981, Sch 1.
7 1948 Act, s 39(1).
8 Ibid, s 39(2).

to be notified. On summary conviction for failure to comply the party may be fined[1].

DEER[2]

(a) General

Deer as game are in general reserved to the landlord. Deer are game for which the tenant is entitled to claim damages[3].

The close season for red deer stags runs from 21 October to 30 June and for hinds from 16 February to 20 October. The Secretary of State may by order fix a close season for deer other than red deer[4]. Night shooting between the expiration of the first hour after sunset and the commencement of the last hour before sunrise is prohibited except in limited circumstances[5].

(b) Occupier's right to take and kill deer

The limited rights conferred on a tenant as occupier of the land to kill deer under the 1948 Act have been replaced by rights under the Deer (Scotland) Act 1959[6].

Notwithstanding anything in any agreement between the occupier of agricultural land or enclosed woodland and the owner thereof, it is lawful for the occupier and any servant of the occupier authorised by him in writing to take or kill deer and sell or otherwise dispose of the carcasses of any deer found on arable land, garden grounds or land laid down in permanent grass (other than moorland and unenclosed land) provided:

'that the occupier has reasonable grounds for believing that serious damage will be caused to crops, pasture, trees or human or animal foodstuffs on that land if the deer are not killed.'[7]

Notwithstanding any agreement to the contrary between the occupier of agricultural land or enclosed woodland and the owner thereof, it is lawful for the occupier in person[8] to carry out night shooting of red or sika deer on agricultural land or enclosed woodland, provided the occupier has reasonable

---

1 1948 Act, s 39(5); Pests Act 1954, ss 1(1) and 14(a); 11 *Stair Memorial Encyclopaedia* paras 862 and 868–871.
2 See 2 *Stair Memorial Encyclopaedia* paras 296-301 ('Animals'); 11 *Stair Memorial Encyclopaedia* paras 926–1001 ('Game') for a fuller treatment of the law relating to deer. Cf p 113.
3 1991 Act, s 52(1) and (5).
4 Deer (Scotland) Act 1959, s 21. Cf s 33(3) which authorises certain persons to take or kill deer and sell carcasses if the occupier has reasonable grounds for believing that serious damage will be caused to crops etc if the deer are not killed.
5 1959 Act, s 23(1); see also s 33(1), (2), (4) and (4A).
6 The right to kill deer under the 1948 Act, s 43(1) was abolished by the Deer (Amendment) (Scotland) Act 1982, s 15 which substituted rights under the amended 1959 Act.
7 1959 Act, s 33(3). NB certain other persons are covered by this section but their right relates to taking or killing in the close season.
8 The occupier cannot authorise any other person to take or kill on his behalf. If anyone else is to act this has to be authorised by the Red Deer Commission; see below.

grounds for believing serious damage will be caused to crops, pasture, trees or human or animal foodstuffs on that land if the deer are not killed[1].

The Red Deer Commission may authorise in writing any person nominated by the occupier of agricultural land or enclosed woodland to shoot deer of any species at night provided that it is satisfied (a) that the shooting is necessary to prevent serious damage to crops, pasture, trees or human or animal foodstuffs, (b) that no other method of control which might reasonably be adopted would be adequate and (c) that the person concerned is a fit and competent person to receive such authorisation[2].

The owner of the agricultural land or enclosed woodland may require the occupier to inform him of the number of red or sika deer shot under the 1959 Act, s 33(3), (4) or (4A) in the 12 months preceding the request[3].

A person who kills or injures deer without legal right or permission from a person who has such a right is guilty of the offence of poaching[4].

The Red Deer Commission may by agreement with any owner or occupier of land assist in or undertake, whether in pursuance of a deer control scheme or otherwise, the taking or killing of red or sika deer and with the disposal of carcasses. The agreement may make provision for the providing of equipment by the Commission. The agreement must make provision for the payment of the Commission's expenses unless the Commission with the approval of the Secretary of State decides otherwise[5].

Any person authorised in writing by the Red Deer Commission has power, at all reasonable time, of entry upon any land in pursuance of their functions or to determine which of their functions should be exercised or to ensure that any requirements placed upon any person has been complied with or to take a census of red or sika deer[6].

(c) Deer returns

The Red Deer Commission may, for the purposes of its functions, by written notice served on the owner or occupier of any land, require him to make a return showing the number of red or sika deer of each sex which to his knowledge have been killed on the land during such period (not exceeding five years) as may be specified. Owners and occupiers should therefore keep a note of all such deer killed.

(d) Marauding deer

Where the Red Deer Commission[7] is satisfied:

'(a) that red deer or sika deer are, on any agricultural land, woodland or garden ground—
  (i) causing serious damage to forestry or to agricultural productions, including any crops or foodstuffs;

1 1959 Act, s 33(4).
2 Ibid, s 33(4A).
3 Ibid, s 33(4C).
4 Ibid, s 22.
5 Ibid, s 12.
6 Ibid, s 15(1).
7 The functions of the Red Deer Commission under the 1959 Act, s 6 may be delegated to a panel established under s 2 in respect of the locality of that panel; s 2(3) and (4).

    (ii) causing injury to farm animals (including serious overgrazing of pastures and competing with them for supplementary feeding); and

(b) that the killing of deer is necessary to prevent further such damage or injury,[1]

it is required to authorise in writing, subject to such conditions as are specified, any person[2] competent to do so to follow and kill such red and sika deer on any land mentioned in the authorisation as appear to be causing the damage and injury[3]. An authorisation remains in force for such period, not exceeding 28 days, as may be specified in the authorisation[4].

Where the deer causing the damage appear to be coming from particular land and the person having the right to kill the deer there appears willing to undertake the killing, the Red Deer Commission is required to make a request to that effect to that person[5]. Where a request has been made the Red Deer Commission cannot grant an authorisation to kill deer until it appears that the person requested to kill deer has become unable or unwilling to comply with the terms of the request[6].

The Red Deer Commission is under a duty to give the owner of any land which is mentioned in an authorisation such notice of its intention to issue an authorisation as may be reasonably practical[7].

Where the Commission intends to issue an authorisation, it is under a duty to give as soon as reasonably practical to any person who is likely to be on the land such warning as it considers necessary to prevent danger to that person[8].

Notices require to be in writing and may be served by delivery to the person or by leaving it at his proper address or by posting it[9]. A notice relating to marauding deer may be served on an agent or servant who is responsible for the management or farming of the land[10].

Where the Commission is satisfied that deer other than red or sika deer are causing serious damage to agricultural land or to woodland and that the killing of the deer is necessary to prevent further such damage, it is entitled, with the consent of the occupier of the agricultural land or woodland, to kill such deer as its servants may encounter in the course of their duties[11].

(e) Deer control schemes

Where the Red Deer Commission is satisfied that red or sika deer have caused damage to agriculture or forestry in any locality, and that for the prevention of further damage the red or sika deer should be reduced in number or exterminated, having regard to the nature and character of the land in that area, it may determine what measures should be taken to reduce or exterminate the deer[12].

---

1 1959 Act, s 6(8).
2 If the person is not a servant of the Commission he may be paid.
3 Ibid, s 6(1).
4 Ibid, s 6(4).
5 Ibid, s 6(2).
6 Ibid, s 6(3).
7 Ibid, s 6(6).
8 Ibid, s 6(5).
9 Ibid, s 16(1).
10 Ibid, s 6(7).
11 Ibid, s 6A(1).
12 Ibid, s 7(1).

The Commission is required to consult with the owners and occupiers of the land to secure agreement on the carrying out of the measures determined[1]. Where agreement cannot be secured or the measures agreed upon are not being carried out, the Commission is required to make a control scheme[2] for carrying out the measures, which requires to be confirmed by the Secretary of State before it comes into operation[3].

Where notice is served on an owner or occupier on whom the control scheme proposes to impose any requirement, that person has 28 days in which to object to the Secretary of State[4]. Any other person has a right to object within 28 days of the publication of the statutory notice to the press[5]. Where an objection is not frivolous[6] or not withdrawn the Secretary of State is required to hold a public inquiry[7].

Any person aggrieved by a control scheme or by any variation or revocation thereof who wishes to question its validity on the grounds that it is not within the powers of the Act or that any requirement of the Act has not been complied with may appeal to the Court of Session within six weeks of the date from which the notice appears in the Edinburgh Gazette stating that the scheme has been approved[8].

Where a deer scheme has been confirmed it is the duty of every owner and occupier of land to take such measures as the scheme may require. Any person who refuses or wilfully fails to comply with any requirement of the scheme is guilty of an offence[9].

Where any owner or occupier fails to carry out any requirement of a scheme, the Red Deer Commission, if satisfied that it is still necessary so to do, is under a duty to carry out the requirement[10]. Where the expenses of carrying out the requirement exceed the proceeds of the sale of carcasses of any deer killed, the excess is recoverable from the owner or occupier concerned[11]. The Commission is required to furnish any owner or occupier with a statement of expenses and amounts received from the sale of carcasses and if the owner or occupier is aggrieved by the statement he may appeal to the Land Court within one month after being notified of the statement[12]. The Land Court may, if it appears to be equitable so to do, vary the amount recoverable from the owner or occupier[13].

(f) Deer farming

The close season does not apply to the killing of deer by any person[14] who keeps deer by way of business on land enclosed by a deer-proof barrier for the

1 1959 Act, s 7(2).
2 See ibid, s 8 for details of what has to be contained in a control scheme.
3 Ibid, s 7(3). The procedures for making, confirming, varying or revoking a control scheme are set out in ibid, Sch 2.
4 Ibid, Sch 2, para 1(a).
5 Ibid, Sch 2, para 1(b).
6 Ibid, Sch 2, para 10.
7 Ibid, Sch 2, para 8.
8 Ibid, Sch 2, para 13.
9 Ibid, s 9.
10 Ibid, s 10.
11 Ibid, s 11(1).
12 Ibid, s 11(2). See also the provisions of the Scottish Land Court Act 1993 and the Rules of the Scottish Land Court 1992.
13 1959 Act, s 11(2).
14 Including the servant or agent authorised by that person for that purpose.

production of meat or foodstuffs or skins or other by-products or as breeding stock, provided the deer are conspicuously marked to demonstrate that they are so kept[1].

DEER (AMENDMENT) (SCOTLAND) BILL 1996

The Deer (Amendment) (Scotland) Bill[2] proposes substantial changes to the Deer (Scotland) Act 1959 (as amended). The principal amendments include:

Clause 1 – (1) changes the name of the 'Red Deer Commission' to the 'Deer Commission for Scotland'; (2) imposes a duty on the Commission to take into account the size and density of the deer population and its impact on the natural heritage, the needs of agriculture and forestry and the interests of owners and occupiers of land, and (3) widens the scope and expertise of persons from whom members of the Commission may be appointed.

Clause 4 – amends s 6 of the 1959 Act (marauding deer) to provide that where satisfied that deer are causing serious damage to woodland or to agricultural production (including crops or foodstuffs), or are causing injury to livestock (by serious overgrazing or competing for supplementary feeding), or constitute a danger to the public, the Commission may authorise the killing of those deer if considered necessary to prevent further damage.

Clause 5 – introduces a new s 6AA, applying the 1959 Act to natural heritage as it applies to woodland if the Commission is satisfied that the deer are causing serious damage to the natural heritage, with different provisions for enclosed and unenclosed land.

Clause 6 – substitutes s 7 of the 1959 Act (control schemes). If satisfied that, for the prevention of further damage or potential damage to woodland, agricultural production, natural heritage or injury to livestock etc, the Commission may introduce a control scheme, excluding deer from a particular area. The new section provides for consultation with owners and occupiers. Where there is agreement, a 'control agreement' is drawn up; where agreement cannot be reached the Commission may make a 'control scheme'. The procedure for making a control scheme remains basically unchanged.

Clause 7 – removes the Commission's general power to dispose of carcasses of deer killed under its authority.

Clause 8 – substitutes s 21 of the 1959 Act (closed seasons), empowering the Secretary of State for Scotland, after consultation, to fix the period in each year during which stags and hinds may be lawfully killed.

Clause 10 – inserts a new s 33A into the 1959 Act giving the Commission power to authorise certain acts which, in other circumstances, would be an offence, including (1) the power to take or kill deer at night contrary to s 23; (2) authority to use a vehicle (not including a helicopter or hovercraft) to drive deer in order to take or kill them for the purposes of deer management (not including driving for any sporting activity), and (3) the power to take or kill deer during the close season.

Clause 11 – inserts a new s 34A to the 1959 Act, disapplying the Act to farmed deer which are defined as 'deer of any species which are on agricultural land enclosed by a deerproof barrier and are kept on that land by any person as livestock'.

1 1959 Act, s 21(5A).
2 As remitted from the House of Lords to the Commons – printed 2 April 1996.

CHAPTER 12

Succession planning

Key points

- Tenants should pre-plan who is to be the successor in a lease and the assets necessary for continuing the farming operation, so as to minimise the prospects for a successful objection by a landlord to the legatee or acquirer of a lease (1991 Act, ss 11(4) or 12(2)) or serving a successful notice to quit under s 25.

- Pre-planning who is to be the successor in a lease and the assets necessary for continuing the farming operation should include a consideration of the tax implications of the succession and whether these could be mitigated by appropriate PETs and tax planning.

- The tenant should establish whether the succession to his lease is governed by a special destination, may be bequeathed or can only be transferred by his executors.

- A lease may only be bequeathed or transferred to one person.

- Where the farming assets are those of a partnership the partnership's heritable assets, such as the lease, the tenant's fixtures, fitting and improvements, milk quota, are moveable in the deceased's succession: Partnership Act 1890, s 22.

- Provision should be made to avoid the situation where an executor might be *auctor in rem suam* in transferring the lease to himself.

- Consideration may have to be given to the timing of the transfer of a tenancy to coincide with the timing provisions for the transfer of sheep and livestock annual premium quota.

Time limits

- None

GENERAL

In every succession, whether of the family farm or a tenanted farm, the tax implications of the succession to, or transfer of, the farm or tenancy should be borne in mind.

It is important that tenant farmers plan the succession to their tenancies and the assets necessary to continue the farming operation. Failure to plan the succession might result in the successor being a person who will have no defence or a doubtful defence to an objection by the landlord[1] or to notice to quit[2]. This puts the tenancy at risk as the successor will be left with insufficient capital to continue the farming operation.

As leases and quotas now have a value in succession it is important to provide in a will for the successor to those assets and what credit is to be given for them in the succession.

The rights and claims which arise on intestacy can be such as to make it impossible for a successor tenant to be left with sufficient capital to continue the tenancy if it is claimed.

THE VALUE OF A DECEASED'S INTEREST IN A LEASE

An executor transfers the deceased's interest in the tenancy to a successor 'in or towards satisfaction of that person's entitlement or claim'[3]. The proper basis for valuing this interest has not been judicially determined.

What is being transferred and valued for the purposes of s 16(2)[4] is the interest of the deceased under the lease and not the lease itself. It would appear that the value of the interest in the lease is the value of the deceased's rights, such as to compensation for improvements or milk quota and the value of the deceased's fittings and fixtures. This approach is consistent with s 12(5) which treats the termination of the lease upon an objection by the landlord as a termination of the acquirer's tenancy, giving him the right to claim compensation under the 1991 Act, Pts IV and V, but not compensation for disturbance[5].

Where a tenancy passes by bequest or under a special destination[6] it is not being transferred 'in or towards satisfaction of that person's entitlement or claim'[7]. It might be said that different valuation principles should be applied to such a transfer because there is no statutory provision as to value. In a bequest, as the lease is just as vulnerable to termination, the same principles probably should apply as in a transfer by an executor[8]. The right to claim compensation for milk quota[9] will pass as one of the deceased's interest's under the lease or

---

1 Under the 1991 Act, ss 11(4) or 12(2).
2 Ibid, s 25 – 'Termination of tenancies acquired by succession'.
3 Succession (Scotland) Act 1964, s 16(2).
4 1964 Act.
5 *Gill* para 592.
6 1964 Act, s 36(2)(a). A lease subject to a special destination is not available for transfer by an executor under s 16.
7 Ibid, s 16(2).
8 This may also apply to a succession under a special destination; *Halliday* vol 4, para 49-16.
9 If the deceased was allocated milk quota or paid for transferred quota or was the successor to such a tenant; 1986 Act, Sch 2, paras 2 and 3.

with the lease or land, which will require to be valued. Sheep and suckler cow premium quotas are not linked to the land and are therefore able to pass as a separate asset from any land or lease of land.

The tenancy and the right to use the milk quota may have a value for the acquirer, but that value will depend on his personal circumstances and whether or not the lease may be terminated by the landlord.

METHOD OF TRANSMISSION OF LEASE

It must first be ascertained that the lease is a lease capable of transmission on the death of the tenant. A liferent lease expires on the death of the tenant[1]. *Gill* notes[2] that it has not been decided whether or not parties can contract out of the transfer provisions of the 1964 Act, s 16. If there is doubt as to whether or not a lease is transmissible on the death of the tenant, the tenant might be wise to raise proceedings *inter vivos* to have it settled that the lease is transmissible.

Issues in relation to the transmission of a lease also apply to the transmission of a *pro indiviso* share of a lease where there are joint tenants[3].

There are three methods by which a lease may be transmitted on death from the tenant to a successor. It should be noted that a lease may only be bequeathed or transferred to one person[4].

The first is by a special destination in the lease[5]. Where there is a special destination, the lease is not available to the executors to transfer to any person other than the heir in terms of the destination[6]. A special destination in a lease probably cannot be evacuated without the consent of the landlord[7].

The second is by a valid bequest by the deceased of his interest under the lease[8]. The tenant may bequeath his lease to his son-in-law or daughter-in-law or to any one of the persons entitled to succeed to his estate on intestacy[9].

The normal destination of an agricultural lease was to the tenant and his heirs, probably also excluding heirs-portioner, the eldest heir-female succeeding without division. Where there is no destination, a destination to the tenant and his heirs was implied at common law[10].

Where there is a right to bequeath the lease, it should be done as a specific bequest. While the right to the lease may in certain circumstances be carried by

---

1 13 *Stair Memorial Encyclopaedia* para 393; *Cormack v McIldowie's Exrs* 1974 SLT 178 at 183 per Lord Cameron.
2 Para 585.
3 Properly tenants in common. See *Smith v Grayton Estates Ltd* 1960 SC 349; *Coats v Logan* 1985 SLT 221.
4 *Kennedy v Johnstone* 1956 SC 39; 1964 Act, s 16(2) 'transfer the interest to any *one* of the persons entitled . . .'.
5 *Halliday* vol 4, para 49-04 'Special destinations'. The decisions in *Cormack v McIldowie's Exrs* 1975 SC 161 and *Reid's Trs v Macpherson* 1975 SLT 101, which decided that a destination to 'A and B and to the survivor, but excluding heirs-portioner (the eldest heir-female always succeeding without division) . . .' was not a special destination are criticised by *Halliday*.
6 1964 Act, ss 18 and 36(2).
7 *Halliday* vol 4, para 49-16.
8 See chap 14.
9 1991 Act, s 11.
10 *Rankine* p 157 ff.

a general bequest of the residue if the will can be so construed[1], it would be safer to make a specific bequest of the lease[2].

It is permissible to contract out of the power to bequeath a lease. An exclusion of 'legatees and assignees' or of 'assignees' alone is sufficient to exclude the right of bequest[3].

Most post-1948 leases exclude transmission by bequest, so the only option available will be a transfer of the deceased's interest in the lease by the deceased's executors to one of the persons entitled to succeed to the deceased's intestate estate, or to claim legal or prior rights of a surviving spouse[4].

It is important to establish whether the deceased's interest under the lease can transmit on death or under a special destination, or can be transferred by a valid bequest or can only be transferred by the executors. Once it has been established that the lease can transfer on death and if so, by what method, the tenant can be advised as to an appropriate will to settle the succession to his tenancy.

Where a tenant has no right to bequeath a lease it is still open to him to nominate in his will to whom the executors should transfer his interest in the lease and what credit, if any, the successor should give to the estate for that succession. Such a direction probably does not override the discretion given to the executors by the 1964 Act, s 16 but an executor would be expected to give effect to such a direction, unless there were strong reasons for departing from the direction.

AUCTOR IN REM SUAM

It is important to provide in the will for the appropriate executor and the person who is to succeed to a tenancy, so that an executor who transfers a lease to himself may not be *auctor in rem suam*.

A transfer of a lease *auctor in rem suam* can be reduced by the other beneficiaries or persons having a right in the estate of the deceased[5]. In order to avoid a challenge the executor must either decline or resign office[5], or obtain the consents of all interested parties to the transfer to himself[6] or he may have a defence to the plea if the deceased expressly or by implication authorised the transfer by the executor to himself[7]. Express authorisation is to be preferred.

It is accordingly important to consider appointing a person to be executor who will have no interest in the succession to the lease or else to specifically provide in the will that the executor should or at least might transfer the lease to himself.

1 *Hardy's Trs v Hardy* (1871) 9 M 736; *Edmond v Edmond* (1873) 1 M 348; *Rankine* p 164 and *Lindsay's Trs v Welsh* (1890) 34 Journal of Jurisprudence 165 (Sh Ct).
2 *Reid's Trs v Macpherson* 1975 SLT 101 at 109 (Lord Ordinary held that a general bequest of the residue did not carry the lease and that the latter must be specifically bequeathed).
3 *Kennedy v Johnston* 1956 SC 39.
4 See chap 13 and 1964 Act, s 16.
5 *Inglis v Inglis* 1983 SC 8.
6 *Phipps v Boardman* [1967] 2 AC 46 at 106.
7 *Johnston v MacFarlane* 1987 SLT 593. Authorisation by the testator may be express or implied, particularly where the testator foresaw a potential conflict of interest and yet appointed the particular executor; *Sarris v Clark* 1995 SLT 44.

HERITABLE OR MOVEABLE?

Whether a right relating to the tenancy is heritable or moveable in succession is of importance in determining what parts of the tenant's estate might be vulnerable to a claim for *jus relictae* or *legitim*.

A lease is heritable in succession[1].

The tenant's fittings, fixtures and improvements and the right to claim compensation due in respect thereof are heritable in succession[2]. Where a lease terminates before the death, the compensation will be moveable even if paid after the death.

The growing crops are heritable in succession until harvested, when they become moveable[3].

The value of the claim for compensation for milk quota[4] is heritable in succession because it is connected to the land[5].

In contrast, sheep and suckler cow annual premium quota, because they are allocated to the producer, are moveable in succession[6].

PARTNERSHIP ASSETS

If the tenant is farming through the medium of a partnership heritable assets owned by the partnership are, 'unless the contrary intentions appear', moveable in a particular partner's succession[7].

The first issue to be resolved is whether or not the lease, albeit it may be in the name of a particular tenant, is partnership property or not. If it is a partnership asset then the individual partner's share in the value of the lease falls to be treated as moveable in his succession. This will be the value of the lease[8] rather than the value of the deceased's interest under the lease[9].

The value of rights in milk quota, because connected to the land, will probably only be a partnership asset if the lease is a partnership asset[10].

WHO SHOULD BE THE SUCCESSOR?

Consideration will have to be given as to the most appropriate person to be the successor under the lease, otherwise the landlord may obtain either an

---

1 Erskine *Institutes* II, 2, 6; Bell's *Principles* 1478.
2 *Brand's Trs v Brand's Trs* (1876) 3 R (HL) 16; *Miller v Muirhead* (1894) 21 R 658.
3 *Chalmer's Trs v Dick's Trs* 1909 SC 761.
4 1986 Act, Sch 2, paras 2 and 3. The right to claim compensation is restricted to the tenant who had milk quota allocated to him or who acquired transferred quota for value and his successors. Note that the successor to a lease under a special destination does not appear to qualify under para 3 for compensation.
5 *Faulks v Faulks* (1992) 15 EG 82; J Murray 'Milk Quotas' 1986 SLT (News) 153.
6 EC Regulation (EEC) No 3567/92; Sheep Annual Premium and Suckler Cow Premium Quotas Regulations 1993, SI 1993/1626 (as amended by SI 1993/3036).
7 Partnership Act 1890, s 22.
8 See chap 24.
9 See p 115.
10 *Faulks v Faulks* (1992) 15 EG 82.

indefensible right to terminate the tenancy or a good prospect for terminating the tenancy.

Forward planning may be necessary to ensure that the prospective legatee or acquirer obtains or is obtaining the relevant training to equip him to take over the lease, to ensure objection cannot be taken to his personal skill or training for farming[1].

Consideration will have to be given to the most appropriate way for that person to be made the successor in the lease.

The executor's powers of transfer are different from the powers of bequest. A power of bequest includes a brother or sister-in-law, but not a spouse. The executors can transfer to a spouse[2].

Where a lease is bequeathed or the deceased's rights under the lease are transferred by the executors, the landlord has the right to object to receiving the prospective person as tenant[3].

It is an open question as to whether a landlord may object to a successor to a lease under a special destination. It has been suggested that the successor to a lease under a special destination should intimate the acquisition as a bequest under the 1991 Act, s 11, which will give the landlord a right to serve a counternotice objecting to the succession[4].

The question of the prospective tenant's suitability is then referred to the Land Court for a decision. The grounds of objection must be personal to the legatee[5].

There is a significant distinction between ss 11 and 12. If the landlord successfully objects to a person bequeathed the lease under s 11, the bequest is declared null and void. The lease then falls to be treated as intestate estate available for transfer by the executors[6]. In contrast, if the landlord makes a successful objection to a person to whom the lease was transferred by the executors, then the lease is terminated[7].

If there is doubt about the acceptability of the person whom the tenant would like to succeed to the lease, and if the power of bequest is available, then the lease should be bequeathed so that if the landlord successfully objects there is a second chance to nominate an alternative successor.

The landlord has a right to serve a notice to quit on certain grounds where a tenancy has been acquired by succession[8]. Where the legatee or acquirer of the lease is a near relative successor[9] the lease can only be terminated with the Land Court's consent if the notice to quit falls within one of the cases set out in Sch 2[10]. Where the legatee or acquirer is not a near relative successor then the landlord can serve a notice to quit to which there is no defence[11].

---

1 See pp 120 and 122; 1991 Act, Sch 2, cases 1 and 5.
2 Ibid, s 11(1) and 1964 Act, s 16(2).
3 1991 Act, ss 11(3) and 12(2).
4 *Halliday* vol 4, para 49-16.
5 See p 120.
6 1991 Act, s 11(8).
7 Ibid, s 12(3).
8 Ibid, s 25.
9 The spouse or child or adopted child (as defined by the 1964 Act, s 23(5)) of the deceased tenant; 1991 Act, Sch 2, Pt III, para 1. A person brought up as 'a child of the family' would not qualify as a near relative; see *Scammell and Densham* p 229, n 2.
10 1991 Act, s 25(2)(c).
11 Ibid, s 25(2)(d).

Consideration will have to be given to choosing a near relative as successor to the lease and who will not be vulnerable to a notice to quit, relying on cases 1, 3, 4, 5 or 6 in Sch 2, as described below[1].

GROUNDS OF OBJECTION TO LEGATEE OR ACQUIRER

Where the landlord objects[2] to the legatee or acquirer of the lease he requires to establish before the Land Court 'a reasonable ground of objection'. The ground of objection must be personal to the legatee or acquirer. It usually relates to his agricultural skill and knowledge; or his financial resources in relation to the holding; or to his personal character and reputation.

The legatee or acquirer does not become tenant of the holding until after this objection has been rejected by the Land Court. This is in contrast to a notice to quit founded on s 25, where the legatee or acquirer has become tenant before the notice to quit is served.

The landlord cannot found on his own needs or requirements, such as his wish for vacant possession, or a belief that he could farm the lands better himself[3].

Common grounds of objection include:

(1) *Agricultural skill.* The Land Court does not require of the legatee or acquirer the same standards which a landlord might require of an open market applicant[4]. Practical experience is accepted as a substitute for formal training[5]. Even if the tenant lacks personal skill and experience the Land Court may not sustain the objection if the court is satisfied that the tenant will have access to the appropriate skill and advice[6]. This is in contrast to the situation under case 1 or 5[7], where the court is concerned with the personal expertise of the tenant.

When dealing with the objection the Land Court is concerned with the performance of the prospective tenant and not with what has gone before. The court is entitled to take into account his performance since taking entry, pending a resolution of the objection[8]. Breaches of the tenancy persisted in by the legatee or acquirer after entry can be taken into account[9]. Neglect or breaches of the tenancy by the former tenant should not be held against the legatee or acquirer if he intends to follow a better course of husbandry[10].

---

1 See p 121.
2 Under the 1991 Act, ss 11(4) or 12(2).
3 *Howie v Lowe* 1952 SLCR 14; *Fraser v Murray's Trs* 1954 SLCR 10.
4 *Gill* para 580.
5 *Dunsinnan Estate Trs v Alexander* 1978 SLCR App 146; *Bell v Forestry Commission* 1980 SLCR App 116.
6 *Service v Duke of Argyll* 1951 SLT (Sh Ct) 2 (employment of competent manager); *Macrae v Macdonald* 1986 SLCR 69 (help from neighbours); *Dunsinnan Estate Trs v Alexander* 1978 SLCR App 146.
7 1991 Act, Sch 2.
8 *Dunsinnan Estate Trs v Alexander* above; *Service v Duke of Argyll* 1951 SLT (Sh Ct) 2; *Sloss v Agnew* 1923 SLT (Sh Ct) 33; *Reid v Duffus Estate* 1955 SLCR 13; *Fraser v Murray's Trs* 1954 SLCR 10; *Anderson v Bennie* 1958 SLCR 34; *Morefarm Property Co v Forbes* 1958 SLCR App 225.
9 *Sloss v Agnew* above.
10 *Service v Duke of Argyll* above.

(2) *Financial resources.* The adequacy of the financial resources of the tenant are a matter for the judgment of the Land Court[1]. The financial resources must be related to the size of the farm[2]. An absolute guarantee of solvency is not required[3]. If the legatee or acquirer of the lease does not have sufficient skill, capital or equipment, or access to such capital or equipment, the objection may be sustained[4].

(3) *Personal character.* The landlord is entitled to a tenant of good character[5]. Alcoholism, dependence on drugs or convictions for serious offences would be reasonable grounds for objection. Isolated minor offences may be overlooked by the Land Court[6]. Spent convictions should not be taken into account[7].

NOTICES TO QUIT UNDER S 25[8] – NEAR RELATIVE SUCCESSOR[9]

A landlord may seek to recover possession of a holding after succession by a near relative by service of a notice to quit under the 1991 Act, s 25. This section is not concerned with the requirements of such a notice to quit, but with the cases under Sch 2 to which a legatee or acquirer may be vulnerable.

In planning the succession to a tenancy consideration will have to be given to the vulnerability of the near relative to a notice to quit under one of the cases. Forward planning in the training undertaken by the prospective legatee or acquirer, or with regard to the succession to or occupation of land by the prospective legatee or acquirer or the bequest of assets may be necessary to reduce the vulnerability to such a notice to quit.

Cases 1 to 3 relate to tenancies let before 1 January 1984 and cases 4 to 7 to tenancies let after that date.

The onus of proof is on the landlord to establish cases 1 to 3 and 6, but it is incumbent on the tenant to satisfy the Land Court that the circumstances are not as specified in cases 4, 5 and 7[10].

There is a potential defence to cases 1, 2, 3, 6 and 7, that the Land Court

---

1 *Dunsinnan Estate Trs v Alexander* 1978 SLCR App 146; *Bell v Forestry Commission* 1980 SLCR App 116.
2 *Macrae v Macdonald* 1986 SLCR 69.
3 *Reid v Duffus Estate* 1955 SLCR 13; *Chalmers Property Investment Co v Bowman* 1953 SLCR App 214.
4 *Sloss v Agnew* 1923 SLT (Sh Ct) 33.
5 *Howie v Lowe* 1952 SLCR 14; *Marquis of Lothian's Trs v Johnstone* 1952 SLCR App 233.
6 *Luss Estates Co v Campbell* 1973 SLCR App 96.
7 Rehabilitation of Offenders Act 1974, s 4.
8 A question has been raised as to whether or not the consolidated provisions of s 24 as read with s 25 remove the right of a landlord to serve both a s 25 and a s 21(1) notice to quit on a near relative successor; see R D Sutherland 'Unforeseen Consequences of the Agricultural Holdings (Scotland) Act 1991' 1993 SLT (News) 351; per contra D G Rennie 'Getting Rid of a Near Relative' 1993 SLT (News) 375. Probably the landlord can serve both notices; see *Gill* para 377.
9 See chap 16 for a fuller consideration of the operation of the cases in the 1991 Act, Sch 2.
10 Ibid, s 25(3); *Macdonald v Macrae* 1987 SLCR 72 at 81.

can withhold consent on the ground that it appears to it that a fair and reasonable landlord would not insist on possession[1].

(a) *Cases 1 and 5*. The ground for the notice is that the tenant has neither sufficient training in agriculture nor sufficient experience in the farming of land to enable him to farm the holding with reasonable efficiency. Training envisages a formal training course, but adequate farming experience may suffice[2]. Unlike an objection under ss 11 or 12, where the tenant may rely on advice from other sources, these cases relate to the personal qualifications of the legatee or acquirer.

While case 1 relates to the training or experience at the date of death, case 5 has the proviso that the case does not apply if the legatee or acquirer, throughout the period from the date of death, has been engaged in a course of relevant training in agriculture which he is expected to complete satisfactorily within four years from the date of death and has made arrangements to secure that the holding will be farmed with reasonable efficiency until he completes the course. The proviso does not benefit a tenant who has taken up the training course after the death.

It is therefore important to try to plan that the prospective legatee or acquirer obtains the necessary training and experience prior to the death of the tenant.

(b) *Cases 2 and 6*. These cases are identical. They relate to the question of whether or not the holding, or any agricultural unit of which it forms part, is a two-man unit[3], coupled with the landlord's intention to amalgamate[4] the holding within two years of the termination of the tenancy with other specified land.

It has been held that these cases relate to the agricultural unit as the land actually occupied, whether formally or informally, by the tenant as a single unit[5].

Consideration will have to be given as to how the holding or the agricultural unit of which it forms part can be made into a two-man unit, by acquiring or renting, including regular grazing lets, land to make it a two-man unit.

(c) *Cases 3 and 7*. While these cases relate to the near relative successor who occupies agricultural land[6] either as owner or tenant[7], which is a two-man unit distinct from the holding, there are significant differences.

1 1991 Act, s 25(3). Cf *Altyre Estates v McLay* 1975 SLT (Land Ct) 12; *Earl of Seafield v Currie* 1980 SLT (Land Ct) 10; *Carnegie v Davidson* 1966 SLT (Land Ct) 3; *Trs of the Main Calthorpe Settlement v Calder* 1988 SLT (Land Ct) 30.
2 *Macdonald v Macrae* 1987 SLCR 72.
3 1991 Act, Sch 2, Pt III, paras 1 and 2; see *Jenners Princes Street Edinburgh v Howe* 1990 SLT (Land Ct) 26 (it was held that any land occupied by the tenant, whether formally or informally, as a single unit was included in the agricultural unit). Cf cases 3 and 7 which relate to the tenant as a formal 'occupier (either as owner or tenant)'.
4 1991 Act, Sch 2, Pt III, para 1. The landlord must show a genuine proposal for amalgamation; *Mackenzie v Lyon* 1984 SLT (Land Ct) 30; *Earl of Seafield v Currie* 1990 SLT (Land Ct) 10.
5 *Jenners Princes Street Edinburgh v Howe* 1990 SLT (Land Ct) 26.
6 *Gill* para 395 raises the question of whether or not the occupied land must be in Scotland or if occupation in England will suffice. It has been suggested that land in England would qualify as agricultural land so occupied; see notes to the 1991 Act, Sch 2, case 3 in *Current Law Statutes* .
7 The tenancy probably need not be a secure tenancy under the 1991 Act; see notes in *Current Law Statutes* to the 1991 Act, Sch 2, case 3.

In case 3 the agricultural land has to be distinct from the holding and any other agricultural unit of which the holding forms part. In case 7 the agricultural land merely has to be distinct from the holding.

In case 3 the occupation by the tenant has to be from before the death of the person from whom he acquired the right to the lease. In case 7 the occupation has to be throughout the period from the date of giving the notice to quit. Case 7 therefore covers the situation where the tenant succeeds to other land from the deceased tenant on that death. It may be advisable to plan, perhaps by way of a trust, that the prospective tenant does not inherit additional agricultural land until after the period during which a notice to quit can be given. A person who inherits two tenancies on one death would be vulnerable to a notice to quit in respect of both tenancies. The termination of one tenancy would mean that there was no continuing occupation for the purposes of case 7 to terminate the other tenancy. The problem is that the tenant could not control which application went ahead first and if both were heard together both tenancies might be vulnerable.

Further it should be noted that under case 3 the occupation has to be by the tenant personally, whereas under case 7 occupation through the medium of a company controlled by the tenant or by a partnership of which the tenant is a partner is deemed to be occupation by the tenant[1].

In planning the succession consideration requires to be given to what other land is occupied at, or is to be occupied after, the relevant death by the prospective tenant. The tenancy can then be transmitted to a person who will not be vulnerable to a case 3 or 7 notice to quit.

(d) *Case 4.* This case relates to the sufficiency of the tenant's financial resources. The criteria set out at p 121 above are relevant to this question, but the difference is that the onus is on the tenant to establish that the case does not apply.

Consideration will have to be given to leaving the prospective tenant sufficient financial resources to allow him to continue with the tenancy.

QUOTAS

The right to use the milk quota and to claim compensation on the termination of the tenancy[2] pass to the person succeeding to land or acquiring the tenancy.

Sheep and livestock annual premium quotas are personal to the producer and will pass in terms of his will or by intestate succession. Where there is an intestacy the executors will have to determine to whom the quota is transferred.

---

1   1991 Act, Sch 2, Pt III, para 3; *Haddo House Estate Trs v Davidson* 1982 SLT (Land Ct) 14, where it was held that for the precursor of case 3 occupation by a partnership was not occupation for the purposes of the case. In *Jackson v Hall; Williamson v Thompson* [1980] AC 854 it was held in England that sole occupancy was not required for the purposes of the equivalent English case - now the Agricultural Holdings Act 1986, s 36(3)(b).

2   If the tenant is a successor in terms of the Agriculture Act 1986, Sch 2, para 3.

It may be important to time the transfer of a lease by the executors[1] so as to coincide with the transfer periods for sheep and livestock annual premium quotas, otherwise the new tenant may find himself without those quotas in the year of the transfer[2]. Unless the quota is transferred with the holding it will be subject to siphon[3]. Further, those quotas can only be transferred to producers.

TAX PLANNING[4]

As this book is not a book on tax law it is not intended to give detailed consideration to the tax implications relating to agricultural land.

As agricultural land, an agricultural tenancy[5] and the assets of an agricultural business or partnership have a substantial value; in planning the succession to farm land, the tenancy or the assets of the business, consideration requires to be given to the tax implications and how they may be mitigated, to ensure that the successor to the farm or tenancy and the business assets is not precluded from continuing to farm by tax demands settled on him by virtue of the succession.

Consideration will have to be given to the use of and impact of:

(a) Inheritance tax relief available on transfers of agricultural property[6]. To qualify for agricultural property relief the transferor must either (i) have occupied[7] the property for the purposes of agriculture throughout the period of two years ending with the date of transfer or (ii) have owned the property for seven years ending with the date of transfer where it has been occupied by himself or another throughout the period for agricultural purposes[8]. The relief is 100 per cent if the transferor has vacant possession or the right to obtain it within twelve months of the transfer otherwise 50 per cent applies[9].

(b) Consideration should be given to whether there are tax advantages in the letting of a farm to a family partnership or otherwise at full market rent, for the purpose, after three years (to avoid the associated operations provisions), of transferring the freehold at a reduced value burdened by the tenancy. The delay of three years may be critical in looking for a seven-year PET and if the transferor dies within three years then the granting of the tenancy and the subsequent transfer are associated operations, which means the property will be valued at open market valuation. The break-even point at which the project might be worthwhile is where the tenanted value will be about 70 per cent of the vacant possession value[10].

---

1 Or to seek an extension of time for the transfer under the 1964 Act, s 16(3)(b); or to renounce a bequest so that the lease can be transferred by the executors.
2 SI 1993/1626, regs 5 and 7.
3 SI 1993/1626, reg 6.
4 See *Rodgers* chap 10 for a fuller discussion of this issue.
5 *Baird's Executors v IRC* 1991 SLT (Lands Tr) 9; *Walton's Exrs v IRC* [1994] 2 EGLR 217.
6 Inheritance Tax Act 1984, ss 115-117.
7 Occupied and not owned, so an occupier who purchases at the end of the two years can make a relevant transfer on that date.
8 Eg a landlord not in possession.
9 In general, the working farmer gets 100 per cent relief and the landlord 50 per cent.
10 *Rodgers* p 215.

(c) The effect of a partnership will have to be considered. In general the death of a partner terminates the lease, so a farm tenanted by a partnership which it was planned to be subject to a tenancy at the death will have a vacant possession value[1]. Occupation of property by a Scottish partnership is treated for inheritance tax purposes as occupation by the partners[2]. Where the land is owned by the partnership a gift of a share in the partnership does not attract agricultural property relief, but may qualify for business property relief[3]. The relief provisions relating to the value of agricultural tenancies provided by the Inheritance Tax Act 1984, s 177 apply only where the deceased 'was the tenant' and do not apply if the lease was to a partnership of which the deceased was a member.

(d) Where the agricultural interest is held as shares in or securities of a company and the share value is attributable to the agricultural value of the property, and if the transferor had control of the company then agricultural reliefs will be available[4].

(e) The valuation of leases for inheritance tax purposes poses problems. An agricultural tenancy, because of the protection afforded to successors, has a quantifiable value as part of the deceased's estate[5]. Provision has been made that any value associated with the prospect of renewal by tacit relocation should be left out of account, where the deceased was tenant for two years continuously before the death or had become tenant by succession[6]. There may be a value in the remainder of a term of a fixed duration lease, because the relief only excludes the value associated with the right to transfer the lease on a death. Where the lease is running on tacit relocation the value is left out of account where the deceased was tenant for two years continuously before the death or had become tenant by succession[7]. There is 100 per cent relief for a tenancy acquired by succession on or after 1 September 1995[8].

(f) Where there is milk quota consideration will have to be given to whose asset the milk quota is. If a farm is being farmed by a partnership, the allocated milk quota will probably not be an asset of the partnership if the farm is neither owned nor tenanted by the partnership[9]. If the farm is tenanted, but the milk quota is registered in the name of the partnership, while the tenant might have an interest in the asset of the quota, the asset may be valueless to him as he probably has no claim for compensation on termination of the tenancy as the quota was not registered as his[10].

(g) In considering whether or not to make a lifetime transfer of a farm and farming assets consideration will have to be given to the interrelationship

1 *IRC v Graham's Trs* 1971 SC (HL) 1; see 19 *Stair Memorial Encyclopaedia* paras 1568–1571 for a fuller discussion of this topic.
2 Inheritance Tax Act 1984, s 119(2).
3 19 *Stair Memorial Encyclopaedia* paras 1559 and 1589. In England, because a partnership does not have a legal persona, the converse applies.
4 Inheritance Tax Act 1984, s 122.
5 19 *Stair Memorial Encyclopaedia* para 1564.
6 Inheritance Tax Act 1984, s 177(1) and (3).
7 Ibid, s 177(2).
8 Inheritance Tax Act 1984, s 116(2)(c) and (3).
9 *Faulks v Faulks* [1992] 1 EGLR 9.
10 See p 226 'Entitlement'.

between capital gains tax[1] and its associated roll-over, hold over and business retirement reliefs, and the impact of inheritance tax if the assets are retained until death[2]. A lifetime transfer of a tenancy may be liable for capital transfer tax if the transfer is a transfer of value[3].

---

1 See 19 *Stair Memorial Encyclopaedia* paras 1524–1529.
2 *Rodgers* p 219.
3 *Baird's Executors v IRC* 1991 SLT (Lands Tr) 9.

CHAPTER 13

Succession to a lease

Key points

(1) *Testate succession*

- Where there is a bequest of the lease confirm that (a) the testator had the power to bequeath the lease and (b) there was a valid bequest of the lease.

- Check whether or not the legatee is defenceless or vulnerable either to an objection by the landlord or to a notice to quit under the 1991 Act, s 25. Consider whether or not the bequest should be accepted if there is a less vulnerable successor to whom the lease might be transferred by the executor.

- Intimation of the bequest has to be made to the actual landlord (and not deemed landlord under s 84(4)). The legatee should confirm who is the landlord before serving the notice.

- Late intimation of an acceptance after 21 days will bring the lease to an end. If *per incuriam* the lease is not accepted timeously the lease should be left to the executor to transfer.

- Non-acceptance of the bequest, or if the Land Court declares the bequest null and void, leaves the lease available for transfer by the executor.

- Intimation of acceptance to which the landlord serves counternotice, followed by a failure to apply to the Land Court for an order declaring the legatee tenant, probably brings the lease to an end.

- The landlord should consider whether to serve a counternotice (s 11(3)) objecting to receiving the legatee as tenant and/or later to serve a notice to quit under s 25.

(2) *Intestate succession*

- The executor requires to confirm to the lease within one year, prior to transferring it.

- The executor may transfer the lease to 'any one of the persons entitled to succeed to the deceased's intestate estate, or to claim legal rights or the prior rights of a surviving spouse': 1964 Act, s 16(1).

- The executor requires to transfer the lease within one year of the deceased's death or such other period fixed by agreement with the landlord or by the sheriff on a summary application. An application to the sheriff for an extension has to be made before the expiry of the year: 1964 Act, s 16(3)(b).

- Where there is a lease to joint tenants, the executors should ascertain whether or not there is a survivorship clause and if not, then the deceased joint tenant's share requires to be transferred by them, otherwise the lease, if running on tacit relocation, will be terminable by the landlord.

- Intimation of the acquisition of the lease by the acquirer has to be made to the actual landlord. The acquirer should confirm who is the landlord before serving the notice.

Time limits

(1) *Testate succession*

- Acceptance of the bequest has to be given to the landlord within 21 days after the death, unless prevented by some unavoidable cause, in which case notice must be given as soon as practical thereafter: s 11(2).

- The landlord's counternotice objecting to the tenant has to be given within one month of the receipt of the legatee's intimation of acceptance: s 11(4).

- If the landlord serves a counternotice the tenant requires to make application to the Land Court expeditiously for an order declaring him to be tenant under the lease.

(2) *Intestate succession*

- Where the executor has the power to transfer the lease he has one *year* (or such longer period as is agreed or granted by the sheriff prior to the end of the year) from the date of death or determination by the Land Court that the bequest was null and void in which to transfer the lease: 1964 Act, s 16(3)(b).

- The person to whom the lease is transferred has to give notice to the landlord of the acquisition within 21 days or if he is prevented by some unavoidable cause from giving such notice within that period as soon as practical thereafter, of the acquisition by him of the lease: s 12(1).

- Where the landlord objects to the acquirer of the lease he has one month in which to give counternotice intimating that he objects to the acquirer of the lease: s 12(2).

- Not before the expiry of one month from giving the counternotice the landlord requires to apply to the Land Court for an order terminating the lease: s 12(2).

GENERAL

On the death of an agricultural tenant it is important to ascertain immediately[1]:

(1) whether there is a will;

(2) if there is a will did it purport to bequeath the lease;

(3) if there was a bequest of the lease (a) did the lease give the deceased the power to bequeath it and (b) if it did, was the bequest a valid bequest of the lease to a person entitled to have the lease bequeathed to him.

If there is no bequest or valid bequest of the lease, then the lease is available to be transferred by the executors.

A VALID BEQUEST OF THE LEASE

At common law the succession to the lease was governed by the destination in the lease or by the implied destination to the tenant and his heirs[2]. The tenant did not have the power to bequeath his lease unless the lease included the power of bequest.

Where the lease gives the power to bequeath it, the bequest has to be to the deceased's 'son-in-law or daughter-in-law or to any one of the persons who would be, or would in any circumstances have been, entitled to succeed to the estate on intestacy by virtue of' the 1964 Act[3]. The range of persons is very much wider than that available to the executor[4], although a spouse is not included.

Notwithstanding the wording of s 11(1) a tenant may contract out of his power of bequest[5]. The power of bequest is excluded by the words 'excluding legatees and assignees' or 'excluding assignees' alone[6].

If the deceased has purported to bequeath his lease a check should be made of the lease to confirm that the power of bequest has not been excluded. If it has been excluded the 'bequest' should probably be treated by the executors as a direction as to the person to whom they should consider transferring the lease.

As a general rule there should be a specific bequest of the lease *nominatum*. This is the safest course. It has been held that a general bequest of the residue does not carry the lease in circumstances where there was an express exclusion

---

1 Intimation of acceptance of a valid bequest of the lease has to be made within 21 days of the death; 1991 Act, s 11(2).

2 *Rankine* p 157.

3 1991 Act, s 11(1).

4 Ie it includes the whole range of relatives even if the particular relative is not actually entitled to suceed on intestacy because there is nearer kin actually entitled to share in the intestate estate. In contrast, the executor may only transfer the lease to the persons actually entitled to succeed to the intestate estate and to a spouse.

5 *Kennedy v Johnstone* 1956 SC 39. It would appear that this decision is not affected by the subsequent enactment of the 1964 Act, s 29(1) which refers to 'an *implied* condition prohibiting assignation'; see *Gill* para 573.

6 *Kennedy v Johnstone* above.

of assignees and the will conveyed to the executors the whole estate 'over which I have power of control or disposal at the time of my death'[1]. If there is no exclusion of assignees it may be a matter of construction of the will to see if a general bequest or a bequest of the residue carries the lease[2].

Whether or not the bequest is a valid one is not within the jurisdiction of the Land Court, but would require to be determined in the ordinary courts[3].

OPTIONS OPEN TO LEGATEE AND TIME LIMITS

The legatee may accept the bequest by giving notice to the landlord within 21 days[4].

A legatee may choose not to accept the bequest of the lease[5]. The lease is then available for transfer by the executors[6].

If there is doubt as to whether or not there has been a valid bequest, the safest course would be not to accept the purported bequest and leave the lease to be transferred by the executor.

A legatee may after accepting the bequest renounce his claim to the lease expressly or impliedly. In such a situation the lease is no longer available for transfer by the executor and the tenancy will cease[7].

Where a legatee who has accepted the lease is served with a counternotice by the landlord objecting to receiving him as a tenant, the legatee requires to apply to the Land Court for an order declaring him tenant[8].

If the legatee decides not to continue to contest the application he should consent to the Land Court declaring the lease null and void, otherwise the lease may not be available for the executor to transfer[9].

Where a legatee accepts the lease and no counternotice is served by the landlord, the legatee becomes tenant from the date of the deceased's death[10].

---

1 *Reid's Trs v Macpherson* 1975 SLT 101 at 109. Cf *Lindsay's Trs v Welsh* (1890) 34 Journal of Jurisprudence 165 (Sh Ct) where the contrary argument was upheld.
2 *Hardy's Trs v Hardy* (1871) 9 M 736; *Edmond v Edmond* (1873) 11 M 348; *Rankine* p 164.
3 *Garvie's Trs v Still* 1972 SLT (Land Ct) 29.
4 See 1991 Act, s 11(2).
5 1991 Act, s 11(2) – 'if he accepts the bequest'.
6 1964 Act, s 16(2)(b).
7 After acceptance the interest under the lease no longer falls within the criteria of the 1964 Act, s 16(2), leaving it available for transfer by the executor.
8 1991 Act, s 11(4) and (5).
9 1964 Act, s 16(2) gives the executor a right to transfer the lease in three situations: (a) where the interest is not subject to a valid bequest; (b) where a bequest is not accepted; and (c) where the bequest is declared null and void by the Land Court under the 1991 Act, s 11. Section 16(3)(b)(i) suggests that an executor has one year to transfer the lease 'from the date of determination or withdrawal of . . . the application'. This might suggest that the executor has one year from the date of withdrawal of an application to the Land Court by the legatee, but this is not consistent with the clear words of s 16(2)(c). It would therefore be safer to consent to a decree that the bequest is null and void, rather than to agree to withdraw the application.
10 Ibid, s 11(3).

LEASE AVAILABLE FOR TRANSFER BY EXECUTOR

The lease is vested in the executor and available for transfer if the 'deceased immediately before his death held the interest of a tenant under a tenancy or lease, which was not expressed to expire on his death'[1].

Where there is a special destination[2] or valid bequest of the lease to which no objection is made or sustained, the executor will be obliged to transfer the lease to the heir in the destination or the legatee.

Where there is a joint tenancy, on the death of one of the tenants it is important to ascertain if there is a survivorship clause in favour of the other joint tenant or if the interest of the deceased joint tenant requires to be transferred by his executors. Failure to transfer the interest of a deceased joint tenant, where there is no survivorship clause, when the lease is running on tacit relocation will mean that the lease will terminate[3]. A deceased joint tenant's pro indiviso interest in the lease is treated in the same way as a deceased tenant's interest in the whole lease.

Where the interest in the lease is:

(a) not the subject of a valid bequest by the deceased; or

(b) is the subject of such a bequest, but the bequest is not accepted by the legatee; or

(c) the bequest is declared null and void by the Land Court under the 1991 Act, s 11

then the lease is available to the executor to transfer 'to any one of the persons entitled to succeed to the deceased's intestate estate, or to claim legal rights or the prior rights of the spouse'[4]. As noted above the range of transferees is more restrictive than the range of persons to whom a lease may be bequeathed.

The executor has an uncontrolled discretion as to the selection of the acquirer, where there is more than one person eligible to receive the transfer. An executor must avoid being *auctor in rem suam* by transferring the lease to himself, unless specifically authorised to do so[5]. A tenant may assist his executor by suggesting the nominee in his will, although such a nomination by the deceased will not override the executor's discretion.

The power of transfer exists even if there is an express or implied prohibition against assignation in the lease[6].

CONFIRMATION AND TRANSFER OF LEASE BY EXECUTOR

The executor only has title to transfer the lease if he has confirmed to it within one year[7]. The lease should be entered as a specific item in the inventory for

---

1 1964 Act, ss 14(1) and 36(2).
2 Ibid, s 36(2)(a).
3 *Smith v Grayston Estates Ltd* 1960 SC 349; *Coats v Logan* 1985 SLT 221.
4 1964 Act, s 16(2).
5 See p 117.
6 1964 Act, s 16(2).
7 *Morrison-Low v Paterson* 1985 SC (HL) 49; *Rotherwick's Trs v Hope* 1975 SLT 187.

confirmation by an appropriate conveyancing description or other description 'sufficient to identify it'[1]. While confirmation should precede the transfer it has been held that the executor may validly transfer the lease prior to confirmation, provided both transfer and confirmation were within the year[2].

The executor requires to make a formal transfer of the lease. A docket in the form (suitably modified) to the 1964 Act, Sch 1 is suggested.

TIME LIMITS FOR TRANSFER BY EXECUTOR

Normally the lease has to be transferred by the executor within one year of the death of the deceased[3].

Where there has been an accepted bequest of a lease, which the Land Court has declared to be null and void[4], the executor has one year from the 'determination or withdrawal of the ... application' in which to transfer the lease[5].

The executor may obtain an extension of the time in which to transfer the lease, fixed either by agreement with the landlord or, failing agreement, by the sheriff on a summary application[6]. The landlord's agreement has to be obtained[7] or the application to the sheriff has to be made before the expiry of the anniversary of the death[8]. The sheriff on such a summary application has no power to rule on the competency of the application, which is a matter for the ordinary courts[9].

INTIMATION OF BEQUEST OR TRANSFER OF LEASE

If the legatee accepts the bequest he is required to intimate the bequest to the landlord of the holding 'within 21 days after the date of the death of the tenant, or, if he is prevented by some unavoidable cause[10] from giving such notice within that period, as soon as practical thereafter'[11].

1 Act of Sederunt (Confirmation of Executor's Amendment) 1966, SI 1966/593; *Currie on Confirmation of Executors* (8th edn, 1995) p 438.
2 *Garvie's Trs v Garvie's Tutors* 1975 SLT 94.
3 1964 Act, s 16(3)(b)(ii).
4 1991 Act, s 11(6).
5 1964 Act, s 16(3)(b)(i). The use of the word 'withdrawal' causes difficulties in that it suggests that one year might run from a withdrawal of an application by the legatee to be declared tenant, whereas the 1964 Act, s 16(2)(c) and the 1991 Act, s 11(8) suggest that the executor only has the right to transfer if the bequest is declared null and void. Where the application has been withdrawn there will be no declaration that the bequest was null and void; see p 130.
6 Ibid, s 16(3)(b).
7 It is clearly competent for the landlord to grant an extension of time after the anniversary, but there would be no *compulsitor* on him to grant the extension. If an extension was refused it would be too late to apply to the sheriff.
8 *Gifford v Buchanan* 1983 SLT 613. It would appear that the extension period does not have to be fixed by the sheriff within the one year; *Cunninghame DC v Payne* 1988 SCLR 144.
9 *Gill* para 593.
10 A phrase used in the Crofters Acts; see *Budge v Gunn* 1925 SLCR 74; *Murray v Lewis Island Crofters Ltd* 1930 SLCR 12; *MacKinnon v MacSween* 1939 SLCR 28; *MacAskill v MacDonald* 1945 SLCR 3.
11 1991 Act, s 11(2).

Similarly an acquirer of a lease 'shall give notice of the acquisition to the landlord of the holding within 21 days after the date of acquisition, or if he is prevented by some unavoidable cause from giving such notice within such period, as soon as practical thereafter'[1].

The notice should be expressed in terms sufficiently clear to bring home to the landlord that the legatee or acquirer is exercising his rights under the bequest or transfer to take up the lease[2].

The notice of bequest or acquisition should be given by the legatee or acquirer himself. If the executor is also the acquirer, the acquirer should make it clear that the intimation is given in his personal capacity. It has been held that a notice of transfer given by the executors and not the acquirer was valid as there was no prejudice to the landlord because the executors and the acquirer were the same[3].

It has been held that a failure to find a will in a search a few days after the death, but which was discovered by chance after the 21 days was an unavoidable delay[4]. Unawareness of the statutory time limits is not an unavoidable cause[5], although mental strain and distress suffered by a legatee may be[6]. Probably only the ordinary courts have the jurisdiction to determine whether or not there was an unavoidable cause[7].

A late intimation by a legatee of an acceptance, if an unavoidable cause cannot be established, has the effect of terminating the tenancy if the landlord does not accept the late intimation. It remains a valid acceptance so as to preclude the executor from transferring the lease thereafter[8].

If the bequest is not intimated within the 21 days it would be safer not to rely on establishing an unavoidable cause, but to leave the lease to be transferred by the executor.

The intimation has to be to the 'landlord', so the legatee or acquirer will have to confirm who is the actual landlord. As neither the legatee nor the acquirer is the tenant they cannot rely on the deemed landlord provisions of s 84(4).

The safe and proper way of giving intimation of a bequest or acquisition is in writing by recorded delivery to the landlord. Verbal intimation by telephone has been conceded to be effective[9].

EFFECT OF INTIMATION

The effect of intimation by the legatee or the acquirer, if the landlord does not object, is to make the lease binding between landlord and tenant in the case of

1 1991 Act, s 12(1).
2 *BSC Pension Funds Trs Ltd v Downing* [1990] 1 EGLR 4; *Lees v Tatchell* [1990] 1 EGLR 10.
3 *Garvie's Trs v Garvie's Tutor* 1975 SLT 94. The import of this case is that in the normal situation an intimation by the executors would be invalid.
4 *MacKinnon v Martin* 1958 SLCR 19.
5 *Wight v Marquis of Lothian's Trs* 1952 SLCR 25.
6 *Thomson v Lyall* 1966 SLCR App 136 (a widow pled mental strain and distress as the unavoidable cause. The Land Court expressed sympathy, but ruled against her as no evidence was led for her).
7 *Garvie's Trs v Still* 1972 SLT (Land Ct) 29; *Gill* para 605.
8 See *Reid's Trs v Macpherson* 1975 SLT 101.
9 *Irvine v Church of Scotland Trs* 1960 SLCR 16.

(1) a legatee, from the date of the death of the deceased tenant[1] and (2) an acquirer, from the date of the transfer to him by the executor[2]. The importance of this distinction is that the fruits of the holding belong to the executory until the date of transfer, whereas under a bequest the fruits belong to the legatee from the date of death[3].

OPTIONS OPEN TO LANDLORD ON INTIMATION OF BEQUEST OR TRANSFER OF LEASE

Upon receipt of an intimation of a bequest or transfer of a lease the landlord has three options:

(1) The landlord may do nothing, in which case the lease becomes binding between the landlord and new tenant in the case of (a) a legatee, from the date of the death of the deceased tenant and (b) an acquirer, from the date of the transfer to him by the executor.

The same result occurs if the landlord per incuriam fails to serve a counternotice timeously.

(2) The landlord may serve a counternotice within one month of the receipt of the intimation objecting to receiving the legatee or acquirer as tenant under the lease[4].

The counternotice need not specify the grounds of the landlord's objections, which may be kept for the landlord's pleadings in the subsequent Land Court application.

The grounds of objection must be personal to the legatee or acquirer and cannot relate to the personal circumstances of the landlord[5].

(3) Accept the legatee or acquirer as tenant, but thereafter serve a notice to quit under s 25. Where the new tenant is not a 'near relative'[6] there will be no defence to the notice to quit. In the case of a near relative the notice will only succeed if the particular cases set out in Sch 2 are established[7].

This option remains available if the landlord omits to serve a counternotice.

Gill[8] suggests that if an acquirer intimates the acquisition out of time, the landlord should not serve a counternotice but should sue in the ordinary courts for possession of the holding, leaving it to the acquirer to establish by way of defence an 'unavoidable cause'. The risk for the landlord is that if the defence succeeds he will have lost his right to serve a counternotice. A landlord might safeguard his position by serving a counternotice on a without prejudice basis and making application to the Land Court, which can then be sisted pending the outcome of the court proceedings.

1 1991 Act, s 11(3).
2 Ibid, s 12(1).
3 *Macdonald v Macdonald* 1995 GWD 34-1761.
4 Ibid, s 11(4) (legatee) and s 12(2) (acquirer).
5 See p 120.
6 1991 Act, Sch 2, Pt III, para 1 – 'surviving spouse or child of the tenant including a child adopted by him in pursuance of an adoption order (as defined in section 23(5) of the Succession (Scotland) Act 1964)'.
7 See p 121.
8 Para 599.

Response to counternotice

After service of a counternotice on a legatee, the legatee has to apply to the Land Court 'for an order declaring him to be tenant under the lease as from the date of the death of the deceased tenant'[1].

No time limit is given for such an application, but if it is not made expeditiously it may be held that the tenant has waived his right to have it declared that he is the tenant under the lease. Any action for declarator that the tenant had lost his right would have to be in the ordinary courts[2].

In the case of an acquirer, after service of the counternotice, the landlord requires to take the matter further by applying to the Land Court, not before the expiry of one month from the giving of the counternotice, for an order terminating the tenancy[3]. Although there is no time limit for an application, a landlord should apply immediately to the Land Court for the order terminating the lease, otherwise he may be held to have waived his right to make the application.

Effect of counternotice

Pending any proceedings following upon a counternotice the legatee with the consent of the executor[4] or the acquirer with the consent of the executor[5] shall have possession of the holding. On cause shown by the landlord, the Land Court may otherwise direct that the legatee or acquirer should not have possession.

During this interim period the lease is vested in the executor[6], who should pay the rent and fulfil the other prestations due under the lease. While the landlord will have no recourse against the person in possession of the holding under the lease, because he is not the tenant, he may have a remedy against the executor.

The landlord should not accept rent from the legatee or acquirer in possession, otherwise this might have the effect of constituting a new lease[7].

THE EFFECT OF A SUCCESSFUL OBJECTION

Where a landlord succeeds in his objection to a legatee the Land Court declares the bequest to be null and void[8]. The lease is then available for transfer by the executor as intestate estate[9]. The unsuccessful legatee has no waygoing rights. The whole interest in the lease vests in the executor by virtue of his confirmation.

1 1991 Act, s 11(5).
2 See *Gill* para 578 regarding the problems raised by a failure to apply expeditiously.
3 1991 Act, s 12(2).
4 Ibid, s 11(7).
5 Ibid, s 12(4).
6 1964 Act, s 14(1). The executor should in the meantime have confirmed to the lease.
7 *Morrison-Low v Paterson* 1985 SC (HL) 49.
8 1991 Act, s 11(6).
9 Ibid, s 11(8); 1964 Act, s 16(2).

Where the landlord succeeds in his objection to an acquirer of the lease, the Land Court pronounces 'an order terminating the lease, to take effect as from such term of Whitsun or Martinmas as they may specify'. The lease is not thereafter available for further transfer by the executor.

Although the acquirer only had possession of the holding pending the outcome of the objection application, s 12(5) provides that the termination of the lease shall be treated for the purposes of compensation as a termination of the acquirer's tenancy of the holding, but nothing in the section entitles him to compensation for disturbance.

It is doubtful that the acquirer would have any right to claim compensation for milk quota because he was never a person who acquired a right to the tenancy[1]. Milk quota compensation should probably be claimed by the executor.

1 1986 Act, Sch 2, para 3.

CHAPTER 14

Termination and part termination of tenancies

Key points

- When negotiating a variation or modification of a lease care requires to be taken to avoid implied renunciation of the old lease and the creation of a new one.

- A conventional irritancy cannot be purged.

- Where a landlord purports to irritate or otherwise terminate a lease, even if the tenant has a prospective defence, the tenant should intimate his waygoing claims and have arbiters appointed timeously, on a without prejudice basis.

- A resumption notice can only be given in terms of a reserved right to resume in a written lease. The amount of land sought to be resumed must not amount to a fraud on the lease.

- A resumption of land for agricultural purposes cannot be done by resumption notice but must be by a notice to quit: s 21(7)(a).

- Where an executor fails to transfer a lease within the time limits, then it requires to be terminated by the landlord giving notice of termination: 1964 Act, s 16(3) and (4).

Time limits

- Waygoing claims following an irritancy or other termination of a lease require to be intimated within two months of the date of the termination. The parties have four months from the irritancy or other termination (or any extensions – two two-month periods granted by the Secretary of State) in which to settle the claims in writing. If the claims cannot be settled within that period, arbitration has to be invoked within one month of the end of the four-month (or any extension thereof) period: s 62.

- A notice terminating a lease, where the executor has failed to transfer the lease within the time limits, requires to be a period of not less than one year and not more than two years ending at such term of Whitsunday or Martinmas as the notice stipulates: 1964 Act, s 16(4).

- A tenant wishing to contest a notice to quit part of a holding under s 29(1) requires to serve a counternotice within one month: s 21(1).

- A tenant wishing to treat a notice to quit part of a holding as a notice to quit the whole holding has to give written notice to the landlord within

28 days of (1) the giving of the notice or (2) the determination of Land Court proceedings on a counternotice: s 30.

- A resumption notice must give at least a two-month period of notice, in order to allow the tenant to make claims on partial dispossession which require two months' notice before the termination of the tenancy.

GENERAL

While the tenant of an agricultural holding is given security of tenure under the 1991 Act, the tenancy can still terminate or be terminated in a number of different ways.

'Termination, in relation to a tenancy, means the termination of the lease by reason of effluxion of time or from any other cause'[1]. The phrase is used extensively throughout the 1991 Act and in other related Acts[2]. It is important to note that a time limit related to the 'termination of the tenancy' refers to the date of the termination of the contract and not to the termination of possession of the subjects by the tenant, which in terms of the contract may be at later dates[3]. The different ways in which a lease may be terminated are discussed below.

RENUNCIATION

A tenant may renounce a lease impliedly by abandoning possession of the subjects[4]. It may be difficult for a landlord to determine when a tenant has in fact abandoned the subjects.

Where a tenant inverts the possession of the holding, by abandoning wholly or substantially agricultural activity on the holding and substituting a different activity, whether with or without the landlord's consent, the lease may cease to be the lease of an agricultural holding subject to the security of tenure provisions of the 1991 Act[5].

A lease may be renounced when the parties enter into a new lease of the subjects differing in material respects from the former lease[6]. It is a question of fact and circumstances whether the modification or variation is sufficiently material so as to constitute a new lease. In view of the differing statutory provisions that relate to leases starting at different dates[7], parties seeking to agree a modification or variation of a lease require to be careful not to constitute a new lease.

Section 16 of the 1991 Act provides that a lease is not terminated because

---

1 1991 Act, s 85(1).
2 Eg the 1986 Act.
3 Eg where the lease provides for multiple handover dates in respect of different parts of a farm. See *Earl of Hopetoun v Wight* (1864) 2 M (HL) 35 and p 157.
4 *Rankine* p 600; *Paton and Cameron* p 239.
5 *Wetherall v Smith* [1980] 1 WLR 1290.
6 Erskine *Institutes* II, 4, 44; Hunter *Law of Landlord and Tenant* (4th edn, 1876) vol 2, p 110; *Morrison v Rendall* 1986 SC 69; *Mackie v Gardner* 1973 SLT (Land Ct) 11; *Tufnell and Nether Whitehaugh Co Ltd, Applicants* 1977 SLT (Land Ct) 14; *Jenkin R Lewis & Son v Kerman* [1971] Ch 477; cf *McCowan's Trs v Wilson-Walker* 1956 SLCR 40.
7 Eg landlord's fixed equipment liability in post-1 November 1948 leases (1991 Act, s 5); near relative successor termination provision for pre and post-1 January 1983 leases (1991 Act, Sch 2); sheep stock valuations in pre or post-6 November 1946 or 1 December 1986 leases. NB with regard to improvements s 34(5) preserves the right of a tenant who has remained in occupation of the holding during two or more tenancies to claim compensation for improvements.

new terms have been added to the lease or that any terms 'have been varied or revised in pursuance of this Act'[1].

A lease may be effectively terminated if the tenant unilaterally renounces the lease and acts upon that renunciation by vacating the subjects[2]. An oral renunciation is not effective, unless it is accompanied by actings[3], although such a renunciation probably cannot be enforced if the tenant does not vacate the subjects[4].

Where a tenant renounces a lease it has been held that he has impliedly renounced all rights arising out of the lease, which would include waygoing rights, unless these are expressly reserved[5]. Renunciation does not affect statutory waygoing claims[6].

AGREEMENT

An agreement to terminate a tenancy other than in terms of s 21 is, in general, unenforceable[7].

Clearly, if parties agree to the termination of the tenancy and the tenant acts upon it by vacating the subjects, the tenancy has been terminated by agreement.

If an agreement is made and acted upon by one party 'the other party may be personally barred from founding on the provisions of the subsection'[8]. The question of whether or not personal bar can operate in the face of the 1991 Act was not considered, and it may be that personal bar cannot be founded upon[9].

It has been held that a landlord can waive his objection to a tenant's invalid notice of intention to quit, and having accepted it, enforce the notice to quit when the tenant later refused to remove[10]. Gill[11] suggested that in light of Morrison v Rendall it is doubtful whether a party on whom an invalid notice of intention to quit is served may waive his objection and enforce the notice.

There was an underlying assumption in Elsden v Pick[12] that an agreement to surrender the tenancy would have been perfectly valid notwithstanding the statutory provisions[13].

Parties may contractually agree to terminate a tenancy after the service of a

1 See p 28.
2 Morrison v Rendall 1986 SC 69.
3 Requirements of Writing (Scotland) Act 1995, s 1(3) and (4). Formerly there required to be rei interventus.
4 Morrison v Rendall above. While the question of waiver or personal bar was raised in that case no consideration was given to whether or not personal bar could operate in the face of the 1991 Act. Cf Kok Hoong v Leong Cheong Kweng Mines Ltd [1964] AC 993; Guthrie v MacLean 1990 SLCR 47 at 32.
5 Lyon v Anderson (1886) 13 R 1020; cf Strang v Stuart (1887) 14 R 637.
6 Strang v Stuart above.
7 Morrison's Exrs v Rendall 1986 SC 69. See chap 4.
8 Morrison's Exrs v Rendall above at 74 per LJC.
9 See n 4 above.
10 Elsden v Pick [1980] 1 WLR 898.
11 Para 257.
12 [1980] 1 WLR 898.
13 Kildrummy (Jersey) Ltd v Calder 1994 SLT 888. Elsden v Pick [1980] 1 WLR 898 was not cited in Morrison's Exrs v Rendall 1986 SC 69.

notice to quit and counternotice as part of the settlement of Land Court proceedings for consent to the operation of a notice to quit[1].

Parties cannot contract in advance either in the lease or by other agreement to serve a notice of intention to quit at a particular date or not to serve a counternotice[2].

BANKRUPTCY

Section 21(6) preserves 'the right of the landlord of an agricultural holding to remove a tenant whose estate has been sequestrated under the Bankruptcy (Scotland) Act 1985 or the Bankruptcy (Scotland) Act 1913. . .'. This provision applies only where bankruptcy is fenced by an irritancy clause. The section amplifies the common law under which sequestration alone was not sufficient to found a legal irritancy and could found a removal only if it was subject to a conventional irritancy[3].

The irritancy clause can make provision that property rights in, for example, the growing crops should vest in the landlord[4].

Where bankruptcy is not fenced by an irritancy clause the tenant may be removed by an incontestable notice to quit[5].

If the lease is not terminated upon bankruptcy either the landlord can leave the tenant in place or circumstances might show the lease to have been adopted by the trustee in bankruptcy[6].

DISSOLUTION OF THE TENANT

The lease terminates if the tenant is an artificial person and ceases to exist. Thus on the dissolution of a partnership or a change in the constitution of the partnership, the lease terminates because the tenant ceases to exist[7]. A lease terminates in such circumstances by the death, bankruptcy[8] or retiral of a partner because the *persona* of the partnership to whom the lease was granted has ceased to exist.

If the deed of partnership provides for a continuing partnership and the lease specifically provides that the lease is to the continuing partnership or to 'the house', then the lease may endure, notwithstanding changes in the constitution of the partnership. It is a matter of construction of the lease and not of the

1 *Kildrummy (Jersey) Ltd v Calder* above.
2 *Johnson v Moreton* [1980] AC 37; *Featherstone v Staples* [1986] 1 WLR 861; cf *Duguid v Muirhead* 1926 SC 1078.
3 *Gill* para 271.
4 *Chalmer's Trs v Dick's Trs* 1909 SC 761.
5 1991 Act, s 22(2)(f).
6 Cf *Rankine* p 698.
7 *IRC v Graham's Trs* 1971 SC (HL) 1; *Jardine-Paterson v Fraser* 1974 SLT 93. NB *William S Gordon & Co Ltd v Mrs Mary Thomson Partnership* 1985 SLT 122 in which the only issue raised was the terms of the partnership agreement and no consideration was given to the terms of the lease.
8 *Rankine* p 694.

deed of partnership whether the lease is to the partnership as then constituted or to the continuing partnership or 'house'. For the lease to continue to the partnership there must be at least two surviving partners[1].

Similarly if a limited company is wound up or struck off the Register of Companies the lease will terminate.

A limited partnership device is a means whereby the landlord can seek to circumvent the security of tenure provisions and to obtain possession upon the termination of the limited partnership. Whether or not the courts will look behind the 'sham' of a limited partnership to give security of tenure to the tenant has not yet been determined[2].

TERMINATION OF TENANCIES BY IRRITANCY

While ss 3 and 21(1) provide that a tenancy may only be terminated by notice to quit, the right to terminate a lease by way of legal or conventional irritancy remains.

(a) The legal irritancy

Section 20 preserves the landlord's right to have a tenancy terminated where six months' rent of the holding is due and unpaid. This is a legal irritancy[3]. The irritancy is purgeable at any time before decree is extracted[4] upon payment of the arrears or upon caution being found for the arrears and one further year's rent.

The section empowers the landlord to raise an action of removing in the sheriff court, concluding for removal from the holding at the term of Whitsunday or Martinmas next ensuing after the action is raised. It has been suggested that there may be practical difficulties where an action is defended and the next ensuing term day has passed before decree is granted[5]. Decree has to be pronounced in terms of the crave, which has to be for the term next ensuing after the action is raised.

It is suggested that this difficulty does not exist. Section 20(2) provides that the sheriff may decern the tenant to remove and may eject him at the term craved 'unless the arrears of rent then due are paid or caution is found to his satisfaction for them, and for one year's rent further'. The objective of this provision appears to be that immediately the action is raised the tenant is required either to pay the arrears or, if he disputes them, to find caution for the arrears and one further year's rent[6].

---

1 *Wallace v Wallace's Trs* (1906) 8 F 558.
2 See p 37 'Sham transactions' and p 37 'Limited partnerships'.
3 *M'Douall's Trs v MacLeod* 1949 SC 593.
4 *Fletcher v Fletcher* 1932 SLT (Sh Ct) 10; *Westwood v Keay* (1950) 66 Sh Ct Rep 147 (purged on appeal).
5 *Gill* para 272.
6 There are similarities with the right to seek caution for violent profits in an action of removing; see Macphail *Sheriff Court Practice* (W Green, 1988) para 23–12.

Failure to pay the arrears or find caution would allow the sheriff to pronounce immediate decree. The sheriff has a discretion, as he 'may' decern. He would require to be satisfied that prima facie there were arrears or that there was a dispute about arrears, where the landlord had a prima facie claim.

If there is a dispute about arrears during the further year, then the parties will require to have the question or difference between them as to quantum or liability for the arrears determined by arbitration or other judicial process[1]. Once the arrears are paid or caution found the right to remove under s 20 is lost.

Gill[2] notes that as the landlord's remedy is easily defeated by the tenant, the section is nowadays seldom used. Clearly a conventional irritancy is preferable. A s 20 application may have advantages over a written demand for payment of rent within two months as a prelude to a notice to quit under s 22(2). For example, if the landlord knows that such a demand may be disputed by arbitration under s 23(2), but seeks caution for rent while the dispute is resolved, if he has reason to suspect the tenant's solvency, then a s 20 application may be more appropriate.

The Act of Sederunt, December 14, 1756[3] provides for a legal irritancy restricted to agricultural leases, where one year's rent is unpaid or where the tenant has deserted the holding[4]. An application in respect of arrears of rent is no longer competent under the Act of Sederunt[5]. While the process is unheard of in modern practice[6], it might be appropriate in cases where the tenant appears to have deserted the holding, but the landlord wants a legal warrant to repossess it.

(b) The conventional irritancy[7]

The right to remove a tenant on the occurrence of a conventional irritancy is preserved by s 21(6). The notice provisions relating to irritancies of commercial leases introduced by the Law Reform (Miscellaneous Provisions) (Scotland) Act 1985 do not apply to leases of agricultural holdings[8].

The right to a conventional irritancy only arises if there is a provision in the lease permitting irritancy. While in earlier leases conventional irritancies tended to be confined to specific breaches of the more important of the tenant's obligations, modern leases tend to have an all-embracing irritancy clause relating to any breach by the tenant.

As a conventional irritancy, unlike a legal irritancy, cannot be purged, it is a most potent weapon in the hands of the landlord. It can be exercised at any time during the duration of the lease, whereas s 22(2)(d) and (e) notices can only be given at a break in the lease.

1 See p 276 'Any question or difference'.
2 Para 272.
3 CASL, xv.
4 *Rankine* pp 533–539.
5 1991 Act, s 20(4).
6 *Gill* para 273.
7 For a review of the general law on conventional irritancies, see *Rankine* pp 532–668; *M'Douall's Trs v MacLeod* 1949 SC 593; *Lucas Exr v Demarco* 1968 SLT 89; *Dorchester Studios (Glasgow) Ltd v Stone* 1975 SC (HL) 56; *CIN Properties Ltd v Dollar Land (Cumbernauld) Ltd* 1992 SC (HL) 104.
8 1985 Act, s 7(1)(b).

Where a conventional irritancy merely repeats a legal irritancy, that irritancy can be purged, unless it is being enforced oppressively[1].

Although an irritancy clause usually provides that on the occurrence of the breach the lease is to be held as null and void, it is accepted that irritancy is an option open to the landlord. If the landlord exercises such an option the tenant cannot thereafter seek to purge the irritancy. If the landlord does not exercise the option or, for example, accepts rent after an irritancy for a period thereafter, unless the lease specifically provides for such acceptance in the case of irritancy, he may be deemed to have departed from the right to irritate[2].

The benefit of an irritancy has to be specifically assigned to a successor landlord before that landlord can seek to enforce the irritancy[3].

Where an irritancy has been incurred no period of notice is required to be given. The landlord should serve on the tenant a formal notice stating that the lease has been irritated coupled with a demand that the tenant remove. If the tenant does not remove the landlord should raise an action of declarator and removing immediately as any delay might give rise to a plea of waiver of the right to irritate[4].

If the action is defended on the grounds that an irritancy has not been incurred or the tenant has one of the other defences available to him the action requires to be sisted for arbitration under s 60[5].

There are four principal defences available to a tenant upon receipt of a notice of irritancy. He can deny that the purported breach amounts to breach of the particular clause in the lease[6]. He can plead that the enforcement of the irritancy amounts to oppression, which has to amount to impropriety of conduct by the landlord resulting in an abuse or misuse of the irritancy[7]. As contractual obligations are mutual it is a defence that the landlord is in breach of his obligations under the lease[8]. A landlord may stipulate in the lease that he is not to be prevented from enforcing an irritancy by reason of any breach by him of his obligations under the lease. If the landlord has acquiesced in a past breach he may be barred from seeking to irritate the lease[9], unless he gives warning that he will not permit a continuance of the particular breach[10]. The landlord may have waived his right to enforce an irritancy[11].

1 See *British Rail Pension Trustee Co Ltd v Wilson* 1989 SLT 340 and *CIN Properties Ltd v Dollar Land (Cumbernauld) Ltd* 1992 SC (HL) 104 at 121 per Lord Jauncey of Tullichettle.
2 *Central Estates (Belgravia) Ltd v Woolgar (No 2)* [1972] 1 WLR 1048; *Cayzer v Hamilton (No 2)* 1995 SLCR 13.
3 *Life Association of Scotland v Blacks Leisure Group* 1989 SC 166.
4 *Rankine* pp 512, 546–548. An action of ejection is not competent; *Alongi v Alongi* 1987 GWD 1–27.
5 See eg *BTC v Forsyth* 1963 SLT (Sh Ct) 32.
6 *BTC v Forsyth* above (purported inversion of possession by using 2 out of 100 acres for a scrap business); *Stuart v Warnocks* (1883) 20 SLR 863 and *Blair Trust Co v Gilbert* 1940 SLT 322 (breaches of residence clause).
7 For a summary of the law on the defence of oppression see *CIN Properties Ltd v Dollar Land (Cumbernauld) Ltd* 1992 SC (HL) 104 at 108 per Lord Jauncey of Tullichettle.
8 *MacNab of MacNab v Willison* 1960 SLT (Notes) 25. Cf *Edmonstone v Lamont* 1975 SLT (Sh Ct) 57 where it was held that a reserved right such as resumption was independent of obligations under a lease and accordingly mutuality did not apply.
9 *Lamb v Mitchell's Trs* (1883) 10 R 640.
10 Gloag *The Law of Contract* (2nd edn, 1929) pp 281–282; *Gatty v Maclaine* 1921 SC (HL) 1; *Lucas Exrs v Demarco* 1968 SLT 89 and *Morrison-Low v Howison* 1961 SLT (Sh Ct) 53.
11 See eg *HMV Fields Properties Ltd v Tandem Shoes Ltd* 1983 SLT 114 (plea unsuccesful); see *Armia Ltd v Daejan Developments Ltd* 1979 SC (HL) 56; *Morrison v Rendall* 1986 SC 69; *Lousada Ltd v JE Lesser Properties Ltd* 1990 SLT 823 and *Cayzer v Hamilton (No 2)* 1995 SLCR 13 for a general consideration of 'waiver'.

A lease is terminated on the date that the irritancy takes place. The rights and obligations of the parties on the occurrence of an irritancy fall to be determined primarily from the lease. Where a lease is terminated by an irritancy the tenant loses his right to compensation for disturbance and in consequence his right to an additional payment[1]. A question of damages may arise. A landlord who enforces an irritancy has no claim for damages for the tenant's failure to complete the lease[2].

Where a landlord purports to irritate a lease and the tenant considers he has a defence the tenant should, in order to protect his position, intimate his waygoing claims within the statutory time limits and invoke the necessary arbitrations[3], without prejudice to his defence. The claims can then be sisted pending the outcome of defence to the irritancy. If the claims are not made timeously and the defence fails, the tenant will lose his waygoing claims.

TERMINATION BY OBJECTION TO ACQUIRER OF LEASE

A lease may be terminated if the Land Court upholds a landlord's objection to the acquirer of a lease[4].

TERMINATION BY FAILURE OF SUCCESSION

The lease may terminate after the death of the tenant if (1) a legatee accepts a bequest and fails to intimate it to the landlord timeously[5], (2) if the acquirer of the lease fails to intimate the acquisition timeously to the landlord[6], (3) if the executor fails to confirm to the lease within a year[7], (4) if the executor fails to transfer the lease within a year of the death or the Land Court declaring the bequest null and void[8] or such other period of time as is agreed by the landlord or authorised by the sheriff[9].

If the interest of a deceased under a lease is not disposed of lawfully within the specified time limits by the executor, the lease does not automatically terminate, but must be terminated by the landlord, giving the executor a period of notice of not less than and not more than two years, ending with such term of Whitsunday or Martinmas as the notice may stipulate[10].

---

1 1991 Act, ss 43 and 54.
2 *Walker's Trs v Manson's Trs* (1886) 13 R 1198; *Chalmer's Trs v Dick's Tr* 1909 SC 761.
3 Under the 1991 Act, s 62.
4 1991 Act, s 12(2); see chap 12 and p 134 'Options open to landlord'.
5 *Reid's Trs v Macpherson* 1975 SLT 101.
6 1991 Act, s 12(1); see p 132 'Intimation of bequest or transfer of lease'.
7 *Rotherwick's Trs v Hope* 1975 SLT 187.
8 Under the 1991 Act, s 11(6).
9 See chap 13.
10 1964 Act, s 16(3) and (4). This is not a formal notice to quit in terms of the 1991 Act but a notice under the 1964 Act. If the executors are allowed to remain on the holding this may create a new lease with the executors; *Morrison-Low v Paterson* 1985 SC (HL) 49.

NOTICES TO QUIT[1]

(a) Notices to quit requiring Land Court consent

A notice to quit, seeking to terminate the tenancy, can be served[2]. If the tenant serves a counternotice[3], the lease can only be terminated with the consent of the Land Court. The Land Court shall only grant its consent if it is satisfied by the landlord on one or more of the matters specified in s 24(1), namely:

'(a) that the carrying out of the purpose for which the landlord proposes to terminate the tenancy is desirable in the interests of good husbandry as respects the land to which the notice relates, treated as a separate unit;
(b) that the carrying out thereof is desirable in the interests of sound management of the estate of which that land consists or forms part;
(c) that the carrying out thereof is desirable for the purposes of agricultural research, education, experiment or demonstration, or for the purposes of the enactments relating to allotments, smallholdings or such holdings as are referred to in section 64 of the Agriculture (Scotland) Act 1948;
(d) that greater hardship would be caused by withholding than by giving consent to the operation of the notice;
(e) that the landlord proposes to terminate the tenancy for the purpose of the land being used for a use, other than for agriculture, not falling within section 22(2)(b) of this Act.'

If the tenant serves a counternotice on the landlord and there is a sub-tenant in occupation, the tenant requires to serve notice on the sub-tenant that he has served the counternotice[4]. The sub-tenant is entitled to be a party to the Land Court proceedings[4].

(b) Notices to quit not requiring Land Court consent

In certain circumstances the landlord may serve an incontestable notice to quit on the tenant[5] where:

'(a) the notice to quit relates to land being permanent pasture which the landlord has been in the habit of letting annually for seasonal grazing or of keeping in his own occupation and which has been let to the tenant for a definite and limited period for cultivation as arable land on the condition that he shall, along with the last or waygoing crop, sow permanent grass seeds;
(b) the notice to quit is given on the ground that the land is required for use, other than agriculture, for which permission has been granted on an application made under the enactments relating to town and country planning, or for which (otherwise than by virtue of any provision of those enactments) such permission is not required;
(c) the Land Court, on an application in that behalf made not more than 9 months before the giving of the notice to quit, were satisfied that the tenant was not fulfilling his responsibilities to farm the holding in accordance with the rules of good husbandry, and certified that they were so satisfied;

---

1 See chap 15 for a consideration of the form, style and content of notices to quit.
2 Under the 1991 Act, s 21.
3 Under ibid, s 22(1).
4 Ibid, s 23(8).
5 Ibid, s 22(2).

(d) at the date of the giving of the notice to quit the tenant had failed to comply with a demand in writing served on him by the landlord requiring him within 2 months from the service thereof to pay any rent due in respect of the holding, or within a reasonable time to remedy any breach by the tenant, which was capable of being remedied, of any term of condition of his tenancy which was not inconsistent with the fulfilment of his responsibilities to farm in accordance with the rules of good husbandry;

(e) at the date of the giving of the notice to quit the interest of the landlord in the holding had been materially prejudiced by a breach by the tenant, which was not capable of being remedied in reasonable time and at economic cost, of any term or condition of the tenancy which was not inconsistent with the fulfilment by the tenant of his responsibilities to farm in accordance with the rules of good husbandry;

(f) at the date of the giving of the notice to quit the tenant's apparent insolvency had been constituted in accordance with section 7 of the Bankruptcy (Scotland) Act 1985[1];

(g) section 25(1) of this Act applies, and the relevant notice complies with section 25(2)(a), (b) and (d) of this Act.'

In order to rely on s 22(2) the notice to quit requires to state in the notice the ground upon which it proceeds[2].

Section 22(2) (c) and (d) are mutually exclusive. It is safer to serve a notice to quit separately in respect of each subsection, but if a combined notice is to be used then the notice must carefully discriminate in its terms between the breaches which are relied upon in respect of each subsection[3].

(c) Notice to quit tenancies acquired by succession[4]

The landlord may serve an incontestable notice to quit upon a successor by bequest or transfer to a tenancy[5], where the successor is not a near relative successor[6].

Where the successor is a near relative the Land Court's consent is required to the notice to quit[7] provided it complies with one of the statutory cases[8].

(d) Notice of intention to quit

The tenant may serve a notice of intention to quit[9] giving not less than one and not more than two year's notice of his intention to quit the tenancy.

It is doubtful whether a landlord can waive his rights to object to an invalid notice of intention to quit and then seek to enforce the notice[10].

---

1 Ibid, s 22(2).
2 See p 160 'Statements in notice to quit'.
3 *Macnabb v Anderson* 1957 SC 213; *Budge v Hicks* [1951] 2 KB 335.
4 See chap 13.
5 1991 Act, ss 25(1) and 22(2)(g).
6 Defined by ibid, Sch 2, Pt III, para 1 - 'means a surviving spouse or child of that tenant, including a child adopted by him in pursuance of an adoption order (as defined in section 23(5) of the Succession (Scotland) Act 1964)'.
7 1991 Act, s 25(3).
8 Ibid, Sch 2; see p 121 'Notice to quit under s 25(1)'.
9 Ibid, s 21(2).
10 *Gill* para 257; *Elsden v Pick* [1980] 1 WLR 898; cf *Kildrummy (Jersey) Ltd v Calder* 1994 SLT 888.

(e) Notice to quit part of holding

In general a notice to quit part of a holding is incompetent[1]. Section 29(1) provides limited exceptions, where the lease is a tenancy from year to year or running on tacit relocation.

Where there is a resumption clause in a lease it should be used rather than a notice to quit under this section, because the tenant cannot serve a counternotice to a resumption notice. A resumption clause can be operated during the period of the lease and does not require to wait until the lease is running on tacit relocation.

The Land Court[2] has left open the question of whether or not a notice to quit part of a holding should be viewed like a resumption notice and only permitted if the area sought to be recovered by the notice does not amount to a fraud on the lease. Such an approach does not appear consistent with the right to serve a counternotice, accepting the partial notice as a notice to quit the entire holding[3] and the provisions for a disturbance payment, which restrict the right to that of part of the subject of the notice if it relates (a) less than a quarter or a quarter in rental value of the farm and (b) the diminished holding is reasonably capable of being farmed as a separate holding[4]. This later provision suggests that a partial notice could leave less than a viable holding and still be valid, because the tenant has the option of treating the whole lease as terminating or accepting the lesser part.

A notice to quit part of a holding is not invalid if it is given:

'(a) for the purpose of adjusting the boundaries between agricultural units or amalgamating agricultural units or parts thereof', or
(b) with a view to the use of the land to which the notice relates for any of the purposes mentioned in subsection (2)....[6]'

The notice must accurately identify the part of the holding to which it relates. It requires to state that it is given for the purpose of adjusting boundaries or amalgamating units or with a view to the particular use specified in s 29(2)[7].

The authorised purposes are[8]:

'(a) the erection of farm labourers' cottages or other houses[9] with or without gardens;
(b) the provision of gardens for farm labourers' cottages or other houses[9];
(c) the provision of allotments;
(d) the provision of small holdings under the Small Landholders (Scotland) Acts 1886 to 1931, or of such holdings as are referred to in section 64 of the Agriculture (Scotland) Act 1948;
(e) the planting of trees;
(f) the opening or working of coal, ironstone, limestone, brickearth, or other minerals,

1 *Gates v Blair* 1923 SC 430.
2 *Hamilton v Lorimer* 1959 SLCR 7.
3 1991 Act, s 30.
4 Ibid, s 43(7).
5 Cf *Hamilton v Lorimer* 1959 SLCR 7.
6 1991 Act, s 29(1)(a) and (b).
7 Ibid, s 29(1).
8 Ibid, s 29(2).
9 'Other houses' is qualified by 'farm labourers'; *Paddock Investments v Lory* (1975) 236 EG 803. Cf *Connell* p 148 which appears to be wrong to suggest that 'other houses' includes housing developments.

or of a stone quarry, clay, sand, or gravel pit, or the construction of works or buildings to be used in connection therewith;

(g) the making of a watercourse or reservoir;

(h) the making of a road, railway, tramroad, siding, canal or basin, wharf, or pier, or work connected therewith.'

A tenant wishing to contest a notice to quit part of the holding should serve a counternotice, within one month[1].

Where a valid notice to quit part of the holding is given[2], the tenant may, within 28 days of either the giving of the notice or the determination of Land Court proceedings on a counternotice, give the landlord a counternotice in writing that he accepts the notice to quit as a notice to quit the entire holding[3].

RESUMPTIONS

The right to resume land[4] from a lease is a matter of contract. If there is no contractual provision permitting resumption, the landlord has no right to resume.

A notice of resumption requires to give at least two months' notice (unless the lease stipulates for a longer period), so that the tenant can give the appropriate notices before the resumption date to allow him to exercise his statutory rights on the partial termination of the tenancy[5]. Any term in the lease providing for notice of resumption on a shorter notice will be invalid[6].

The reserved purposes for which resumption is permitted under the lease will be closely scrutinised to see if the purpose specified in the resumption notice falls within the reserved purpose and in particular within words of general purpose following emuneration of specific purposes[7].

To reserve a right to resume for agricultural purposes can 'only be done by clear and express terms, for such a provision would go far to negative the security of tenure which the lease is designed to give'[8].

A resumption for agricultural purposes can only be done by notice to quit and not by a resumption notice[9].

It may be a question of fact whether the resumption is for agricultural or non-agricultural purposes[10]. Subject to the *de minimus* rule, where the proposed use includes an agricultural use, it does not matter that the area might

---

1 1991 Act, s 22(1).

2 Under ibid, s 29.

3 Ibid, s 30.

4 'Land' includes buildings; *Glencruitten Trs v Love* 1966 SLT (Land Ct) 5.

5 Eg 1991 Act, s 18 (right to remove fixtures and buildings) and s 44 (high farming).

6 *Re Disraeli's Agreement* [1939] Ch 382; *Coates v Diment* [1951] 1 All ER 890.

7 Cf *Admiralty v Burns* 1910 SC 531; *Crichton-Stuart v Ogilvie* 1914 SC 888; *Coates v Diment* [1951] 1 All ER 890.

8 *Pigott v Robson* 1958 SLT 49 (1st Div); cf *Crichton-Stuart v Ogilvie* above; *Turner v Wilson* 1954 SC 296; *Admiralty v Burns* 1910 SC 531.

9 1991 Act, s 21(7)(a).

10 Cf *Sykes and Edgar, Applicants* 1974 SLT (Land Ct) 4 where the tenant suggested that the landlord's proposed resumption was for 'shelter belts', which might be an agricultural purpose (see s 85(1) – 'agriculture . . . land used for woodlands where the use is ancillary to the farming of the land' and 'fixed equipment . . . (h) shelter belts'), whereas the landlord contended the resumption was for forestry.

be used for a non-agricultural purpose, whether or not of more substance or importance[1].

The landlord may not resume land for opencast coal mining[2].

The right to resume is a reserved right by the landlord at the formation of the contract. It is independent of the landlord's obligations under the lease. Accordingly the landlord may resume land even if he is in breach of other obligations under the lease[3].

A resumption notice requires to specify the purpose for which the landlord intends to resume the land. The tenant can challenge the existence of the purpose, but the onus would be on the tenant to prove that the purpose does not exist[4].

A resumption will only be permitted provided the purpose specified in the notice falls within the terms of the reservation and the extent of the resumption is not contrary to the good faith of the lease or amount to a fraud on the lease. The test of good faith was set in *Admiralty v Burns*[5]:

'...it must be a question of circumstances whether the land sought to be resumed forms so material a part of the subjects let that it cannot be reasonably regarded as within the contemplation of and would be against the good faith of the bargain embodied in the lease.'

The Land Court has suggested[6] that the resumption should not have the effect of changing the character of the farm as contemplated in the lease. In that case the landlord proposed to resume the hill land, thus turning a hill sheep farm with some arable low ground into a primarily low ground farm, albeit with the farm remaining viable.

The resumption of a building alone, while competent, may amount to a fraud, on the lease, if the building

'forms so material a part of the subjects let that its resumption could not reasonably be regarded as within the contemplation of and would be against the good faith of the bargain embodied in the lease.'[7]

The Land Court has recognised a paradox which remains unresolved[8]. If the test is leaving a viable unit after resumption, what happens in the case of the efficient farmer who, having lost part of the farm leaving him with a just viable unit, then improves his efficiency, so that a few years later he could lose further acres and still be left with a viable unit. Probably this paradox can be resolved if one looks to the situation at the commencement of the lease and what was in

---

1 *Crawford v Dunn* 1981 SLT (Sh Ct) 66 where it was held that a resumption for use of a field for riding and teaching riding and for grazing horses was an agricultural purpose. This decision does not appear to have considered the inter-relationship between 'agricultural purpose' in the 1991 Act, s 21(7)(a) and 'agricultural land' in s 1(2) which defined agricultural land as land used for a trade or business. If the proposed resumption is for amenity land to a house on which there will be some grazing of an entirely non-business character, it is arguable that under the scheme of the Act such a purpose is non-agricultural in that it is not for the purpose of a trade or business.
2 Opencast Coal Act 1958, s 14A(10).
3 *Edmonstone v Lamont* 1975 SLT (Sh Ct) 57.
4 *Gill* para 235.
5 1910 SC 531 at 542 per Lord Salveson.
6 *Fothringham v Fotheringham* 1987 SLT (Land Ct) 10 at 17K.
7 *Glencruitten Trs v Love* 1966 SLT (Land Ct) 5 at 7.
8 *Fothringham v Fotheringham* above.

the contemplation of the parties at that date. This would properly ignore tenant's improvements to the farm and in management techniques.

In *Edinburgh Corporation v Gray*[1] the court recognised that a resumption clause might provide for the resumption of the whole farm in successive parcels. Lord President (Cooper) said[2]:

'The tenant has chosen to remain year after year on tacit relocation in the occupation of an arable farm in the suburbs of a large city, and he must long have known perfectly well that the risk of resumption for housing purposes was real and increasingly imminent; for successive resumptions have been in progress since 1935, the present landlords are the local housing authority, and the entire farm has eventually been absorbed in a housing scheme.'

The *Edinburgh Corporation* case predated the introduction of statutory security of tenure. The Land Court has commented[3] that a resumption clause now requires to be construed to the effect that it cannot be used to override the statutory security of tenure provisions[4].

The amount of land sought to be resumed is either a fraud on the lease or it is not. An arbiter has no power to substitute a smaller area for the area specified in the notice[5].

A tenant may have a claim for damages if the landlord does not use the land for the purpose expressed in the resumption notice[6].

RENT REDUCTION AND WAYGOING CLAIMS ON PARTIAL DISPOSSESSION

Where a tenant is partially dispossessed from the holding by either a notice to quit under s 29 or a resumption, he is entitled to a rent reduction.

The rent reduction is calculated 'proportionate to that part of the holding'[7] taken back, 'together with an amount in respect of any depreciation in value to him of the residue of the holding caused by the severance or by the use to be made of the part severed'[8].

In fixing a rent reduction following upon a resumption (but not in respect of a partial notice to quit) the arbiter is required to take into account any benefit or relief allowed to the tenant under the lease in respect of the part whose possession is resumed[9].

---

1 1948 SC 538.
2 *Edinburgh Corp v Gray* 1948 SC 538 at 545.
3 *Fothringham v Fotheringham* 1987 SLT (Land Ct) 10 at 14L.
4 Per contra D G Rennie in 'Resumptions', a lecture to PQLE Advanced Agricultural Law Course, 5/7 November 1992. Cf *Strathern v MacColl* 1993 SLT 301, a crofting case where the court held that a landlord could not exercise his reserved right to extract minerals in such a way as to render the croft incapable of crofting tenure. The ratio supports the principle that a resumption clause should not be used in such a manner as to evade the rights of security of tenure conferred by the 1991 Act.
5 *Fothringham v Fotheringham* 1987 SLT (Land Ct) 10.
6 *Gill* para 235.
7 *Hoile and Sheriffs* 1948 SLCR 24 (the Land Court took the productive capacity of the part of the holding resumed in relation to the whole 'as giving the criterion for measuring the reduction of rent to which the tenant is entitled').
8 1991 Act, s 31(1).
9 Ibid, s 31(2).

If the area from which the tenant is dispossessed is of no rental value, there may be no rental reduction[1].

A resumption clause can provide for the rent reduction or other payment to be paid upon a resumption, in which case it is binding[2]. However parties cannot contract out of the tenant's rights to statutory claims on the partial termination of a tenancy[3].

1 *Sinclair v Secretary of State for Scotland* 1966 SLT (Land Ct) 2.
2 *Edinburgh Corp v Gray* 1948 SC 538.
3 1991 Act, s 53; *Coates v Diment* [1951] 1 All ER 890; see p 236 'Claims on partial dispossession'.

CHAPTER 15

Notices to quit and of intention to quit

Key points

- For the purposes of serving a notice to quit on a person acquiring a lease on succession, the date of acquisition is (1) in respect of a bequest the date of death of the previous tenant or (2) in respect of a transfer from the executors, the date of transfer.

- A check should be made to confirm whether the lease has a single ish or multiple ishes or a single ish with a number of handover dates.

- Where a holding has a number of landlords the notice must be served by all the landlords.

- The notice to quit requires to be in the form prescribed by OCR 1993, Form H2, specifying the subjects, the lease and the ish date.

- The notice to quit should be served by one of the methods prescribed in OCR 1993, r 34.8. It is not competent to serve a notice to quit in terms of the 1991 Act, s 84.

- When serving a notice to quit a check should be made as to whether an additional statement is required or should be made in the notice: see p 160.

- There is no statutory form for a notice of intention to quit.

- A notice of intention to quit given by any one of a number of joint tenants terminates the lease, even without the consent of the other tenants.

- A landlord should avoid concluding a contract to sell the holding where there is an effective notice to quit in place. A notice to quit ceases to be effective if the landlord concludes missives for the sale of the subjects after service of the notice and prior to the ish date under the notice, except in special circumstances: s 28(3). A disposition of part of the subjects during the currency of a notice to quit precludes a landlord insisting in an application to the Land Court for consent to the operation of the notice.

Time limits

- A notice to quit or of intention to quit requires to be served more than one year, but less than two years, before ish or the anniversary of the ish and in the case of multiple ishes before the earliest ish: s 21(3)(c) and (d).

- A notice to quit given to terminate a tenancy acquired by succession, where the unexpired period of the lease is two years or less or the lease is

153

running on tacit relocation, has to specify as its effective date an ish that falls during the period not less than one year nor more than three years after the acquisition: s 25(2)(b)(ii).

- Where a notice to quit is served founding on a Land Court certificate of bad husbandry the notice has to be served not more than nine months after the application was made to the Land Court. The notice cannot be served until the certificate has been granted: s 22(2)(c).

- Where a landlord has concluded a contract for the sale of the holding, unless the tenant and the landlord have agreed within three months before missives were concluded that the notice will continue to be effective, he is required to serve on a tenant a notice within 14 days of conclusion of missives that he has made the contract: s 28(2)(a).

- A tenant has one month in which to respond to a notice under s 28(2)(a) if he elects that the notice shall continue to have effect: s 28(2)(b).

GENERAL

This chapter deals with the essential requirements for notices to quit, notices of intention to quit, counternotices and demands to remedy.

A lease cannot be brought to an end by notice to quit or notice of intention to quit procedure except at the end of the stipulated endurance of the lease or at stipulated breaks[1] in the lease[2].

When the lease is on tacit relocation it can only be brought to an end at a stipulated anniversary of the lease, provided proper notice is given.

It is not competent to contract out of the notice to quit provisions of the 1991 Act[3], ss 21(1) or 22(1).

FORMAL VALIDITY OF AND TIME LIMITS FOR A NOTICE TO QUIT OR OF INTENTION TO QUIT

To be formally valid a notice requires to comply with the provisions of s 21(3).

It requires to be in writing[4]. It requires to be a notice of intention to bring the tenancy to an end[5]. A notice to quit expressed in conditional terms is invalid[6].

The notice requires to comply with the specified time limits[7]. A notice must be served not less than one year nor more than two years before the termination of the stipulated endurance of the lease or the anniversary of that date if the lease is running on tacit relocation[8].

A notice cannot be served before the commencement of the tenancy[9].

In the case of a notice to quit served on a tenant who acquired right to the lease by succession or transfer[10] there are time windows within which the ish stipulated in the notice to quit must fall, if the landlord is to rely on the provisions of s 25. The notice still requires to comply with s 21[11] as to formal validity and time limits.

Where the unexpired period of the lease exceeds two years from the date the tenant acquired[12] the lease, the notice to quit requires to stipulate the term of outgo in the lease. In other words the landlord requires to bring the lease to an end by a valid notice to quit at the outgo stipulated in the lease.

---

1 Some leases stipulate for breaks for the purposes of rent review only. The lease cannot be terminated at these breaks. There is now doubt as to whether it is competent to seek a rent review at such breaks; *Moll v MacGregor* 1990 SLT (Land Ct) 59.
2 *Macnabb v A and J Anderson* 1955 SC 38 at 44 per Lord Patrick; *Alston's Trs v Muir* 1919 2 SLT 8; *Strachan v Hunter* 1916 SC 901; *Edell v Dulieu* [1924] AC 38; *Beck v Davidson* 1980 Lothian RN 30.
3 *Morrison v Rendall* 1986 SC 69; see chap 4.
4 1991 Act, s 21(3)(a).
5 Ibid, s 21(3)(b).
6 *Murray v Grieve* (1920) 36 Sh Ct Rep 126.
7 1991 Act, s 21(3)(c) and (d).
8 For the relevant date where there are multiple ishes, see p 157 'The ish date'.
9 *Lower v Sorrell* [1963] 1 QB 959.
10 See 1991 Act, s 25.
11 Ibid, s 25(2)(a).
12 'Acquired' means in relation to (1) a bequest, the date of death of the tenant unless the will provided for another date and (2) a transfer under the 1964 Act, s 16, the date of transfer, which may be up to a year or possibly more after death.

Where the unexpired period of a lease is less than two years from the date of acquisition or the lease is running on tacit relocation the notice to quit must specify as its effective date an ish that falls during the period of not less than one year nor more than three years from the date of the acquisition[1].

Where a notice to quit is served founding on a Land Court certificate of bad husbandry[2] the notice has to be served not more than nine months after the application was made to the Land Court[3]. The notice cannot be served until the certificate has been granted, which means that there is a tight time schedule in which to lodge the application with the Land Court, obtain a decision and then serve the notice to quit, before the nine months expire[4].

FORM OF A BASIC NOTICE TO QUIT

A notice to quit[5] requires to conform to the provisions of the 1907 Act, subject to s 21[6]. The relevant provisions of the 1907 Act are ss 34 to 38 and OCR 1993, rr 34.5[7] to 34.9.

The notice requires to be in the form prescribed by the 1907 Act[8], which is set out in OCR 1993, Form H2 as follows:

FORM H2 Rule 34.6(1)

Form of notice of removal

To (insert name, designation, and address of party in possession)

You are required to remove from (describe subjects) at the term of (or if different terms, state them and the subjects to which they apply[9]), in terms of lease (describe it) [or in terms of your letter of removal] dated (insert date) [or otherwise as the case may be].

Date (insert date)   Signed
                       (add designation and address)

It is essential that the requirements of the form are strictly complied with. There are three cardinal elements in the form, namely the subjects, the lease and the ish. Failure to include any of the cardinal elements renders the notice invalid even if the tenant suffers no prejudice[10].

The legal position was that if the notice did not follow the statutory form,

---

1 1991 Act, s 25(2)(a).
2 Ibid, s 22(2)(c).
3 *Cooke v Talbot* (1977) 243 EG 831 (the time runs from the date when the application was lodged).
4 Cf *Macnabb v A and J Anderson* 1955 SC 38 at 44 per Lord Patrick.
5 But not a notice of intention to quit; see 1991 Act, s 21(4) which relates to 'removings'.
6 Ibid, s 21(4).
7 OCR, r 34.5 is expressed as being '(1) Subject to section 21 of the Agricultural Holdings (Scotland) Act 1991 (notice to quit and notice of intention to quit)'.
8 1991 Act, s 21(5)(b). The reference to the Removal Terms (Scotland) Act 1886 in s 21(5)(a) relates only to the manner of service; *Rae v Davidson* 1954 SC 361.
9 This insertion relating to multiple ish dates amends the form from that contained in earlier OCRs and deals with problems highlighted in *Milne v Earl of Seafield* 1981 SLT (Sh Ct) 37; see p 157 'The ish date'.
10 *Rae v Davidson* 1954 SC 361; *Scott v Livingstone* 1919 SC 1; *Taylor v Brick* 1982 SLT 25.

but included the cardinal elements, it might be held valid if the tenant suffered no prejudice[1]. It is not clear that this is still the position. The earlier sheriff court rules[2] provided that 'Notices ... shall be as nearly as may be in terms of Form L...', whereas OCR 1993, r 34.6 provides '(1) A notice ... shall be in Form H2'. The early form of words allowed some latitude in the style of the notice, which appears to have been removed by the new rule.

Dealing with each of the requirements in turn:

(1) *The party in possession.* The notice requires to be served on the tenant[3], who will require to be accurately described.

If there are a number of joint tenants or tenants in common, the notice has to be served on all the tenants.

A notice may be served on a tenant's agent[4], but it is safer to serve the notice on the actual tenant, otherwise the landlord may be put to proof on the question of whether or not the agent had authority to accept a notice to quit.

(2) *The subjects.* A failure to refer to the subjects of the lease is fatal[5]. It is best to describe the subjects in the terms set out in the lease itself. If there is no written lease then the common name of the subjects or a description of the fields and their acreage will suffice[6]. What is required is some reference sufficient to identify the subjects in a reasonable sense. The question is 'Does the notice contain a reasonable description of the subjects in question'[7]. A description of only part of the subjects is also fatal[8]. It is not clear whether a description of more than the subjects is fatal.

(3) *The ish date.* The notice requires to specify '*the term of (or if different terms, state them and the subjects to which they apply)*' at which the tenant is required to remove.

The 'term' as applicable to the 1991 Act is either 'the termination of the stipulated endurance of the lease' or an anniversary of that date, if the lease is on tacit relocation[9].

It is important to distinguish between leases which include multiple ish dates and leases with a single ish and a number of handover dates. In multiple ish leases the contract runs, for example, from Whitsunday to Whitsunday as to the house and grass and from Martinmas to Martinmas as to the arable land, etc[10]. In such a case the notice must specify the separate ish dates and give not less than one year's notice in respect of the earliest date[11].

---

1 *Campbell's Trs v O'Neill* 1911 SC 188 at 198, 200; *Callander v Watherston* 1970 SLT (Land Ct) 13; *Graham v Lamont* 1970 SLT (Land Ct) 10; *Earl of Seafield v Cameron* (1979, unreported) Elgin Sh Ct; *Milne v Earl of Seafield* 1981 SLT (Sh Ct) 37.
2 Latterly OCR 1983, r 104.
3 See the extended definition of 'tenant' in the 1991 Act, s 85(1).
4 Ibid, s 85(5).
5 *Scott v Livingstone* 1919 SC 1.
6 *Taylor v Brick* 1982 SLT 25.
7 *Scott v Livingstone* above at 5 per Lord Guthrie.
8 *Scott v Livingstone* above.
9 1991 Act, s 21(3)(c) and (d); cf *Watters v Hunter* 1927 SC 310.
10 See eg *Strang v Stuart* (1887) 14 R 637 (double ish); *Black v Clay* (1894) 21 R (HL) 72 (treble ish); *Hendry v Walker* 1924 SC 757; *Montgomerie v Wilson* 1924 SLT (Sh Ct) 48 (double ish).
11 *Earl of Hopetoun v Wight* (1863) 1 M 1074, (1864) 2 M (HL) 35; 1907 Act, s 34; *Rankine* pp 566–567.

More commonly a lease has one ish, qualified by various handover dates. Such a lease usually runs from Whitsunday to Whitsunday with handover arrangements running from March through to the separation of crops with, for example, different dates for the gardens, fallow land, houses, grass and arable land[1]. A notice should be served specifying and against the single ish date[2].

A formal lease usually specifies Whitsunday or Martinmas as the date of entry and of ish. Prior to the passing of the Term and Quarter Days (Scotland) Act 1990 these terms, if unqualified as to date, caused problems[3]. At common law these terms were 15 May and 11 November respectively. For certain limited purposes of the 1949 Act they were defined as 28 May or 28 November.

These terms are now defined[4], 'unless the context otherwise requires' and are now 28 May and 28 November in (1) any lease entered into after 13 July 1990[5] and (2) in any lease entered into before 13 July 1990 if the term is used 'without further specification as to the date or month'[6]. This provision does not apply to pre 13 July 1990 leases if the sheriff, on a summary application made within 12 months of the coming into force of the Act, was satisfied that the date intended in the lease was a specific date other than the date so prescribed and made a declaration accordingly[7].

A notice served against an ish of Martinmas (28 November), where the actual ish date was 11 November, has been held valid because the tenant was not prejudiced[8].

Where a lease runs from a particular date it is safer to serve the notice against the same date in a following year, although a notice against the date before the anniversary date has been held valid[9].

Where the landlord does not know the ish date problems arise. The safest course would be to establish the ish date either by agreement with the tenant or by demanding a written lease[10] before the notice to quit is served. If this is not possible, a notice to quit specifying a Whitsunday or Martinmas or other specific date falling after the probable date of entry may be valid, as the tenant is not prejudiced by the longer time in the holding[11].

1 See eg *Earl of Hopetoun v Wight* (1863) 1 M 1074; *Gatherer v Cumming's Exr* (1870) 8 M 379; *Waldie v Mungall* (1896) 23 R 792; *Millar v M'Robie* 1949 SC 1; *Coutts v Barclay-Harvey* 1956 SLT (Sh Ct) 54; *Milne v Earl of Seafield* 1981 SLT (Sh Ct) 37.
2 *Gill* para 304. *Fairfax-Lucy v Macdonald* 1935 SLCR 13 (Div Ct) at 32 to the opposite effect was wrongly decided; *Gill* para 304, n 2. It has been held that a notice which specified the handover dates rather than the single ish date was valid; *Earl of Seafield v Cameron* (1979, unreported) Elgin Sh Ct.
3 See *Gill* para 302.
4 Term and Quarter Days (Scotland) Act 1990.
5 Ibid, s 1(1) and (2).
6 Ibid, s 1(4).
7 Ibid, s 1(5).
8 *Callander v Watherston* 1970 SLT (Land Ct) 13.
9 *Morrison's Exrs v Rendall* 1989 SLT (Land Ct) 89; *McGill v Bichan* 1970 SLCR App 122; cf D C Coull 'Termination Date in a Notice to Quit' 1989 SLT (News) 431.
10 1991 Act, s 4.
11 Cf *Callander v Watherston* 1970 SLT (Land Ct) 13; *Addis v Burrows* [1948] 1 KB 444; cf *Scammell and Densham* p 143 'running notice'. The notice should probably provide 'You are required to remove from (subjects) at Whitsunday (1995), being the term next after your date of entry in about April (1956), the exact date being unknown to your landlord, in terms of lease (describe)'.

(4) *The lease.* The notice must specify 'the warrant upon which the notice is based'[1] at the very least by 'some reference sufficient (in a reasonable sense) to identify'[2] the lease. Failure to refer to the lease or to sufficiently identify it is fatal[3].

Where there is a written lease it should be described by its conveyancing description giving the parties, date and a specification of any minutes varying the lease. If an earlier superseded lease is specified in error the notice is invalid[4]. A misdescription of the lease in the notice (ie erroneous date) will not invalidate the notice if the tenant is not prejudiced[5].

Where there is only an oral lease the landlord's position may be more difficult, particularly where he is not the original landlord and does not know when the lease was entered into or what was its ish date. In such circumstances it may be necessary to have a written lease entered into[6], as a prelude to being able to serve a valid notice to quit.

Where there was an oral lease, reference to an arbiter's award determining the terms of a lease was held a sufficient description of the lease[7].

(5) *The signatory.* A notice to quit may be signed and served by the landlord or by his agent[8]. A notice by an agent even if the agency is not disclosed on the face of the notice may be valid[9].

Where there are a number of landlords of the holding they must all join in the service of a single notice to quit[10]. A notice by one of the landlords in relation to the whole holding is invalid[11]. A notice by one landlord in relation to part of the holding is incompetent[12]. If a landlord dispones part of the holding after serving the notice to quit, it is incompetent for him to proceed with an application for the Land Court's consent to the operation of the notice[13].

1 *Watters v Hunter* 1927 SC 310.
2 *Watters v Hunter* above; *Morrison's Exrs v Rendall* 1989 SLT (Land Ct) 89.
3 *Rae v Davidson* 1954 SC 361.
4 *Mackie v Gardiner* 1973 SLT (Land Ct) 11; in *Morrison's Exrs v Rendall* 1989 SLT (Land Ct) 89 the court upheld a notice which referred to missives of let, but did not date them. In that case there were a series of missives of let for periods of less than one year after a number of years. It was in dispute as to whether the lease should be ascribed to the first missives, which took effect from year to year in terms of s 3, or whether each successive missive was deemed to be the acceptance of a new let where the last missive took effect from year to year.
5 *Earl of Seafield v Cameron* (1979, unreported) Elgin Sh Ct.
6 Under 1991 Act, s 4.
7 *Gemmell v Andrew* 1975 SLT (Land Ct) 5.
8 1991 Act, s 85(5).
9 *Graham v Stirling* 1922 SC 90; *Walker v Hendry* 1925 SC 855; *Combey v Gumbrill* [1990] 2 EGLR 7 (wife deemed to have husband's authority to serve notice by virtue of court orders in matrimonial proceedings).
10 *Walker v Hendry* above; *Stewart v Moir* 1965 SLT (Land Ct) 11.
11 *Stewart v Moir* above; *Secretary of State v Prentice* 1963 SLT (Sh Ct) 48; *Gordon v Rankin* 1972 SLT (Land Ct) 7.
12 *Secretary of State v Prentice* above.
13 *Gordon v Rankin* above.

STATEMENTS IN NOTICES TO QUIT

(a) Circumstances in which a statement is required

A basic notice to quit permits the tenant to serve a counternotice requiring the Land Court to consent to the notice[1].

Where the landlord seeks to serve an incontestable notice to quit excluding the tenant's right to serve a counternotice[2] 'the ground under the appropriate paragraph on which the notice to quit proceeds' requires to be 'stated in the notice'[2].

Where a landlord serves a notice to quit on a near relative successor to a tenancy the notice requires to specify 'the Case set out in Schedule 2 to this Act under which it is given'[3].

Where a landlord serves a notice to quit part of a holding[4] the notice requires to specify the purpose for which it is served or the use to which the land is to be put[5].

The landlord may avoid payment of compensation for disturbance, where the notice to quit relies upon and specifies that it proceeds upon s 22(2)(a), (c) or (f)[6].

The landlord may also avoid payment of an additional payment[7] where the notice to quit relies upon and specifies that:

(1) it proceeds upon s 22(2)(a), (c) or (f); or

(2) the notice contains a statement that the purpose for which the landlord proposes to terminate the tenancy is desirable on any grounds referred to in s 24(1)(a) to (c)[8] or that the landlord will suffer hardship unless the notice has effect[9] and the Land Court, in granting its consent to the notice to quit, states in the reasons for its decision that it is satisfied as to the statutory tests[10].

If the reasons include that the Land Court is satisfied that the land will be used for a purpose other than agriculture, the additional payment is not avoided, even if the application does not mention s 24(1)(e)[11] and the Land Court states in its decision that it would have been satisfied that the use was to be other than agriculture[12]; or

(3) the notice contains a statement required by s 25(2)(c) (near relative successor) and the Land Court consents to the notice[13]. Where the Land Court's consent proceeds on cases 1, 3, 5 or 7 an additional payment is not avoided if the court states in its decision that it would have been satisfied if the case had

---

1 1991 Act, s 22(1).
2 Under ibid, s 22(2).
3 Ibid, s 25(2)(c).
4 Ibid, s 29.
5 Ibid, s 29(1).
6 Ibid, s 43(2).
7 Ibid, s 54.
8 (a) husbandry, (b) sound estate management and (c) agricultural research.
9 1991 Act, s 24(1)(d).
10 Ibid, s 55(1)(a) and (b).
11 The land being used for a purpose other than agriculture.
12 1991 Act, s 55(2)(a).
13 Ibid, s 55(1)(c).

been brought under s 24(1)(e) that the land will be used for a purpose other than agriculture[1].

(b) Form of the statement

A statement in a notice to quit must be made unequivocally leaving the tenant in no doubt as to what is being relied upon[2]. No set form is prescribed, but the substance of the statement must be clear[3]. It is wise to make particular reference to the statutory provision although a bare reference to the section or subsection may be sufficient[4].

Where the notice contains the statutory statement and an additional and erroneous statement it does not necessarily invalidate the notice to quit[5].

A statement given as a matter of courtesy does not necessarily make the statement into a statutory statement[6]. If the statement leaves it in doubt whether the notice is served under s 22(1) or (2) the notice will be invalid[7], because the tenant is left in doubt as to his proper response.

(c) Statements in accompanying documents

Notwithstanding the statutory provisions which appear to require the statement to be in the notice itself, it has been held that a statement in an accompanying letter complies with the statutory provisions[8]. A statement in a subsequent letter or notice cannot be said to be a statement in the notice to quit[9].

NOTICE TO QUIT PART OF HOLDING

In addition to the requirements listed above, a valid notice to quit part of the holding under s 29 requires:

(1) to specify accurately the part to which it relates and

(2) to contain a statement in the notice specifying either that the notice is served for the purpose of adjusting boundaries betweeen agricultural units or

---

1 1991 Act, s 55(2)(b).
2 *Macnabb v Anderson* 1957 SC 213; *Cowan v Wrayford* [1953] 1 WLR 1340; *Mills v Edwards* [1971] 1 QB 379.
3 *Graham v Lamont* 1971 SC 170; *Budge v Hicks* [1951] 2 KB 335; *Macnab v Anderson* above.
4 Cf *Re Digby and Penny* [1932] 2 KB 491 (it was held sufficient to refer to the paragraphs without setting out the reasons).
5 *French v Elliott* [1960] 1 WLR 40 at 50 per Paull J; *Milne v Earl of Seafield* 1981 SLT (Sh Ct) 37.
6 *Hammon v Fairbrother* [1956] 1 WLR 490.
7 *Mills v Edwards* [1971] 1 QB 379.
8 *Turton v Turnbull* [1934] 2 KB 197; *Graham v Lamont* 1970 SLT (Land Ct) 10; *Copeland v McQuaker* 1973 SLT 186. It is said that the Agricultural Holdings Act 1986 now requires the statement to be made 'in the notice'; *Scammell and Densham* p 167.
9 *Connell* p 140; *Budge v Hicks* [1951] 2 KB 335.

of amalgamating agricultural units or parts thereof or with a view to one of the specified uses[1].

NOTICES OF INTENTION TO QUIT

There is no statutory form for a notice of intention to quit[2].

The notice of intention to quit requires to be in writing[3]. It requires to be a clear notice of intention to bring the tenancy to an end[4]. It requires to be served more than one year and less than two years before the termination of the stipulated endurance of the lease or an anniversary thereof[5]. The notice must be 'such as to convey to the party to whom it is addressed, in the circumstances in which it is so addressed, a clear intimation that tacit relocation is not consented to by the [tenant]'[6].

The notice may be given by an agent[7]. Where there are a number of joint tenants, then the service of a notice of intention to quit by any one of the tenants, even without the consent of the others, is sufficient to bring the tenancy to an end[8].

A tenant who vacates the holding without serving a notice of intention to quit leaves himself open to an action of damages for breach of contract.

SERVICE OF NOTICES TO QUIT

A notice to quit requires to be served in terms of the 1991 Act, s 21(5) either '(a) in the same manner as a notice of removal under section 6 of the Removal Terms (Scotland) Act 1886; or (b) in the form and manner prescribed by the Sheriff Courts (Scotland) Act 1907'. The provisions of the 1886 Act have effectively been superseded by service under the 1907 Act.

Notice, under the 1907 Act, may be given by (a) a sheriff officer, (b) the person entitled to give such notice or (c) the solicitor or factor of such a person. Service has to be by

'posting the notice by registered post or first class recorded delivery service at any post office within the United Kingdom in time for it to be delivered at the address on the notice before the last date on which by law such notice must be given, addressed to the person entitled to receive such notice, and bearing the address of that person at the time, if known, or, if not known, to the last known address of that person.'[9]

---

1 1991 Act, s 29(1). The specified uses are listed at s 29(2)(a)–(h).
2 Ibid, s 21(4) and (5) only applies to notices to quit.
3 Ibid, s 21(3)(a).
4 Ibid, s 21(3)(b).
5 Ibid, s 21(3)(c) and (d).
6 *Graham v Stirling* 1922 SC 90 at 105-106.
7 1991 Act, s 85(5).
8 *Smith v Grayston Estates Ltd* 1960 SC 349, resolving the doubts expressed in *Graham v Stirling*. This would not apply if one of the joint tenants was the landlord or his nominee under an arrangement which was designed to avoid the normal security of tenure given by the 1991 Act; cf *Featherstone v Staples* [1984] 1 WLR 861; see chap 4.
9 OCR 1993, r 34.8(1).

A sheriff officer may also serve the notice 'in any manner in which he may serve an initial writ'[1].

Service by any other means is fatal to the validity of the notice to quit even if the tenant suffers no prejudice, because the rule is mandatory[2].

A tenant cannot avoid the service by refusing to sign the delivery slip[3].

It should be noted that the provisions of the 1991 Act, s 21(5) override the provisions regarding the service of general notices provided for in s 84(1). It is not competent to serve a notice to quit by giving it or delivering it to the tenant or by just leaving it at his address.

Notices of intention to quit are not covered by s 21(5), so a notice of intention to quit can be delivered or served in the manner provided for in s 84(1).

CONSTRUCTION OF NOTICES TO QUIT AND OF INTENTION TO QUIT

The rule as to the inadmissibility of extrinsic evidence to explain and construe a document does not apply to notices to quit[4].

A notice is to be construed along with any accompanying letter[5] in the light of the relevant surrounding circumstances and by reference to its effect on a reasonable tenant reading it who is taken to have the knowledge of the surrounding facts and circumstances which he and the landlord enjoy[6]. It has been said that:

'The notice must be such as to convey to the party to whom it is addressed, in the circumstances in which it is so addressed a clear intimation. . . .'[7]

SUB-TENANTS[8]

Where a notice to quit is served on a tenant, the tenant may, but need not, serve a notice to quit on the sub-tenant[9]. If the tenant serves a notice to quit on the sub-tenant, the sub-tenant cannot serve a counternotice and has no right to invoke the consent of the Land Court[10]. The sub-tenancy terminates with the tenancy.

Where a notice to quit does not have effect, then any notice to quit given by the tenant to the sub-tenant does not have effect either[10].

1 OCR 1993, r 34.8(2).
2 *Dept of Agriculture v Goodfellow* 1931 SC 556; *Watt v Findlay* (1921) 37 Sh Ct Rep 34.
3 *Van Grutten v Trevenen* [1902] 2 KB 82. Cf Citation Amendment (Scotland) Act 1882, s 4.
4 *Cayzer v Hamilton (No 1)* 1995 SLCR 1.
5 *Graham v Lamont* 1970 SLT (Land Ct) 10.
6 *Lands v Sykes* (1992) 03 EG 115; *Cayzer v Hamilton* above; *Graham v Stirling* 1922 SC 90.
7 *Graham v Stirling* above at 105 per Lord President. Cf *Bury v Thompson* [1895] 1 QB 696 per Lopes LJ - 'a notice to quit is a good notice if it be so expressed that a person of ordinary capacity receiving the notice cannot very well mistake its nature; it must be clear and unambiguous . . .'.
8 See p 22.
9 Cf 1991 Act, s 23(6).
10 Ibid, s 23(6).

Where a tenant serves a counternotice on the landlord in response to a notice to quit, he requires to serve a notice in writing on the sub-tenant that he has served such a counternotice and the sub-tenant is entitled to be a party to the Land Court proceedings[1].

A pre-existing tenant of an agricultural holding who has become a sub-tenant by virtue of a later interposed lease cannot be removed as a sub-tenant with no right to serve a counternotice[2].

Where the sub-tenancy is a sham arrangement entered into with the view to defeating the tenant's statutory protection, the sub-tenant may have the right to serve a counternotice and to enjoy the statutory protection of the 1991 Act[3].

FRAUDULENT STATEMENTS IN A NOTICE TO QUIT

A notice to quit which contains a statement which is false and made fraudulently by the landlord by reason of knowing the statement to be untrue or being reckless, whether it is true or false, is invalid and unenforceable, even if the misrepresentation did not deceive[4].

EFFECT ON NOTICE TO QUIT OF SALE OF ALL OR PART OF HOLDING

A notice to quit a holding ceases to be of any effect if the landlord concludes a contract for the sale[5] of his interest in all or part of the holding after[6] service of the notice to quit and before it takes effect[7] unless:

(1) the landlord and tenant agree in writing within the period of three months ending with the date on which the contract is made, that the notice shall continue to have effect[8];

(2) the landlord serves a notice on the tenant within 14 days after making the contract and the tenant, not later than one month after the service of the landlord's notice, gives the landlord written notice that he elects that the notice to quit shall continue to have effect[9]; or

(3) the landlord fails to give notice of the making of the contract and the tenant quits in consequence of the notice[10].

1 1991 Act, s 23(8).
2 Land Tenure Reform (Scotland) Act 1974, s 17; cf *Kildrummy (Jersey) Ltd v Calder* 1996 GWD 8–458.
3 *Gisbourne v Burton* [1989] QB 390.
4 *Earl of Stradbroke v Mitchell* [1991] 1 WLR 469, sub nom *Rous v Mitchell* [1991] 1 All ER 676; *Omnivale v Boldan* [1994] EGCS 63; *Jones v Gates* [1954] 1 WLR 222 '. . . a fictitious notice'.
5 1991 Act, s 28 applies to the conclusion of the contract for sale (ie missives) and not to the implementation of the contract by the delivery of the disposition.
6 NB 'after' – s 28 does not apply to missives concluded before the notice to quit is served. A notice should be served by the selling landlord because the purchaser under the missives has no right to serve such a notice; *Waddell v Howat* 1925 SC 484.
7 1991 Act, s 28
8 Ibid, s 28(2).
9 Ibid, s 28(3)(b).
10 Ibid, s 28(3)(c).

This provision is a leftover from the 1923 Act, which is of no practical effect, beyond being a trap for the unwary landlord who concludes missives after serving an effective notice to quit.

Further, if the landlord dispones[1] part of the holding after the service of a notice to quit, he may no longer insist upon an application to the Land Court under the 1991 Act[2], s 23(1) because he is no longer the landlord of the whole holding.

1 Ie delivers a disposition rather than merely concluding missives.
2 *Gordon v Rankin* 1972 SLT (Land Ct) 7.

CHAPTER 16

Response to notices to quit

Key points

- The counternotice procedure applies to (1) notices to quit to which s 22(2) does not apply, (2) notices to quit served on a near relative successor under s 25(1) (s 25(3)) and (3) a notice to quit part of a holding under s 29.

- The tenant has to serve a counternotice in writing within one month of the service of the notice to quit requiring s 22(1) to apply to the notice so that it does not have effect unless the Land Court consents to the operation of the notice: s 22(1).

- The landlord requires to respond to a counternotice by applying to the Land Court within one month for consent to the operation of the notice to quit. The application requires to specify the grounds upon which the consent is sought: ss 23(1) and 24(1).

- Where an application has been made to the Land Court, the operation of the notice to quit is suspended until the issue of the Land Court's decision: s 23(4).

- Where the Land Court's decision is issued at a date later than six months from the date at which the notice to quit is to take effect, the tenant may apply, within one month of the issue of the decision, to the Land Court to have the operation of the notice to quit postponed for a period not exceeding 12 months: s 23(5).

- A tenant who intends to invoke the Land Court's jurisdiction to refuse a notice to quit on the grounds that 'a fair and reasonable landlord would not insist upon possession' must give notice of this in the pleadings: s 24(2).

- If a tenant succeeds in an arbitration on any question under a s 22(2) notice to quit, that notice becomes a s 22(1) notice to quit, to which the tenant must serve a counternotice.

- A tenant can either challenge a notice to quit part of the holding by serving a counternotice, or by serving a counternotice accepting the notice to quit part of the holding as a notice to quit the entire holding. The option to serve a counternotice accepting the notice to quit as a notice to quit the entire holding can be exercised either upon receipt of the notice to quit or upon determination of the Land Court proceedings following a counternotice: ss 29 and 30.

- Where a notice to quit is served under threat of a compulsory purchase order, the tenant should consider whether or not to take notice of entry compensation in lieu of compensation on the termination of the tenancy: see p 237 'Compulsory acquisition'.

Time limits

- The tenant has to serve the counternotice on the landlord within one month of the date of (1) the receipt of the notice to quit (s 22(1)) or (2) within one month of the issue of an arbiter's award, where s 23(3) applies.

- The landlord has to apply to the Land Court for consent within one month after service on the landlord of the counternotice.

- Where the Land Court's decision is issued at a date later than six months from the date at which the notice to quit is to take effect, the tenant has to apply to the Land Court, within one month of the Land Court's decision on a notice to quit, for a postponement of the operation of a notice to quit: s 23(5).

- Where a notice to quit part of the holding is served and the tenant accepts the notice as a notice to quit the entire holding the tenant has 28 days from either (1) the giving of the notice or (2) the date of determination of Land Court proceedings on a counternotice, in which to serve a notice in writing accepting the notice to quit as a notice to quit the entire holding: s 30.

- Where a notice to quit is served under threat of a compulsory purchase order, and the tenant has decided to elect to take notice of entry compensation in lieu of compensation on the termination of the tenancy, the time limits for making the election require to be observed: see p 237 'Compulsory acquisition'.

GENERAL

Section 22(1) procedure applies to all notices to quit to which s 22(2) does not apply; to notices to quit served on a near relative successor[1]; and in respect of notices to quit part of the holding[2].

It should be noted that if a tenant is successful in an arbitration on any question in a s 22(2) notice, then upon the issue of the arbiter's award the notice becomes a s 22(1) notice. The tenant requires to serve a counternotice within one month of the issue of the arbiter's award[3].

ACTION BY TENANT ON RECEIPT OF NOTICE TO QUIT THE WHOLE HOLDING

(1) The tenant must first decide whether or not he accepts the notice to quit. If the notice is accepted, then the tenant need make no response. He will then require to quit the holding at the ish date given in the notice to quit. He should make the appropriate claims for waygoings and compensation timeously[4].

(2) Next the tenant will have to determine whether or not the notice to quit is one to which ss 22(1) or 25(2)(c) apply and to which a counternotice should be served, or if the notice is served under s 22(2) to which the correct response is a reference to arbitration[5].

It should be clear that s 22(2) applies if there is a statement making reference to s 22(2)(a) to (f).

A difficulty arises where the notice is served under s 25(1) (termination of tenancy acquired by succession), because such a notice does not require a statement, except in the case of a near relative successor, where it requires to specify the case in Sch 2 under which it is given.

The correct response to a notice to quit served on a near relative successor under s 25(1) is a counternotice[6].

Where the tenant is not a near relative successor, s 22(2)(g) applies to a notice to quit served under s 25(1). The tenant will be presumed to know that this subsection applies from the time limits under which the notice is served. The notice may also contain a statement made to avoid liability for a reorganisation payment.

If the notice to quit purportedly served under s 25(1) does not comply with s 25(2), the tenant should serve a counternotice to protect his position.

(3) A tenant is not entitled to serve a counternotice to a notice to quit served by the Secretary of State under the Agriculture Act 1967, s 29(4). The Secretary of State requires to certify in writing that the notice is given in order that he may use or dispose of the land for the purpose of effecting any amalgamation or reshaping of the agricultural unit, and the tenancy agreement contains an acknowledgement by the tenant that the tenancy is subject to the provisions of the 1967 Act, s 29.

1 1991 Act, s 25(3).
2 See p 181.
3 1991 Act, s 23(3).
4 See chaps 18 and 19.
5 See chap 17.
6 1991 Act, s 25(3).

(4) Where a tenant makes the wrong response to the notice to quit he loses his rights under either s 22(1) or s 23(2) to require arbitration[1]. The only remaining defence will be to challenge the validity of the notice to quit or its service[2]. Similarly, if the counternotice is defective or out of time the tenant loses his rights[3].

THE COUNTERNOTICE

No particular form of counternotice is laid down. It requires to be in writing[4], and to be served on the landlord within one month of the giving of the notice to quit[5] preferably by registered or recorded delivery.

Failure to serve timeously is fatal to the counternotice[6].

The counternotice has to require 'that this subsection shall apply to the notice to quit'. It is therefore important that the counternotice actually provides *in germino* that the tenant requires that s 21(1) should apply to the notice to quit[7].

In *Mountford v Hodkinson*[8] a tenant replied to a notice to quit saying: 'I don't intend to go. I shall appeal against it and take the matter up with [the Agricultural Executive Committee]'[9].

The letter was held not to be a counternotice, because it did not require that the equivalent subsection should apply to the notice to quit. In *Taylor v Brick*[10] a letter in which the tenant intimated a refusal to remove was held insufficient as a counternotice, because it made no reference to (now) s 22(1).

An appropriate form of counternotice addressed to the landlord might be:

'With reference to your notice to quit the holding of . . . dated . . . , I hereby require section 22(1) of the Agricultural Holdings (Scotland) Act 1991 to apply to the said notice.

This notice is served without prejudice to my whole rights and pleas to object to the competence, validity or enforceability of the said notice in any subsequent proceedings.'

RESPONSE TO A DEFECTIVE NOTICE

If the notice to quit is defective in the manner of its form or the way in which it was served, the tenant may ignore it. Subsequently he may defend any action of removing on the ground that the notice was null and void.

1 *Morris v Muirhead* 1969 SLT 70; *Magdalen College Oxford v Heritage* [1974] 1 WLR 441; *Harding v Marshall* (1983) 267 EG 161.
2 See *Scammell and Densham* pp 165-166; *Earl of Stradbroke v Mitchell* [1991] 1 WLR 469; *Jones v Gates* [1954] 1 WLR 222.
3 *Mountford v Hodkinson* [1956] 1 WLR 422; *Luss Estates Co v Colquhoun* 1982 SLCR 1 (failure to serve counternotice timeously).
4 1991 Act, s 22(1).
5 Ibid, s 22(1). See p 6 for meaning of 'within one month'.
6 *Luss Estates Co v Colquhoun* 1982 SLCR 1.
7 In *Secretary of State v Fraser* 1954 SLCR 24 the Land Court held that a notice which referred to a repealed, but predecessor, subsection was invalid.
8 [1956] 1 WLR 422.
9 The committee in England was the equivalent of the Land Court, so there was a clear reference to the correct tribunal for determining whether or not consent should be given to a notice to quit.
10 1982 SLT 25.

It is safer to serve a counternotice[1], but one expressly reserving the right to object to the validity of the notice in any subsequent proceedings[2].

Where there is doubt as to whether the notice to quit is a s 22(1) or a s 22(2) notice, it is probably safer to serve both a counternotice and a demand for arbitration, under express reservation of right to object to the validity of the notice and without prejudice to either the counternotice or reference to arbitration. The landlord will then have to elect on what basis he claims the notice was served.

The fact that there is doubt as to the basis upon which the notice is served may be sufficient to render the notice invalid.

THE LANDLORD'S RESPONSE TO A COUNTERNOTICE

On receipt of a counternotice the landlord requires to apply to the Land Court within one month of the service of the counternotice for its consent to the notice to quit, otherwise the notice does not take effect[3]. The application requires to be a formal application to the court. A letter asking for an application form to be sent is not an application for the purposes of s 22(1)[4].

The application has to specify the ground or grounds under which the consent is sought. The application is incompetent if it does not specify the grounds under s 24(1) upon which the landlord seeks the Land Court's consent, because:

'it is a prerequisite of further procedure before the Court that the landlord's application should specify the matter or matters upon which they seek to satisfy the Court that consent to the operation of the notice to quit is proper.[5]'

Prior to serving a s 22(1) notice to quit the landlord should have made his preparations for presenting a case to the Land Court for consent on one or more of the grounds set out in s 24(1). Once the Land Court procedure is underway the landlord may find himself in difficulties qua time for adjustment etc, if his experts have not prepared the groundwork before service of the notice.

THE LAND COURT APPLICATION

Competency and relevancy

The Land Court may exculpate its own jurisdiction and dismiss an incompetent application[6].

---

1 If it is a s 22(2) notice to quit it requires arbitration under s 23(2).
2 Appropriate words might be 'This counternotice is served without prejudice to the tenant's whole rights and pleas to object to the competence, validity or enforceability of the notice in any subsequent proceedings'.
3 1991 Act, s 23(1). The Land Court has no power to extend the time limit; *Gemmell v Hodge* 1974 SLT (Land Ct) 2; *Still v Reid* 1957 SLCR 16.
4 *Still v Reid* above.
5 See 1991 Act, s 24(1); *Bennington-Wood's Trs v Mackay* 1969 SLT (Land Ct) 9; *O'Donnell v Heath* 1995 SLT (Land Ct) 15.
6 *Garvie's Trs v Still* 1972 SLT 29; *Luss Estates Co v Colquhoun* 1982 SLCR 1.

It is competent to present an application notwithstanding the fact that the parties are at issue as to whether or not there is a tenancy[1]. The application should probably be sisted pending the outcome of the proceedings to determine whether or not there is a tenancy[2], although with the consent of the parties the court may determine the substantive issues, while proceedings in the ordinary courts are pending[3].

An application is incompetent if it relates to land which the applicant does not own in whole or in part[4].

An application cannot relate to more than one notice to quit[5].

The Land Court may, in exceptional circumstances, dismiss an action as irrelevant[6].

In *Leask v Grains*[7] the Land Court held that the plea of *res judicata* did not apply to successive applications based on different grounds, but commented that it was desirable that alternative grounds should be tabled in one case. The court left open the question of whether or not *res judicata* might apply to successive applications on the same ground and observed that in that case a plea of competent but omitted had not been taken. In England it has been held[8] that successive applications on the same ground are competent and the landlord does not require to show a change of circumstances.

General

On receipt of a counternotice, a landlord may apply to the Land Court for consent to the notice to quit on (1) any of the grounds specified in s 24(1) or (2) in respect of a near relative successor, on the basis of one of the cases set out in Sch 2.

In reaching a decision the court takes into account not only the circumstances prevailing at the date of the notice to quit, but also the circumstances at the date of the decision[9].

The onus of proof is on the landlord to establish the grounds in applications under ss 24(1)[10] and 25(3)(a) and (c)[11]. Where the application is under s 25(3)(b)[12] the onus is on the tenant to satisfy the court 'that the circumstances are not as specified in that case (provided that for the purposes of Case

1 *Allan-Fraser's Trs v Macpherson* 1981 SLT (Land Ct) 17. The parties must resolve the question of whether or not there is a tenancy in the ordinary courts; *Garvie's Trs v Still* 1972 SLT (Land Ct) 29; *Allan-Fraser's Trs v Macpherson* above.
2 *Allan-Fraser's Trs v Macpherson* 1981 SLT (Land Ct) 17; *Eagle Star Insurance Co Ltd v Simpson* 1984 SLT (Land Ct) 37. Cf *O' Donnell v Heath* 1995 SLT (Land Ct) 15.
3 *Prior v J & A Henderson Ltd* 1984 SLT (Land Ct) 51; *O'Donnell v Heath* above.
4 *Stewart v Moir* 1965 SLT (Land Ct) 11; *Gordon v Rankin* 1972 SLT (Land Ct) 7; *Secretary of State v Prentice* 1963 SLT (Sh Ct) 48.
5 *Beck v Davidson* 1980 Lothian RN 30.
6 *Smoor v Macpherson* 1981 SLT (Land Ct) 25.
7 1981 SLT (Land Ct) 11.
8 *Wickington v Bonney* (1984) 47 P & CR 655.
9 *Eagle Star Insurance Co Ltd v Simpson* 1984 SLT (Land Ct) 37 at 41.
10 *McLaren v Lawrie* 1964 SLT (Land Ct) 10; *Edinburgh University v Craik* 1954 SC 190. *M'Callum v Arthur* 1955 SC 188 was superseded, so far as it commented upon onus, by amendment of the 1949 Act, s 26(1) by the 1958 Act.
11 1991 Act, Sch 2, cases 1–3 and 6.
12 Ibid, Sch 2, cases 4, 5 and 7.

7, the tenant shall not be required to prove that he is not the owner of any land)'[1]. Where the onus is on the landlord, the Land Court requires to be 'satisfied' that the ground has been made out. An application cannot be granted *de plano* in the absence of the tenant without proof being led[2].

The Land Court's consent has to be to the whole notice to quit and it is not competent for the court to grant consent in respect of part of the holding[3].

In *Edinburgh University v Craik*[4] the Lord President suggested that where the Land Court was likely to find against a landlord applicant on the merits of an application, it was competent for the Land Court to give the landlord an opportunity to offer another holding to the tenant as a means of tilting the balance in the landlord's favour for obtaining consent in relation to the notice to quit of the particular holding. This would suggest that it would be competent for the Land Court to take into account as one of the factors that a tenant has been offered an alternative holding.

Consents under s 24(1)

Ground a

'(a) that the carrying out of the purpose for which the landlord proposes to terminate the tenancy is desirable in the interests of good husbandry as respects the land to which the notice relates, treated as a separate unit'

This ground requires that the holding be 'treated as a separate unit'[5] when the application is considered by the Land Court. This is in contrast to ground b where consideration is given to the sound management of the estate of which the land forms part. This is an important distinction.

This ground contemplates a comparison between the tenant's regime and the landlord's proposed new regime[6]. It is not enough to criticise the tenant's farming methods; the landlord has to go on and prove that significantly better husbandry will result from his proposed regime[7]. The landlord does not have to prove bad husbandry, although he may found on criticisms of the tenant's standard of husbandry to contrast with his proposals[8].

The landlord's purpose, which by implication must be agricultural[9], can include a proposal to relet[10], to take the land in hand or to amalgamate the holding[11]. Where amalgamation is proposed, consideration must still be to the holding as a separate unit, so it might be wise to make the application under ground b as well, because resulting economies are disregarded[12].

---

1 1991 Act, s 25(3)(b).
2 *McLellan v McGregor* 1952 SLCR 3; *Buchanan v Buchanan* 1983 SLT (Land Ct) 31.
3 *Stewart v Moir* 1965 SLT (Land Ct) 11; *McBay v Birse* 1965 SLT (Land Ct) 10.
4 1954 SC 190.
5 *Austin v Gibson* 1979 SLT (Land Ct) 12; *Clarke v Smith* 1981 SLCR App 84.
6 *Davies v Price* [1958] 1 WLR 434; *Clarke v Smith* above.
7 *Clarke v Smith* above at 92; *Prior v J & A Henderson* 1984 SLT (Land Ct) 51.
8 *Davidson v Barrowman* 1978 SLCR App 155; *Austin v Gibson* 1979 SLT (Land Ct) 12.
9 *Clarke v Smith* 1981 SLCR App 84.
10 *Prior v J & A Henderson* 1984 SLT (Land Ct) 51.
11 *Gill* para 379.
12 *Austin v Gibson* 1979 SLT (Land Ct) 12.

Hardship to the tenant is probably not a relevant consideration[1], although it may arise under the fair and reasonable landlord defence[2].

Ground b

'(b) that the carrying out thereof is desirable in the interests of sound management of the estate of which that land consists or forms part'

This ground includes consideration of the sound management of the estate of which the land forms part. The holding may constitute the whole estate[3].

It has been held that the estate must consist of land owned by the landlord, although the court did not consider whether or not 'estate' could include land tenanted or occupied by the landlord along with land owned by himself[4]. The 'estate' could not comprise land in the ownership of others, where the proposed amalgamation was to be with land not owned by the landlord, even if that land was to be farmed as one unit in partnership by those owners.

There would appear to be a distinction between sound estate management and good estate management[5]. *Scammell and Densham*[6] suggests that good estate management is directed towards farming the land, whereas sound estate management embraces a wider conception and the possible use of the land for other purposes[7].

The obligation of the landlord is to manage the estate properly[8]. Therefore sound estate management relates to the actual management of the land, rather than to the personal financial interests of the landlord[9]. The court makes a comparison between the existing system of farming by the tenant and a consideration of the management situation in relation to the rest of the landlord's estate[10]. It would appear that the court may take into account the landlord's past management record as well as proposals for the future[11].

Sound estate management might include amalgamation to make two uneconomic units into one economic unit[12] or to amalgamate to upgrade a farming enterprise on other estate land[13]; improving the estate income[14] or the return from the holding; or selling the holding to finance other development on the estate[15] or to reduce the estate debts[16].

1 *Evans v Roper* [1960] 1 WLR 814, which refers to ground b, although the principles should also apply to ground a.
2 1991 Act, s 24(2).
3 *NCB v Naylor* [1972] 1 WLR 908; *Greaves v Mitchell* (1972) EG 440.
4 *Smoor v Macpherson* 1981 SLT (Land Ct) 25; *Beck v Davidson* 1980 Lothian RN 30.
5 Cf 1948 Act, Sch 5 'Rules of Good Estate Management'.
6 Page 155.
7 Cf *Gill* para 380, which queries if there is such a distinction.
8 *NCB v Naylor* [1972] 1 WLR 908.
9 *Gill* para 380.
10 *Peace v Peace* 1984 SLT (Land Ct) 6; *Leask v Grains* 1981 SLT (Land Ct) 11.
11 *Gibson v McKechnie* 1961 SLCR 11 at 14 and 15.
12 *Scammell and Densham* p 154; *Leask v Grains* above.
13 *Gemmell v Andrew* 1975 SLT (Land Ct) 5.
14 *McLaren v Lawrie* 1964 SLT (Land Ct) 10, but the question of increased income may be more relevant to ground d (hardship).
15 *Lewis v Moss* (1962) 181 EG 685.
16 *Forsyth-Grant v Wood* 1963 SLCR App 170.

This ground does not permit termination for the purposes of renegotiating a lease on more favourable terms for the landlord[1] or for switching the tenancy from one farmer to another[2]. It is essential the proposal involves a change to the management of the landlord's estate.

Hardship to the tenant is not a relevant consideration[3], although it might be invoked under the fair and reasonable landlord proviso[4].

Ground c

'(c) that the carrying out thereof is desirable for the purposes of agricultural research, education, experiment or demonstration, or for the purposes of the enactments relating to allotments, smallholdings or such holdings as are referred to in section 64 of the Agriculture (Scotland) Act 1948'

Gill comments[5] that this paragraph has not been invoked for many years.

The Land Court has to balance the desirability of the proposed use from a public interest in agricultural production against the interests of the tenant[6].

Ground d

'(d) that greater hardship would be caused by withholding than by giving consent to the operation of the notice'

The phrase 'greater hardship' is unqualified and has been given a wide construction to include not only the respective financial circumstances of the parties, which is usually the principal issue, but also the family[7] or personal[8] circumstances of the parties. The court may take into account the circumstances of third parties[9].

Where the landlord has acquired the holding by gift or succession it is difficult to establish hardship[10]. Similarly a landlord who has purchased a holding with a sitting tenant will find it difficult to establish hardship[11].

The fact that the tenant has security of tenure and the landlord cannot

---

1 *NCB v Naylor* [1972] 1 WLR 908.
2 *Prior v J & A Henderson Ltd* 1984 SLT (Land Ct) 51.
3 *Evans v Roper* [1960] 1 WLR 814; *Leask v Grains* 1981 SLT (Land Ct) 11.
4 1991 Act, s 24(2).
5 Para 383.
6 *Edinburgh University v Craik* 1954 SC 190.
7 Eg a son wanting to return to farm – *Graham v Lamont* 1970 SLT (Land Ct) 10; *Edmonstone v Smith* 1986 SLCR 97.
8 Eg age and health of the landlord – *McRobie v Halley* 1984 SLCR 10; member of the family homeless – *Lindsay-Macdougall v Peterson* 1987 SLCR 59; per contra consent would make the tenant homeless – *Reid v MacLeod* 1956 SLCR 47.
9 Eg intentions of children of the landlord – *Graham v Lamont* 1971 SC 170; an employee who might be made redundant – *Benington-Wood's Trs v Mackay* 1967 SLCR App 135; impact on partnership – *Edmonston v Smith* 1986 SLCR 97; but cf *Prior v J & A Henderson Ltd* 1984 SLT (Land Ct) 51 at 54 where the third party interest was considered too remote.
10 Eg *Davidson v Barrowman* 1978 SLCR App 155; but cf *Lindsay-Macdougall v Peterson* 1987 SLCR 59.
11 *McGill v Bichan* 1970 SLCR App 20; *Mackenzie v Tait* 1951 SLCR 3; *Copeland v McQuaker* 1973 SLT 186 at 190 and 191.

recover vacant possession is not hardship[1]. The mere inconvenience of having an agricultural tenant is not hardship[2]. The depreciation of the capital value of the holding by reason of the security of tenure does not in general constitute hardship[3].

The parties require to make specific averments about the hardship upon which they rely[4].

The onus is on the landlord to satisfy the Land Court that greater hardship would be caused by withholding than by giving consent[5]. The landlord requires to prove material, although not necessarily severe, hardship before the tenant's hardship needs to be considered[6].

Incidental proof of the tenant's hardship rests on the tenant[7]. After proof it is then for the court to weigh and balance the resultant hardships on either side in reaching a decision, but if the hardship is equal the landlord fails[8].

In considering financial hardship it is essential for the landlord to calculate and take into account the financial impact on him of such a termination. This might include the capital that he would have to find to farm the holding and any payments that may be due by him to the tenant upon the termination of the tenancy. If such payments in fact increase the hardship the Land Court may refuse the application[9].

Similarly the value of the tenant's rights to compensation and the cessation of the obligation to pay rent are factors to be weighed in the balance[10].

Ground e

'(e) that the landlord proposes to terminate the tenancy for the purpose of the land being used for a use, other than for agriculture, not falling within section 22(2)(b) of this Act'

This section applies where planning permission is not required by reason of a provision in the planning Acts; eg forestry, which does not require planning permission by reason of the Town and Country Planning (Scotland) Act 1972, s 19(2)(e)[11].

This is in contrast to a s 22(2)(b) notice to quit, which applies where

---

1 *Hutchison v Buchanan* 1980 SLT (Land Ct) 17; *Somerville v Watson* 1980 SLT (Land Ct) 14; *Ritson v McIntyre* 1982 SLCR 213; *McRobie v Halley* 1984 SLCR 10; *Edmonston v Smith* 1986 SLCR 97.
2 *Geddes v Mackay* 1971 SLCR App 94.
3 *Clamp v Sharp* 1986 SLT (Land Ct) 2.
4 *Beck v Davidson* 1980 Lothian RN 30.
5 Eg *Somerville v Watson* 1980 SLT (Land Ct) 14; *Edmonstone v Smith* 1986 SLCR 97.
6 *Clamp v Sharp* 1986 SLT (Land Ct) 2; *Edmonstone v Smith* 1986 SLCR 97; *McRobie v Halley* 1984 SLCR 10.
7 *McLaren v Lawrie* 1964 SLT (Land Ct) 10.
8 *McLaren v Lawrie* above.
9 *Graham v Lamont* 1970 SLT (Land Ct) 10 at 13; *Macdonald's Exrs v Taylor* 1984 SLT (Land Ct) 49.
10 *Dickson v Allison* 1976 SLCR App 108.
11 See *Ministry of Agriculture, Fisheries and Food v Jenkins* [1963] 2 QB 317 where it was held that as the Crown was not bound by the Town and Country Planning Act 1947 its plans for afforestation were not plans excluded from the Act by a provision of the Act. Therefore, in relation to the 1991 Act, s 22(2)(b) applied rather than s 24(1)(e).

planning permission has already been granted or is not required, other than by reason of a provision in the Act.

Development authorised by a development order falls under this ground and not under s 22(2)(b)[1].

This section does not apply where the proposed non-agricultural use is opencast mining, or where planning permission has been granted subject to a restoration condition and an aftercare condition in which the use specified is for agriculture[2].

The landlord has to prove that he 'proposes' to terminate the tenancy for the particular purpose, but does not have to prove the practicability or otherwise of the purpose[3]. Accordingly a tenant will usually have to rely on s 24(2) – that a fair and reasonable landlord would not insist upon possession if the proposed purpose is not practical[4].

It should be noted that a consent under s 23(1)(e), or where the Land Court states in its reasons that it is satisfied that the subsection would apply, will give rise to a reorganisation payment[5].

Consents under Sch 2

Under Sch 2, cases 1, 2 and 3 apply to a tenancy let before 1 January 1984 and cases 4, 5, 6 and 7 apply to a tenancy let after 1 January 1984.

(a) Cases 1 and 5

'The tenant has neither sufficient training in agriculture nor sufficient experience in the farming of land to enable him to farm the holding with reasonable efficiency.

[Case 5 only] Provided that this Case shall not apply where the tenant has been engaged, throughout the period from the date of death of the person from whom he acquired right to the lease, in a course of relevant training in agriculture which he is expected to complete satisfactorily within 4 years from the said date, and has made arrangements to secure that the holding will be farmed with reasonable efficiency until he completes that course.'

Both these cases refer to the tenant's personal training and experience, which are treated disjunctively so that either a training in agriculture or sufficient experience in the farming of land can be sufficient to resist the operation of the notice to quit[6].

A tenant cannot rely, as he may in opposition to an objection by the landlord under ss 11 or 12, on access to skilled advice[7].

1 Town and Country Planning (Scotland) Act 1972, s 21; *Muir Watt* p 96 (special development order); *Ministry of Agriculture v Jenkins* [1963] 2 QB 317 (general development order).
2 Agriculture Act 1958, s 14A inserted by the Housing and Planning Act 1986, Sch 8, para 5.
3 *Carnegie v Davidson* 1966 SLT (Land Ct) 3.
4 *Carnegie v Davidson* above; *Arbroath TC v Carrie* 1972 SLCR App 114.
5 1991 Act, s 55(2).
6 *Macdonald v Macrae* 1987 SLCR 72.
7 See p 120 'Agricultural skill'; cf *Macdonald v Macrae* above at 82.

The phrase 'training in agriculture' refers to a formal course, but a tenant who has personal farming experience may succeed in holding the tenancy[1]. The Land Court may take into account the standard of the tenant's farming since succeeding to the tenancy, and take note of his standard of farming before the succession if he was in fact farming the land for the tenant[2].

It should be noted that case 5 differs from case 1 in three significant respects:

(1) the onus is on the tenant to satisfy the Land Court that the circumstances are not as specified in the case[3];

(2) the tenant cannot invoke the fair and reasonable landlord exception[4];

(3) the tenant can invoke the proviso to case 5. The tenant has to be engaged in the relevant training before acquiring the lease. As the proviso requires that the tenant 'throughout the period from the date of the death' should have been engaged in the relevant training it may not be possible to invoke the proviso by joining such a course after the date of death but before acquiring the lease by transfer.

(b) Cases 2 and 6

'(a) The holding or any agricultural unit of which it forms part is not a two-man unit;
(b) the landlord intends to use the holding for the purpose of effecting an amalgamation within 2 years after the termination of the tenancy; and
(c) the notice specifies the land with which the holding is to be amalgamated.'

The landlord requires to establish all three elements.

'A two-man unit' means 'an agricultural unit[5] which in the opinion of the Land Court is capable of providing full-time employment for an individual occupying it and at least one other man'[6]. It is further provided that:

'For the purpose of determining whether land is a two-man unit, in assessing the capability of the unit of providing employment it shall be assumed that the unit is farmed under reasonably skilled management, that a system of husbandry suitable for the district is followed and that the greater part of the feeding stuffs required by any livestock kept on the unit is grown there.'[7]

The unit includes land actually occupied by the tenant, whether informally or not. 'The test is simply that of single occupation for agricultural purposes.

1 *Macdonald v Macrae* 1987 SLCR 72.
2 *Macdonald v Macrae* above at 83.
3 1991 Act, s 25(3)(b).
4 Ibid, s 25(3).
5 Ibid, s 85(1) – '"agricultural unit" means land which is an agricultural unit for the purposes of the Agriculture (Scotland) Act 1948'; 1948 Act, s 86(2) – '"agricultural unit" means land which is occupied as a unit for agricultural purposes, including – (a) any dwellinghouse or other building occupied by the same person for the purpose of farming the land and (b) any other land falling within the definition in this Act of the expression "agricultural land" which is in occupation of the same person, being land as to which the Secretary of State is satisfied that having regard to the character and situation thereof and other relevant circumstances ought in the interest of full and efficient production to be farmed in conjunction with the agricultural unit'; see *Jenners Princes Street Edinburgh Ltd v Howe* 1990 SLT (Land Ct) 26 at 27J.
6 1991 Act, Sch 2, Pt III, para 1.
7 Ibid, Sch 2, Pt III, para 2.

There is no specific requirement that the extra land be held under a formal lease or disposition'[1].

The court requires to be satisfied that cases apply 'in the circumstances prevailing at the date of the hearing and not in relation to a different state of affairs existing at the prior date of the notice to quit'[2].

The test is an objective test taking into account the hypothetical occupier and not the actual management regime of the tenant.

The Land Court has said[3]:

'We consider that Parliament must have intended to attach some significance to the phrase "an individual occupying it" instead of merely saying "two full-time employees". We conclude that the words used require us to contemplate a notional occupier who might, depending upon circumstances, have some latitude for working longer hours, but would on the other hand, be involved in the administration of the holding.

In assessing the capability of the holding to provide full-time employment for an individual so occupying it and for at least one other man, the court has also to make the assumptions directed in s 19 (3). The subsection requires us in the first place to assume that the unit is farmed under reasonably skilled management. As this is an assumption and the question is one of capabilities, we are not necessarily concerned with the present state of husbandry, but rather the farm's potential for generating employment. The other assumptions we are required to make are that a system of husbandry suitable to the district is followed and also that the greater part of the feeding stuffs required by any livestock kept are grown on the farm.'

The fair and reasonable landlord exception can be invoked by the tenant.

(c) Cases 3 and 7

'[Case 3] The tenant is the occupier (either as owner or tenant) of agricultural land which -
(a) is a two-man unit;
(b) is distinct from the holding and from any agricultural unit of which the holding forms part; and
(c) has been occupied by him since before the death of the person from whom he acquired right to the lease of the holding;
and the notice specifies the agricultural land.

[Case 7] The tenant is the occupier (either as owner or tenant) of agricultural land which–
(a) is a two-man unit;
(b) is distinct from the holding; and
(c) has been occupied by him throughout the period from the date of giving of the notice;
and the notice specifies the land.'

---

1 *Jenners Princes Street Edinburgh Ltd v Howe* 1990 SLT (Land Ct) 26.
2 *Jenners Princes Street Edinburgh Ltd v Howe* above at 28H. In this case the court took into account land held under temporary grazing lets because such leases had been taken in the past. The court commented that if the land were leased as a contrivance to avoid the case, it would be open to the court to hold that it was not genuinely held as part of the agricultural unit.
3 *Earl of Seafield v Currie* 1980 SLT (Land Ct) 10 at 11; see *Jenners Princes Street Edinburgh Ltd v Howe* 1990 SLT (Land Ct) 26 at 28L.

These two cases, while very similar, have significant differences, which require to be noted.

The onus is on the landlord to prove case 3, but on the tenant to satisfy the Land Court that the circumstances in case 7 are not as specified, although the tenant is not required to prove that he is not the owner of land[1].

To bring a tenant within these cases where he is the owner of agricultural land, it is not enough that he is the owner of tenanted land. In case 3 occupation of land by a partnership or a company does not count[2], but under case 7 occupation by a company controlled by the tenant or by a partnership including the tenant is treated as occupation by the tenant[3]. In England it has been held that the occupation need not be as sole occupier[4].

The cases do not appear to require that the tenancy should be a secure tenancy of other land[5].

It does not appear that the other agricultural land, whether owned or tenanted, has to be in Scotland[6].

What amounts to a 'two-man unit' is considered under cases 2 and 6 above. It should be noted that case 3 (c) refers to 'occupied by him since before the death' so the unit could not be altered by leasings after death.

In contrast, the provision in case 7 relates to occupation from the date of 'giving of the notice'. As the date of assessment of the unit as a two-man unit is the date of the hearing[7], it may be possible for a vulnerable tenant in expectation of the service of a notice to quit to shed land to reduce his unit to less than a two-man unit, although this might then make him vulnerable to a case 2 or 6 notice[8].

The difference between part (b) of the requirements is important in that for case 7 the land owned or occupied need only be 'distinct from the holding', whereas under case 3 it also has to be distinct 'from any agricultural land of which the holding forms part'[9]. This means that under case 7 if the holding is part of a larger agricultural unit, where the other land is objectively a two-man unit on its own, then the lease can be terminated.

Paragraph (c) in relation to the periods of occupation is different. The important point is that under case 3 agricultural land which is inherited by way of ownership or tenancy as a result of the death does not fall to be taken into account, whereas under case 7 land that has been acquired as a result of the death does fall to be taken into account. A landlord should be careful, under case 7, to ensure that any other tenancy which he is to rely on has been acquired by transfer from the executors before service of the notice to quit.

The fair and reasonable landlord exception applies to both cases[10].

1 1991 Act, s 25(3)(b).
2 *Haddo House Estate Trs v Davidson* 1982 SLT (Land Ct) 14.
3 1991 Act, Sch 2, Pt III, para 3.
4 *Jackson v Hall, Williamson v Thompson* [1980] AC 854.
5 *Gill* para 395. This appears to be consistent with the dicta in *Jenners Princes Street Edinburgh Ltd v Howe* that a two-man unit can include land taken on a temporary grazing lease.
6 *Gill* para 395.
7 *Jenners Princes Street Edinburgh Ltd v Howe* 1990 SLT (Land Ct) 26.
8 *Jenners Princes Street Edinburgh Ltd v Howe* above at 28H. A 'mere temporary contrivance' to avoid the case may be ignored by the court.
9 See cases 2 and 6 above and *Jenners Princes Street Edinburgh Ltd v Howe* for a consideration of this phrase.
10 1991 Act, s 25(3).

(d) Case 4

'The tenant does not have sufficient financial resources to enable him to farm the holding with reasonable efficiency.'

The onus is on the tenant to satisfy the Land Court that the circumstances specified in the case do not apply[1].

This test will probably be similar to those considered by the Land Court in objections to a successor under the 1991 Act, ss 11 and 12.[2]

The fair and reasonable landlord exception is not available[3].

The 'fair and reasonable landlord' exception

This exception can be invoked in respect of any of the grounds given in s 24(1)[4] or in respect of cases 1, 2, 3, 6 or 7 of Sch 2[5].

A tenant must give notice in his pleadings if he intends to found on this exception[6]. It is competent to defend on the exception alone[7].

It is rare for this exception to succeed[8]. It is not enough that a plea is *ad misericordiam*. A tenant's case 'is not facilitated when he gives ... unsupported testimony and without detailed information as to his overall financial situation and means of livelihood'[9].

Power to attach condition to consent

Where the Land Court consents to a notice to quit on any of the grounds in s 24(1) it 'may ... impose such conditions as appear to them requisite for securing that the land to which the notice relates will be used for the purpose for which the landlord proposes to terminate the tenancy'[10].

It may not be competent for the Land Court to impose a condition in relation to ground (d) (hardship)[11] unless perhaps the case discloses a purpose[12].

Where a landlord fails to comply with a condition imposed under s 24(3) on an application by the Crown to the Land Court, the Land Court may impose

---

1 1991 Act, s 25(3)(b).
2 See p 121 'Financial resources'.
3 1991 Act, s 25(3).
4 Ibid, s 24(2).
5 Ibid, s 25(3).
6 Eg *Lindsay-Macdougall v Peterson* 1987 SLCR 59.
7 *Trs of the Main Calthorpe Settlement v Calder* 1988 SLT (Land Ct) 30; *Arbroath TC v J Carrie & Son* 1972 SLCR App 114.
8 The plea succeeded in *Carnegie v Davidson* 1966 SLT (Land Ct) 3; *Altyre Estate Trs v McLay* 1975 SLT (Land Ct) 12; *Mackenzie v Lyon* 1984 SLT (Land Ct) 30; *Trs of the Main Calthorpe Settlement v Calder* 1988 SLT (Land Ct) 30.
9 *Arbroath TC v J Carrie & Son* 1972 SLCR App 114.
10 1991 Act, s 24(3); see eg *Robertson v Lindsay* 1957 SLCR 3; *Shaw-Mackenzie v Forbes* 1957 SLCR 34; *Graham v Lamont* 1970 SLT (Land Ct) 10.
11 *Gill* para 407; per contra *Muir Watt* p 98.
12 *Robertson v Lindsay* 1957 SLCR 3; *Shaw-Mackenzie v Forbes* 1957 SLCR 34; *Graham v Lamont* 1970 SLT (Land Ct) 10.

on the landlord a penalty of up to two years' rent[1]. The penalty is payable to the consolidated fund and is not available for the benefit of the tenant[2].

Similar powers are available for consents given in respect of a near relative successor under cases 2 or 6 to ensure that the proposed amalgamation takes place within two years of the termination of the tenancy[3].

A condition for any other purpose would be *ultra vires*[4]. The Land Court could not impose a condition which gave the tenant further occupation to a date in the future[5].

Upon an application by a landlord the Land Court may vary or revoke a condition, if satisfied by reason of any change of circumstances or otherwise that the condition imposed ought to be varied or revoked[6].

NOTICE TO QUIT PART OF THE HOLDING

Where the landlord serves a valid notice to quit part of the holding under s 29, the tenant has three options:

(1) To accept the notice to quit part of the holding.

(2) To serve a counternotice under s 22(1) within one month of the service of the notice to quit. The notice to quit is then subject to the Land Court's consent[7].

If the tenant is unsuccessful in the Land Court proceedings option (3) below remains open to him.

(3) To serve a written counternotice on the landlord within 28 days of the receipt of the notice to quit part of the holding stating that he accepts the notice as a notice to quit the entire holding[8]. The notice to quit then operates as a notice to quit the whole holding.

This option can be exercised within 28 days of the Land Court determination[9].

In deciding whether or not to exercise the option to treat the notice to quit as a notice to quit the entire holding, the tenant should bear in mind the provisions relating to disturbance[10].

Where the notice to quit relates to less than a quarter of the area of the holding or an area of less than a quarter of the rental value, and the diminished holding is reasonably capable of being farmed as a separate holding, then disturbance is payable only in respect of the part of the holding to which the notice to quit relates. The consequence of this is that the additional payment[11] will similarly be reduced.

1 1991 Act, s 27(1).
2 Ibid, s 27(2).
3 Ibid, s 25(4).
4 *Earl of Moreton v Hamilton* 1977 SLCR App 136; cf *British Airports Authority v Secretary of State* 1979 SC 200.
5 *Arbroath TC v J Carrie & Son* 1972 SLCR App 114.
6 1991 Act, s 24(4).
7 Ibid, ss 22(1) and 24(1) (grounds for consent).
8 Ibid, s 30(a).
9 Ibid, s 30(b).
10 Ibid, s 43(7).
11 Ibid, s 54.

The tenant does not have the right to elect to treat the notice to quit as a notice to quit the entire holding in a compulsory purchase situation, if he qualifies and has elected to take notice of entry compensation[1].

NOTICE TO QUIT – COMPULSORY PURCHASE

Where a notice to quit founding on s 24(1)(e)[2] is served after an acquiring authority has served a notice to treat on the landlord or, being an authority with compulsory purchase powers, has agreed to acquire his interest and the Land Court has consented, the tenant should consider whether or not to elect to take notice of entry compensation from the acquiring authority in lieu of compensation from the landlord[3].

A similar election can be made where the notice to quit relates to part of the holding.

1 Land Compensation (Scotland) Act 1973, s 55(1) and (2).
2 Land required for non-agricultural purpose where planning permission not required.
3 See p 237 'Compulsory acquisition' for a fuller consideration of such an election.

CHAPTER 17

Response to a s 22(2) notice to quit and demands to remedy

Key points

- The correct response to a s 22(2) notice to quit is to require arbitration except where the notice (1) relates to fixed equipment under s 22(2)(d), or (2) relates under s 22(2)(g) to the succession of a near relative under s 25(3): s 23(2).

- A counternotice in response to a s 22(2) notice to quit is not competent except where the notice (1) relates to fixed equipment under s 22(2)(d), or (2) relates under s 22(2)(g) to the succession of a near relative under s 25(3): s 23(2).

- The operation of a notice to quit is suspended after the tenant requires arbitration until the issue of the arbiter's award: s 23(4).

- Where the arbiter's award is issued less than six months before the ish date in the notice to quit, the tenant may apply to the Land Court to postpone the operation of the notice to quit by up to 12 months: s 23(5).

- A landlord who fails to institute arbitration within a reasonable time of the service of a requirement for arbitration and prior to the ish date under the notice may be held to have waived his rights to found on the notice to quit.

- A tenant served with a demand to pay rent should pay the rent claimed and then require arbitration on the question of whether or not that rent was due, rather than risk an arbitration on the notice to quit challenging accuracy of the demand.

- If a tenant fails to require arbitration, then the intrinsic validity of the notice to quit or a demand to pay rent or remedy cannot be challenged in subsequent proceedings, although the technical validity can be challenged.

- Where a tenant receives a demand to remedy or a notice to quit founded under s 22(2)(d) on a failure to remedy a breach of a term or condition of the lease relating to the provision, repair, maintenance or replacement of fixed equipment, the tenant must carefully consider the additional options open to him under ss 32 and 66.

- Where a counternotice or requirement for arbitration is given in respect of a demand to remedy or a notice to quit founded under s 22(2)(d) on a failure to remedy a breach of a term or condition of the lease relating to the provision, repair, maintenance or replacement of fixed equipment the landlord will require to consider his options under ss 32 and 66.

- Where a tenant is successful in a reference of a question in a s 22(2) notice to quit to arbitration, then the notice becomes one to which s 22(1) applies and the tenant requires to serve a counternotice within one month of the arbiter's award: s 23(3).

- Where a notice to quit is served under threat of a compulsory purchase order, the tenant should consider whether or not to take notice of entry compensation in lieu of compensation on the termination of the tenancy: see p 237 'Compulsory acquisition'.

Time limits

- A notice to quit founded on a certificate of bad husbandry has to be served after the certificate is obtained, but not more than nine months after the application for the certificate is made to the Land Court: s 22(2)(c).

- A notice requiring a s 22(2) notice to quit to be referred to arbitration requires to be served on the landlord within one month of the notice to quit being served on the tenant: s 23(2).

- Following the service of the requirement for arbitration the landlord should institute arbitration proceedings within a reasonable time and prior to the ish date in the notice to quit.

- Where a tenant receives a demand to remedy or a notice to quit founded under s 22(2)(d) on a failure to remedy a breach of a term or condition of the lease relating to the provision, repair, maintenance or replacement of fixed equipment, the tenant must carefully consider the additional options open to him under ss 32 and 66 and serve either a counternotice or a requirement for arbitration within one month from the giving of the notice to quit: s 32(2) and (3).

- Where a tenant succeeds in a reference to arbitration on a s 22(2) notice to quit the notice becomes a s 22(1) notice to quit and the tenant requires to serve a counternotice within one month of the arbiter's award: s 23(3).

- A tenant requires to apply to the Land Court within one month of the issue of the arbiter's award for a postponement of the operation of the notice to quit, if required: s 23(5).

- Where a notice to quit is served under threat of a compulsory purchase order, and the tenant has decided to elect to take notice of entry compensation in lieu of compensation on the termination of the tenancy, the time limits for making the election require to be observed: see p 237 'Compulsory acquisition'.

GENERAL

A notice to quit to which s 22(2)[1] applies is incontestable provided (1) the notice is formally and technically correct and (2) the facts in the statement in the notice are factually accurate.

Where the notice to quit follows upon a demand to remedy in relation to fixed equipment, an arbiter appointed under s 23(2) may have power to modify the demand under s 66(1).

ACTION BY TENANT ON RECEIPT OF NOTICE TO QUIT

The tenant must ascertain whether or not the notice is one to which s 22(2) applies[2].

It is not competent to serve a counternotice to a s 22(2) notice to quit[3]. It is therefore essential to establish that the notice is a s 22(2) notice.

If the notice is a s 22(2) notice to quit the tenant requires to consider whether there are any questions arising as to the accuracy of the statement in the notice which he wishes to challenge. Such questions require to be determined by arbitration.

RESPONSE TO A DEFECTIVE NOTICE

If the notice to quit is defective as to form or service the tenant may ignore it and subsequently defend any action of removing on the ground that the notice was null and void.

It is safer to serve a requirement for arbitration under reservation of the right to object to the notice in any subsequent proceedings[4].

If the tenant is in doubt as to whether the notice is given under s 22(2) or s 22(1) the tenant should also serve a counternotice under reservation of the right (1) to challenge the validity of the notice to quit and (2) to deal with the notice as a notice to which the provisions of s 22(2) apply and similarly serve a demand requiring arbitration under reservation[4].

If there is doubt as to whether or not the notice is a notice under s 22(1) or 22(2) it is probably an invalid notice in any event.

ARBITRATION

The correct response to a s 22(2) notice to quit is to require arbitration under s 23(2), if the tenant 'requires such question to be determined by arbitration'.

---

1 With the limited exception of a s 25(2)(c) notice relating to a near relative successor.
2 See chap 15.
3 s 22(2); *Morris v Muirhead* 1969 SLT 70; *Earl of Stradbroke v Mitchell* [1991] 1 WLR 469. NB the limited exception in relation to s 22(2)(h) notices under s 25(1) relating to near relative successors.
4 See p 169 'Response to a defective notice'.

The tenant has to give notice to the landlord within one month[1] after the notice to quit has been served requiring arbitration[2] on any such questions[3].

Failure to require arbitration

Failure to serve a notice requiring arbitration within one month is fatal[4].

If a tenant fails to require arbitration or there are no questions which can be referred to arbitration, then the only defence available to the tenant is to challenge the validity of the notice to quit on technical grounds in subsequent proceedings. The tenant cannot, in that defence, challenge the validity of the notice on the ground that the reasons in it were defective[5], unless they were fraudulent[6].

Once the challenge is raised in subsequent proceedings, the question of the formal validity of the notice to quit will have to be determined by arbitration.

Landlord's response to requirement for arbitration

Where a tenant requires arbitration under s 23(2), no time limit in which arbitration should be instituted is provided. A landlord who fails to take steps within a reasonable time and before the ish date under the notice, or who fails to have an arbiter appointed or to agree with the tenant a reference to the Land Court[7] may be held to have waived his right to found on the notice to quit[8].

The question referred to the arbiter is usually framed as:

'All questions arising out of and in connection with the notice to quit served by the landlord on the tenant dated . . . .'

A widely framed question allows all issues to be raised in the arbitration.

It would be open to the tenant to have the arbiter appointed and to state the questions that he wishes to have determined in the application.

Effect of requirement for arbitration

Where arbitration has been required by the tenant the operation of the notice to quit is suspended until the issue of the arbiter's award[9].

---

1 Ie to the corresponding date; *Dodds v Walker* [1981] 1 WLR 1027.
2 1991 Act, s 23(2).
3 Style – 'With reference to your notice to quit [specify holding] dated . . . . , I hereby intimate to you that in terms of section 23(2), I require all questions arising out of the reasons stated in the said notice to quit to be determined by arbitration under the Agricultural Holdings (Scotland) Act 1991'. The notice should reserve questions of validity etc; see p 170, n 2.
4 *Gemmell v Robert Hodge & Sons* 1974 SLT (Land Ct) 2.
5 *Magdalen College Oxford v Heritage* [1974] 1 WLR 441; *Morris v Muirhead* 1969 SLT 70.
6 *Earl of Stradbroke v Mitchell* [1991] 1 WLR 469; *Jones v Gates* [1954] 1 WLR 222 ' . . . a fictitious notice'.
7 1991 Act, s 60(2).
8 *Cayzer v Hamilton (No 2)* 1995 SLCR 13.
9 1991 Act, s 23(4).

If the award is issued later than six months before the date on which the notice to quit is expressed to take effect, the Land Court, on an application made to it by the tenant, may postpone the operation of the notice to quit for a period not exceeding twelve months. The tenant requires to make the application not later than one month after the issue of the award[1].

Effect of success in arbitration

If the tenant is successful in an arbitration challenging the notice to quit, on the arbiter issuing his award, the notice to quit then becomes a notice to quit to which s 22(1) applies[2]. A counternotice requires to be served by the tenant within one month of the arbiter's award[3].

GROUNDS UPON WHICH A S 22(2) NOTICE TO QUIT MAY BE SERVED

Ground a

'(a) the notice to quit relates to land being permanent pasture which the landlord has been in the habit of letting annually for seasonal grazing or of keeping in his own occupation and which has been let to the tenant for a definite and limited period for cultivation as arable land on the condition that he shall, along with the last or waygoing crop, sow permanent grass seeds.'

This provision allows a landlord who runs grazing parks to let the land for the limited purpose of running a crop or crops through the land prior to reseeding as grazing.

The lease has to be of 'permanent pasture' let 'for a definite and limited period'[4] for the purpose of 'cultivation' on condition the tenant 'shall, along with the last or waygoing crop, sow permanent grass'. A lease from 'year to year' would not qualify for this exception[5].

Further if the original lease runs onto tacit relocation the landlord loses the protection of this subsection.

It is therefore essential that a notice to quit is served under this provision to terminate the lease at its ish, because if a notice to quit is not served then the lease does not terminate, but goes on to relocation.

---

1 1991 Act, s 23(5).
2 Ibid, s 23(3); *Wilson-Clarke v Graham* 1963 SLT (Sh Ct) 2 at 4. Cf A G M Duncan 'Agricultural Holdings; a Scottish-English Divergence?' 1973 SLT (News) 177 where the author suggests that the notice to quit requires to specify that in the event that s 22(2) is found not to apply, then it is served alternatively as a s 22(1) notice as well; a view disapproved of by *Gill* para 364.
3 1991 Act, s 23(3).
4 *Stirrat v Whyte* 1968 SLT 157 at 163 per Lord Cameron.
5 *Roberts v Simpson* (1954) 70 Sh Ct Rep 159.

Ground b

'(b) the notice to quit is given on the ground that the land is required for use, other than agriculture, for which permission has been granted on an application made under the enactments relating to town and country planning, or for which (otherwise than by virtue of any provision of those enactments) such permission is not required.'

This provision contrasts with the ground under which the Land Court may consent to a s 22(1) notice to quit under s 24(1)(e)[1].

The land must be required at the date of the notice[2] either by the landlord or some other party[3]. The land must be required for 'a use other than agriculture' for which planning permission has already been granted[4] or for which planning permission is not required other than by virtue of any provision of the Town and Country Planning Acts[5].

Development authorised by a development order falls under s 24(1)(e) and not ground b[6].

While it has been held that planning permission need not relate to the whole area[7], it probably must relate to the whole or substantially the whole holding[8].

A tenant should know whether or not planning permission has been granted, because he is entitled to notification of the application to allow him to make representations about it and to be informed of the decision[9].

The ground does not apply to planning permission for opencast coal workings subject to a restoration and after care condition[10].

Ground c

'(c) the Land Court, on an application in that behalf made not more than 9 months before the giving of the notice to quit, were satisfied that the tenant was not fulfilling his responsibilities to farm the holding in accordance with the rules of good husbandry, and certified that they were so satisfied.'

---

1 See p 175 'Ground e'.
2 *Paddock Investments v Lory* (1975) 236 EG 803; *Jones v Gates* [1954] 1 WLR 222. Cf *Scammell and Densham* p 169 for a discussion of whether 'is required' related to the date of service of the notice or the date of challenge or some later date, perhaps the date of decision of the arbiter which appears contrary to the cited cases.
3 *Rugby Joint Water Board v Foottit* [1973] AC 202.
4 Outline planning permission is sufficient; *Ministry of Agriculture v Jenkins* [1963] 2 QB 317.
5 Eg the Crown does not require planning permission by reason of being the Crown, rather than under any provision of the Act; cf *Ministry of Agriculture v Jenkins* above. A landlord's proposed non-agricultural use, if it does not involve a change of use, falls within this provision; *Bell v McCubbin* [1990] 1 QB 976.
6 Town and Country Planning (Scotland) Act 1972, s 21; *Muir Watt* p 96 (special development order); *Ministry of Agriculture v Jenkins* [1963] 2 QB 317 (general development order).
7 *Crawley v Pratt* [1988] 2 EGLR 6.
8 *Scammell and Densham* p 170. In *Crawley v Pratt* above, which relates to a very small holding, while the planning permission did not relate to the whole holding, the small area excluded from the actual permission appears to have been an area that would be included in the amenity area of the actual development.
9 Town and Country Planning (Scotland) Act 1972, ss 24 and 26.
10 Opencast Coal Act 1958, s 14A(6).

This provision makes effective a certificate of bad husbandry obtained under s 26.

The nine-month period starts with the lodging of the application in the Land Court, but the notice to quit cannot be served until after the certificate is issued[1]. It is therefore important to ensure that the Land Court decision is available in time to serve the notice within the nine-month period.

A certificate is applied for under s 26 by lodging an application with the Land Court, wherein the landlord requires to specify the particular rules of good husbandry upon which the breach is founded[2].

The Land Court normally holds an inspection of the holding as soon as the application is lodged[3].

Certificates of bad husbandry are only granted where there is proof of the tenant's failure to comply with the rules of good husbandry[4]. The landlord has to discharge the onus of proof to a very high standard[5]. The Land Court cannot grant a certificate without proof, even if the application is unopposed[6].

The Land Court 'if satisfied that the tenant is not fulfilling his said responsibilities' to farm in accordance with the rules of good husbandry 'shall grant the certificate'[7]. If satisfied, the Land Court has no discretion but to grant the certificate, even if the failure is through no fault (for example, illness) of the tenant[8].

The court has taken into account mitigating factors excusing the tenant's failures under 'other relevant circumstances'[9].

The landlord does not have to prove a breach over the whole holding. The breach

'must significantly affect the holding so that it can broadly be said that a reasonable standard of efficient production has not been maintained nor the unit kept in such a condition to maintain such a standard in the future.[10]'

A minor and temporary breach of the rules will not justify the grant of a certificate of bad husbandry[11].

The rules in r 2

'are subsidiary to and without prejudice to the generality of Rule 1 so any minor breach of any particular item in Rule 2 which does not also involve the matters mentioned in Rule 1 will not justify the grant of a Certificate of Bad Husbandry.[12]'

In considering whether or not the tenant is in breach of the rules of good husbandry, the court may take into account the standard of management of the

---

1 *Cooke v Talbot* (1977) 243 EG 831.
2 *McGill v Bichan* 1982 SLCR 33; *Austin v Gibson* 1977 SLT (Land Ct) 12.
3 *Ross v Donaldson* 1983 SLT (Land Ct) 26 at 26.
4 1948 Act, Sch 6; *McGill v Bichan* above.
5 *McGill v Bichan* above; *Austin v Gibson* 1979 SLT (Land Ct) 12.
6 *Buchanan v Buchanan* 1983 SLT (Land Ct) 31.
7 1991 Act, s 26(1).
8 *Cambusmore Estate Trust v Little* 1991 SLT (Land Ct) 33 at 39E and 40A.
9 *Luss Estate Co v Firkin Farm Co* 1985 SLT (Land Ct) 17; *Dalgleish v Dalgleish* 1952 SLCR 7. The approach in these cases is perhaps difficult to reconcile with *Cambusmore Estate Trs v Little* above.
10 *Ross v Donaldson* 1983 SLT (Land Ct) 26 at 27.
11 *Ross v Donaldson* above.
12 *McGill v Bichan* 1982 SLCR 33 at 40; *Cambusmore Estate Trs v Little* 1991 SLT (Land Ct) 33 at 39J.

owner and his compliance with the rules of good estate management, in so far as they affect the holding[1]. The landlord does not have to prove that he is complying with the rules of good estate management[2].

Where the landlord has the power under the lease to remedy the tenant's breaches, the fact that he does not invoke this power does not bar him from applying for a certificate, because the remedy was additional to the other remedies available to him[3].

In order to comply with the rules the tenant is 'normally only obliged to husband or farm his holding as equipped by the landlords in accordance with the lease'. Failure to follow a more intensive system of farming requiring capital expenditure not available to a tenant does not amount to a breach of the rules of good husbandry[4].

Leasing out milk quota cannot be

'regarded as the equivalent of farming in accordance with the rules of good husbandry or of maintaining a reasonable standard of production ... it is not a farming operation or a "kind of produce" to obtain an income from the leasing out of milk quota to a lessee who runs a dairy herd on other land.[5]'

A similar *ratio* would apply to leasing out sheep and suckler cow premium quota.

Rule 2(e) provides for detailed rules in regard to hill sheep farming[6]. In hill farms the court is more likely to find a breach of the rules relating to stock management rather than land management[7].

Any practice adopted by a tenant in compliance with any obligation imposed by or accepted under the Control of Pollution Act 1974, s 31B is disregarded by the Land Court in considering any question of bad husbandry[8].

Where a landlord applies for a certificate of bad husbandry, he would be well advised to serve concurrent notices to quit seeking the Land Court's consent under s 24(1)(a) and (b), where the standards of proof in relation to poor husbandry are not so strict.

Ground d

'(d) at the date of the giving of the notice to quit the tenant had failed to comply with a demand in writing served on him by the landlord requiring him within 2 months from the service thereof to pay any rent due in respect of the holding, or within a reasonable time to remedy any breach by the tenant, which was capable of being remedied, of any term or condition of his tenancy which was not inconsistent with the fulfilment of his responsibilities to farm in accordance with the rules of good husbandry.'

---

1 Cf r 1 – 'having regard to . . . the standard of management thereof by the owner'; *Cambusmore Estate Trs v Little* 1991 SLT (Land Ct) 33 at 41K; *Ross v Donaldson* 1985 SLT (Land Ct) 26; *Sinclair v Mackintosh* 1983 SLT (Land Ct) 29; *McCrindle v Andrew* 1972 SLCR App 117.
2 *Dalgleish v Dalgleish* 1952 SLCR 7 at 9.
3 *Halliday v Ferguson* 1961 SC 24.
4 *McGill v Bichan* 1982 SLCR 33.
5 *Cambusmore Estate Trust v Little* 1991 SLT (Land Ct) 33 at 41H.
6 Cf *Luss Estate Co v Firkin Farm Co* 1985 SLT (Land Ct) 17.
7 *Jedlischka v Fuller* 1985 SLCR 90 at 91.
8 1991 Act, s 26(2).

A notice to quit can be based on one of three types of demand to remedy, which need to be considered separately, namely (1) demand to pay rent due, (2) demand to remedy a breach of the terms of the lease (except in relation to maintenance or repair of fixed equipment) and (3) a demand relating to the maintenance or repair of fixed equipment for which there are additional special provisions[1].

The ground must be distinguished from ground e, which relates to breaches of conditions of the lease which are not capable of being remedied in a reasonable time and at an economic cost[2]. It is incompetent to rely on ground d if the breach cannot be remedied in a reasonable time and at an economic cost.

As a prelude to serving an incontestable notice to quit under s 22(2)(d) the landlord requires to serve either a rent demand or a demand to remedy.

Where a demand contains a time limit within which performance must be completed, with the exception of specialities relating to demands to repair or maintain fixed equipment, a demand for arbitration at that stage of whether or not the tenant is in breach of the terms of the lease does not appear to interrupt the running of the time limit[3]. A tenant would therefore be wise to comply with the demand timeously and then require arbitration on the question of whether or not he was in arrears of rent or in breach of any term or condition of the lease.

(a) Form of the demand

There is no laid down format for a demand in Scotland, although a statutory form is provided for in England[4].

The Land Court has recently summarised the law as to the requirements of a notice to remedy[5]:

'a number of requirements seem to us to have been established. The Notice need not be in any particular form, *Budge v Hicks* [1951] 2 KB 335 approved in *Macnabb v Anderson* 1957 SC 213. The landlord must make his position clear as to what the nature of the breach is and as to what requires to be done in order to remedy it, *Morris v Muirhead, Buchanan & Macpherson* 1969 SLT 70. The tenant must not be misled or left in any doubt as to what is required of him, *Carradine Properties Ltd v Aslam* [1976] 1 WLR 442, approved in *Land v Sykes* (1992) 03 EG 115.'

1 See 1991 Act, ss 32 and 66.
2 *Macnabb v A & J Anderson* 1955 SC 38 at 43 per Lord Russell.
3 Cf 1991 Act, s 23(4) where the operation of a notice to quit is suspended where arbitration is demanded or an application is made to the Land Court for consent; and s 66 where an arbiter has the power to vary time limits in a demand to remedy breaches of repair and maintenance obligations.
4 See forms AH60 'Notice to tenant to pay rent'; AH61 'Notice to tenant to remedy breach of tenancy agreement by doing work of repair, maintenance or replacement'; AH62 'Notice to tenant to remedy breach of tenancy agreement' (not being a notice requiring the doing of any work, repair, maintenance or replacement); Agricultural Holdings (Forms of Notice to Pay Rent or to Remedy) Regulations 1987, SI 1987/711. In *Morris v Muirhead* 1969 SLT 70 at 74 the Lord Ordinary said that he was sympathetic to the argument that the English statutory forms laid down what in fact was the law in Scotland as to what should appear in a demand.
5 *Cayzer v Hamilton (No 1)* 1995 SLCR 1.

The demand must be in writing. It must run in the name of the correct land-lord[1] to the correct tenant[2] at his proper address[3]. It may be sent by an agent or to an agent[4], although it is safer to address a demand to the tenant personally.

A demand to pay rent or to remedy must give a clear and unambiguous statement of the breach to be remedied. If the demand is ambiguous the notice to quit will likewise be invalid[5].

(b) Rent demand

A rent demand is a written demand addressed by the landlord to the tenant requiring the tenant within two months of the service of the demand to pay any rent due in respect of the holding.

A demand for rent must contain an explicit demand for payment within two months[6]. The sum due must be accurately stated[7], and it must be due at the date of the demand[8]. It would be wise to specify in the demand the date at which each sum and the amount thereof of the arrears became due[9]. If the sum claimed is inaccurate then the demand is invalid[10].

If the arrears of rent are *de minimus* the court may not allow the landlord to rely on s 22(2)(d), unless the *de minimus* arrears are part of a course of conduct to irritate the landlord[11].

The tenant requires to make payment within the two-month period. A cheque received within the two months, provided it is honoured thereafter, is timeous payment[12]. If the parties have agreed that the tenant may pay rent by

---

1 *Pickard v Bishop* (1975) 32 P & CR 108 (a demand in the name of the previous landlord was held invalid); cf *Frankland v Capstick* [1959] 1 WLR 204 (the misdescription of a landlord in a notice claiming dilapidations was held to be a technical error not invalidating the notice because it in no way affected the tenant, her rights, or her understanding of what the proceedings were and her obligations thereunder).
2 *Jones v Lewis* (1973) 25 P & CR 375 (a demand to one of two joint tenants was held invalid).
3 1991 Act, s 84(3).
4 Ibid, s 85(5).
5 *Cayzer v Hamilton (No 1)* 1995 SLCR 1; *Macnabb v A & J Anderson* 1955 SC 38.
6 *Magdalen College Oxford v Heritage* [1974] 1 WLR 441; *Official Solicitor v Thomas* (1986) 279 EG 407.
7 *Dickinson v Boucher* (1984) 269 EG 1159.
8 *Urwidk v Taylor* (1969) EGD 1106; *Harding v Marshall* (1983) 267 EG 161; *Dallhold Estates Ltd v Lindsey Trading Properties Inc* [1992] EGLR 88. Payment date for rent is usually stipulated in the lease; if not the common law terms of Whitsunday and Martinmas will apply; see *Gill* para 336.
9 Cf the Agricultural Holdings Act 1986, Form 1 'Notice to Tenant to Pay Rent'.
10 *Dickinson v Boucher* (1984) 269 EG 1159.
11 *Luttenberger v North Thoresby Farms Ltd* [1993] 1 EGLR 3.
12 *Balfour's Exrs v Inland Revenue* 1909 SC 619; *Lord Herries v Maxwell's Curator* (1873) 11 M 396; *Luttenberger v North Thoresby Farms Ltd* above. In *Hannaford v Smallacombe* [1994] 1 EGLR 9 it was held that upon the dishonouring of a cheque on first presentation the tenant had failed to comply with the demand to pay rent. The fact that the cheque was honoured on the second presentation by the landlord before the service of the notice to quit did not waive the landlord's right to serve the notice.

cheque sent in the post to the landlord, then the court might accept that the date of payment was the date of posting[1].

The notice to quit cannot be served until the two-month period has expired[2]. The tenant cannot retrieve the situation by paying the rent after the expiry of the two months, whether before or after service of the notice to quit[3].

If the demand is invalid, the notice to quit is also invalid[4]. If the tenant fails to require arbitration on the question of whether or not the rent is due, he cannot challenge the reasons stated in the demand in later proceedings[5], unless the demand was fraudulently or recklessly made[6].

(c) Demand to remedy (except in relation to fixed equipment)

This part of the section relates to the tenant's failure to comply with a written demand requiring him within a reasonable time to remedy a breach of any term or condition of the lease, which was capable of being remedied, which was not inconsistent with the fulfilment of his responsibilities to farm in accordance with the rules of good husbandry.

The demand can relate to the 'breach by the tenant of *any* of the terms or conditions of the tenancy'[7], including a residence clause[8], stocking obligation[9] or the payment of a sum of money due under the lease[10], inversion of possession[11], provided the term or condition is not inconsistent with the responsibilities to farm in accordance with the rules of good husbandry.

It should be clear from the terms of the demand what provision of the lease the tenant is said to have breached. The breach may be a breach of a specific

---

1 *Beevers v Mason* (1978) 37 P & CR 452 (CA) where the court held that the payment was made on the day that the cheque was posted, even though it was received after the expiry of two months, because the tenant had been in the habit of paying by cheque posted to the landlord. The decision accords with the line of authority that when a creditor authorises payment by cheque to be sent throught the post, that payment is made when the cheque is posted; *Norman v Ricketts* (1886) 3 TLR 182; *Thairwall v Great Northern Ry Co* [1910] 2 KB 509; *Thorey v Wylie & Lockhead* (1890) 6 Sh Ct Rep 201. The rule in *Beevers v Mason* does not apply where the posted cheque is not one which the bank is bound to honour – eg where a signature was missing; *Luttenberger v North Thoresby Farms Ltd* [1993] 1 EGLR 3.

2 *Macnabb v A & J Anderson* 1955 SC 38 at 45; cf *French v Elliott* [1960] 1 WLR 40 where a notice to quit posted before, but received after, the expiry of the two months was held valid.

3 *Stoneman v Brown* [1973] 1 WLR 459; *Price v Romilly* [1960] 1 WLR 1360 at 1363; cf *Hannaford v Smallacombe* [1994] 1 EGLR 9.

4 *Morris v Muirhead* 1969 SLT 70; *Magdalen College Oxford v Heritage* [1974] 1 WLR 441; *Harding v Marshall* (1983) 267 EG 161; *Official Solicitor v Thomas* (1986) 279 EG 407.

5 *Morris v Muirhead* above; *Magdalen College Oxford v Heritage* above; *Harding v Marshall* above; *Parrish v Kinsey* (1983) 268 EG 1113.

6 *Rous v Mitchell* [1991] 1 All ER 676; *Luttenberger v North Thoresby Farms Ltd* [1993] 1 EGLR 3; cf *Jones v Gates* [1954] 1 WLR 222.

7 *Macnabb v A & J Anderson* 1955 SC 38 at 45 per Lord Patrick; *Fane v Murray* 1995 SLT 567. *Morris v Muirhead* 1969 SLT 70 at 74 is wrong in so far as it suggests that the subsection relates only to the repair, maintenance and replacement of fixed equipment.

8 *Morrison-Low v Howison* 1961 SLT (Sh Ct) 53; *Lloyds Bank Ltd v Jones* [1955] 2 QB 298. Cf *Blair Trust Co v Gilbert* 1940 SLT 322.

9 *Pentland v Hart* 1967 SLT (Land Ct) 2.

10 *Official Solicitor v Thomas* (1986) 279 EG 407.

11 *Cayzer v Hamilton (No 2)* 1995 SLCR 13.

condition of a written lease or term implied into a written lease, or of any terms implied at common law into an oral lease[1].

The 'reasonable time' must have expired before the notice can be served[2]. While the demand should specify a time limit, it has been held sufficient to require the breach to be remedied in a 'reasonable time'[3]. The notice to quit can then only be served upon the expiry of a reasonable time. If the tenant contends that he was not given reasonable time, that will be a question of fact for arbitration under s 23(2).

In determining what is a reasonable time an arbiter should not consider each item in isolation, but should have regard to what is a reasonable time to comply with the whole demand including those demands that are ill-founded[4].

Where the demand relates to a series of breaches, if only one breach remains unremedied at the expiry of the time limit, that breach alone, if material, justifies the notice to quit[5]. Further, even if an arbiter finds that part of the demands in the notice was not justified, but the substantial part was and the tenant has not remedied them, the landlord may rely on those breaches in serving a notice to quit[6].

Even if the landlord is in breach of some of the conditions of the lease, this does not preclude him serving a demand under this subsection[7], unless the tenant's breach has been caused by the landlord's breach[8].

A tenant fails to comply with a demand at his peril, even if he intends to challenge the validity of the demand. Unlike demands in relation to the repair or maintenance of fixed equipment, an arbiter does not have power to vary the demand or to extend the time limits. If a demand is not complied with and the challenge fails, then the tenant will lose the tenancy.

(d) Demands in relation to repair or maintenance of fixed equipment

Sections 32 and 66 introduce restrictions on the operation of notices to quit, including a statement that the notice relates, under s 22(2)(d), to the tenant's failure to remedy 'a breach of any terms or condition of his tenancy by doing work of provision, repair, maintenance or replacement of fixed equipment'.

The most common demand to remedy relates to the provision, repair, maintenance or replacement of fixed equipment. Prior to the amendments introduced by the 1976 Act, a tenant faced considerable difficulties and awkward choices in challenging a demand to remedy relating to fixed equipment[9]. The amendments introduced additional options and defences to such a demand to remedy.

1 *Cayzer v Hamilton* above where it was said that it would be best if the notice specified the clause founded upon, although this was not essential if the breach founded on was clear to the tenant from the demand, as construed in the light of the surrounding circumstances.
2 *Macnabb v A & J Anderson* 1955 SC 38 at 45 per Lord Patrick.
3 *Morrison-Low v Howison* 1961 SLT (Sh Ct) 53; *Stewart v Brims* 1969 SLT (Sh Ct) 2.
4 *Nicholls Trs v McLarty* 1971 SLCR App 85; *Wykes v Davis* [1975] QB 843.
5 *Price v Romilly* [1960] 1 WLR 1360; *Pentland v Hart* 1967 SLT (Land Ct) 2.
6 *Shepherd v Lomas* [1963] 1 WLR 962.
7 *Wilson-Clarke v Graham* 1963 SLT (Sh Ct) 2. This decision seems inconsistent with the mutuality of obligations in a contract.
8 *Shepherd v Lomas* above.
9 Cf *Gill* para 342 for a summary of the problems.

The general law relating to demands to remedy, considered in the previous section, also applies to demands in relation to maintenance or repair of fixed equipment, except where specifically altered by ss 32 and 66.

Of particular importance is the requirement that the demand should specify in detail the exact works that the tenant is required to undertake, otherwise the demand will be held invalid because it is vague[1].

Care should be exercised in considering English authority in relation to demands to remedy relating to fixed equipment, because the law in England has been substantially altered by the Agricultural Holdings Act 1986 and statutory instruments made thereunder, so that it differs from the Scottish provisions[2].

When a tenant receives a demand to remedy relating to fixed equipment his options are:

(1) If he disputes the need for the work he should do nothing and immediately invoke arbitration under s 60(1) on the question of whether or not the demand is justified. If the arbiter holds the work is justified, then the arbiter can be invited to exercise his powers under s 66 (1)(a) to extend the time for compliance.

The tenant can require arbitration after the service of the notice to quit in terms of s 32(3) and at that stage invoke the provisions of s 66[3]. However the arbiter's rights to extend the period at that stage are more restricted, if he holds that the original period was reasonable[4].

If only some of the items in the demand are disputed, then time runs on the remainder in accordance with the demand[5].

(2) If he merely disputes the adequacy of the time given for the work, the tenant should start the work and then require arbitration asking the arbiter to extend the time limit[6]. Provided the tenant has made reasonable progress with the work, the arbiter is likely to extend the time.

The tenant should invoke the provisions of s 66 to have the demand modified by an arbiter, unless he accepts that it is reasonable.

Where an arbiter is appointed, he may use his powers to:

(a) provide for such period as he considers reasonable if no period is specified in the demand or substitute a period for one already specified as he considers reasonable[6]. The onus will be on the tenant to persuade the arbiter to substitute a different period.

(b) delete from the demand any item or part of an item which, having regard to the interests of good husbandry as respects the holding and sound management of the estate of which the holding forms part or constitutes, the arbiter is

---

1 Cf the Agricultural Holdings Act 1986, Form 2 'Notice to Tenant to Remedy Breach', which requires a schedule headed 'Terms or Condition of Tenancy – Particulars of Breach and Work Required to Remedy it'. The form, as an example of what should be included, was approved in *Morris v Muirhead* 1969 SLT 70.
2 Cf *Fane v Murray* 1995 SLT 567.
3 *Fane v Murray* above.
4 See below.
5 *Ladds Radio v Docker* (1973) 226 EG 1565.
6 s 66(1)(a).

satisfied is unnecessary or unjustified[1]. The onus will be on the tenant to satisfy the arbiter that the deletions should be made.

(c) substitute in the case of any item or part of an item specified in the demand a different method or material from that which the demand would otherwise require, having regard to the purpose which that item or part is intended to achieve, where the arbiter is satisfied that (i) the specified method or material would involve undue difficulty or expense, (ii) the alternative method would be substantially as effective for the purpose and (iii) in the circumstances the substitution is justified. The onus will be on the tenant to satisfy the arbiter that he should make the substitution[2]. In considering whether or not to make a substitution the arbiter must be satisfied on the three grounds which are cumulative.

Where the arbiter extends the time for compliance, which has the effect that the landlord will miss an ish date for the purposes of serving a notice to quit if the tenant fails to comply with the demand, then the arbiter or the landlord can make application to the Land Court to specify a date for the termination of the tenancy by notice to quit. The specified date has to be the later of (a) the date at which the tenancy could have been terminated by notice to quit upon expiry of the original period or, if no period was specified, (b) the date of the demand of 6 months after the period specified or extended by the arbiter[3].

The arbiter has additional powers under s 66(4) after the service of the notice to quit.

The options open to a tenant upon receipt of a notice to quit to which s 32 applies are:

(a) The tenant may serve a counternotice requiring that s 32(2) should apply to the notice[4]. The notice to quit then cannot have effect without the Land Court's consent[5]. The landlord requires to apply to the Land Court for consent to the operation of the notice to quit, although no time limit is given for that application to be made[6]. The Land Court 'shall consent ... to the operation of the notice to quit unless in all the circumstances it appears to them that a fair and reasonable landlord would not insist upon possession'[7].

The onus is on the tenant to establish that a fair and reasonable landlord would not insist upon possession[7].

(b) Alternatively the tenant can require arbitration under s 23(2) by giving the appropriate notice to the landlord requiring arbitration within one month after the notice to quit has been served.

If the tenant had first served a counternotice under s 32(2) that counternotice is rendered of no effect by the subsequent service of a requirement for arbitration[8].

---

1 s 66(1)(b).
2 s 66(1)(c).
3 s 66(2).
4 See p 169 regarding the requirement for a counternotice.
5 s 32(2).
6 s 32(5).
7 NB 'unless in all the circumstances', which contrasts with the 1991 Act, s 24(2) where the fair and reasonable landlord provision is invoked 'if in all the circumstances it appears' to the Land Court.
8 s 32(3). Cf Fane v Murray 1995 SLT 567.

Where arbitration is required after the service of a notice to quit, the tenant can invoke the provision of s 66[1], but the arbiter's powers are more restricted if he holds that the original period was reasonable[2].

Where the arbiter is appointed after the service of a notice to quit, given because of the tenant's failure to remedy within the specified period work in relation to the provision, repair, maintenance or replacement of fixed equipment[3], the arbiter's powers of extension are more limited if he finds that the period originally given or extended was reasonable.

In such circumstances the arbiter may only extend the period if 'it would in consequence of any happening before the expiry of that period have been unreasonable to require the tenant to remedy the breach within that period'[4].

Although 'happening' is not defined it appears to cover personal disasters to the tenant or some external happening such as failure of contractors, industrial disputes and the weather[5]. The word does not appear to cover refusal by the tenant to do the work, because he considers he is not liable.

Where the arbiter finds that the period became unreasonable by reason of a happening he may extend or further extend the period as he considers reasonable 'having regard to the length of period which has elapsed since the service of the demand'[6]. The later provision perils the tenant if he has done no work under the demand, because even if he can rely on a happening, the arbiter may not be able to extend the period long enough to complete the works, where he has taken into account a period during which works should notionally have been undertaken.

If the arbiter extends the time then any notice to quit that may have been served is invalidated; but the landlord can serve a second notice to quit, based on the tenant's failure to complete the works within the extended time, within one month of the period specified or the extended time allowed by the arbiter[7].

When the period was extended, the landlord should have made application to the Land Court for a specified date on which the tenancy could have been terminated[8].

(c) If the arbiter finds against the tenant and the notice to quit accordingly is to have effect, the tenant may serve a counternotice within one month of the arbiter's award requiring that s 32(4) shall apply to the notice to quit[9]. The notice to quit then cannot take effect unless the Land Court consents. The Land Court shall consent to such a notice 'unless in all the circumstances it appears to them that a fair and reasonable landlord would not insist upon possession'[10].

---

1 *Fane v Murray* 1995 SLT 567. In this case it was argued for the landlord that if the tenant sought to invoke the provisions of s 66 he required, as in England, to do that before service of the notice to quit. The First Division, in answering the questions posed in a special case from the Land Court, rejected this submission.
2 s 66(4).
3 s 66(4)(a).
4 s 66(4)(b).
5 *Duncan,* note to s 66(4) in *Current Law Statutes.*
6 s 66(4).
7 s 66(3).
8 See above; s 66(2).
9 s 32(4).
10 s 32(5).

It should be noted that the tenant can, again, require arbitration under s 32(3) to a second notice to quit, with a view to relying on a 'happening'[1] for the purpose of asking the arbiter to extend the period.

Where a notice to quit is served under s 66(3) the tenant may serve a counternotice in writing on the landlord within one month of receipt of the notice. Where that is done the notice to quit shall not have effect unless the Land Court consents[2].

Ground e

'(e) at the date of the giving of the notice to quit the interest of the landlord in the holding had been materially prejudiced by a breach by the tenant, which was not capable of being remedied in reasonable time and at economic cost, of any term or condition of the tenancy which was not inconsistent with the fulfilment by the tenant of his responsibilities to farm in accordance with the rules of good husbandry.'

This ground is mutually exclusive of ground d in that ground d contemplates that the breach is remedial, whereas under ground e the breach is 'not capable of being remedied in reasonable time and at economic cost'. It is not competent to include in one notice to quit references to both grounds d and e unless the notice makes it absolutely clear which breach relates to which ground. If a landlord seeks to serve a notice to quit relying separately on breaches of grounds d and e it is preferable that two separate notices to quit are served without prejudice to each other[3].

In order to succeed with a notice to quit under this ground the landlord will have to establish (1) that his interest in the holding has been *materially* prejudiced, (2) by a breach by the tenant of any term or condition of the lease[4], (3) which, and this is crucial, was not capable of being remedied in reasonable time and at an economic cost and (4) which was not inconsistent with the fulfilment by the tenant of his responsibilities to farm in accordance with the rules of good husbandry. Where such a breach is established there is a presumption that the landlord's interest has been prejudiced[5].

It should be noted that while sub-letting in England is an irremedial breach[6], this is not the case in Scotland as a subtenancy granted without the landlord's consent does not give the sub-tenant any right to remain on the holding[7].

As with a demand under ground d the notice to quit under ground e should specify, preferably by reference, the term or condition which is said to have been breached and how it has been breached[8].

The fact that the terms of the lease permit the landlord to effect repairs or

1   s 66(4).
2   s 32(6).
3   *Macnabb v A & J Anderson* 1955 SC 38.
4   This can be a breach of a written term, a term implied into the written lease by law or otherwise, or the terms or implied terms of an oral lease.
5   *Gill* para 354.
6   Cf eg *Troop v Gibson* [1986] 1 EGLR 1.
7   *Rankine* p 192 – 'The landlord may then remove [the sub-tenant] as possessing without title'.
8   *Cayzer v Hamilton* 1994 Tayside RN 66.

maintenance if the tenant fails to do so, does not prevent the landlord serving a notice to quit on ground e[1].

The special position of an executor[2] should be noted. On the reference of a s 22(2)(e) notice to quit to arbitration[3] 'the arbiter shall not make an award in favour of the landlord, unless ... the arbiter is satisfied that it is reasonable having regard to the fact that the interest is vested in the executor in his capacity as executor[4], that it should be made'. This provision probably also applies where the notice was served on the tenant who died and was succeeded as tenant by his executor, prior to the arbiter issuing his award.

Ground f

'(f) at the date of the giving of the notice to quit the tenant's apparent insolvency had been constituted in accordance with section 7 of the Bankruptcy (Scotland) Act 1985.'

The ground is concerned with 'apparent' insolvency. Where a tenant is sequestrated the landlord's rights to an irritancy are protected by s 21(6).

At common law a lease could not be ended by the landlord on the tenant's sequestration or insolvency etc, but it could be adopted by the trustee in bankruptcy[5]. If there is no irritancy provision in the lease the landlord will have to rely on ground f to remove a trustee in bankruptcy.

Apparent insolvency can be constituted in a a number of ways[6]. If the tenant is apparently insolvent it will be up to him to prove that he was willing and able to pay the debts as they came due. In *Murray v Nisbet*[7] the Land Court allowed the tenant a proof restricted to two aspects, (1) to rebut the resumption of insolvency and (2) that at the material time the tenant had realisable assets to meet his current obligations A tenant who required at the material time to negotiate a financing deal probably would not be held to have had realisable assets to meet the obligations[8].

This ground can be invoked on a tenant who is found to have been apparently insolvent at ingo[9].

Where the Land Court has directed that the holding or part of the holding be treated as a market garden[10], then on the termination of the tenancy under ground f the tenant is not entitled to compensation unless he has produced a statutory offer, which is not accepted by the landlord[11].

1 *Halliday v Ferguson* 1961 SC 24, disapproving *Forbes-Sempill's Trs v Brown* 1954 SLCR 36.
2 Cf s 85(1) – 'tenant . . . includes executor'.
3 Under s 23(2).
4 NB an executor can sometimes become a tenant in his own right, in which case this provision will not apply; *Morrison-Low v Paterson* 1985 SC (HL) 49.
5 *Chalmer's Tr v Dick's Tr* 1909 SC 761; *Dobie v Marquis of Lothian* (1864) 2 M 788.
6 Bankruptcy (Scotland) Act 1985, s 7, including (1) giving creditors written notice that he has ceased to pay his debts in the ordinary source of business, (2) the granting of a trust deed, (3) failing to pay on a duly executed charge, (4) sequestration or receivership, (5) failure to pay following a statutory demand etc.
7 1967 SLT (Land Ct) 14.
8 *R A Cripps & Son Ltd v Wickenden* [1973] 2 All ER 606.
9 *Hart v Cameron* (1935) 51 Sh Ct Rep 166.
10 s 41(1).
11 s 41(3) and (4).

Ground g

'(g) section 25(1) of this Act applies, and the relevant notice complies with section 25(2)(a), (b) and (d) of this Act.'

This ground applies only to a tenant who has succeeded to the lease either as an acquirer[1] or as a legatee[2]. It does not apply to a near relative successor[3].

The specialities of notices to quit under this ground are considered at chapters 15 and 16.

NOTICE TO QUIT - COMPULSORY PURCHASE

Where a notice to quit founding on s 22(2)(b)[4] is served after an acquiring authority has served a notice to treat on the landlord or, being an authority with compulsory purchase powers, has agreed to acquire his interest and the Land Court has consented, the tenant should consider whether or not to elect to take notice of entry compensation from the acquiring authority in lieu of compensation from the landlord[5].

A similar election can be made where the notice to quit relates to part of the holding.

---

1 s 16.
2 s 12.
3 s 25(2)(c) and 'near relative' as defined by Sch 2, Pt III, para 1.
4 Land required for non-agricultural purpose where planning permission not required.
5 See p 237 'Compulsory acquisition'.

CHAPTER 18

Rights and claims on termination of tenancy or partial dispossession

Key points

- Except as expressly provided in the 1991 Act, it is not competent to contract out of statutory rights to compensation including compensation for milk quota.

- Non-statutory claims are only enforceable under an agreement in writing.

- Before making a claim on the termination of the tenancy a careful check should be made of what claims are available at common law and under statute and if any of those claims have been varied by an agreement in writing.

- If a tenant does not have a claim for fixtures or buildings, because compensation is not payable for them as an improvement, a check should be made as to whether or not the tenant has a right to remove the fixture or building: s 18.

- A tenant's right to compensation for milk quota depends on (1) the tenant having milk quota registered as his, (2) milk quota having been allocated to the tenant or (3) in respect of transferred quota, that milk quota was allocated to the tenant or that he was in occupation of the holding on 2 April 1984: 1986 Act, Sch 2, para 2(1).

- A landlord's right, on the termination of the tenancy, to make a statutory claim for dilapidations, deterioration or damage to the holding, in respect of a lease entered into after 31 July 1931, is conditional upon a record of the holding having been made: s 47(3)(a).

- A landlord's right, on the termination of the tenancy, to make a claim under a written lease, in lieu of the statutory claim, for dilapidations, deterioration or damage to the holding, in respect of a lease entered into on or after 1 November 1948 is conditional upon a record of the holding having been made: s 47(3)(b).

- A claim on the termination of the tenancy has to be in writing. It has to specify the nature of the claim and it is sufficient specification if the notice refers to the statutory provision, custom, or term of any agreement under which it is made: s 62(3); 1986 Act, Sch 2, para 11.

- Unless early agreement on a claim is likely, it is to the claiming party's advantage to seek early arbitration, as an arbiter cannot award interest from a date earlier than his award.

- Where a tenant disputes that the tenancy has been terminated, claims for compensation should be made 'without prejudice' timeously and the arbiter appointed within the time limits.

- A tenant dispossessed from part of the holding by a notice to quit under s 29 or a resumption notice is entitled to the usual claims, and some particular claims, on the termination of the tenancy restricted to that part of the holding: ss 49 and 58.

Time limits

- In general, for the purposes of calculating time, 'termination of the tenancy' means the date at which the contract comes to an end and does not include periods of occupation, lawful or otherwise, after that date.

- Written notice of all claims including claims for milk quota compensation (except claims for high farming or deterioration and dilapidations) have to be given before the expiry of two months from the termination of the tenancy: s 62(2); 1986 Act, Sch 2, para 11(1).

- Written notice of claims for high farming have to be given not later than one month before the termination of the tenancy: s 44(2).

- Written notice of claims by the landlord for dilapidations, deterioration or damage under s 45 have to be given not later than three months before the termination of the tenancy: s 47(1).

- Where a landlord makes a claim for dilapidations, deterioration or damage under a written lease, in lieu of the statutory claim, then the normal notice applies, namely before the expiry of two months from the termination of the tenancy: ss 45(3) and 62.

- For claims under the 1991 Act the parties have a four-month period for negotiation, which may be extended before the expiry of the period for a further two months and again for a further two: s 62(4).

- For claims for milk quota compensation parties have an eight-month negotiating period, which cannot be extended: 1986 Act, Sch 2, para 11(2).

- For claims under the 1991 Act the arbiter has to be appointed before the expiry of one month after the end of the negotiating, or extended negotiating, period or such longer time as the Secretary of State may in special circumstances allow: s 62(5).

- For claims for milk quota compensation the arbiter has to be appointed within the eight-month negotiating period: 1986 Act, Sch 2, para 11(1) and (2).

- A tenant wishing to exercise his rights under s 18 to remove fixtures and buildings on the termination of the tenancy requires to give at least one month's notice prior to the termination of the tenancy: s 18(2)(b).

GENERAL

On the termination[1] of a tenancy or part of a tenancy certain potential claims become open to the landlord and to the tenant. In general it is the tenant who has the right to make the majority of the claims, although the landlord has limited claims for dilapidations and deterioration of the holding.

The extent of the claims that might be open to the tenant depend on the type of farming being carried out and the way in which the tenancy was terminated.

The common law and some of the statutory rights to waygoing valuations and compensation may have been altered by agreement, either in the lease or by a separate agreement. Before a claim is made, the tenant will have to ascertain the basis upon which any claim is being made.

Special provisions relate to compensation for improvements for market gardens[2].

This chapter aims to outline the claims that may be available to landlord and tenant and to detail the procedure for making the claim effective.

TENANT'S CLAIMS

Broadly the tenant's claims fall under three categories, (1) waygoing valuations, which relate to the sale by the tenant to the landlord or the incoming tenant of the produce of the holding and any bound stock, (2) compensation for improvements, milk quota and high farming and (3) statutory rights to disturbance and additional payments.

In addition the tenant may have a right to remove fixtures and buildings (other than ones in respect of which the tenant is entitled to compensation)[3].

On termination of the tenancy the claims which may be open to the tenant include:

(a) A payment for manure, compost, hay, straw or roots grown in the last year of the tenancy[4].

(b) A payment for implements of husbandry, fixtures, farm produce or farm stock[5].

(c) Bound sheep stock valuations[6].

(d) A payment for waygoing valuations including ploughing and sowing, white crops, grass, green crops, fallow land and unexpired manurial valuations.

(e) Compensation for improvements[7].

---

1 s 85(1) – 'termination . . . means the termination of the lease by effluxion of time or from any other cause'.
2 See chap 21.
3 1991 Act, s 18.
4 Ibid, s 17.
5 Ibid, s 19.
6 Ibid, ss 68–72.
7 Ibid, ss 33–39.

(f) Compensation for milk quota[1].

(g) Compensation for high farming[2]. A tenant is only entitled to a payment for high farming if a record has been made of the holding[3].

(h) Compensation for disturbance[4].

(i) An additional payment, sometimes called a reorganisation payment, of four times the annual rent[5].

Where the tenancy is terminated by compulsory purchase additional payments may be applicable[6].

On partial dispossession

Where a tenant is dispossessed from part of the holding under a notice to quit[7] or a resumption notice:

'... the provisions of this Act with respect to compensation shall apply as if that part of the holding were a separate holding which the tenant had quitted in consequence of a notice to quit.[8]'

The tenant's right to claim compensation for milk quota on a partial dispossession is preserved[9].

In addition, on a resumption the tenant can claim compensation in respect of the early resumption of part of the holding[10].

LANDLORD'S RIGHTS AND CLAIMS

The landlord's rights and claims may be stipulated for or modified by the lease or other written agreement[11].

A landlord's right to make a statutory claim for dilapidations[12], deterioration or damage to the holding on the tenant quitting the holding on the termination of the lease, where the lease was entered into after 31 July 1931, is conditional upon a record of the holding having been made[13].

It is doubtful whether a landlord can now make a claim for damages and deterioration of the holding during the currency of the lease, because s 100 of the 1949 Act had not been continued in this Act[14].

1 1986 Act, Sch 2.
2 1991 Act, s 44.
3 Ibid, s 44(2)(b).
4 Ibid, s 43.
5 Ibid, s 54.
6 Ibid, ss 56–59.
7 Ibid, s 29.
8 Ibid, s 49(1).
9 1986 Act, Sch 2, para 1 – 'termination means the resumption of possession of the whole or part of the tenancy'.
10 1991 Act, s 58.
11 Page 243 'Contractual rights'.
12 1991 Act, s 45.
13 Ibid, s 47(3)(a).
14 See p 9 '1949 Act, s 100'.

On termination of the tenancy the rights and claims which may be open to the landlord include:

(a) Rights and claims provided for in the lease or written agreement[1].

(b) Compensation for the reduction in value of the holding by dilapidation, deterioration or damage[2].

(c) Compensation for the cost of making good dilapidation, deterioration or damage to the holding[3].

(d) In lieu of (b) and (c), if the lease in writing so provides, claims under the lease[4].

(e) A claim for damages for any injury to or deterioration of the holding attributable to the exercise by the tenant of his rights to dispose of the produce of the holding (other than manure) and to practise any system of cropping of the arable land[5].

A landlord is precluded from claiming compensation in excess of the value of damage actually suffered by him in consequence of the breach or non-fulfilment of any condition in the lease. Any provision in the lease to the contrary is null and void[6].

NOTICE OF CLAIM

A landlord or tenant making a claim under the 1991 Act has to 'give notice in writing ... of his intention to make the claim'[7]. The notice 'shall specify the nature of the claim, and it shall be sufficient specification thereof if the notice refers to the statutory provision, custom, or term of an agreement under which the claim is made'[8].

Claims for milk quota compensation[9] are unenforceable 'unless ... the tenant has served notice in writing on the landlord of his intention to make the claim specifying the nature of the claim'. Note that there is no statutory definition of what is sufficient specification[10].

The test of whether or not a notice is valid is probably whether the terms of the

---

1 Page 243 'Contractual rights'.
2 1991 Act, s 45(1)(a) and (2)(a).
3 Ibid, s 45(1)(b) and (2)(b).
4 Ibid, s 45(3).
5 Ibid, s 7(3)(b); p 76 'Freedom of cropping'.
6 Ibid, s 48.
7 Ibid, s 62(3).
8 Ibid, s 62(3). It should be noted that the wording in the Agricultural Holdings (Scotland) Act 1923, s 15 provided that the claim was unenforceable 'unless particulars thereof have been given ... by the tenant to the landlord'. The present wording comes from the 1949 Act. Cases under the pre-1949 Act legislation (eg *Duke of Montrose v Hart* 1925 SC 160; *Simpson v Henderson* 1944 SC 365) are therefore not of assistance in construing the phrase 'shall specify the nature of the claim'.
9 1986 Act, Sch 2, para 11(1).
10 *Walker v Crocker* [1992] 1 EGLR 29 where the judge commented that in view of this difference in wording a claim for milk quota compensation might be valid in circumstances where a claim for compensation under the 1991 Act might be invalid.

purported notice are sufficiently clear to bring home to the ordinary landlord and/or his professional advisers that the tenant intends to make a claim for compensation and to indicate clearly the heads of claim or statutory provision under which the claim is made[1]. A letter making a claim may be construed in light of the surrounding circumstances and the contemporary correspondence[2].

A letter stating that 'We give you prior warning that it is our present intention to serve a notice under section 70 of the Act in due course' was held not to be a valid notice, because the words only intimated an intention to serve a notice[3].

While the statutes provide for the minimum specification required in such a notice it is good practice to have the claim fully valued before the notice is given. The claim can then be detailed item by item and by value, with reference to each head of claim, in the notice[4].

TIME LIMITS FOR CLAIMS

There are strict time limits within which claims on the termination of a tenancy have to be intimated. The time limit in each case relates to 'the termination of the tenancy'[5]. Termination 'means the termination of the lease by reasons of effluxion of time or from any other cause'[6]. This relates to the termination of the lease rather than to the handover dates, which might be after that date[7].

Except for claims in respect of high farming[8] and claims by the landlord for deterioration and dilapidations[9], 'where the tenant lawfully remains in occupation of part of an agricultural holding after the termination of the tenancy...'[10] the time limit runs from the termination of the occupancy rather than from the termination of the tenancy[11]. It would appear that these provisions only apply where the tenant remains in lawful occupation under a new agreement and do not apply where the tenant remains in occupation under a provision in the lease that he may remain in occupation of part of the farm, for example until the separation of the crops[12].

1 *Walker v Crocker* [1992] 1 EGLR 29 at 30F and 31B; *Amalgamated Estates Ltd v Joystretch Manufacturing Ltd* (1980) 257 EG 489; *Nunes v Davies, Laing and Dick* [1986] 1 EGLR 106. Cf *ED & AD Cooke Bourne (Farms) Ltd v Mellows* [1983] QB 104.
2 *Walker v Crocker* [1992] 1 EGLR 29 at 31B; cf *Duke of Montrose v Hart* 1925 SC 160 at 166 per LJC (it was considered relevant that particulars of a claim, which the tenant later challenged as insufficient notice, had been intimated to the tenant, who then made no suggestion that he was puzzled by the letter).
3 *Lady Hallinan v Jones* (1984) 272 EG 1081 (County Ct) approved in *Walker v Crocker* [1992] 1 EGLR 29.
4 *Williams* para 11–3 regarding dilapidations claims.
5 Eg 1991 Act, ss 44, 47 and 62; 1986 Act, Sch 2, para 11(1).
6 1991 Act, s 85(1).
7 *Waddell v Howat* 1925 SC 484; *Coutts v Barclay–Harvey* 1956 SLT (Sh Ct) 54; see p 157 'The ish date'.
8 1991 Act, s 44(2)(a).
9 Ibid, s 47(1).
10 Ibid, s 62(6).
11 Ibid, s 62(6); 1986 Act, Sch 2, para 11(3) where the wording, although slightly different, is to the same effect.
12 *Coutts v Barclay–Harvey* 1956 SLT (Sh Ct) 54 following *Swinburne v Andrews* [1923] 2 KB 483 at 489 and *Arden v Rutter* [1923] 2 KB 865.

The relevant time limits are:

(a) The time limit for all claims, except those noted below, requires that notice be given 'before the expiry of 2 months after the termination of the tenancy'[1]or the lawful continued occupation[2]. There is nothing to prevent the notice being given before the date of the termination of the tenancy[3].

This includes claims by the landlord for dilapidations etc if made under a written lease, rather than a statutory claim in terms of s 45(1)[4].

(b) The time limit for a high farming claim requires it to be made 'not later than one month before the termination of the tenancy'[5]. Section 62(2) cannot be invoked to extend this period[6].

(c) A claim by the landlord for dilapidations, deterioration or damage to the holding under s 45 requires to be made 'not later than 3 months before the termination of the tenancy'[7]. Section 62(2) cannot be invoked to extend this period[8].

This time limit precludes a claim for dilapidations etc occurring within the last three months of the tenancy under s 45.

Where the landlord makes a claim under a written lease in terms of s 45(3) in lieu of a statutory claim under s 45(1)(b), then the time limit is two months from the termination of the tenancy[9].

PROCEDURE AFTER MAKING A CLAIM

Where a claim under the 1991 Act has been properly and timeously made the parties have four months from the termination of the tenancy in which to settle the claim in writing[10]. The Secretary of State may, on the application of either the landlord or tenant, made within the four-month period, extend the period by two months and again on a further application for a further two months[10].

If a claim under the 1991 Act has not been settled by agreement in writing within the four months or the extended period, the claim ceases to be enforceable 'unless before the expiry of one month after the end of the said period and any such extension or such longer time as the Secretary of State may in special circumstances allow, an arbiter has been appointed'[11]. The Secretary of State

---

1 1991 Act, s 62(2); 1986 Act, Sch 2, para 11(1) where the wording, although slightly different, is to the same effect.
2 1991 Act, s 62(2); 1986 Act, Sch 2, para 11(3).
3 *Moffat v Young* 1975 SLCR App 98; *Lady Hallinan v Jones* (1984) 272 EG 1081 (County Ct); cf *Earl of Moreton's Trs v Macdougall* 1944 SLT 409.
4 1991 Act, ss 45(3), 62(1) and 66(2).
5 Ibid, s 44(2)(a).
6 *Coutts v Barclay–Harvey* 1956 SLT (Sh Ct) 54.
7 1991 Act, s 47(1).
8 *Coutts v Barclay–Harvey* above.
9 1991 Act, s 62(1) and (2).
10 Ibid, s 62(4).
11 Ibid, s 62(5).

has a discretion 'in special circumstances' to appoint an arbiter after the expiry of the time limit[1].

With regard to milk quota compensation the negotiating period is eight months from the termination of the tenancy[2]. There is no provision for the extension of the period. The arbiter requires to be appointed before the expiry of the eight months[3]. There is no discretion to the Secretary of State[4] for appointing the arbiter after the expiry of that period.

To satisfy the provisions relating to the appointment of an arbiter by agreement of the parties, there must be a delivered document in writing in the hands of the landlord and tenant or their agents appointing a named arbiter to determine the specified dispute[5]. Where the Secretary of State is asked to appoint an arbiter s 62(5) requires 'that an application for the appointment of an arbiter ... has been made'. The Secretary of State accepts applications by fax.

Unless agreement is likely, it is to the tenant's advantage to seek the appointment of an arbiter in early course, because an arbiter cannot award interest on his award from a date earlier than the date of the award. A tenant who is not allowed his compensation during a lengthy period of negotiation and then arbitration is not entitled to any interest on that money for the time prior to the arbiter's award[6].

CLAIMS WHERE TENANT DISPUTES TERMINATION OF TENANCY

Where a tenant disputes the termination of the tenancy, and to preserve the position in case the tenant fails to establish that the tenancy was not terminated, the appropriate claims for compensation should be made timeously 'without prejudice' to the dispute. An arbiter will have to be appointed timeously 'without prejudice'. The arbitration should be sisted until the outcome of any litigation relating to the dispute anent termination.

If a tenant fails to take these precautionary steps and the courts hold that the tenancy was lawfully terminated, then occupation after the date of termination will have been unlawful[7]. If, following the termination of the tenancy, the landlord agreed that the tenant could remain lawfully in occupation of part of the holding until the resolution of the dispute, then the date might run from the termination of the occupation[8].

The landlord's remedies arising from such a period of unlawful occupation fall to be determined by the ordinary courts[9].

---

1 Cf *Crawford's Trs v Smith* 1952 SLT (Notes) 5. Where a Secretary of State exercises this discretion it is only open to challenge by judicial review. It would appear that the Secretary of State is not required to give reasons for the exercise of this discretion; *Crawford's Trs* above.
2 1986 Act, Sch 2, para 11(2).
3 Ibid, Sch 2, para 11(1) and (3).
4 Cf the discretion given to the Secretary of State under the 1991 Act, s 62(5).
5 *Chalmers Property Investment Co v McColl* 1951 SC 24.
6 See p 282.
7 *Hendry v Walker* 1926 SLT 679, 1927 SLT 333.
8 1991 Act, s 62(6); 1986 Act, Sch 2, para 11(3).
9 *Hendry v Walker* above.

VALUATION

The sums due to the landlord or tenant under any of the claims on termination of the tenancy are matters of valuation for an expert[1].

CONTRACTING OUT

It is not competent to contract out of the right to statutory compensation under the 1991 Act, 'Unless this Act makes express provision to the contrary'[2].

A tenant who 'has milk quota registered as his ... shall be entitled, on quitting the tenancy, to obtain from his landlord a payment'[3] for milk quota[4]. Parties therefore cannot contract out of the statutory provisions on compensation. The tenant may elect to claim compensation under an agreement in writing 'in lieu of payment provided for by this paragraph'[5].

GENERAL CONSIDERATIONS

(1) *Sale of holding.* If the proprietor of the holding changes hands at the date of the termination of the tenancy or thereafter, the tenant's claim remains against the outgoing proprietor[6]. The new proprietor should ensure that the outgoing proprietor settles the claims, otherwise the holding might be made the subject of a charge[7].

Similarly the new proprietor can only enforce dilapidations etc claims against the outgoing tenant upon an assignation from the outgoing proprietor[8].

(2) *Landlord's interest divided.* Unless the rent has been apportioned by agreement or under statute, the tenant is entitled to require an arbiter to determine the compensation as if the holding had not been divided. The arbiter may apportion the amount awarded between the joint landlords, and may direct that any additional expense caused by the apportionment shall be paid by the joint landlords in such proportions as he may determine[9].

If the rent has been apportioned then the compensation falls to be paid by the joint landlords in respect of that apportionment.

(3) *Non-statutory claims.* Non-statutory claims are only enforceable under an agreement in writing[10].

---

1 See *Marshall* and *Williams*.
2 1991 Act, s 53; *Coates v Diment* [1951] 1 All ER 890. Express provision to the contrary is made in ss 34(4), (7), 37(2), 38(5), 42, 45(3) and 53(2).
3 1986 Act, Sch 2, para 2(1).
4 The right to compensation depends on the quite complex provisions of ibid, Sch 2, para 2; see p 262 'Entitlement'.
5 Ibid, Sch 2, para 2(4).
6 *Waddell v Howat* 1925 SC 484.
7 1991 Act, s 75(1); 1986 Act, Sch 2, para 12.
8 *Waddell v Howat* above.
9 1991 Act, s 50. Liability is not joint and several; *Weston v Duke of Devonshire* [1923] 12 LJCCR 74.
10 1991 Act, s 53(3).

(4) *Enforcement of agreement or award of compensation.* If the sum payable as to compensation, expenses or otherwise under any agreement or award is not paid within one month after the date it becomes payable it may be recorded in the Books of Council and Session or the sheriff court books. The agreement or award is then enforceable in like manner as a decree arbitral[1].

While a warrant for registration and execution is probably not required in an agreement it would be wise to include one[2].

A tenant also has the option of obtaining a charge over the holding where the money remains unpaid for one month after the date it was due[3]. As the annuity period is currently 30 years[4], this is an unattractive option, seldom used.

A landlord who is not the owner of the *dominium utile*, and who has paid the compensation, may, likewise, obtain a charging order against the absolute owner[5].

1 1991 Act, s 65; 1986 Act, Sch 2, para 12.
2 *Duncan*, note to s 65 in *Current Law Statutes*.
3 1991 Act, s 75(1).
4 Ibid, s 75(4).
5 Ibid, s 75(3).

CHAPTER 19

Tenant's claims on termination of tenancy, partial dispossession and compulsory purchase

Key points

- After a notice to quit is given the tenant may not sell or remove from the holding any manure or compost, hay, straw or roots grown in the last year of the tenancy, unless and until he has given the landlord or the incoming tenant a reasonable opportunity of agreeing to purchase them on the termination of the tenancy: s 17.

- A check should be made to see if the common law waygoing valuations have been varied by the terms of the lease or any agreement.

- Bound sheep stock valuations are carried out under different statutory provisions depending on the date at which the lease was entered into: s 68.

- Where the lease provides that the tenant shall sell to the landlord or incoming tenant the implements of husbandry, fixtures, farm produce or stock, the property in the goods does not pass until payment is made. If payment is not made within one month the tenant may sell and claim compensation for loss or expense: s 19.

- The right to compensation for improvements only arises if the improvements are suitable and appropriate to the holding, and the statutory provisions such as consents, notice etc have been complied with: see chapter 9 and p 222 'Compensation for improvements'.

- Compensation for high farming is only available if a record has been made of the holding: s 44(2)(b).

- The tenant has a right to compensation for disturbance where the tenancy is terminated by notice to quit (unless the notice to quit is given in terms of s 22(2)(a) or (c)–(f)) and where the tenant gives a counternotice under s 30.

- The tenant has a right to a reorganisation payment where he is entitled to compensation for disturbance, unless the right has been excluded: ss 54 and 55.

- Where the tenant is dispossessed from part of the holding he has the normal claims on the termination of the tenancy, treating that part of the tenancy as a separate holding. On a resumption the tenant may have an additional claim in respect of the effect of early resumption: ss 49 and 58.

- Where agricultural land is acquired compulsorily special provision is made for additional compensation in respect of the loss of the farm or agricultural holding. The tenant is also given the right to elect whether to claim against his landlord or the acquiring authority: Land Compensation (Scotland) Act 1973 (1973 Act), s 56 and ss 31–33, 44–45, 49–53, 55–58.

- Where notice to treat or notice of entry is served in respect of part of an agricultural unit or holding the owner and/or the tenant have the right to serve a counternotice claiming that the remainder of the unit cannot be farmed on its own or in conjunction with other relevant land. If the counternotice is disputed the question of its validity is referred to the Lands Tribunal and if upheld then the notice to treat is held to relate to the whole land: 1973 Act, ss 49–53.

- The special provisions which apply between landlord and acquiring authority where part of an agricultural unit is acquired, but the tenant gives up possession of the whole holding, should be noted: 1973 Act, s 52(3) and (4).

- Where a tenant is given notice to quit a holding or part of a holding after notice to treat has been served on the landlord or the landlord has agreed to sell his interest to an acquiring authority the tenant may elect to take notice of entry compensation in lieu of the normal compensation on the termination of the tenancy: 1973 Act, ss 55–58.

Time limits

- See chapter 18.

- Where the lease provides that the tenant shall sell to the landlord or incoming tenant the implements of husbandry, fixtures, farm produce or stock, payment must be made within one month of the tenant quitting the holding or one month after delivery of the valuation award: s 19.

- Not less than one month's notice has to be given to the landlord of the sale of any goods, implements, fixtures, produce, stock etc, if the tenant intends to claim compensation for disturbance in a sum exceeding one year's rent: s 43(4)(b).

- A claim for a farm loss payment has to be made before the expiration of the period of one year beginning with the date on which the claimant began to farm another farm: 1973 Act, s 33(1).

- A counternotice has to be served against a notice to treat in respect of part of an agricultural unit or holding by the owner or tenant within two months of the date of the service of the notice to treat: 1973 Act, ss 49(1) and 51(1).

- An acquiring authority has two months in which to decide whether or not to accept a counternotice as valid. Thereafter both parties have two months in which to refer the counternotice to the Lands Tribunal for Scotland to determine whether or not it is valid.

- Where an acquiring authority acquires part of an agricultural unit, but the tenant gives up possession of the whole holding, the landlord has three months from the giving up of the possession to intimate his intention to the acquiring authority to make a claim for dilapidations and deterioration under s 45: 1973 Act, s 52(4).

- A tenant has to make an election on whether or not to take notice of entry compensation in lieu of compensation on the termination of his tenancy by serving notice in writing on the acquiring authority not later than the day on which possession is given up: 1973 Act, s 55(4).

- A tenant intending to claim notice of entry compensation in respect of a notice to quit part of a holding has to (1) make the election within two months of the date of service of the notice to quit or, if later, the Land Court's decision and (2) within the two months give notice to the acquiring authority claiming that the remainder of the holding cannot itself or with other relevant land be farmed as a separate agricultural unit.

- Following service of a notice claiming that the remainder of the unit cannot be farmed itself or with other relevant land as a separate agricultural unit, the acquiring authority has two months in which to accept the notice as valid and if that has not been done, the parties have two months thereafter to apply to the Lands Tribunal for Scotland to determine the validity or otherwise of the notice: 1973 Act, s 57(1).

- Where a notice claiming that the remainder of the unit cannot be farmed itself or with other relevant land as a separate agricultural unit is accepted or declared valid, then the tenant requires to give up possession before the end of twelve months in order to qualify for notice of entry compensation: 1973 Act, s 57(5).

GENERAL

As indicated at page 203 ('Tenant's claims') certain claims may be open to the tenant on the termination of the tenancy. This chapter aims to consider those claims in greater detail.

MANURE, COMPOST, HAY, STRAW, OR ROOTS[1]

The 1991 Act, s 17 provides:

'Where ... notice to quit is given by the landlord or notice of intention to quit is given by the tenant, the tenant shall not, subject to any agreement to the contrary, at any time after the date of the notice, sell or remove from the holding any manure or compost, or any hay, straw or roots[2] grown in the last year of the tenancy, unless and until he has given the landlord or the incoming tenant a reasonable opportunity of agreeing to purchase them on the termination of the tenancy at their fair market value, or at such other value as is provided by the lease.'

The section permits contracting out. Contracting out is likely to be in the lease, which should therefore be consulted to see if its terms differ from the statutory provision.

The phrase 'last year' of the tenancy is difficult to construe in relation to a Whitsunday lease, which provides for the separation of the crops thereafter. In such a situation it has been suggested that 'last year' should be construed to mean the last year of the tenant's agricultural operations[3].

The tenant has to give either the landlord or the incoming tenant a reasonable opportunity to purchase them, but he need not give an opportunity to both[4]. What is a reasonable opportunity 'must depend on the circumstances of the case'[5]. In *Barber v M'Douall*, while the tenant had not communicated with the landlord, the landlord's factor had seen notices of an impending displenishing sale. On that basis it was held that the landlord had a 'reasonable opportunity' to value the machinery on sale.

While there is no obligation on the tenant to notify either party of the opportunity to purchase, it is prudent for the tenant to make a written offer to sell at the fair market value or at the valuation provided for in the lease[6]. The valuation date is the termination of the tenancy or, under a Whitsunday lease, the date of the separation of crops.

A fair market value, ie the value which the tenant could obtain by removing the items and selling in the open market, may be greater than the value provided for in the lease or greater than the value to the incoming tenant[7]. In the former case the tenant may have to consider whether it would be appropriate

---

1 1991 Act, s 17.
2 Including mangolds, swedes, turnips and cabbages, but not potatoes; *Connell* p 122. The Agricultural Holdings Act 1986, s 15(3) and (7) defines roots to mean 'the produce of any roots crop of a kind normally grown for consumption on the holding'.
3 *Gill* para 440; *Connell* p 122.
4 *Gill* para 442.
5 *Barber v M'Douall* 1914 SC 844 at 851 per Lord Mackenzie.
6 *Gill* para 442; *Connell* p 122.
7 Cf *Williamson v Stewart* 1912 SC 235.

to make the offer to the party least likely to purchase. In the latter case the landlord or incoming tenant will have to consider whether it is worthwhile purchasing the items.

Where the lease provides for takeover at a valuation, the arbitration will be a non-statutory one, even if the landlord is the party taking over[1]. Where the matter is not dealt with in the lease an arbitration on valuation between the landlord and the tenant will be a statutory arbitration under the 1991 Act, s 60. An arbitration between the tenant and the incoming tenant should be a non-statutory arbitration[2].

In any arbitration in regard to a valuation the parties should state clearly in their submission the basis upon which they contend the valuation is to be carried out. If no basis is agreed in the submission, the arbiter requires to establish the 'agreed basis' by reference to the lease (if any), by proof (eg of the custom of the district) or by making the valuation in terms of the section 'at their fair market price'[3].

WAYGOING VALUATIONS

Waygoing valuations at common law may include the right to a payment for ploughing and sowing, undersown grass, straw, dung and fallow land[4]. Bound sheep stock valuations are dealt with below.

The tenant loses his freedom to crop in the last year of the tenancy or during any period after he has received notice to quit or given notice of intention to quit[5]. In the last year he requires to husband the ground and re-establish the proper rotations in accordance with the provisions of the lease or custom of the country. This means that the tenant will have to plough and sow for the benefit of his successor in the holding.

The tenant's obligation to consume the straw on the holding re-emerges in the last year[6], unless the holding is not equipped to store the straw[7].

In general a written lease will provide for the waygoing valuation payments to which the tenant will be entitled at his waygoing. Where the terms of the lease settle the parties' waygoing rights, in general the tenant cannot claim additional compensation at common law[8].

---

1 1991 Act, s 61(7).
2 *Gill* para 443, but cf *Connell* p 122 where it is suggested that the 'point is not . . . clear'; *Roger v Hutcheson* 1922 SC (HL) 140 (it was held that there was nothing illegal under the Agricultural Acts for an outgoing and an incoming tenant to agree to a non–statutory arbitration. The case did not hold that the outgoing and incoming tenants could not arbitrate under the statutory provisions).
3 *Couper v Anstruther* 1949 SLCR 37.
4 See *Williams* chap 4 – 'Growing Crops'; D L Laird 'Waygoing Claims and Valuations' in *Aspects of Agricultural Law* (Law Society of Scotland, 1981).
5 1991 Act, s 7(5).
6 Ibid, s 7(1)(a); see *Rankine* pp 421–422.
7 *Mitchell v Adam* (1866) 1 SLCR 247.
8 *Shireff v Lovat* (1854) 17 D 177; *Allan v Thomson* (1829) 7 S 784 at 785 per Lord Pitmilly.

The lease may provide for the tenant to reap the growing crops after the termination of the tenancy or provide for them to be taken over at valuation. It may provide that items such as straw and dung are to be left 'steelbow'[1].

An irritancy clause may have the effect of depriving the tenant of his rights to the growing crops[2].

The right to waygoing valuations should be considered along with the provisions relating to prohibition from removal of manures or compost, hay, straw or roots on the termination of the tenancy without offering them for sale to the landlord or incoming tenant[3].

Where there are no written stipulations in the lease, then the common law regulates the tenant's waygoing rights. The common law may be varied or regulated by the custom of the district[4].

'The general rule of law is that he who sows a crop is entitled to reap it, but that right may be modified or taken away altogether by contract ...'[5] so at common law the tenant may reap the crop he has sown and remove it with its straw[6]. This includes the green crop (turnips and the like) which the tenant is entitled to sell or consume off the ground[7].

Where the rotation so requires, the tenant has to undersow his waygoing white crop with grass, unless the incoming tenant is allowed to come in and sow the grass. The outgoing tenant is entitled to the value of the costs of the seed and sowing[8]. The tenant may be liable in damages if the failure of the grass crop can be traced to his negligence[9].

The tenant is obliged at common law to apply the dung made on the holding to the land[10]. This does not apply to the surplus[11] for which the tenant is entitled to payment at market value[12].

The tenant is entitled to a payment if he is required to leave land fallow, which is based on the difference between the return on the fallow and the return he would have made from a crop[13]. An outgoing tenant is bound to leave the same amount fallow and duly manured as he received at ingo and is only entitled to payment for the surplus[14].

1 'Steel–bow goods: consist in corn, cattle, straw, implements of husbandry, delivered by the landlord to his tenant, by means of which the tenant is enabled to stock and labour the farm, and in consideration of which he becomes bound to return articles in equal quantity and quality at the expiration of the lease'; *Bell's Dictionary*. See *Chapman v Dept of Agriculture for Scotland* 1938 SLCR 47 for a consideration of the law relating to steelbow.

2 Cf *Moncrieff v Hay* (1842) 5 D 249 and *Chalmers' Tr v Dick's Trs* 1909 SC 761 as examples of where the lease provided that the landlord was entitled to the growing crops on bankruptcy of the tenant.

3 1991 Act, s 17.

4 *Allan v Thomson* (1829) 7 S 784; *Coupar v Anstruther* 1949 SLCR 37.

5 *Chalmer's Tr v Dick's Trs* 1909 SC 761 at 769 per Lord Low; *Edinburgh Corp v Gray* 1948 SC 538; *McKinley v Hutchison's Trs* 1935 SLT 62.

6 *Lord Elibank v Scott* (1884) 11 R 494.

7 *Rankine* p 424; cf *Cameron v Nicol* 1930 SC 1.

8 *Gordon v Hogg* 1912 SC 986; *Simson's Trs v Carnegie* (1870) 8 M 811.

9 *Rankine* p 428.

10 *Reid's Exrs v Reid* (1890) 17 R 519; *Dowall v Milne* (1874) 1 R 1180.

11 *Allan v Thompson* (1829) 7 S 784.

12 *Marshall* pp 22–23; *Herriot v Halket* (1826) 4 S 452; cf *Williams* p 32.

13 *Marshall v Walker* (1869) 7 M 833 at 834 per Lord President.

14 *Brown v College of St Andrews* (1851) 13 D 1355.

In assessing compensation to an outgoing tenant, where land has been ploughed in pursuance of an order by the Secretary of State to plough up permanent pasture[1], the value per hectare of any tenant's pasture comprised in the holding shall be taken not to exceed the average value per hectare of the whole of the tenant's pasture comprised in the holding on the termination of the lease[2]. A similar rule applies to assessing compensation, where an arbiter has given a s 9(2) direction modifying the provisions as to permanent pasture in a lease[3].

Arbitrations to settle waygoing valuations provided for in the lease are common law arbitrations[4].

SHEEP STOCK VALUATIONS

Where, under the terms of the lease, there is a bound sheep stock[5] the tenant is obliged to maintain the flock in regular ages and to return at the termination of the tenancy no less than the number he took from the landlord and for which the tenant is entitled to a payment[6].

The method of valuation is statutorily determined by the date on or before which the lease was entered into[7]. None of the amending legislation, with the exception of the initial amendments introduced in 1937[8], are retrospective in effect.

Bound sheep stock valuations are governed by the 1991 Act, ss 68–72, which

'apply where under the lease of an agricultural holding[9], the tenant is required at the termination of the tenancy to leave the stock of sheep on the holding to be taken over by the landlord or the incoming tenant at a price or valuation to be fixed by arbitration .... [10]'

The arbiter includes an oversman and any person required to determine the value or price of the sheep stock in pursuance of any provision in the lease[11].

The statutory provisions only apply where (1) the stock is bound under the lease and (2) the price is to be fixed by arbitration. They do not apply if the

1 1948 Act, s 35 and Sch 3; see p 75 'Permanent pasture'.
2 Ibid, Sch 3, para 3(1). 'Tenant's pasture' means pasture laid down at the expense of the tenant or paid for by the tenant on entering the holding; para 3(2).
3 1991 Act, s 51(1)(b); see p 75 'Permanent pasture'.
4 Ibid, s 61(7).
5 Cf *Rankine* p 434 and *Gill* para 454 for a discussion of the historical reasons, which have led to bound sheep stocks, and the problems which arose from arbiters adding to the value to cover factors such as acclimatisation and hefting etc.
6 *Luss Estates Co v Firkin Farm Co* 1985 SLT (Land Ct) 17; *Duke of Argyll v MacArthur's Trs* (1889) 17 R 135.
7 See below.
8 Sheep Stocks Valuation (Scotland) Act 1937.
9 Specifically defined by the 1991 Act, s 72(a) to mean 'a piece of land held by a tenant which is wholly or in part pastoral, and which is not let to a tenant during and in connection with his continuance in any office, appointment, or employment held under the landlord'. This definition is different to the definition in ibid, s 1(1) and means that there cannot be a bound sheep stock on an agricultural holding which is wholly arable.
10 Ibid, s 68(1).
11 Ibid, s 72(b).

lease fixes the price to be paid or if the agreement to take over the sheep stock at the termination of the tenancy is not one contained in the lease.

Any arrangement by the outgoing tenant, outwith the terms of the lease, with the landlord or the incoming tenant to take over the sheep stock at valuation has to be a common law arbitration[1].

Where the lease of an incoming tenant provides that the sheep stock shall be bound stock, the landlord and the tenant may agree that the sheep will be valued at ingo on the statutory basis, which will apply at the outgo[2].

Where any question as to the value of sheep stock has been submitted for determination to the Land Court or to the arbiter, the outgoing tenant is required, not less than 28 days before the determination of the question, to provide a statement of the sales of sheep from his stock[3]. In respect of Whitsunday waygoings the statement of sales are for the preceding three years; and for Martinmas waygoings the current year and the two preceding years[4]. The documents are open to inspection by the other party to the valuation[5].

In any lease entered into after 10 June 1937[6], the arbiter may at any stage in the proceedings, and shall if so directed by the sheriff on the application of either party, state a case for the opinion of the sheriff on any question of law arising in the course of the arbitration[7]. The sheriff's decision is final, unless either party appeals to the Court of Session from whose decision there is no appeal[8].

Where a case has been stated for the opinion of the sheriff, and if the arbiter is satisfied that whatever the decision, the sum that will ultimately be found due will not be less than a particular amount, the arbiter may make an interim order for payment of a sum not exceeding that amount[9].

The tenant has the protection of the 1991 Act, s 19(1) if the sheep stock is not paid for within one month of the date of the delivery of the valuation award[10].

(a) Leases entered into on or before 6 November 1946

The arbiter[11]:

'shall in his award show the basis of the valuation[12] of each class of stock[13] and state separately any amounts included in respect of acclimatisation or hefting or any other consideration or factor for which he had made a special allowance.[14]'

1 *Bell v Simpson* 1965 SLT (Sh Ct) 9; *Toms and Parnell* 1948 SLCR 8.
2 *Secretary of State v White* 1967 SLCR App 133.
3 1991 Act, s 71(1).
4 Ibid, s 71(1)(a) and (b).
5 Ibid, s 71(2).
6 The date of coming into force of the Sheep Stocks Valuation (Scotland) Act 1937.
7 1991 Act, s 69(1).
8 Ibid, s 69(2).
9 Ibid, s 69(3).
10 See p 221 'Payment for implements etc'.
11 See 1991 Act, s 72(b).
12 'Basis of the valuation' – cf *Williamson v Stewart* 1912 SC 235.
13 See the Hill Farming Act 1946, Sch 2.
14 1991 Act, s 68(2). This provision derives from the Sheep Stocks Valuation (Scotland) Act 1937 as amended by the 1946 Act; cf *Dunlop v Mundell* 1943 SLT 286 for a consideration of the operation of this provision.

The statutory provision is mandatory and the award requires to show the basis of the valuation[1].

Where an arbiter fails to comply with any requirement of s 68(2) his award may[2] be set aside by the sheriff[3]. The jurisdiction of the Court of Session to reduce an award of an arbiter is not ousted by s 68(4)[4].

Any other consideration could now include the availability or otherwise of sheep annual premium quota if this affects the value of the stock.

The arbiter has an unrestricted discretion in fixing the value of the stock[5], unless the lease makes provision for the basis of the valuation[6].

On a joint application by both parties the Land Court may be asked to determine the value of sheep stock[7]. Appeal lies by way of special case to the Court of Session[8].

(b) Leases entered into after 6 November 1946

The Hill Farming Act 1946 introduced a new method of valuation, which was based on average sales prices. This severely restricted the arbiter's discretion in valuing sheep stocks. The Act provided for different valuations to be applied for Whitsunday and Martinmas waygoings. The provisions of the Act gave rise to difficulties in application. Further, it led to considerable injustice in the valuation of sheep stocks[9].

In consequence there were amendments to the Act, which took effect in relation to leases entered into on or after (1) 15 May 1963[10] and (2) 1 December 1986[11].

Sheep stock valuations from post-6 November 1946 leases are now governed by the 1991 Act, s 68(3).

The 1991 Act, Sch 9 provides for valuation of sheep stocks for leases entered into before 1 December 1986. It makes provision to distinguish between leases entered into before and on or after 15 May 1963 and between Whitsunday[12] and Martinmas[13] waygoings.

Schedule 10 provides for the valuation of sheep stocks for leases entered into on or after 1 December 1986. It makes provision for valuations for Whitsunday[14] and Martinmas[15] waygoings. The Secretary of State may vary the provisions of Sch 10 by statutory instrument[16].

1 *Dunlop v Mundell* 1943 SLT 286 at 290.
2 The use of the word 'may' in s 68(3) suggests that the sheriff has a discretion; *Dunlop v Mundell* above.
3 1991 Act, s 68(4); cf *Dunlop v Mundell* 1943 SLT 286.
4 *Dunlop v Mundell* above.
5 *Duncan*, note to the 1991 Act, s 68(2) in *Current Law Statutes*.
6 *Stewart v Watter's Reps* 1955 SLCR 27.
7 1991 Act, s 70(1)(a).
8 *Gill* para 457, n 13.
9 Cf *Pott's JF v Johnstone* 1951 SLCR 22; *Garrow and anr* 1958 SLCR 13; and *Tufnell and Nether Whitehaugh Co Ltd, Applicants* 1977 SLT (Land Ct) 14.
10 Agriculture (Miscellaneous Provisions) Act 1963, s 21.
11 Law Reform (Miscellaneous Provisions) (Scotland) Act 1985, s 32; Hill Farming Act 1946 (Variation of Second Schedule) (Scotland) Order 1986, SI 1986/1823.
12 1991 Act, Sch 9, Pt I.
13 Ibid, Sch 9, Pt II.
14 Ibid, Sch 10, Pt I.
15 Ibid, Sch 10, Pt II.
16 Ibid, s 68(5).

While it is competent to make reference to similar stocks in the district under Pt I of Schs 9 and 10, it is not competent to look at similar stocks for a Martinmas waygoing valuation under Pt II[1].

Where parties seek to rely on valuations from similar stocks they are required to specify in their statement of case the stocks in the district which they regard as similar and kept under similar conditions[2].

The arbiter shall in his award state separately the particulars set forth in Pt III of Schs 9 or 10 as the case may be[3].

Where an arbiter fails to comply with any requirement of the 1991 Act, s 68(4) his award may[4] be set aside by the sheriff[5]. The jurisdiction of the Court of Session to reduce an award of an arbiter is not ousted by s 68(4)[6].

On the application of either party the sheep stock may be valued by the Land Court in lieu of the manner provided for in the lease[7]. In determining the value the Land Court is required to proceed under Schs 9 or 10 depending on whether the lease was entered into before, on or after 1 December 1986[8].

The Land Court has a particular jurisdiction under para 4 of both Schedules, where the number of ewes or lambs sold off the hill at the autumn sales during the preceding three years has been less than half the total number of ewes or lambs sold to determine prices by reference to prices realised at such sales from similar stocks kept in the same district under similar conditions[9].

Appeal will lie from the Land Court to the Court of Session by way of special case.

In both Schs 9 and 10 provision is made that the arbiter may take into account 'the profit which the purchaser may reasonably expect it [the stock] to earn'[10]. The profit that a purchaser may reasonably expect to earn may be dependent on whether or not the hypothetical purchaser is deemed to have sheep quota available for no consideration or whether it is deemed that the quota has to be bought or leased in. Similarly in both Schedules[11] tups are valued 'having regard to ... any other factor for which he thinks it proper to make an allowance', where the other factor could be the impact of sheep annual premium quota. Accordingly the introduction of sheep annual premium quota is likely to have an effect on bound sheep stock valuations.

---

1 *Macpherson v Secretary of State for Scotland* 1967 SLT (Land Ct) 9.
2 *Macpherson v Secretary of State for Scotland* above at 10.
3 1991 Act, s 68(3).
4 The use of the word 'may' in s 68(3) suggests that the sheriff has a discretion; *Dunlop v Mundell* 1943 SLT 286.
5 1991 Act, s 68(4); cf *Dunlop v Mundell* above.
6 *Dunlop v Mundell* above.
7 1991 Act, s 70(1)(b).
8 Ibid, s 70(2).
9 Ibid, Sch 9, para 4 and Sch 10, para 4. This applies only to Whitsunday waygoings.
10 See eg ibid, Sch 9, Pt I, para 5 and Pt II, para 8; and Sch 10, Pt I, para 5 and Pt II, para 8 'Profit' will be dependent on the availability of subsidies; see chap 25 – SAPS, HLCA etc.
11 See eg ibid, Sch 9, Pt I, para 6(c) and Pt II, para 9(c); and Sch 10, Pt I, para 6(c) and Pt II, para 9(c).

PAYMENT FOR IMPLEMENTS ETC SOLD ON QUITTING HOLDING[1]

The right to a payment for implements etc sold to the landlord or incoming tenant only arises under s 19(1) if there is an agreement or a term of the lease to that effect:

'Where a tenant . . . has entered into an agreement or it is a term of the lease ... that the tenant will on quitting[2] the holding sell to the landlord or to the incoming tenant any implements of husbandry, fixtures, farm produce or farm stocks on or used in connection with the holding . . . '

If there is such an agreement or term of the lease it is not competent to contract out of the provisions of s 19[3].

The lease will usually specify whether the sale is to be to the landlord or to the incoming tenant, but if it does not the tenant may choose between them.

The property in the goods does not pass to the buyer until the price is paid[4]. The price has to be paid within one month of the tenant quitting the holding or, if the price is to be ascertained by valuation, within one month of the delivery of the award[5].

Where payment of the price is not made within the one month, the outgoing tenant is entitled to sell or remove the goods and to receive from the landlord or the incoming tenant by whom the price was payable:

'compensation of an amount equal to any loss or expense unavoidably incurred by the outgoing tenant upon or in connection with any such sale or removal, together with any expenses reasonably incurred by him in the preparation of his claim for compensation.[6]'

The measure of loss and damage is specified by s 19(2), but probably includes the loss that might be incurred on a forced disposal[7]. The provision is difficult to apply in the case of bound sheep stock valuations[8].

Where the price of the goods, under s 19(1), is to be ascertained by valuation, this will be a common law valuation[9].

Any question arising as to the amount of compensation payable for loss, where the price has not been timeously paid, requires to be determined by arbitration[10]. This will be a statutory arbitration between landlord and tenant or a common law arbitration between outgoing and incoming tenant[11].

---

1 1991 Act, s 19.
2 NB 'quitting the holding' and not termination of the lease.
3 1991 Act, s 19(1) which provides 'notwithstanding everything in the agreement or lease to the contrary'.
4 Ibid, s 19(1). Cf Sale of Goods Act 1979, s 18, r 1 where property passes when the contract is made irrespective of when the price falls to be paid. This rule is overridden by the statutory provision.
5 1991 Act, s 19(1).
6 Ibid, s 19(2).
7 Cf *Williamson v Stewart* 1912 SC 235; *Keswick v Wright* 1924 SC 766.
8 *Connell* p 133.
9 1991 Act, s 61(7).
10 Ibid, s 19(3).
11 *Duncan*, note to s 19(3) in *Current Law Statutes*.

COMPENSATION FOR IMPROVEMENTS

Improvements are classified into (1) 'old' and 'new' improvements and (2) improvements (a) which require consent, (b) which require notice in writing to the landlord and (3) which require no consent or notice[1].

A tenant is not entitled to compensation for an old improvement

'carried out on land which, at the time the improvement was begun, was not a holding within the meaning of the Agricultural Holdings (Scotland) Act 1923 as originally enacted[2], or land to which provisions of that Act relating to compensation for improvements and disturbance were applied by section 33[3] of that Act.[4]'

The tenant only has a right to claim compensation at the termination of the tenancy if the statutory requirements as to consent and notice have been complied with.

In certain circumstances[5], the statutory right to compensation may have been varied by written agreement to a right to claim either no compensation or substitute compensation.

Before a claim is made for compensation for improvements on the termination of the tenancy a check will have to be made that (1) the appropriate consents or notices have been given and (2) whether or not an agreement in writing exists varying the statutory rights.

The amount of any compensation payable to the tenant 'shall be such sum as fairly represents the value of the improvement to an incoming tenant'[6]. The value is the value of the improvement to a hypothetical incoming tenant to the whole holding intending to use it agriculturally[7]. It does not matter if there is no incoming tenant.

---

1 See chap 9 and 1991 Act, Schs 3, 4 and 5.
2 1923 Act, s 49(1) – '"Holding" means any piece of land held by a tenant which is either wholly agricultural or wholly pastoral, or in part agricultural and as to the residue pastoral, or in whole or in part cultivated as a market garden, and which is not let to the tenant during his continuance in any office, appointment, or employment held under the landlord'. See *Green's Encyclopaedia of the Law of Scotland* (1926) vol 1, para 593 and cases there cited.
3 1923 Act, s 33 provides:
'(1) Where the land comprised in a lease is not a holding within the meaning of this Act by reason only of the fact that the land so comprised includes land (hereinafter referred to as "the non–statutory land") which, owing to the nature of the buildings thereon or the use to which it is put, would not, if it had been separately let, be a holding within the meaning of this Act, the provisions of this Act relating to compensation for improvements and disturbance shall, unless otherwise agreed in writing, apply to the part of the land exclusive of the non–statutory land as if that part were a separate holding.
(2) This section shall not apply in relation to a lease entered into before the first day of January, nineteen hundred and twenty one.'
4 1991 Act, s 34(3).
5 Considered in chap 9.
6 1991 Act, s 36(1). The principle of valuation embodied in the phrase 'the value . . . to an incoming tenant' has appeared in all the Agricultural Holdings Acts and in the Crofting Acts, until the principle was changed in 1961 by the Crofters (Scotland) Act 1961, s 6; viz Crofters Holdings (Scotland) Act 1886, s 10; Crofters (Scotland) Act 1955, s 14(4). Cases under the pre-1961 Crofting Acts are relevant to a consideration of the meaning of the phrase, but need to be treated with caution; see *McEwen and Low* 1986 SLCR 109 at 120.
7 Cf *MacMaster v Esson* 1921 SLCR 18; *Mackenzie v Macgillivray* 1921 SC 722 at 730 per the Lord President; *Hannan v Dalziel* 1923 SLCR 15; *Strachan's Trs v Harding* 1990 SLT (Land Ct) 6 at 8B and C.

The improvement must be suitable and appropriate to the holding as an agricultural subject[1]. It is only when the arbiter has found that the improvement is suitable and appropriate to the holding that the question of valuation arises[2]. The question of suitability refers to the class and character of the improvement and if it reasonably fulfils the requirements of the holding[2]. The test is whether the improvement is of a class that increases or enhances the value of the land to an incoming tenant[3].

Once the arbiter has held that the improvement is suitable then the question of valuation arises. The method by which the value may be ascertained is a matter for the expert, because 'it is impossible to affirm that there is only one legitimate principle by which a just result can be reached'[4]. The court has approved (1) a capitalisation of the additional rental value of the improvement[5] and (2) the cost of erecting the improvement by the incoming tenant, discounted for age, state of repair etc and excluding unsuitable features[6]. The historic cost is of little or no relevance[7].

The Land Court has said:

'If rent producing capacity is used as the method of valuation, this cannot therefore be calculated at such a level as is beyond the capacity of the holding to sustain. Likewise a value based on what it would cost an incoming landholder to make the improvement himself must be circumscribed by the limits of what an incoming tenant would regard as economic.[8]'

If the buildings are unsuitable they may have little or no value[9]. If a building is excessive or too extravagantly constructed for the holding it may be held unsuitable to the extent that it is excessive or too extravagant[10].

In calculating the measure of compensation payable:

'In a straightforward case the proper approach is for the valuer to decide whether the improvement is of a kind which suitably enhances the productive capacity of the holding, and whether the benefit which it effects fully enures to the holding considered on its own. If he is satisfied as to those matters, the value to the hypothetical tenant should be assessed at the price which such a tenant would have to pay to have the improvement erected at his entry were it not already there. From this, of course, there must be deducted a discount in respect of the age and the existing state of repair of the improvement and any obsolescence in its design or fittings.[11]'

In ascertaining the amount of compensation payable for an old improvement

1 *MacMaster v Esson* 1921 SLCR 18.
2 *Wight v Morison* 1922 SLCR 53 at 56, 1922 SLCR 91 (IH).
3 Cf *Lord Advocate v Earl of Home* (1891) 18 R 397 at 402 and 403; *Hannan v Dalziel* 1923 SLCR 15.
4 *Wight v Morison* 1922 SLCR 91 (IH) at 96 per the Lord President.
5 *Wight v Morison* above at 96 per the Lord President; *Strachan's Trs v Harding* 1990 SLT (Land Ct) 6 at 9F and K.
6 *Strachan's Trs v Harding* above at 9L and 10B; *MacEwen and Low* 1986 SLCR 109.
7 *MacEwen and Low* above at 122; cf *Frier v Earl of Haddington* (1871) 10 M 118.
8 *Strachan's Trs v Harding* 1990 SLT (Land Ct) 6 at 8B.
9 *Lord Advocate v Home* (1891) 18 R 397 (the question was whether military barracks, because of their peculiar character, enhanced the value of the land).
10 *Hannan v Dalziel* 1923 SLCR 15; *Stornoway Trs v Jamieson* 1926 SLCR 88; *MacEwen and Low* 1986 SLCR 109.
11 *Gill* para 478; approved by the Land Court in *MacEwen and Low* 1986 SLCR 109 at 120.

'there shall be taken into account any benefit which the landlord has given or allowed to the tenant (under the lease or otherwise) in consideration of the tenant carrying out the improvement.[1]'

The benefit given or the agreement under which the benefit is given does not require to be documented[2]. It may be given expressly or by implication[3]. It is for the landlord to prove that there is a causal connection between the alleged benefit and the execution of the improvement[4]. It will be for an arbiter to determine whether or not a benefit was given which was causally connected to the execution of the improvement.

There is no express provision that grants are to be taken into account[5] in valuing old improvements In a case under the Landholders Acts the court left open the question whether or not the availability of grants should be taken into account in assessing the value to an incoming tenant, where the statutory provision made no reference to grants[6]. Probably a hypothetical incoming tenant would take into account the current availability of grants.

The statutory deductions under the 1949 Act[7] for a contribution to liming under the Agriculture Act 1937 and for grant aid for drainage under the Defence Regulations have not been incorporated into the 1991 Act.

In the ascertainment of the amount of compensation payable for a new improvement

'there shall be taken into account –
(a) any benefit which the landlord has agreed in writing to give the tenant in consideration of the tenant carrying out the improvement; and
(b) any grant out of moneys provided by Parliament which has been or will be made to the tenant in respect of the improvement.[8]'

'Benefit' in the section is

'a particular benefit in consideration whereof the tenant has executed the specific improvements that are in question.[9]'

It must be a benefit given voluntarily. The benefit is usually made by way of a financial contribution or by the supply of materials It must be measurable in financial terms[10].

In determining the value of the benefit or grant;

'The arbiter should ... determine the historic cost and assess what proportion of that cost was met by the benefit or grant. He should assess the current value of the improvement, with a deduction if appropriate for depreciation, and thereafter deduct from the net figure the same proportion, and not the original amount, on account of the original benefit or grant.[10]'

---

1 1991 Act, s 36(2).
2 This is in contrast to new improvements; see below and ibid, s 36(3)(a).
3 *Earl of Galloway v McClelland* 1915 SC 1062.
4 *McQuater v Fergusson* 1911 SC 640; *Earl of Galloway v McLelland* 1915 SC 1062 and *Mackenzie v Macgillivray* 1921 SC 722.
5 See the 1991 Act, s 36(3)(b) in respect of new improvements.
6 *Strachan's Trs v Harding* 1990 SLT (Land Ct) 6 at 9F. The Crofters Holdings (Scotland) Act 1886, s 10 requires improvements to be valued at the value to an incoming tenant, but makes no reference to grants.
7 s 44(2) and (3).
8 1991 Act, s 36(3).
9 *McQuater v Fergusson* 1911 SC 640 at 644 per Lord Kinnear.
10 *Gill* para 479.

The Hill Farming Act 1946 provides:

'In assessing the amount of compensation payable, whether under the Act of [1991] or under custom or agreement, to the tenant of an agricultural holding, if it is shown to the satisfaction of the person assessing the compensation that the improvement or cultivations in respect of which compensation is claimed was or were wholly or in part the result of or incidental to work in respect of the cost of which an improvement grant has been paid or will be payable, the amount of the grant shall be taken into account as if it had been a benefit allowed to the tenant in consideration of his executing the improvement or cultivations, and the compensation shall be reduced to such extent as that person considers appropriate.[1]'

COMPENSATION FOR MILK QUOTA[2]

(a) General

A tenant's right to compensation from his landlord for milk quota[3] allocated to his holding was conferred by the Agriculture Act 1986, s 14. Compensation is assessed in terms of Sch 2.

Tenants whose leases terminated between the introduction of milk quota on 2 April 1984 and 25 September 1986 had no statutory right to claim compensation[4].

Landlord and tenant are defined by reference to the 1911 or 1991 Acts[5]. Limited owners are made absolutely liable for compensation[6]. Schedule 2 applies where the landlord is the Crown or a government department[7].

The tenant may elect to claim compensation for milk quota either under an agreement in writing or under Schedule 2[8].

In considering compensation for milk quota references to 'holding'[9] means:

'all production units operated by the single producer and located within the geographical territory of the community[10]'.

The quota holding to which the milk quota attaches is therefore all the land from which the quota holder produces milk. This may include land other than that included in the agricultural tenancy.

---

1 s 9(4).
2 1986 Act, Sch 2.
3 Milk quota is defined to mean the amount of quota registered in the tenant's name in either the direct sales or the wholesale register – 1986 Act, Sch 2, para 1(1).
4 Cf *R v MAFF ex p Bostock* [1991] 2 EGLR 1, 1994 ECJ (C–2/92) 955 (it was held that the UK government was under no obligation to introduce such a compensatory scheme).
5 1986 Act, Sch 2, para 1(1).
6 Ibid, Sch 2, para 13.
7 Ibid, Sch 2, para 15.
8 Ibid, Sch 2, para 2(4).
9 Called 'quota holding' in the section.
10 1986 Act, Sch 2, para 1; DPQRs 1994, reg 2; and EC Council Regulations 3950/92, para 9.d.

(b) Entitlement

The right to compensation for milk quota only arises on 'the termination of the lease' of the whole or a part of the tenancy[1]. Compensation is only payable 'on quitting the tenancy'[2].

Compensation can be claimed on renunciation or resumption of part of the holding or if a notice to quit is served in respect of part of the holding[3].

In order to have a claim for compensation the tenant has to qualify under at least three of four heads:

(1) the tenancy has to be of either an agricultural holding, a croft, a landholding or a statutory small tenancy under the 1911 Act[4].

(2) It is only the tenant who 'has milk quota registered as his in relation to a holding consisting of or including the tenancy' who has a right to claim compensation[5].

The effect of this provision would appear to bar a tenant who farms through the medium of a partnership or limited company from claiming compensation for milk quota, if the quota is registered in the name of the partnership or limited company as producer[6].

(3) The milk quota had to be allocated to the tenant in relation to the holding consisting of or including the tenancy[7].

Allocated quota will include quota allocated on the introduction of quota in 1984 and quota, such as hardship, special or additional milk products quota allocated thereafter.

Where the tenant was farming and producing milk through the medium of a partnership or limited company and the quota was allocated to the partnership or limited company as producer, the tenant probably has no claim for compensation. It is arguable that the restrictive provisions at (2) and (3) are contrary to EC law in that they discriminate against the actual producer in a common farming situation. Alternatively if the provisions are construed liberally along teleological lines, then it might be said that 'tenant' must be taken to include anyone producing milk under his interest.

A difficulty may arise where the milk quota was allocated to one tenant, but subsequently another tenant was added to the lease. The lease will then terminate in the names of the two tenants, but only one will have had milk quota allocated to him. Probably the one tenant who had milk quota allocated to him can claim the whole compensation, but he is under an obligation to account to the additional tenant for his share[8].

(4) Where the tenant had quota allocated to him or was in occupation of the holding on 2 April 1984 (whether or not under the lease which is terminating)

---

1 1986 Act, Sch 2, para 2(1). Termination is defined to mean 'the resumption of possession of the whole or part of the tenancy by the landlord by virtue of any enactment, rule of law or term of the lease'.
2 Ibid, Sch 2, para 2(1).
3 1991 Act, s 29.
4 1986 Act, Sch 2, para 2(1) – 'Tenancy'.
5 Ibid, Sch 2, para 2(1).
6 *Scammell and Densham* p 320, n 4; *Rodgers* p 333.
7 1986 Act, Sch 2, para 2(1)(a) – 'Allocated quota'.
8 Cf *Rodgers* p 333.

COMPENSATION FOR MILK QUOTA   227

the tenant is entitled to compensation for so much of the relevant quota as consists of transferred quota[1], the cost of which was borne wholly or partly by him[2].
The problems noted in (3) above also apply here.

A successor who has acquired right to the tenancy as an acquirer[3] or legatee[4] or lawful assignee is entitled to compensation for milk quota provided that he and his predecessor fulfil the statutory requirements[5].

Sub-tenants, provided that they are otherwise qualified[6], are entitled to obtain payment of compensation from the head tenant for milk quota on the termination of the sub-tenancy.

If the head tenant takes over the tenancy and continues in occupation he 'shall be deemed to have had the relevant quota allocated to him, and to have been in occupation of the tenancy as tenant on 2nd April 1984' for the purpose of claiming compensation for milk quota on the termination of the tenancy[7].

If the head tenant does not take up occupation of the tenancy when the sub-tenant quits he is treated as if he had quit the tenancy when the sub-tenant quitted for the purposes of claiming compensation from the landlord[8].

Provision is made for the outgoing tenant's dairy improvements and fixed equipment to be treated as those of the successor for the purposes of calculating compensation[9].

(c) The amount of compensation

The tenant's compensation is made up of the value of[10]:

(1) the tenant's fraction of so much of the allocated quota as does not exceed the standard quota[11]. Where the allocated quota is less than the standard quota, such proportion of the tenant's fraction of the allocated quota as the allocated quota bears to the standard quota[12];

(2) the amount of the excess of allocated quota over standard quota, if any[13]; and

(3) the transferred quota if the tenant has borne the whole cost of the transfer[14]. Where the tenant has borne only part of the cost of the transfer, the corresponding part of the transferred quota[15].

---

1 'Transferred quota' means milk quota transferred to the tenant by virtue of the transfer to him of the whole or part of a holding – 1986 Act, Sch 2, para 2(2). Cf chap 24 – 'Transfer of quota'.
2 Ibid, Sch 2, para 2(1)(b) – 'Transferred quota'.
3 1964 Act, s 16.
4 1991 Act, s 11; 1886 Act, s 16 (landholdings).
5 1986 Act, Sch 2, para 3.
6 Ibid, Sch 2, para 2.
7 Ibid, Sch 2, para 4(b)(i).
8 Ibid, Sch 2, para 4(b)(ii).
9 Ibid, Sch 2, para 7(3) and (5)(c).
10 Ibid, Sch 2, para 5.
11 Ibid, Sch 2, para 5(2).
12 Ibid, Sch 2, para 5(2)(c).
13 Ibid, Sch 2, para 5(2)(a)(ii).
14 Ibid, Sch 2, para 5(3)(a).
15 Ibid, Sch 2, para 5(3)(b).

The various components of this calculation are considered in (d) below.

(d) Preliminaries to calculating compensation

In order to determine the amount of payment to which the tenant is entitled in accordance with the provisions of Sch 2, para 5, certain preliminary criteria have to be established. These include:

(1) The relevant quota

The tenant may only claim compensation in respect of 'relevant quota' which pertains to or has been apportioned to the tenancy on the termination of the lease[1].

Where the tenancy and the quota holding are the same, the quota registered in relation to the quota holding is the relevant quota[2].

Where a tenancy, which forms only part of the quota holding, terminates then the milk quota requires to be apportioned between the tenancy land and the other land of the quota holding. The apportionment will have to be carried out in terms of the statutory procedure, by reference to the period before the termination of the tenancy[3]. The milk quota apportioned to the tenancy is then the relevant quota.

Apportionment probably cannot be carried out in the milk quota compensation arbitration, particularly if other parties have an interest in the apportionment.

(2) The relevant period[4]

The majority of calculations which have to be carried out to determine the amount of compensation to which the tenant is entitled have to be carried out by reference to 'the relevant period'.

In order to make the calculations regard has to be had to land quality, the climate, land use, the rent paid, the value of dairy improvements and fixed equipment and compensation, benefits allowed in respect of dairy improvements and fixed equipment during or at the end of the relevant period.

The 'relevant period' is either:

(i) the period in relation to which the allocated quota was determined. This is usually the calender year 1983 in respect of the initial allocation of quota[5]; or

(ii) where allocated quota was determined in relation to more than one period, the period in relation to which the majority was determined or if equal amounts were determined in relation to different periods, the later of the periods[6].

Parties to a lease should therefore make sure that adequate records are retained regarding the relevant information relating to the relevant period or invoke the

---

1 1986 Act, Sch 2, para 2(2)(b).
2 Ibid, Sch 2, para 2(2)(a).
3 DPQRs 1994, reg 7(2) and Sch 3, para 2(2); see p 306 'Apportionment'.
4 1986 Act, Sch 2, para 8.
5 Ibid, Sch 2, para 8(a).
6 Ibid, Sch 2, para 8(b).

procedure to have the standard quota and tenant's fraction determined before the end of the lease[1].

(3) The standard quota[2]

The importance of the standard quota lies in the fact that the higher the 'standard quota' the less the compensation that the tenant will probably be entitled to.

The standard quota is calculated by multiplying the relevant number of hectares by the standard yield[3] per hectare[4].

Where

'by virtue of the quality of the land in question or of climatic conditions in the area the amount of milk which could reasonably be expected to have been produced[5] from one hectare of the tenancy during the relevant period ('the reasonable amount') is greater than the average yield per hectare [then the standard quota is calculated] by multiplying the relevant number of hectares by such proportion of the standard yield per hectare as the reasonable amount bears to the average yield per hectare.[6]'

Paragraph 6(2) only comes into play if the arbiter determines that the quality of the land[7] or the climatic conditions in the area[8], during the relevant period, have had the effect of increasing or decreasing the milk yield. An arbiter is not bound to do a para 6(2) calculation

'... if he is satisfied, on the evidence without more, that the quality of the land and the climatic conditions are not so unorthodox that they would affect the milk yield either way.[9]'

The 'relevant number of hectares' means:

'the average number of hectares of the tenancy used during the relevant period for the feeding[10] of dairy cows[11] kept on the tenancy or, if different, the average number of

1 1986 Act, Sch 2, para 10.
2 Ibid, Sch 2, para 6.
3 The 'standard yield per hectare' is prescribed by the Secretary of State for Scotland – ibid, Sch 2, para 6(5)(b). Cf the Milk Quota (Calculation of Standard Quota) (Scotland) Order 1986, SI 1986/1475 (as amended by the Milk Quota (Calculation of Standard Quota) (Scotland) Amendment Order 1992, SI 1992/1152). The SI is regularly updated.
4 1986 Act, Sch 2, para 6(1).
5 In calculating the reasonable amount that could be produced from one hectare the arbiter should take into account the actual practice of reasonably skilled farmers in normally feeding concentrated foodstuffs to their animals; Grounds v AG for Duchy of Lancaster [1989] 1 EGLR 6.
6 1986 Act, Sch 2, para 6(2).
7 Eg particularly rich or poor land.
8 Eg exceptionally good or poor weather during the relevant period.
9 Surrey CC v Main [1992] 1 EGLR 26 at 27L.
10 The areas used for feeding dairy cows does 'not include references to land used for growing cereal crops for feeding to dairy cows in the form of loose grain' – 1986 Act, Sch 2, para 6(6)(a). As dairy cows are not fed loose grain it has been held that this includes 'grain grown on the land which is then processed and fed to the cattle' – Grounds v AG for Duchy of Lancaster [1989] 1 EGLR 6.
11 Means milking cows and calved heifers – 1986 Act, Sch 2, para 6(6)(b). Cf ibid, Sch 1, para 6(5)(b) (the English provision) where dairy cows are defined as 'cows kept for milk production (other than uncalved heifers)'. Although the two definitions probably mean the same, it can be argued that the Scottish definition, because it refers to 'milking cows', does not include dry cows.

hectares of the tenancy which could reasonably be expected to have been so used (having regard to the number of grazing animals other than dairy cows kept on the tenancy during the period).''

This provision is stated in the alternative with the effect that the second alternative has to be used unless it is the same as the first alternative. The reason for the provision is to allow for an abnormal use during the relevant period.

It should be noted that definition of 'the relevant number of hectares' is different from the meaning of 'areas used for milk production'[2], used in the apportionment of milk quota[3]. Further, in the former, the arbiter is concerned with the relevant period, whereas in an apportionment he is concerned with the period before the termination of the lease. The effect of this can be that milk quota is apportioned to part of the quota holding, but that no compensation is payable in respect of that apportionment if the area was not used for feeding milking cows and calved heifers during the relevant period[4].

Where the relevant quota includes quota awarded by the Dairy Produce Quota Tribunal for Scotland, but which has not been allocated in full, then the standard quota falls to be reduced by the amount of milk quota allocated under the award that falls short of the amount awarded, or where only part of the milk quota allocated under an award is included in the relevant quota, by the corresponding proportion of the shortfall[5].

(4) The tenant's fraction[6]

The tenant's fraction determines the amount of allocated quota in respect of which the tenant is entitled to compensation[7].

The tenant's fraction, being the 'numerator' divided by the 'denominator', is expressed as:

$$\frac{\text{Numerator}}{\text{(Numerator + rent attributable to land used for dairying)}}$$

The 'numerator' is the annual rental value at the end of the relevant period[8] of the tenant's dairy improvements and fixed equipment[9].

The tenant's dairy improvements[10] and fixed equipment[11] are those 'relevant to the feeding, accommodation or milking of dairy cows kept on the tenancy'[12]. Where an improvement or item of fixed equipment is used for more than one purpose the arbiter will have to apportion the rental value between the different uses.

---

1 1986 Act, Sch 2, para 6(5).
2 *Puncknowle Farms Ltd v Kane* [1985] 3 All ER 790.
3 DPQRs 1994, Sch 3, para 2(2).
4 Eg it was used for young dairy stock or bulls.
5 1986 Act, Sch 2, para 6(3).
6 Ibid, Sch 2, para 7.
7 Cf ibid, Sch 2, para 5(2).
8 Ibid, Sch 2, para 8 – usually 31 December 1984.
9 Ibid, Sch 2, para 7(1)(a).
10 Cf ibid, Sch 2, para 7(4)(a) – a new or old improvement defined by the 1991 Act, s 33 or a permanent improvement under the 1886 Act, s 34.
11 Cf 1986 Act, Sch 2, para 7(4)(b) – fixed equipment as defined by the 1991 Act, s 85(1).
12 1986 Act, Sch 2, para 7(2).

All of the tenant's dairy improvements and fixed equipment are taken into account, except for those for which the tenant has, before the end of the relevant period, received full compensation directly related to their value[1]. Any allowance or benefit given or compensation paid after the end of the relevant period is disregarded[2].

The annual rental value of the tenant's dairy improvements and fixed equipment is the amount of rental that would be disregarded in a rental arbitration under the 1991 Act, s 13.

The 'denominator' is the sum of the numerator plus

'such part of the rent payable by the tenant in respect of the relevant period as is attributable to the land used in that period for the feeding, accommodation or milking of dairy cows kept on the tenancy.[3]'

It should be noted that the area referred to in this provision is different from the area in 'the relevant number of hectares'. Further, as it relates to 'rent payable' it does not include the rental value of tenant's dairy improvements and fixed equipment. The rent payable is the actual rent paid and not the rental value of the land[4].

Where there is a joint use of land or buildings with non-dairy cows, then an arbiter will have to apportion rent between those uses.

Where the relevant period is less or greater than 12 months or the tenant paid rent for only part of the relevant period, then the rent payable is calculated by reference to the average rent payable in respect of one month during the period in which rent was paid multiplied by twelve[5].

(5) Valuation of milk quota[6]

Milk quota for the purpose of compensation is valued at the time of the termination of the lease[6].

The value is determined by taking into account such evidence as is available, including evidence of the price being paid for interests in land (a) where milk quota is registered in relation to land and (b) where no milk quota is so registered[7].

As milk quota has acquired a sale value of its own in most land sales, the price is separately specified in relation to the land and to the quota. This means that the later provision is difficult to apply, although the actual price paid for quota is part of 'such evidence as is available'.

The value of the quota at the termination of the tenancy may be affected by factors such as (i) its butter fat content, (ii) the quantum of unused quota at

---

1 1986 Act, Sch 2, para 7(4)(c).
2 Ibid, Sch 2, para 7(5)(a) and (b).
3 Ibid, Sch 2, para 7(1)(b).
4 *Crerar v Fearon* [1994] 2 EGLR 12, where the actual rent was unusually low, which had a distorting effect when considered in relation to the annual rental value of the dairy improvements, which were considered in relation to the figure that would be disregarded in a hypothetical rental arbitration.
5 1986 Act, Sch 2, para 7(3).
6 Ibid, Sch 2, para 9.
7 Ibid, Sch 2, para 9(a) and (b).

that date, (iii) the current open market value of quota, (iv) the current leasing value of quota, and (v) the value attributed to quota on the leasing of dairy farms.

In *Carson v Cornwall CC*[1] the arbitrator's decision that the quota should not be valued by reference to its open market value on sales by transfer, but by reference to a capitalisation of the annual leasing value of the quota over ten years was upheld. Burges R said[2]:

'From this it is clear that the compensation one is considering is not concerned with sums paid by farmers or graziers, possibly in urgent need of additional grazing or pasture, who may be prepared to pay very much "above the odds" for short term licences or grazing agreements.... The compensation is to go to those tenants who have been concerned with the building-up of milk production on the holding. ... it is clear that the arbitrator is not to be concerned with valuations which have been found by method A [open market transfer value] type valuation.... He was therefore right to disallow this method of valuation ... there is no doubt that the law is there to compensate for what is being surrendered "the building-up of the holdings" – not to give ... "unjust enrichment"'.

The argument in *Carson* was that the tenant should be compensated for the increase in value of the landlord's holding attributable to the tenant's share of the milk quota. The value was reflected in increased rental value of the holding to the landlord. Such an approach appears unfair to the tenant, in that on the termination of the tenancy the landlord could immediately sell the quota on the open market.

Suspended quota falls to be valued, if it has a value. This may depend on the likelihood of the suspension being reversed.

COMPENSATION FOR HIGH FARMING[3]

A claim for compensation for high farming is only available if a record of the holding has been made[4]. Any high farming prior to the date of the record, if made after the commencement of the tenancy, does not qualify for compensation[5].

In order to substantiate this claim the tenant requires to prove

'that the value of the holding to an incoming tenant has been increased during the tenancy by the continuous adoption of a standard of farming or a system of farming which has been more beneficial to the holding than –'

the standard or system required by the lease[6] or, in so far as the lease makes no provision, the system of farming normally practised on comparable holdings in the district[7].

1 [1993] 1 EGLR 21. The report is difficult to understand without access to the arbitrator's award and his legal advisor's opinion.
2 [1993] 1 EGLR 21 at 22K and L.
3 This is a rare claim; *Gill* para 501.
4 1991 Act, s 44(2)(b).
5 Ibid, s 44(2).
6 Ibid, s 44(1); the lease may stipulate a standard system of farming thereby depriving the tenant of a claim which would otherwise be available by reference to the normal practice on comparable holdings.
7 1991 Act, s 44(1).

The section requires 'the continuous adoption' of high farming. A tenant may be precluded by the terms of the lease from continuing to adopt that system in the last year of the lease, when his freedom of cropping stops[1]. The accepted view is that 'the continuous adoption' applies up until the period at which the system has to be stopped in order to comply at the ish with the terms of the tenancy[2].

The compensation is such 'as represents the value to an incoming tenant of the adoption of that more beneficial standard or system'[3]. Gill[4] suggests that

'the tenant's claim must be capitalised at a sum representing the additional annual rental values which the high farming would generate until its effects had been worn through.'

In assessing the compensation for high farming 'due allowance shall be made for any compensation agreed or awarded to be paid to the tenant under Part IV of this Act for any improvement which has caused or contributed to the benefit'[5].

The tenant cannot recover compensation for any improvement in respect of high farming which he would not have been entitled to recover, except for the provisions included in s 44[6].

COMPENSATION FOR DISTURBANCE

(a) Entitlement

Compensation for disturbance is additional to any other compensation to which the tenant might be entitled[7]. It is not subject to capital gains tax[8].

The right to compensation for disturbance arises where the tenancy terminated by reason of:

(1) a notice to quit given by the landlord, unless the notice to quit is given in terms of the 1991 Act, s 22(2)(a) or (c) to (f); or

(2) where the tenant gives a counternotice under the 1991 Act, s 30[9].

and in consequence the tenant quits the holding[10].

There is no right to compensation for disturbance if the tenant gives notice of intention to quit; is removed under an irritancy; or if an executor is given notice to quit under the 1964 Act, s 16[11].

---

1 1991 Act, s 7(5).
2 *Marshall* pp 101–102; *Gill* para 500.
3 1991 Act, s 44(1). See p 222 'Compensation for improvements' regarding 'value to incoming tenant'.
4 Para 501.
5 1991 Act, s 44(3).
6 Ibid, s 44(4); eg because the appropriate consent or notice had not been given.
7 Ibid, s 43(8).
8 *Davis v Powell* [1977] 1 WLR 258; *Drummond (Inspector of Taxes) v Brown* [1986] Ch 52.
9 Where the tenant gives notice that he requires a notice to quit part of the holding under s 29 to take effect as a notice to quit the whole holding.
10 1991 Act, s 43(1).
11 *Duncan*, note to ibid, s 43(2) of *Current Law Statutes*.

The right to compensation for disturbance, when the tenant gives a counternotice under s 30, is restricted if the area subject to the notice is either less than a quarter of the area or of a rental value of less than one quarter of the rental value, and the balance of the holding is reasonably capable of being farmed as a separate holding[1]. In such circumstances compensation for disturbance is not payable except in respect of the part of the holding to which the partial notice to quit relates[1].

Where the tenant has lawfully[2] sub-let the whole or part of the holding and in consequence of a notice to quit given by the landlord, becomes liable to pay the sub-tenant compensation for disturbance, then the tenant may claim compensation from the landlord notwithstanding the fact that he is not in occupation of all or part of the holding and so does not quit the holding or part thereof[3].

The tenant has to quit 'in consequence' of the notice to quit or counternotice[4]. A tenant quits 'in consequence' of the notice even if he is unsuccessful in proceedings under a counternotice under s 21(1) or arbitration proceedings under s 23(2). Where a tenant unsuccessfully challenges the validity of a notice to quit in an action of removing it can probably be said that the tenant quit in consequence of the notice to quit[5]. A protective claim must be made timeously, even if the tenant intends to challenge the removing[6].

A tenant quitting in consequence of an invalid notice to quit, which the tenant accepts, probably has a right to compensation for disturbance[7].

(b) Quantification

The amount of compensation payable

'shall be the amount of the loss or expense directly attributable to the quitting of the holding which is unavoidably incurred by the tenant upon or in connection with the sale or removal of his household goods, implements of husbandry, fixtures, farm produce or farm stock on or used in connection with the holding, and shall include any expenses reasonably incurred by him in the preparation of his claim for compensation (not being expenses of an arbitration[8] to determine any question arising under this section).'

The loss has to be 'directly attributable to the quitting of the holding' and 'unavoidably incurred'. This is a question of fact and there has to be a causal connection between the loss and the quitting[10].

---

1 1991 Act, s 43(7).
2 Ie with the landlord's consent.
3 1991 Act, s 43(6).
4 It is a question of fact for the arbiter whether or not the tenant quit 'in consequence of the notice' or for some other reason; *Johnston v Malcolm* 1923 SLT (Sh Ct) 81.
5 *Hendry v Walker* 1927 SLT 333 (it was held that on the special facts of the case that quitting was not in consequence of the notice to quit); *Preston v Norfolk CC* [1947] KB 775.
6 See p 222 'Claims where tenant disputes . . .'.
7 *Westlake v Page* [1926] 1 KB 298; *Kestell v Langmaid* [1950] 1 KB 233; *Forbes v Pratt* 1923 SLT (Sh Ct) 91; per contra *Earl of Galloway v Elliot* 1926 SLT (Sh Ct) 123 (a sheriff held that the notice to quit had to be a notice that complied with the Act).
8 The arbiter will have a discretion to deal with these expenses as appropriate.
9 1991 Act, s 43(3).
10 *Keswick v Wright* 1924 SC 766; *Macgregor v Board of Agriculture* 1925 SC 613 (the tenant's loss arose from an error by an arbiter in estimating crops and this was held to be not attributable to quitting the holding).

The loss can include the loss caused by selling (eg sheep stock) at a displenishing sale as opposed to selling the stock as a going concern[1].

The tenant is entitled to compensation 'equal to one year's rent of the holding at the rate at which rent was payable immediately before the termination of the tenancy' without proof of loss or expense[2].

With proof of loss and expense as defined in the 1991 Act, s 43(3), the tenant is entitled to claim compensation up to a sum not exceeding two years' rent[3].

The tenant is not entitled to claim a greater amount than one year's rent unless he has given the landlord not less than one month's notice of the sale of any goods, implements, fixtures, produce or stock and has afforded the landlord a reasonable opportunity[4] of making a valuation thereof[5].

There is no statutory definition of 'rent'. The statutory deductions provided for in s 43(5) are presently not applicable[6]. The rent is the net rent paid in the last year of the tenancy, whether the claim is for one or two years' rent[7]. Where there are abatements from rent allowed by the landlord, those abatements probably fall to be taken into account[8]. Rent probably includes additional payments made by the tenant, such as insurance premiums and interest on improvements[9].

ADDITIONAL OR REORGANISATION PAYMENT

An additional payment 'to assist in the reorganisation of the tenant's affairs' of four times the annual rent[10], payable immediately before the termination of the tenancy, is payable where compensation for disturbance is payable[11], unless the payment can be excluded[12].

The payment is an automatic right, unless excluded. There is no onus on the tenant to prove that he requires to reorganise or that he has incurred any loss.

Payment may be excluded if terms of the statutory exceptions given in s 55 apply[13].

---

1 *Keswick v Wright* 1924 SC 766.
2 1991 Act, s 43(4)(a).
3 Ibid, s 43(4)(b) and (c).
4 What is a 'reasonable opportunity' is a question of fact; *Dale v Hatfield Chase Corp* [1922] 2 KB 282. In *Barbour v M'Douall* 1914 SC 844 it was held that the tenant did not have to give the landlord written notice of the opportunity, provided the landlord knew about it. However, it would be good practice to give written notice of the opportunity.
5 1991 Act, s 43(4)(b).
6 Cf owner's rates abolished in 1956 and agricultural land and farm buildings exempt from rates. The situation may change.
7 *Connell* p 154; cf *Copeland v McQuaker* 1973 SLT 186.
8 *Connell* p 154.
9 *Connell* p 154 citing *Callander v Smith* (1900) 8 SLT 109; *Duke of Hamilton's Trs v Fleming* (1870) 9 M 329; *Bennie v Mack* (1832) 10 S 255; *Clark v Hume* (1902) 5 F 252; per contra *Marquis of Breadalbane v Robertson* 1914 SC 215.
10 'Rent' – see above.
11 1991 Act, s 54.
12 Ibid, s 55.
13 See p 160 'Statements in notices to quit'.

CLAIMS ON PARTIAL DISPOSSESSION

Where part of the tenancy terminates under a notice to quit[1] or the landlord resumes possession of part of the holding in terms of a provision in the lease

'the provisions of this Act with respect to compensation shall apply as if that part of the holding were a separate holding which the tenant had quitted in consequence of a notice to quit.[2]'

This preserves the tenant's right to make all the claims to which he would normally be entitled on a s 21(1) notice to quit. The right to claim compensation for milk quota relating to part of the holding is also preserved[3].

In assessing the amount of compensation payable to the tenant where the land is resumed, the arbiter is required to

'take into account any benefit or relief allowed to the tenant under the lease in respect of the land possession of which is resumed by the landlord.[4]'

With regard to compensation for disturbance[5], if the tenant has served a counternotice accepting the notice to quit as applying to the entire holding, then the right to compensation for disturbance, and in consequence the right to an additional payment[6], can be restricted to the part referred to in the notice to quit, if the area of land was less than one quarter of the area of the holding or of a rental value of less than one quarter of the rental value of the holding and the diminished holding is capable of being farmed as a separate holding[7].

Where the landlord resumes land under a provision of the lease, the tenant is entitled to compensation for the effect of early resumption[8].

The amount of compensation payable

'shall be equal to the value of the additional benefit (if any) which would have accrued to the tenant if the land had, instead of being resumed at the date of the resumption, been resumed at the expiry of 12 months from the end of the current year[9] of the tenancy.[10]'

The length of period will vary depending on when the resumption notice was served. For arable land the compensation will be the value of the crops (less deductions) that might have been taken off the land. For grazing land the compensation is probably the letting value (less deductions) of the grazings let.

---

1 1991 Act, s 29.
2 Ibid, s 49(1).
3 1986 Act, Sch 2, para 1 – 'termination means the resumption of possession of the whole or part of the tenancy'.
4 1991 Act, s 49(2).
5 Ibid, s 43.
6 Ibid, s 54(2).
7 Ibid, s 43(7).
8 Ibid, s 58(1).
9 Defined by ibid, s 58(5) – 'For the purposes of subsection (1) above, the current year of a tenancy for a term of 2 years or more is the year beginning with such day in the period of 12 months ending with a date 2 months before the resumption mentioned in that subsection as corresponds to the day on which the term would expire by the effluxion of time'.
10 Ibid, s 58(2).

It is not competent to contract out of this section[1]. The tenant is also entitled to a seek a rent reduction[2].

ENFORCEMENT

Any award or agreement as to compensation payable, expenses or otherwise on the termination of the tenancy, if not paid within one month after the date it becomes payable, can be recorded in the Books of Council and Session or the sheriff court books, and is enforceable as a recorded decree arbitral[3].

COMPULSORY ACQUISITION

Compensation claims by the owners and occupiers of agricultural land have their claims determined in accordance with the normal rules relating to compensation[4]. Additional special provision is made in respect of agricultural land.

(a) Compensation in respect of agricultural holdings

The owner/occupier will have a claim under the Lands Clauses Consolidation (Scotland) Act 1845, s 61 for the value of the land taken, disturbance and perhaps severance and injurious affection. He may be entitled to a home loss payment[5].

Both landlord and tenant are entitled to compensation when an agricultural holding, landholding or statutory small tenancy is acquired. The landlord's claim is for the value of his interest in the land[6], but he will not have a claim for disturbance, home loss or farm loss He may have a claim in respect of severance or injurious affection. In assessing the value of the landlord's interest the provisions of the Land Compensation (Scotland) Act 1973 (the 1973 Act), s 44 regarding the tenant's security of tenure have to be taken into account[7].

The tenant is compensated for the unexpired term[8]. The claim is assessed on the basis that the tenant has security of tenure, disregarding the service of or the right to serve a notice to quit under the 1991 Act, s 22(2)(b) or 24(1)(e) or any right of resumption under the lease[9].

The claim will include the value of the remaining term of the lease assessed on the basis of when the tenant might have had to vacate in any event, loss of

1 1991 Act, s 58(3).
2 Ibid, s 31.
3 Ibid, s 65; 1986 Act, Sch 2, para 12.
4 5 *Stair Memorial Encyclopaedia* para 105.
5 1973 Act, ss 27–30; 5 *Stair Memorial Encyclopaedia* para 186.
6 Lands Clauses (Consolidation) (Scotland) Act 1845, s 61.
7 See below; *Anderson v Moray DC* 1977 SLT (Lands Tr) 37.
8 Lands Clauses (Consolidation) (Scotland) Act 1845, s 114.
9 1973 Act, s 44(1).

profits, the normal claims on termination of a tenancy including improvements, waygoings, loss caused by forced sale and removal expenses[1].

Any claim under the 1973 Act, s 44 is reduced by any payment due under the 1991 Act, s 56[2].

If compensation calculated under s 44 is less than it would have been, but for that section, it is increased accordingly[2].

Special provision is made for landholdings and statutory small tenancies[3].

(b) Farm loss payment

Owner/occupiers and tenants who have to move to a new farm as a result of the compulsory acquisition are entitled to a farm loss payment[4]. To qualify the claimant (i) must have been in occupation of the agricultural unit as owner or tenant with an unexpired term of at least three years to run, or as a landholder[5] (ii) be displaced[6] from the whole of the unit and (iii) within three years of the displacement begin to farm another agricultural unit within Great Britain as owner or tenant.

A claim for a farm loss payment has to be made before the expiration of the period of one year beginning with the date on which the claimant began to farm another farm[7].

The amount of any farm loss payment 'shall be equal to the average annual profit derived from the use for agricultural purposes of the agricultural land comprised in the land acquired'[8]. Section 32 of the 1973 Act makes provision for the formula of calculating the average annual profit and the sum that might ultimately be due.

A person who serves a blight notice is not entitled to a farm loss payment[9].

A tenant entitled to a payment under the 1991 Act, s 56 is not entitled to a farm loss payment[10].

Any dispute as to the amount of a farm loss payment is to be referred to the Lands Tribunal for Scotland for determination[11].

(c) Severance of land

Where a notice to treat or a notice of entry is served in respect of agricultural land which is part of an agricultural unit or an agricultural holding, the person having an interest in that land may serve a counternotice within two months

---

1 *Anderson v Moray DC* 1978 SLT (Lands Tr) 37; *Wakerley v St Edmundsbury BC* (1978) 38 P & CR 551.
2 1973 Act, s 44(5).
3 Ibid, s 45.
4 Ibid, ss 31–33.
5 Ibid, s 31(2).
6 An occupier who leaves in advance of displacement is not entitled to the payment; J Rowan–Robinson *Compulsory Purchase and Compensation* p 277.
7 1973 Act, s 33(1).
8 Ibid, s 32(1).
9 Ibid, s 31(6).
10 Ibid, s 31(5).
11 Ibid, s 32(7).

from the date of the service of the notice to treat, claiming that the other land is not reasonably capable of being farmed by itself or in conjunction with other relevant land as a separate agricultural unit, and requiring the acquiring authority to purchase his interest in the whole land. This provision applies both to persons with a greater interest than a tenancy for a year or from year to year, and to those whose interest is that of a tenant for a year or from year to year[1].

On service of the counternotice the acquiring authority has two months in which to accept it as valid. Thereafter either the claimant or the acquiring authority may refer the counternotice to the Lands Tribunal for Scotland for a determination of whether or not the notice is valid[2]. If the counternotice is upheld the notice to treat is held to extend to the land or agricultural holding mentioned in the counternotice[3].

A counternotice by a person with an interest greater than a tenancy for a year or from year to year may be withdrawn at any time before the compensation is determined by the Lands Tribunal for Scotland or at any time up till six weeks after it is determined[4].

Where a tenant gives up possession following upon notice of entry and the acquiring authority is not then authorised to acquire the landlord's interest in that part of the holding subject to the counternotice, then that land must be surrendered to the landlord[5]. The tenancy is treated as having terminated on the date the tenant gave up possession[6]. Any claims by or against the tenant by or against the landlord arising out of or on the termination of the tenancy become rights and liabilities of the acquiring authority and any question as to payment to be made in respect thereof has to be referred to the Lands Tribunal for Scotland for determination[7].

Any increase in the value of the land not subject to the compulsory purchase which is attributable to the landlord taking possession of it falls to be deducted from the compensation payable to the landlord in respect of the acquisition of his interest in the remainder of the holding.

Where a tenancy is terminated by the tenant giving up possession and the landlord has a claim for dilapidations and deterioration under the 1991 Act, s 45, he requires to give notice to the acquiring authority within three months after the termination of the tenancy.

(d) Section 56 reorganisation payment

Section 56[8] of the 1991 Act provides that where an acquiring authority acquires the interest of a tenant in or takes possession of an agricultural holding or any part of an agricultural holding or of the holding of a statutory small

---

1 1973 Act, ss 49 and 51.
2 Ibid, ss 50(1) and 52(1).
3 Ibid, ss 50(2) and 52(2).
4 Ibid, ss 50(3).
5 Ibid, s 52(3)(b).
6 Ibid, s 52(3)(c).
7 Ibid, s 52(3)(d).
8 Introduction to ameliorate the decision in *Rugby Joint Water Board v Shaw-Fox* [1973] AC 202.

tenant, the authority will pay a reorganisation payment under s 54 of four times the rent. This payment is tax free[1].

Where the tenancy is for a term of two years or more the right to compensation under this section is excluded unless the amount of compensation which the tenant receives is less than the amount that would be paid if the tenancy was from year to year[2]. This provides for the situation that a tenant on a longer-term tenancy will be compensated for the loss of the unexpired portion of the tenancy under the normal compulsory purchase provisions.

The tenant has to be in possession or entitled to take possession of the holding. A tenant under a s 2(1) lease which has not yet taken effect as a lease from year to year is excluded[3]. This excludes a tenant claiming where there is a subtenancy.

Section 56(1) does not apply where the land is required for the purposes of agricultural research or experiment or for demonstrating agricultural methods or for the purpose of the Landholders (Scotland) Acts or where the Secretary of State acquires the land under the 1948 Act, ss 57(1)(c) or 64[4]. Where the land is acquired by virtue of the powers conferred by the Town and Country Planning (Scotland) Act 1972, ss 102 or 110, then the acquiring authority is deemed not to have acquired it for any of the excluded purposes[5].

In the event of dispute, compensation falls to be assessed by the Lands Tribunal for Scotland[6]. Provision is made for the Lands Tribunal to assess the proper rental level if the rent is 'substantially higher than the appropriate rent' and has not been fixed by arbitration or the Land Court[7].

(e) Election of notice of entry compensation in lieu of notice to quit compensation

Where a notice to quit is served upon a tenant of an agricultural holding whose interest is that of a tenant for a year or from year to year, after an acquiring authority has served a notice to treat on the landlord or, being an authority with compulsory purchase powers, has agreed to acquire the landlord's interest, then the tenant may elect to take notice of entry compensation in lieu of normal compensation on the termination of the tenancy, if it is either a 1991 Act, s 22(2)(b) notice to quit or the Land Court has consented to a s 24(1)(e) notice to quit.

An election is made by serving a notice in writing on the acquiring authority not later than the date on which possession of the holding is given up[8]. An election cannot be made or ceases to have effect if the acquiring authority takes possession under an enactment providing for taking possession of land compulsorily[9].

---

1 *Anderson v Moray DC* 1977 SLT (Lands Tr) 37 at 42.
2 1991 Act, 56(3). See *Duncan*, note to s 56(3) in *Current Law Statutes*, which points out what appears to be a drafting error.
3 1991 Act, s 57(1).
4 Ibid, s 57(2).
5 Ibid, s 57(3).
6 Ibid, Sch 8, para 1.
7 Ibid, Sch 8.
8 1973 Act, s 55(4).
9 Ibid, s 55(3).

Section 55 also applies to notices to quit part of a holding[1]. A person served with a notice to quit part of a holding cannot elect to serve a notice under s 55 and a counternotice under the 1991 Act, s 30.[2]

Provision is made to elect notice of entry compensation in respect of the whole holding, where the notice to quit relates to only part of the holding[3]. Where an election is made within two months of the service of the notice or, if later, the decision of the Land Court, then the tenant may within that period serve a notice on the acquiring authority claiming that the remainder of the holding is not capable of being farmed either by itself or in conjunction with other relevant land, as a separate agricultural unit.

Following service of a notice claiming that the remainder of the unit cannot be farmed itself or with other relevant land as a separate agricultural unit, the acquiring authority has two months in which to accept the notice as valid, and if that has not been done the parties have two months thereafter to apply to the Lands Tribunal for Scotland to determine the validity or otherwise of the notice[4]. Where the notice is accepted or declared valid, then the tenant requires to give up possession before the end of twelve months in order to qualify for notice of entry compensation[5].

Similar provisions are made in respect of the surrender of, or part of, a land-holding or statutory small tenancy[6].

1 1973 Act, s 55(5).
2 *Dawson v Norwich CC* (1979) 37 P & CR 516. The right of election applies even if the acquiring authority is the landlord.
3 1973 Act, s 57.
4 Ibid, s 57(1).
5 Ibid, s 57(5).
6 Ibid, ss 56 and 58.

CHAPTER 20

Landlord's rights and claims on termination of tenancy

Key points

- On termination of the tenancy a check should be made of the lease and any other written agreement to confirm the landlord's contractual rights on termination of the tenancy.

- In respect of any lease entered into after 31 July 1931 there has to be a record of the holding before the landlord can claim the statutory compensation for dilapidations, deterioration etc on the termination of the tenancy: ss 45(1) and 47(2)(b) and (3).

- To claim compensation under any lease entered into on or after 1 November 1948, in lieu of statutory compensation, on the termination of the tenancy there has to be a record of the holding: s 47(2)(b) and (3).

Time limits

- See chapter 18.

GENERAL

As indicated at p 204 ('Landlord's rights and claims'), on the termination of the tenancy the landlord may have certain rights and claims against the tenant. The aim of this chapter is to consider those rights and claims more fully.

This chapter should be read in conjunction with chapter 19, because many of the landlord's rights and claims are dependent upon the tenant's rights and claims and any modification thereof made in the lease or by written agreement.

CONTRACTUAL RIGHTS

On the termination of the lease it is important to check on the landlord's contractual rights arising out of the lease or other written agreements.

The contractual rights which are commonly found in leases and other written agreements include:

(1) contractual provisions regarding dung and straw, which may provide for those items being left steelbow[1];

(2) contractual provisions regarding waygoing crops and valuations[2];

(3) contractual provisions for taking over bound sheep stock[3];

(4) contractual provisions regarding quotas;

(5) contractual obligations regarding the purchase of implements of husbandry, fixtures, farm produce or farm stocks[4];

(6) agreements regarding compensation for improvements[5];

(7) contractual rights regarding deterioration of holding[6].

STATUTORY RIGHTS AND CLAIMS

The 1991 Act may give the landlord rights and claims on the termination the tenancy. These rights and claims include:

(a) Right to purchase tenant's fixture and buildings

Section 18 gives the tenant the right to remove certain fixtures and buildings (other than those in respect of which the tenant is entitled to compensation)[7].

---

1 See p 214 and the 1991 Act, s 17.
2 See p 215 and ibid, s 19(1).
3 See p 217.
4 See p 221 and the 1991 Act, s 19(1).
5 See p 222.
6 See p 244
7 1991 Act, s 18(1).

The tenant's right is only exercisable if the tenant has paid all the rent owing by him and has performed or satisfied all his other obligations to the landlord in respect of the holding, and the tenant gives the landlord one month's notice of his intention to remove the fixture or building[1].

If before the expiry of the notice the landlord gives the tenant a written counternotice electing to purchase the fixture or building, the landlord acquires a right to purchase[2].

(b) Statutory compensation for deterioration etc of holding

In respect of a lease entered into after 31 July 1931, the landlord only has a right to claim on the termination of the tenancy for dilapidations or deterioration under the Act[3] if a record of the holding[4] was made during the occupancy of the tenant[5]. There is no claim in respect of any matter arising before the date of the record[6] or first record if more than one[7]. The landlord does not lose his right to claim compensation the 1991 Act, s 45 by reason of the fact that the tenant has remained in occupation under two or more tenancies[8].

The time limits for making claims should be observed[9].

Section 45(1) provides:

'(1) The landlord ... shall be entitled to recover from the tenant, on his quitting the holding[10] on termination of the tenancy compensation-
(a) where the landlord shows that the value of the holding has been reduced by dilapidation[11], deterioration[12] or damage caused by;
(b) where dilapidation, deterioration or damage has been caused to any part of the holding or to anything in or on the holding by;
non-fulfilment by the tenant of his responsibilities to farm in accordance with the rules of good husbandry.'

It should be noted that the claim arises where the dilapidation, deterioration or damage is caused by a failure to farm in accordance with the rules of good husbandry[13]. Section 45(3) preserves the landlord's right, in the alternative, to

---

1 1991 Act, s 18(2).
2 Ibid, s 18(3).
3 Ibid, s 45(1).
4 Under ibid, s 8.
5 Ibid, s 47(3)(a). The parties may agree in writing to adopt a record made during a previous tenancy - ibid, s 47(4).
6 Ibid, s 47(3)(b).
7 Ibid, s 47(3)(c).
8 Ibid, s 47(5).
9 See p 206 'Time limits for claims'.
10 A tenant who remains in unlawful occupation after the termination of the tenancy cannot prevent a landlord recovering under this section; *Gulliver v Catt* [1952] 2 QB 308; *Kent v Conniff* [1953] 1 QB 361.
11 Generally 'items such as failure to repair gates, fences, drains, buildings, roads, ditches etc'; *Scammell and Densham* p 341.
12 Includes 'the kind of damage which could not readily be put right for example where the tenant has allowed a serious loss of fertility to occur . . . or where there is an infestation of wild oats'; *Scammell and Densham* p 341, where it is suggested that there might be a claim against a tenant who ceased milk production before the introduction of milk quota, where the lease provided that the farm was let as a dairy farm.
13 See the 1948 Act, Sch 6.

claim under the lease in respect of dilapidation, deterioration or damage caused to any part of the holding or to anything in or on the holding[1].

'Dilapidation' and 'deterioration' are not defined. Dilapidations will generally include items such as 'failure to repair gates, fences, drains, buildings, roads, ditches etc in breach of a term of the tenancy'[2].

Deterioration claims will refer to damage which cannot readily be put right such as serious loss of soil fertility, infestations of weeds etc, which will require considerable time and money to put right[2].

The claims for dilapidations and deteriorations are not exclusive, in that the extent of dilapidations may amount to a general deterioration of the holding[3].

It has been suggested that a landlord might have a deterioration claim for the value of milk quota against a tenant who ceased milk production prior to 2 April 1984 without his landlord's consent, under a lease that provided for the farm being let as a dairy farm[4]. The same principles would apply where a tenant by inefficient farming at the introduction of milk quota had less milk quota allocated to the holding than would have been allocated had the tenant been farming with the competence of an ordinarily competent tenant.

The amount of compensation payable is set in the alternative[5], but is not cumulative[6]. A claim under s 45(1)(a) is therefore an additional claim for the general reduction in value of the holding.

Where the claim is under s 45(1)(a), compensation is an amount equal to the reduction in the value of the holding, in so far as the landlord is not receiving compensation under s 45(1)(b)[7]. *Connell*[8] suggests that the value of the claim is the expenditure required to bring the land back to a proper condition and the loss incurred during the period when the land is not earning a proper rent. *Scammell and Densham* suggests the value of the claim will be an amount to compensate the landlord for loss of rental value during the time that it will take for the defects to be remedied[9]. Where there is serious deterioration the landlord has the option of either keeping the land in hand until the deterioration has been restored or of immediately reletting the land at a lower rent, where any restoration of deterioration will be a tenant's improvement compensatable on the termination of the tenancy. The value of the claim will probably be whichever course minimises the loss.

Where the claim is under s 45(1)(b), compensation is the cost, as at the date of the tenant's quitting the holding, of making good the dilapidation, deterioration or damage[10].

---

1 See section (c) below.
2 *Scammell and Densham* p 341.
3 *Evans v Jones* [1955] 2 QB 58.
4 *Scammell and Densham* p 341. Presumably this would also apply to other quotas introduced in the future, which are linked to the land.
5 1991 Act, s 45(2).
6 Ibid, s 45(2)(a).
7 Ibid, s 45(1)(a); cf *Evans v Jones* [1955] 2 QB 58.
8 Page 172.
9 Page 343.
10 1991 Act, s 45(2)(b).

(c) Compensation under the lease

The landlord may claim compensation under a written lease in lieu of compensation under s 45(1)(b)[1] on the tenant quitting the holding on termination of the lease[2].

The landlord cannot claim under both the lease and statute[3]. The landlord may intimate the claim in the alternative[4], provided one of the heads of claim is later abandoned[5]. If a claim is made in the alternative it is important that the notice does not intermix the claims, but makes it clear that the claims are alternative, otherwise the notice may be void for uncertainty[6].

The concluding words of s 45(3) make it clear that a claim under the lease is to be treated as a claim under s 45(2)(b) for the purposes of set off against a claim for general deterioration under s 45(1)(a).

If the landlord wishes to exercise his rights to claim compensation under the lease in lieu of the statutory compensation under s 45(1)(b), then in respect of any lease entered into on or after 1 November 1948, there has to be a record of the holding[7].

Section 62 of the 1991 Act relating to time limits and notice in writing applies to claims under the lease.

(d) Claim arising from damage caused by the freedom to crop

Where a tenant exercises his freedom of cropping and disposal of produce rights[8], the 1991 Act, s 7(3) provides that:

'the landlord shall have the following remedies, but no other-
(a) . . .
(b) in any case, on the tenant quitting the holding[9] on the termination of the tenancy the landlord shall be entitled to recover damages for any injury to or deterioration of the holding attributable to the exercise by the tenant of his rights . . .'

It is significant that the section uses the words 'recover damages' rather than compensation. *Gill* suggests that it is an open question whether an arbiter could award damages under this section[10], but s 62 provides that any claim 'which arises under this Act . . . on or out of the termination of the tenancy . . . shall . . . be determined by arbitration'[11].

---

1 1991 Act, s 45(3).
2 Ibid, s 45(3)(a).
3 Ibid, s 45(3)(b).
4 Stating a claim in the alternative is useful in circumstances, eg, (1) where the respective values of the claims still have to be calculated or (2) if there is a dispute as to the validity of the lease, so as to preserve the statutory claim if the lease is held not binding (see *Boyd v Wilton* [1957] 2 QB 277).
5 *Boyd v Wilton* above (one of the claims was abandoned during the arbitration).
6 Cf *Boyd v Wilton* above.
7 1991 Act, s 47(2)(b) and (3).
8 Ibid, s 7.
9 A tenant who remains in unlawful occupation after the termination of the tenancy cannot prevent a landlord recovering under this section; *Gulliver v Catt* [1952] 2 QB 308; *Kent v Conniff* [1953] 1 QB 361.
10 Para 161. Cf chap 22.
11 See *Hill v Wildfowl Trust (Holdings) Ltd* 1995 SCLR 778.

CHAPTER 21

Market gardens

Key points

- A holding or part of a holding can only be a market garden (1) if it was in use, on a current lease, with the knowledge of the landlord as a market garden pre-1 January 1898 and a Sch 6 improvement has been carried out after that date without objection in writing by the landlord; or on or after 1 January 1898 (2) there is an agreement in writing that it is let or is to be treated as a market garden or (3) there is a Land Court direction to that effect: s 40(1) and 41.

- Where the Land Court has made a s 40 direction, the tenant's right to claim compensation for market garden improvements on the termination of the tenancy by notice of intention to quit or by reason of the tenant's apparent bankruptcy, is subject to the Evesham custom[1]: s 41(3).

- Where an Evesham custom offer is accepted the tenancy accepted by the incoming tenant is deemed not to be a new tenancy: s 41(7).

- The right to remove fixtures and buildings conferred by s 18 applies to all fixtures and buildings erected by the tenant on the holding or acquired by him since 31 December 1900 for the purpose of trade or business as a market gardener: s 40(4)(a).

- A market garden tenant has the right to remove all fruit trees and fruit bushes planted by him which were not set out permanently before the termination of the tenancy, but if this right is not exercised the fruit trees and fruit bushes become the property of the landlord without compensation: s 40(4)(b).

Time limits

- Where the Evesham custom applies and the tenant has given notice of intention to quit or the tenancy is terminated by reason of the tenant's bankruptcy, the tenant requires to produce an offer complying with

---

1 See the 1991 Act, s 41(3) and (4). Where the Land Court directs that the holding or part of it is to be treated as a market garden, the tenant is not entitled to compensation for market garden improvements on the termination of the tenancy by notice of intention to quit or by bankruptcy of the tenant, unless the tenant first finds a suitable person who offers to take on the tenancy and the landlord declines the offer.

s 41(3) and (4), within one month after the date of the notice to quit or constitution of insolvency or such later date as may be agreed: s 41(4)(c).

- The landlord requires to accept the Evesham custom offer within three months after the production of the offer: s 41(3).

GENERAL

In certain circumstances the whole or part of an agricultural holding may be a market garden. If the holding or part of it is a market garden, the tenant has additional rights in regard to compensation and the removal of fixtures, buildings and fruit trees and bushes.

A tenant's right to freedom of cropping does not apply in respect of a holding or a part of a holding which is a market garden[1].

In market garden tenancies, but probably not where only part of the holding is a market garden, the landlord retains his common law right of hypothec[2].

DEFINITION OF MARKET GARDEN

The 1991 Act, s 85(1) defines 'market garden' to mean:

'a holding, cultivated, wholly or mainly, for the purpose of trade or business of market gardening.'[3]

It is a prerequisite that the holding is an agricultural holding within the meaning of the 1991 Act, s 1. Where the market garden produce is sold occasionally from a garden used mainly for domestic purposes, then the holding will not be treated as a market garden[4].

In *Watters v Hunter*[5] the tenant used the subjects 'as an experimental bulb growing establishment' and claimed they were a market garden. The Lord President said of the claim[6]:

'Bulb growing is extensively carried on in Holland; but I confess it never occurred to me to connect that industry with market gardening as understood in this country. I think the terms "market garden" and "market gardening" as used in the Act of 1923, must be interpreted according to their ordinary meaning in popular language. The trade or business of market gardener is, in my opinion, the trade or business which produces the class of goods characteristic of a greengrocer's shop, and which in ordinary course reaches that shop via the early morning market where such goods are disposed of wholesale. It is no doubt the case that this class of goods includes small fruit, and it may be, flowers. But this does not warrant the extension of "market gardening" so as to cover the highly specialised industry in which the defender is engaged. Marginal cases such as that which was considered in *Grewar v Moncur's Curator Bonis*[7] will always occur in applying such a statute as the Act of 1923. But I see no reason to regard a holding devoted to rearing bulbs as a "market garden"; and I decline to interpret the term by

---

1 *Taylor v Steel-Maitland* 1913 SC 562 at 571 per the Lord President.
2 *Rankine* p 372; *Clark v Keir* (1888) 15 R 458. The Hypothec Abolition (Scotland) Act 1830 applies only to land 'let for agriculture or pasture'.
3 The same definition was in the Agricultural Holdings Act 1923. The definition was not continued in the 1947, 1948 or 1986 Acts. Cf *Hood Barrs v Howard* (1967) 201 EG 768 per Dankwerts J - 'It looked as though the legislature had given up in despair the attempt to define the term'. English cases considering the definition of 'market garden' after 1947 fall to be considered against the background that in England there was no longer a statutory definition.
4 *Bickerdike v Lucy* [1920] 1 KB 707 at 711; *In re Wallis v ex p Sully* (1885) 14 QB 950.
5 1927 SC 310.
6 1927 SC 310 at 317.
7 1916 SC 764.

considering *in abstracto* as it were, whether such a holding might come under the general description of a "garden" and if so, whether, in respect that the bulbs are reared for sale, it might (in that sense) be called a "market garden".[1]

Where a fruit orchard was let with the produce being grown and sold under the trees, the holding was held to be a market garden[2].

Whether or not the definition of market garden extends to a holding growing wholly or mainly flowers remains undecided, although there are *obiter dicta* that it does[3].

CONSTITUTION AS A MARKET GARDEN

A holding or part of a holding may have been or may be constituted as a market garden in three ways.

(a) Pre-1 January 1898

With effect from 1 January 1898 the tenants of market gardens were given special provisions in regard to compensation for improvements[4]. These rights are recognised by the 1991 Act[5]. In order to qualify as a pre-1898 market garden the holding has to qualify in three cumulative respects:

'(a) [the] holding was, on January 1, 1898 under a lease then current, in use or cultivation as a market garden with the knowledge of the landlord; and
(b) an improvement of a kind specified in Schedule 6 to this Act (other than such an alteration of a building as did not constitute an enlargement thereof) has been carried out on the holding[6]; and
(c) the landlord did not, before the improvement was carried out, serve on the tenant a written notice dissenting from the carrying out of the improvement.'[7]

Under the 1949 Act[8] it had been held that the use of the word 'then' in s 65(2) meant 'thereafter' and that the improvement had to have been carried out after 1 January 1898[9]. While the changed wording in the 1991 Act, s 40(2)(b) would appear to be habile to include improvements carried out both before and after 1 January 1898, as a consolidating Act it is to be presumed that there was no intention to change the law.

The retention of the provision in relation to market gardens dating from 1898 is probably of little present importance[10].

1 See *Twygen Ltd v Assessor for Tayside Region* 1991 GWD 4-226 - 'In ordinary language "a market garden" supplied a market for buying and selling produce for consumption'.
2 *Lowther v Clifford* [1927] 1 KB 130.
3 See *Watters v Hunter* 1927 SC 310; *Grewar v Moncur's CB* 1916 SC 764 at 768 per Lord Salveson; *Drummond v Thomson* (1921) 37 Sh Ct Rep 180; cf *Short v Greaves* [1988] 1 EGLR 1.
4 Market Gardeners Compensation (Scotland) Act 1897.
5 s 40(2).
6 1949 Act, s 65(2) provided 'the tenant thereof had then carried out . . . an improvement'.
7 1991 Act, s 40(2)(a)–(c).
8 1949 Act, s 65(2).
9 *Smith v Callander* (1901) 3 F (HL) 28; *Taylor v Steel-Maitland* 1913 SC 562 at 570 per the Lord President.
10 *Gill* para 548.

(b) Constitution by agreement in writing

The market garden provision applies to 'any agricultural holding which, by virtue of an agreement in writing made on or after January 1, 1898, is let or is to be treated as a market garden'[1].

The agreement must be in writing. It will probably appear in the lease, but not necessarily[2].

The agreement can provide for the market garden part of the holding to be rotated through different parts of the land[3]. Where part of the holding is treated as a market garden, then, for the purposes of s 40, it is treated as a separate holding[4].

(c) Constitution by direction of Land Court

Where the tenant intimates to the landlord in writing his desire to carry out on the holding or part of the holding any of the Sch 6[5] improvements, and the landlord refuses or fails within a reasonable time[6] to agree in writing that the holding or part thereof shall be treated as a market garden the tenant may apply to the Land Court for a direction that s 40 shall apply to the holding or part thereof[7].

If the tenant satisfies[8] the Land Court that the holding or part thereof is suitable for the purposes of market gardening[9], the Land Court may direct that s 40 shall apply to the holding or part of it[10]. The direction may be in respect of all Sch 6 improvements or in respect of only some of those improvements[11].

Section 40 then applies in respect of any authorised improvement carried out after the date on which the direction is given[12].

In making a direction the Land Court may impose 'such conditions, if any, for the protection of the landlord as the Land Court may think fit'[13]. Connell[14] suggests that the only or main way of protecting the landlord might be the requirement to pay an increased rent.

Where the direction applies only to part of the holding, the direction may, on the application of the landlord, be given subject to the condition that the tenant shall consent to the division of the holding into two parts, one part being the part to which the direction applies[15]. The landlord would be wise to

---

1 1991 Act, s 40(1).
2 *Saunders-Jacobs v Yates* [1933] 1 KB 392 - 'Let as a market garden' means let for the purpose of a market garden. *Morse v Dixon* (1917) 87 LJKB 1 (a term in the lease permitting a tenant who planted fruit trees to remove them was held not to be an agreement within the subsection).
3 *Taylor v Steel-Maitland* 1912 SC 562 (Lord Johnston dissenting).
4 1991 Act, s 40(6).
5 Sch 6 of the 1991 Act specifies market garden improvements.
6 'Reasonable time' will be a question of fact for arbitration.
7 1991 Act, s 41(1).
8 The onus is on the tenant.
9 1991 Act, s 41(1)(d).
10 Ibid, s 41(1).
11 Ibid, s 41(1)(i) and (ii).
12 Ibid, s 41(1).
13 Ibid, s 41(2).
14 Page 182.
15 1991 Act, s 41(2).

seek such a direction as it would revive the common law right of hypothec in respect of the market garden holding[1].

The divided holdings are to be held at rents agreed by the landlord and tenant or in default of agreement determined by arbitration, but otherwise on the same terms and conditions (so far as applicable) as those on which the holding is held[2].

Where the Land Court has given a s 40 direction, then the tenant's right to compensation for the improvements specified in the direction is subject to the Evesham custom[3].

MARKET GARDEN IMPROVEMENTS

Particular market garden improvements are provided for in Sch 6. They are:

(1) Planting of fruit trees or bushes permanently set out[4];

(2) Planting of strawberry plants;

(3) Planting of asparagus, rhubarb, or other vegetable crops which continue productive for two or more years;

(4) Erection, alteration or enlargement of buildings for the purpose of the trade or business of a market gardener.

The Secretary of State, after consultation, may vary the provisions of Sch 6 by statutory instrument[5]. Such a variation does not affect accrued rights to compensation[6].

RIGHTS ON TERMINATION OF TENANCY

(a) Tenant's right to remove fixtures and fittings

The tenant of a market garden has the right, under the 1991 Act, s 18 to remove every fixture or building affixed or erected by the tenant to or upon the holding or acquired by him since 31 December 1900 for the purposes of his trade or business as a market gardener[7]. An ordinary tenant is only entitled to remove buildings for which he is not entitled to compensation.

Under s 18 the tenant's right to remove the fixtures and buildings is dependent upon (1) the tenant having paid all rent owing by him and having performed or satisfied all his other obligations to the landlord and (2) the tenant

---

1 *Rankine* p 372; *Clark v Keir* (1888) 15 R 458. The Hypothec Abolition (Scotland) Act 1830 applies only to land 'let for agriculture or pasture'.
2 1991 Act, s 41(2).
3 Ibid, s 41(3).
4 'Permanently set out' means 'when having ceased to be a nursery plant, it is planted out in what is intended to be a permanent position for cropping'; *Marshall* p 235.
5 1991 Act, s 73(1) and (4).
6 Ibid, s 73(3).
7 Ibid, s 40(4)(a).

having given the landlord an opportunity to purchase the fixtures and buildings[1].

(b) Tenant's right to remove fruit trees and bushes

The tenant has the right to remove all fruit trees and bushes planted by him on the holding and not permanently set out. If the tenant does not remove such fruit trees and bushes before the termination of the tenancy, they remain the property of the landlord. The tenant is then not entitled to compensation for them[2].

(c) Tenant's right to compensation for improvements

A market garden tenant has the ordinary tenant's right to compensation for improvements, but in addition has a right to claim compensation for Sch 6 market garden improvements, subject to the Evesham custom[3]. As Sch 6 improvements are included as Part 3 improvements[3] it means that neither the landlord's consent nor notice to the landlord is required to effect and claim compensation for an improvement.

Where the market garden tenancy is a pre-1 January 1898 tenancy from year to year, then the compensation payable in respect of a Sch 6 improvement (other than alteration of a building which did not constitute enlargement) is such as could be claimed if the 1949 Act had not been passed[4].

Compensation is assessed in terms of the 1991 Act, s 36, namely the value of the improvement to the incoming tenant less the statutory items that have to be taken into account[5].

The parties may agree in writing for substitute compensation which is fair and reasonable, having regard to the circumstances existing at the time of making the agreement for an improvement for which compensation is payable by virtue of s 40[6].

Where the landlord and tenant have agreed that the holding shall be let or treated as a market garden, they may by written agreement, agree to substitute compensation on the basis of the Evesham custom for compensation due under the Act[7].

An incoming tenant who purchased the whole or part of an improvement from the outgoing tenant may claim compensation in respect thereof, although the landlord has not consented in writing[8].

The tenant's right to compensation for Sch 6, para 4 improvements relating to an alteration of a building (not being an alteration constituting an

1 1991 Act, s 18(2) and (3).
2 Ibid, s 40(4)(b).
3 Ibid, s 40(3).
4 Ibid, s 40(5).
5 See p 222 'Compensation for improvements'.
6 1991 Act, s 42(1).
7 Ibid, s 42(2).
8 Ibid, s 40(4)(c). This is in contrast to an ordinary tenant who may only make such a claim if the landlord consented in writing to the incoming tenant paying compensation to the outgoing tenant; ibid, s 35(4).

enlargement of the building) is restricted to alterations begun on or after 1
November 1948[1].

(d) The Evesham custom

Where the Land Court directs that the holding or part of it is to be treated as
a market garden, then the Evesham custom applies to claims for compensation
where the tenancy is terminated by notice of intention to quit or by reason of
the tenant's apparent insolvency constituted under the Bankruptcy (Scotland)
Act 1985, s 7[2].

In such circumstances the tenant cannot claim compensation unless he produces an offer (1) in writing; (2) made by a substantial and otherwise suitable
person[3]; (3) which is produced by the tenant to the landlord not later than one
month after the date of the notice to quit or constitution of apparent insolvency or such later date as may be agreed; (4) which is an offer to accept a tenancy of the holding from the termination of the existing tenancy on the terms
and conditions of the existing tenancy so far as applicable; (5) which includes
an offer to pay the outgoing tenant all compensation payable under the Act or
the lease; and (6) which is open for acceptance for three months from the date
it is produced[4].

If the landlord accepts an offer complying with s 41(4) the incoming tenant
is required to pay to the landlord on demand all sums payable to him by the
outgoing tenant on the termination of the tenancy in respect of rent or breach
of contract or otherwise in respect of the holding[5]. Any amount paid by the
incoming tenant in respect of such a demand may, subject to any agreement
between the incoming and outgoing tenant, be deducted by the incoming tenant from any compensation payable by him to the outgoing tenant[6].

A tenancy created by the acceptance of such an offer is deemed for the purpose of the 1991 Act, s 13 not to be a new tenancy. This is important, in particular, for rent review periods.

If the landlord does not accept the offer then the tenant is entitled to compensation in terms of the 1991 Act.

1 1991 Act, s 40(7).
2 Ibid, s 41(3).
3 *Connell* pp 181 and 182 – a person of sufficient capital and otherwise capable from experience
  and character of undertaking the obligations of the tenancy.
4 1991 Act, s 41(4).
5 Ibid, s 41(5).
6 Ibid, s 41(6).

CHAPTER 22

Small landholdings and statutory small tenancies

Key points

- Outwith the crofting counties, where a holding is named '.... Smallholding' or 'Croft' or some similar name, or where the holding is less than 50 acres, practitioners should be alert to the possibility that the holding is subject to the Small Landholders (Scotland) Act 1911 (the 1911 Act).

- The distinction between a landholding and a statutory smallholding should be noted: 1911 Act, s 2.

Time limits

- The landlord or landholder has to intimate a renunciation of a tenancy to the Secretary of State within two months of the notice: Crofters Holdings (Scotland) Act 1886 (the 1886 Act), s 7.

- A legatee has to give notice to the landlord within two months that he has been bequeathed the tenancy: 1886 Act, s 16(a).

- Where there is no bequest of the holdings, the time limits in which the executor has to act are the same as those applicable to an agricultural holding: chapter 13.

GENERAL

The Small Landholders (Scotland) Act 1911 extended the Crofters Holdings (Scotland) Act 1886 to the whole of Scotland. The Crofters (Scotland) Act 1955[1] excluded the crofting counties[2] from the provisions of the 1911 Act. The 1911 Act established, and distinguishes, between a landholder and a statutory small tenant[3].

There are still a surprising number of small landholdings and some statutory smallholdings outwith the crofting counties governed by the Small Landholders Acts[4] in Scotland. A large number of landholdings were created under the Land Settlement (Scotland) Act 1919 for returning servicemen, where the landlord was the Secretary of State for Scotland. Most of these holdings were sold off in the late 1970s early 1980s to the sitting tenants.

This chapter aims to outline the legislation relating to small landholdings and statutory smallholdings, so that the practitioner may identify these from an agricultural holding and proceed accordingly. In one chapter it is not possible to examine in any great detail this technical and difficult area of law, which has much in common with the crofting legislation.

If the holding under consideration is called 'smallholding', 'croft' or some similar name, or is less than 50 acres in extent, particularly if it is one of a group of holdings, practitioners should be alert to the possibility that it might be a holding governed by the 1911 Act.

THE LANDHOLDING

Unlike the Crofting Acts, which confer the status of a croft on the holding, the Landholders Acts conferred the status of a landholder on the tenant of a qualifying holding at 1 April 1912, under the Acts[5].

Outwith the crofting counties, the status of landholder could only be acquired on 1 April 1912 at the commencement of the 1911 Act or by registration as a landholder by the Land Court thereafter[6]. A successor of a landholder, being his heir or legatee, is a landholder[7].

A 'landholder' is defined to include (1) every existing yearly tenant, (2) every qualified leaseholder, (3) every new holder and (4) the successors of each and every such person in the holding being his heirs or legatees[8].

---

1 Now the Crofters (Scotland) Act 1993.
2 The former counties of Argyll, Caithness, Inverness, Orkney, Ross and Cromarty, Sutherland and Zetland – Crofters (Scotland) Act 1993, s 61.
3 See below.
4 The Acts are: Crofters Holdings (Scotland) Acts 1886 and 1887; Crofters Common Grazings Regulation Act 1891; Congested Districts (Scotland) Act 1897; Crofters Common Grazings Regulation Act 1908; Small Landholders (Scotland) Act 1911; Small Holdings Colonies (Amendment) (Scotland) Act 1918; Land Settlement (Scotland) Act 1919. The Small Landholders and Agricultural Holdings (Scotland) Act 1931 amended the Landholders Acts.
5 See in general 1 *Stair Memorial Encyclopaedia* paras 836-860; *Paton and Cameron* chap XIX.
6 1911 Act, s 7.
7 Ibid, s 2(2). Post-1964 'heir' will have to be construed in terms of the 1964 Act, s 16 to include a transferee; see the 1911 Act, s 31(1).
8 Ibid, s 2(2).

It is still competent for the landlord and the tenant to agree that a tenant should be registered by the Land Court as a landholder under the Acts, where the holding is otherwise qualified for registration[1].

Once a tenant acquired the status of landholder he came under the protection of the Acts. The holding in the future could, in general, only be let to another landholder under the Acts[2].

A sub-tenant could not become a landholder[3] and land subject to a sub-tenancy at 1 April 1912 could not become a holding or part of a holding[4].

A 'holding' is defined[5] to mean and include:

'(i) [applicable only to the crofting counties]
(ii) ... every holding which at the commencement of this Act is held by a tenant from year to year who resides[6] on or within two miles[7] from the holding and by himself or his family cultivates the holding with or without hired labour ...
(iii) As from the termination of the lease ... every holding which at the commencement of this Act is held under a lease for a term longer than one year by a tenant who resides[8] on or within two miles[9] from the holding and by himself or his family cultivates the holding with or without hired labour ...[10]'

A holding is deemed to include any right in pasture or grazing land held by the tenant or landholder, whether alone or in common with others[11], and any dwellinghouse erected or to be erected on the holding or held or to be held therewith and any offices or other conveniences connected with the dwellinghouse[12].

A person could not be admitted as a new holder in respect of land belonging to more than one landlord or in respect of more than one holding, unless such land or holdings had been worked as one holding[13]. Joint tenants were excluded, except in the case of existing joint tenants[14].

If the leaseholder or his predecessor in the same family[15] had provided or paid for the whole or the greater part of the buildings or other permanent improvements on the holding without receiving from the landlord fair consideration, he was called a qualified leaseholder or 'landholder'[16]. In every other case the tenant was a statutory small tenant[17]. There were important exceptions however, which prevented qualification as a landholder[18].

1 1911 Act, s 7(1).
2 Ibid, s 17.
3 *McDougall v McAlister* (1890) 17 R 555.
4 1911 Act, s 26(6); see Scott *The Law of Smallholdings in Scotland* (W Green, 1933) pp 135 and 136.
5 1911 Act, s 2(1).
6 The residential qualification applied at the commencement of the Act. If the landholder ceased to reside thereafter he did not lose the qualification; *Rogerson v Viscount Chilston* 1917 SC 453.
7 Measured in a straight line; *Simpson v Yool* 1919 SLCR 18.
8 The residential qualification applied at the commencement of the Act. If the landholder ceased to reside thereafter he did not lose the qualification; *Rogerson v Viscount Chilston* 1917 SC 453.
9 Measured in a straight line; *Simpson v Yool* 1919 SLCR 18.
10 The 'qualified leaseholder'.
11 *Scott* p 45; *Macdonald v Prentice's Trs* 1993 SLT (Land Ct) 60.
12 1911 Act, s 26(1).
13 Ibid, s 26(2).
14 Ibid, s 26(8).
15 See *Scott* p 24.
16 1911 Act, s 2(1)(a) and (2).
17 Ibid, s 2(1)(b).
18 See Crispin Agnew of Lochnaw 'When is a Croft not a Croft' (1991) 36 JLSS 115.

A person was excluded from qualifying if the rent of the land on 1 April 1912 exceeded £50, unless the land excluding common grazings or pasture did not exceed 50 acres[1].

Other particular exceptions are provided for by the 1911 Act, s 26(3). These include garden ground or appurtenant to a house[2] or connected with an ancient monument, land within a burgh[3], a market garden[4], glebe land[5], land not an agricultural holding under the Agricultural Holdings (Scotland) Act 1908[6], woodland[7], land forming part of the home farm[8], or any policy, park or pleasure ground of amenity for any residence or farmhouse and permanent grass parks held for a business of calling not primarily agricultural or pastoral[9], land held for public recreation and land compulsorily acquired or by agreement for any public undertaking.

By the Crofters Holdings (Scotland) Act 1886, s 33[10], a holding or building let (1) to a person during his continuance in any office, appointment or employment of the landlord[11], (2) at a nominal rent or without rent as a pension for former service[12], (3) during the tenure of office such as a minister of religion or schoolmaster or (4) let to any innkeeper or tradesman placed in the district by the landlord for the benefit of the neighbourhood, was excluded[13].

Common grazings and other rights

Although grazing rights are deemed to be part of a holding, there are now few common grazings in the non-crofting counties.

The Land Court may prescribe regulations as to the exercise of pasture, grazing or other rights held or to be held in common under the 1911 Act[14].

A grazings committee may be appointed, which may make regulations as to the number of stock that may be carried on the common grazing and as, to other matters affecting the fair exercise of the landholder's joint rights[15]. The Land Court has powers to remove or suspend members of the grazings committee and may appoint or provide for the appointment of other persons, or appoint a grazings constable[16].

Breach of regulations is a criminal offence[17].

---

1 1911 Act, s 26(3)(a).
2 *Malcolm v M'Dougall* 1916 SC 283, 1916 SLCR 40.
3 *Bontein v MacDougall* 1916 SLCR 72.
4 Cf p 249 'Definition of market garden'.
5 *Chisholm v Macdonald* 1913 SLCR 69.
6 *Rankine* pp 609 and 610; *McNeill v Duke of Hamilton's Trs* 1918 SC 221.
7 *Board of Agriculture v Macdonald* 1914 SLCR 43.
8 *Ross v Matheson* 1913 SLCR 15.
9 *MacKay v Countess of Seafield's Trs* 1914 SLCR 37 ('cattle dealer').
10 Applied to the 1911 Act by s 26(7).
11 *Budge v Gunn* 1925 SLCR 74; *Guthrie v MacLean* 1990 SLCR 47.
12 *Stewart v Lewis and Harris Welfare Development Co Ltd* 1922 SLCR 66.
13 *Stormonth Darling v Young* 1915 SC 44; *Macgregor v Milne* 1922 SLCR 32; *Taylor v Fordyce* 1918 SC 824.
14 1911 Act, s 24(1).
15 Crofters Common Grazings Regulations Act 1891.
16 1911 Act, s 24(4).
17 Ibid, s 24(3).

The Land Court may, on the application of the landlord or any landholder, and on such conditions as it considers equitable, apportion a common pasture or grazing into separate parts for the exclusive use of several townships or the persons interested therein, if satisfied that such apportionment is for the good of the estate or estates and the holdings or tenancies concerned[1].

The Land Court may, on the application of the landlord or any landholder or of the Secretary of State, admit new holders to participate in a common pasture or grazing along with existing landholders and others[2].

The landholders interested in a common grazing may appoint two of their number to kill and take ground game under the Ground Game Act 1880[3].

The Land Court may regulate the use of seaweed, peat and thatching materials by landholders on the same estate[4].

The landlord, landholder or the Secretary of State may apply to the Land Court to make a record of the holding when the rent is being fixed or at anytime[5].

The parties cannot contract out of the rights under the Landholders Acts[6].

Vacant holdings

Where a holding has ceased to be held by a landholder, the landlord is required to intimate the fact in writing to the Secretary of State forthwith[7]. The landlord is not entitled to re-let the holding without the consent of the Secretary of State other than to a neighbouring landholder for the enlargement of his holding or to a new holder[7].

Where a holding is vacant the Secretary of State has the power to re-let it, subject to payment of appropriate compensation to the landlord[7].

Where the landlord does not re-let in compliance with s 17, the Secretary of State is entitled to declare the let null and void and to treat the holding, without compensation, as a new holding under the Act[8]. The Secretary of State is required to take action timeously[9].

Ceasing to be a landholding

The landlord and the tenant cannot unilaterally agree to remove the subjects from the jurisdiction of the Acts[10].

A landholding can cease to be a landholding if:

1 1911 Act, s 24(5). NB the application has to be at the instance of the landlord or a landholder, but the apportionment can be to a non-landholder, who may be a 'person interested'.
2 Ibid, s 24(5)(b).
3 1931 Act, s 23.
4 1886 Act, s 12 and 1911 Act, s 28.
5 1931 Act, s 10.
6 Ibid, s 25; cf *Scott* pp 166 and 167.
7 1911 Act, s 17.
8 Ibid, s 17 proviso.
9 *Board of Agriculture v Countess-Dowager of Seafield's Trs* 1916 SLCR 63; *Murray v Anstruther's Trs* 1940 SLCR 59.
10 *Whyte v Garden's Trs* 1925 SLCR 99; *MacNee v Strathearn's Trs* 1970 SLCR App 29.

(1) the landholder purchases the holding[1];

(2) the tenant accepts a new lease outwith the protection of the Act[2];

(3) the landlord let the holding without the consent of the Board of Agriculture and the board did not object timeously[3].

Obligations of landholder

The landholder is obliged to:

(1) comply with the statutory conditions set out in the 1886 Act, s 1[4];

(2) cultivate[5] the holding by himself or his family, with or without hired labour[6]. A landholder is not entitled to sub-let without the consent of the landlord[7]. Granting seasonal grazing lets is not prohibited sub-letting[8].

Rights of a landholder

The rights of a landholder include:

(1) Security of tenure provided the statutory conditions are not breached. The original requirement that the landholder had to reside on the holding is not a continuing statutory condition[9]. If a year's rent remains unpaid or the landholder is in breach of any statutory condition he may be removed by the Land Court[10].

(2) The right to bequeath the holding to one person who may be his son-in-law or any person entitled to succeed to his estate on intestacy under the Succession (Scotland) Act 1964[11]. The bequest has to be intimated to the landlord within two months[12]. The landlord may object to receiving the legatee

---

1 *Scott* p 124.

2 *Kennedy v Marquess of Breadalbane's Trs* 1933 SLCR 3; *McColl v Beresford's Trs* 1921 SLCR 3; *Greg's Exrs v Macdonald* 1991 SLCR 135.

3 *Board of Agriculture v Countess-Dowager of Seafield's Trs* 1916 SLCR 63; *Murray v Anstruther's Trs* 1940 SLCR 59.

4 Including in summary: (1) that the landholder pay rent when due; (2) not to purport to assign the tenancy; (3) not to prejudice the landlord's interest by persistently allowing dilapidations; (4) not to divide or sub-let (this does not include seasonal grazing lets; *Morrison v Nicolson* 1913 SLCR 90); (5) not to persistently violate any written conditions for the protection of the landlord or neighbouring landholders; and (6) not to do any act whereby the landholder becomes bankrupt.

5 'Cultivate' includes 'use of the holding for horticulture or for any purpose of husbandry, inclusive of keeping or breeding of live stock, poultry or bees, and the growth of fruit, vegetables and the like'; 1911 Act, s 10(1).

6 Ibid, s 10(1).

7 1886 Act, s 1(4).

8 *Morrison v Nicholson* 1913 SLCR 90; *Borland v Muir* 1918 SLCR 21.

9 *Rogerson v Viscount Chilston* 1917 SC 453.

10 1931 Act, s 3. See eg *Mackenzie's Trs v MacKenzie* (1905) 7 F 505; *Little v McEwan* 1965 SLT (Land Ct) 3; *Arran Properties Ltd v Currie* 1983 SLCR 92.

11 1886 Act, ss 16 and 34(1); 1964 Act, Sch 2, paras 9-12.

12 1886 Act, s 16(a) (as amended).

within one month[1]. If there is no objection the legatee becomes the tenant from the date of death[2], but if there is an objection the legatee may petition the Land Court for decree of declarator that he is the landholder from the date of death[3]. If there are reasonable grounds for objection the Land Court may declare the bequest null and void[3].

If the bequest is not accepted or it is declared null and void, the tenancy is treated as intestate estate available for transfer by the executors under the 1964 Act, s 16[4].

Where there is no bequest of the tenancy, it requires to be transferred by the executor[5].

(3) Although assignation is prohibited by the statutory conditions[6], where the landholder is unable to work the holding through illness, old age or infirmity, he may apply to the Land Court for leave to assign the holding to his son-in-law or any one of the persons who would be entitled to succeed on intestacy. The Land Court may grant such leave on terms and conditions, if any, as may seem fit[7]. A condition could be that the assignee is to be permitted to live in the house for the rest of his life[8].

A landholder may assign his tenancy with the consent of the landlord[9].

(4) The right to a fair rent is fixed, failing agreement, by the Land Court[10]. A rent fixed by the Land Court applies for seven years[11]. A fair rent is not an open market rent as applies under the 1991 Act, but one which seeks to provide for a fair division of the profits from the land[12].

In a fair rent application the Land Court may deal with arrears of rent by ordering that all or only part be paid off, or by writing them off, or ordering payment by instalments The Land Court has power, in a fair rent application or other application, to sist proceedings in other courts for removal of a landholder for non-payment of rent, until the application is determined[13].

The Land Court has to take into account any milk quota allocated to the holding[14].

(5) The right to use the holding for subsidiary or auxiliary occupations, in the case of dispute, may be found by the Land Court to be reasonable and not inconsistent with the cultivation of the holding[15].

1 1886 Act, s 16(c).
2 Ibid, s 16.
3 Ibid, s 16(d).
4 Ibid, s 16(h).
5 1964 Act, s 16 and 1911 Act, s 31(1). See chap 13 as the law and time limits etc are similar to those applying to agricultural holdings.
6 1886 Act, s 1(2).
7 1911 Act, s 21.
8 *Macdonald v Uig Crofters Ltd* 1950 SLCR 11.
9 *Campbell v Board of Agriculture* 1928 SLCR 27.
10 1886 Act, ss 5 and 6; see the 1911 Act, s 13.
11 1886 Act, s 6(2).
12 *W C Johnston Ltd v Fitzsimon* 1983 SLCR 95. In *Ward v Shetland Islands Council* 1990 SLCR 119 the Land Court reconsidered the basis upon which a 'fair rent' was to be fixed for crofts. The same considerations would seem to apply to a landholding.
13 1886 Act, s 6(4) and (5).
14 1986 Act, s 16(2)(b); see p 66 'Milk quota'.
15 1911 Act, s 10(1); *Taylor v Fordyce* 1918 SC 824.

(6) A landholder may recover compensation for game damage, assessed by the Land Court, failing agreement[1].

(7) A landholder may renounce his tenancy at any term of Whitsunday or Martinmas upon giving one year's notice in writing[2]. A new landholder (ie one who became a landholder by registration after 1 April 1912) or his statutory successor has to give the year's notice to the term of his ingo or entry[3].

A renunciation is of no effect without the consent of the Land Court unless, within two months from the date of the notice, it has been intimated to the Secretary of State[4]. The landholder is not entitled to renounce the tenancy without the consent of the Secretary of State so long as any liability owing by him to SOAEFDS is not wholly discharged[4].

(8) A right to compensation for (1) waygoings under the Agricultural Holdings Acts and (2) improvements, on removal from the holding or renunciation of the lease.

The improvements (1) have to be suitable to the holding, (2) have been executed or paid by the landholder or his predecessors in the same family[5] and (3) have not been executed in virtue of any specific agreement in writing under which the landholder was bound to execute the improvements[6] unless he has received fair consideration for the improvements by way of reduction of rent or otherwise[7].

Improvements are valued at a sum that fairly represents the value of the improvement to an incoming tenant[8]. Failing agreement improvements fall to be valued by the Land Court[9].

Where the Secretary of State has given the landholder a loan the agreement should be recorded in the Landholders Holdings Book, which has the effect of transferring to the Secretary of State all the rights of the landholder to compensation on termination of the tenancy[10].

Rights of the landlord

The statutory conditions reserve to the landlord *inter alia* mining and quarrying, cutting timber, opening roads and drains, access to the sea or any loch to exercise proprietorial rights therein, viewing the holding and sporting rights[11].

The landlord has a right to resume all or part of the holding for some reasonable purpose[12] having relation to the good of the holding or of the estate, with the consent of the Land Court[13]. Resumption for the protection of an

1 1911 Act, s 10(3).
2 1886 Act, s 7.
3 1931 Act, s 22.
4 1886 Act, s 7 proviso; see the 1911 Act, s 18.
5 Cf the 1931 Act, s 9.
6 1886 Act, s 8.
7 1931 Act, s 12.
8 1886 Act, s 10; *Strachan's Trs v Harding* 1990 SLT (Land Ct) 6; cf p 222 'Compensation for improvements'.
9 1886 Act, s 31.
10 1911 Act, s 8.
11 1886 Act, s 1(7).
12 Personal occupation by the landlord is not a reasonable purpose; 1931 Act, s 8(1).
13 1886 Act, s 2.

ancient monument or other object of historical or archaeological interest from destruction or injury is deemed a reasonable purpose[1].

The Land Court can authorise resumption on such terms and conditions as it thinks fit[2]. The tenant is entitled to compensation for loss of tenancy and loss of profits[3].

The landlord has a right, upon payment of compensation for any surface damage, to use spring water, not required by the holding for any estate purpose. Any dispute as to the requirement of the holding or as to compensation requires to be determined by the Land Court[4].

When a landholder renounces the tenancy or is removed from the holding the landlord is entitled to set off any rent due or that may become due against any sums found due to the landholder or to the Secretary of State for improvements on the holding[5].

STATUTORY SMALL TENANCIES[6]

There are very few statutory small tenancies still in existence. If there is a dispute as to whether or not the tenant is a statutory small tenant, it is competent for the Land Court to determine such a question summarily[7].

A tenant, otherwise qualified as a landholder, is a statutory small tenant if he or his predecessors in the same family did not provide or pay for the whole or the greater part of the buildings or other permanent improvements on the holding, or without receiving from the landlord or any predecessor payment or fair consideration therefor[8].

A statutory small tenancy is deemed to include any right in pasture or grazing land held or to be held by the tenant alone or in common with others, as well as the site of any dwellinghouse erected or to be erected on the holding or held or to be held therewith, and of any offices or other conveniences connected with the dwellinghouse[9].

Except as expressly applied by the 1911 Act, the Landholders Acts do not apply to statutory small tenants[10]. The Agricultural Holdings (Scotland) Acts 1908 and 1910 apply to the holdings, except in so far as varied by the 1911 Act, s 32[11].

A statutory small tenancy is passed to the tenant's heirs, legatees[12] or assignees, if permitted under the lease or by implication[13].

1 1911 Act, s 19.
2 1886 Act, s 2 (as amended by the 1911 Act, ss 19, 31(2) and Sch 2 and the 1931 Act, s 8(1) and (2)).
3 *Stornoway Trust v Landholders of North Street, Sandwick* 1948 SLCR 3; *Macdonald v Macdonald* 1950 SLCR 14; *Hay v Matthew* 1967 SLT (Land Ct) 13.
4 1911 Act, s 12.
5 Ibid, s 23.
6 See 1 *Stair Memorial Encyclopaedia* paras 854-860.
7 1911 Act, s 32(13).
8 Ibid, ss 2 and 32(1).
9 Ibid, ss 26(1) and 32(14).
10 Ibid, s 32(2).
11 Ibid, s 32(5).
12 Within the meaning of the 1886 Act, s 16 (as amended).
13 1911 Act, s 32(1); *Reps of Hugh Matheson v Master of Lovat* 1984 SLCR 82 at 92.

Unlike a landholding, when a statutory small tenancy becomes vacant it may be let in any way the landlord chooses[1]. A statutory smallholding cannot be merged into an agricultural holding except with the consent of the Secretary of State[2].

Failing agreement the tenant may apply to the Land Court to fix an 'equitable rent'[3] when the tenancy comes to an end. The Land Court may sist proceedings for the removal of the tenant in respect of non-payment of rent, when fixing the rent[4]. The Land Court has to take into account any milk quota allocated to the holding[5].

Prior to the termination of the tenancy, the tenant may apply to the Land Court to fix the period for which the tenancy is to be renewed[6]. The tenant is entitled to such a renewal unless the landlord can satisfy the Land Court that there is reasonable ground of objection to the statutory small tenant[7].

In the event of the landlord, on the renewal of the lease, failing to provide such buildings as will enable the tenant to cultivate the holding according to the terms of the lease or agreement, or at any time failing to maintain the buildings and permanent improvements required for the cultivation and reasonable equipment of the holding, in so far as the tenant is not obliged to provide them at common law or under an agreement, the tenant may apply to the Land Court to find and declare the tenant to be a landholder[8].

Without prejudice to any agreement between the parties, the Land Court may authorise the landlord to resume the holding or part thereof for building, planting, feuing or some other reasonable purpose having relation to the good of the holding[9]. The tenant is entitled to like compensation as he would be entitled to under the Agricultural Holdings (Scotland) Act 1908, a disturbance payment[10] and compensation, where the resumption is authorised by the Land Court, for early resumption[11].

The landlord, tenant or the Secretary of State may ask the Land Court to make a record of the holding, when fixing a fair rent[12].

On quitting the holding on the termination of the tenancy, a statutory small tenant has a right to claim (1) compensation for improvements, (2) compensation for disturbance on a resumption of a holding[13], (3) an additional payment of four times the rent, where compensation for disturbance is payable[14]

1 *Morrison v McIndoer* 1966 SLCR App 132.
2 1911 Act, s 23(3); *Morrison v McIndoer* above.
3 Ibid, s 32(7) and (8). An 'equitable rent' is different from either a 'fair rent' or an 'open market rent'; *Wilkie v Hill* 1916 SC 892; *Fullarton v Duke of Hamilton's Trs* 1918 SC 292; *W C Johnston v Fitzsimon* 1983 SLCR 95.
4 1886 Act, s 6(4) and 1911 Act, s 32(14).
5 1986 Act, s 16(2)(b); see p 66 'Milk quota'.
6 1911 Act, s 32(4) and (7).
7 Ibid, s 32(4); *Macleod v Christie* 1916 SLCR 94 (it was held that the reasonable objections must relate to personal conduct or character of the tenant connected with the performance of his legal obligations as tenant). Cf *Grant v Seafield's Trs* 1921 SLCR 87.
8 1911 Act, s 32(11).
9 Ibid, s 32(15).
10 1931 Act, s 13 and 1991 Act, s 54.
11 1991 Act, s 58(1)(b).
12 1911 Act, ss 8(4) and 32(14).
13 1931 Act, s 13.
14 1991 Act, s 54(1)(b).

and (4) a payment in respect of the effect of early resumption, where the resumption is authorised by the Land Court[1].

Compensation for improvements falls to be determined by arbitration unless both parties agree to refer the matter to the Land Court[2].

Where the statutory smallholding is acquired in consequence of compulsory acquisition, the statutory small tenant is entitled to an additional payment[3].

The landlord has a right, upon payment of compensation for any surface damage, to use spring water, not required by the holding for any estate purpose. Any dispute as to the requirement of the holding or as to compensation requires to be determined by the Land Court[4].

MILK QUOTA

A landholder or a statutory small tenant is entitled to compensation for milk quota on quitting the holding on the termination of the tenancy[5]. Compensation falls to be determined by the Land Court[6].

Where there requires to be an apportionment or prospective apportionment of milk quota in respect of a landholding or a statutory small tenancy this falls to be carried out by the Land Court[7].

1 1991 Act, s 58; see p 236 'Claims on partial dispossession'.
2 Scott p 37; 1931 Act, s 34.
3 1991 Act, s 56; see p 222 'Compensation for improvements'.
4 1911 Act, s 12.
5 1986 Act, Sch 2; see p 225 'Compensation for milk quota'.
6 Ibid, Sch 2, para 11(1)(b).
7 DPQRs 1994, Sch 3, para 1(2)(b) and (c).

CHAPTER 23

Dispute resolution – courts and arbitration

Key points

- Particular care requires to be taken to determine whether or not a dispute falls to be determined by statutory or common law arbitration, by the Land Court or by one of the other courts or tribunals.

- 'Any question of difference between the landlord and the tenant of an agricultural holding arising out of the tenancy or in connection with the holding' falls to be determined by arbitration, although the courts retain the pre-eminent jurisdiction to determine whether or not the relationship of landlord and tenant of an agricultural holding existed at the relevant time: s 60(1).

- Notwithstanding the terms of a lease any arbitration under the 1991 Act has to be conducted by a single arbiter: s 61(7).

- Parties cannot prorogate the jurisdiction of the courts, if the 1991 Act requires the issue to be determined by arbitration.

- The Land Court's jurisdiction is limited to the specific jurisdictions conferred upon it by statute. When acting on a joint reference in lieu of arbitration, the Land Court's jurisdiction is no wider than that of an arbiter acting under the 1991 Act.

- Unless there are good reasons to the contrary, parties would be well advised to have an arbiter appointed by the Secretary of State, rather than agree upon an appointment themselves. Serious consideration should be given to agreeing to a joint reference to the Land Court.

- Where there are time limits for the appointment of an arbiter, the arbiter has to have been appointed by the particular date, otherwise the right to have the issue determined is lost. Agreement that a particular person should be appointed does not amount to an appointment.

- Delay in demanding arbitration on a claim or issue, even where there are no statutory time limits, may bar a claim on the grounds of waiver or personal bar.

- An arbiter's clerk, if required, should only be appointed after the statement of case has been lodged by the arbiter, otherwise his remuneration and expenses cannot be recovered: Sch 7, para 19.

- It is not competent to request a stated case after the arbiter has signed and delivered his award. An arbiter should be asked to issue a draft award so that parties can consider whether or not to request a stated case.

- With the exception of a right of appeal to the Land Court against an arbitration award in a rental arbitration, under s 13(1), it is not competent to appeal against an arbiter's award.

- Where the Secretary of State is landlord or tenant of a holding, then the Land Court acts in place of him in the appointment of an arbiter in any matter which would otherwise be referred to the Secretary of State under the 1991 Act for a decision or relating to an arbitration concerning the holding: ss 64 and 80.

- Statutory provision is made for reference of disputed compulsory purchase and like compensation claims to the Lands Tribunal for Scotland. It has the jurisdiction to act as arbiters under a reference by consent; a joint reference is accepted where the issues raised are similar to those that fall within the tribunal's normal jurisdiction: Lands Tribunal Act 1949, s 1(5).

- Where the holding is a small landholding (and in some matters a statutory small tenancy) it is the Land Court which has jurisdiction in relation to issues arising between landlord and tenant.

Time limits

- In an arbitration under the 1991 Act, statements of case with all necessary particulars have to be with the arbiter within 28 days 'from the appointment of the arbiter': Sch 7, para 5.

- The arbiter under the 1991 Act requires to sign his award within three months of his appointment or within such longer period as the parties may agree in writing or be fixed by the Secretary of State: Sch 7, para 8; Dairy Produce Quotas Regulations 1994, SI 1994/672 (DPQRs), Sch 3, para 14(1) (apportionment of milk quota).

- An appeal to the Land Court against a statutory arbiter's award in a rental arbitration must be brought within two months of the date of the issue of the award: s 61(3).

- An arbiter's award under the 1991 Act has to be paid within one month of the date it becomes due otherwise it can be enforced by registration in the Books of Council and Session or the sheriff court books as a decree arbitral: 1986 Act, s 65 and Sch 2, para 2.

- An arbiter, to determine the apportionment of milk quota, requires to be appointed within 28 days of the change of occupation, or the arbiter will be appointed by the Secretary of State: DPQRs 1994, Sch 3, para 3(1), (2) and (3).

- Where an arbiter is appointed by the parties or a reference is agreed to the Land Court for an apportionment of milk quota, then the appointment requires to be intimated to the Secretary of State within 14 days: DPQRs 1994, Sch 3, para 3(1), 4(2) and 26.

GENERAL

The respective jurisdictions of the courts, the Land Court, statutory arbitration and common law arbitration may give rise to difficulties in matters relating to an agricultural holding, statutory small tenancy or landholding.

While particular sections of the 1991 Act provide that certain issues should be referred to arbitration, s 60(1) is a catch all section requiring that 'any question or difference' shall be referred to arbitration. The scope of that provision requires to be considered in detail.

Where parties are required to refer a matter to arbitration under the 1991 Act they have the option of (1) selecting an arbiter[1], or (2) having an arbiter appointed by the Secretary of State[1] or (3) agreeing a joint reference to the Land Court[2]. There are advantages and disadvantages to each of these courses.

The powers and duties of an agreed arbiter are different and in general less advantageous to the parties than those of a statutorily appointed arbiter. Where difficult questions of law arise, except in relation to rent arbitrations, parties should consider a joint reference to the Land Court, where the court has substantial experience and legal expertise. The court expenses are substantially less than the costs of an arbiter and his clerk.

The disadvantage of a reference to the Land Court is that a party, if eligible, may apply for legal aid for the reference, in circumstances where legal aid is not available for an arbitration.

Where the holding is a landholding it is the Land Court that has jurisdiction in questions arising between landlord and tenant[3]. With a statutory small tenancy some issues have to be referred to the Land Court, while others are dealt with by arbitration[4].

The aim of this chapter is to outline the respective jurisdictions of arbitration and the courts. The chapter will also deal with relevant procedures.

STATUTORY ARBITRATION

Where statute provides for arbitration, the parties cannot agree expressly or by implication to prorogate the jurisdiction of the courts so as to oust the arbitration provision[5]. The courts, including the Land Court, or the arbiter can, and should, raise the question of jurisdiction *ex proprio motu* as it affects the competency of the application[6].

---

1 1991 Act, Sch 7, para 1.
2 Ibid, s 60(2). For sheep stock valuations the parties may agree a joint reference, where the lease was entered into on or before 6 November 1946, but otherwise either party may refer the matter to the Land Court - ibid, s 70.
3 Small Landholders (Scotland) Act 1911; 1986 Act, Sch 2; DPQRs 1994, Sch 3.
4 See chap 22.
5 *Brodie v Ker, McCallum v McNair* 1952 SC 216; *Taylor v Brick* 1982 SLT 25.
6 *NCB v Drysdale* 1989 SLT 825; *Craig, Applicants* 1981 SLT (Land Ct) 12; *McDiarmid v Secretary of State for Scotland* 1970 SLT (Land Ct) 17.

The Secretary of State for Scotland, on an application to appoint an arbiter, will normally not inquire into jurisdiction, but leave it to the parties to raise the question of jurisdiction before the arbiter[1] or by interdict. It is competent to interdict an arbiter purportedly acting under the 1991 Act, if his remit is out-with the jurisdiction conferred on him by the Act[2]. The Land Court, when acting to appoint an arbiter in place of the Secretary of State, may make such inquiries[3].

Any arbitration between landlord and tenant arising under statute[4] or under the terms of the lease, notwithstanding any agreement to the contrary, now has to be determined by a single arbiter in accordance with the statutory procedures[5].

Waygoing arbitrations are an exception[6]. Questions between an incoming and the outgoing tenant are not governed by the statutory procedures, because they are not questions between landlord and tenant.

The parties are free to agree upon an arbiter[7], or in default of agreement[8], an arbiter[9] can be appointed by the Secretary of State upon a written application[10] by either party[11]. The selection of an arbiter by the Secretary of State[12] is an administrative act[13] and is therefore subject to judicial review, should the Secretary of State act unlawfully or unreasonably in the appointment[14].

If an arbiter dies, or is incapable of acting[15], or for seven days after written notice[16] from either party requiring him to act fails to act, a new arbiter may be appointed as if no arbiter had previously been appointed[17]. Provided the

1 *Christison's Tr v Callender-Brodie* (1908) 8 F 928. Cf *Dundee Corp v Guthrie* 1969 SLT 93 at 97 per LJC and 99 per Lord Milligan, where it is suggested that it is not competent to challenge the appointed arbiter's right to determine the question submitted to him while the appointment remains unreduced. This comment is at variance with *Christison's Tr* and *Cormack v McIldowie's Exrs* 1974 SLT 178.

2 *Cormack v McIldowie's Exrs* above; *Donaldson's Hospital v Esslemont* 1925 SC 199, 1926 SC (HL) 68. Cf *Dundee Corp v Guthrie* in the previous note.

3 *McDiarmid v Secretary of State for Scotland* 1970 SLT (Land Ct) 17 at 20L.

4 1991 Act, and on milk quota compensation under the 1986 Act.

5 1991 Act, s 61(1). The arbitration has to be conducted in terms of ibid, Sch 7. Cf *Cameron v Nicol* 1950 SC 1, which left open the question of whether or not the requirement was peremptory or whether parties who allowed arbitration to proceed outwith the provisions of the Act might not be personally barred from thereafter challenging the procedure. Dicta in the case suggest that the Inner House might have found the provision to be peremptory.

6 1991 Act, s 61(7). See p 284 'Common law arbitrations'.

7 The 'private arbiter', who need not be a member of the Secretary of State's panel of arbiters. 'Agreed upon' means that there must be a delivered document in writing in the hands of the landlord and tenant appointing a named arbiter to determine a specific dispute within the time limits; *Chalmers Property Investment Co Ltd v MacColl* 1951 SC 24.

8 The parties do not have to attempt to agree, but merely do not have to have agreed upon an arbiter; *Chalmers Property Investment Co Ltd v MacColl* above; *F R Evans (Leeds) Ltd v Webster* (1962) 112 LJ 703.

9 The 'statutory arbiter', who must be appointed from the panel of arbiters.

10 Agricultural Holdings (Specification of Forms) (Scotland) Order 1991, SI 1991/2154, Sch 2, Form A. Application is made to SOAEFDS, Pentland House, 47 Robbs Loan, Edinburgh EH14 1TW. Applications are accepted by Fax.

11 1991 Act, Sch 7, para 1.

12 And presumably also by the Land Court.

13 *Ramsay v M'Laren* 1936 SLT 35 at 36 per Lord Mackay.

14 R L C Hunter *The Law of Arbitration in Scotland* (1987) para 19-24.

15 This has a wide meaning; *Dundee Corp v Guthrie* 1969 SLT 93.

16 1991 Act, Sch 7, para 4.

17 Ibid, Sch 7, para 2.

original arbiter was appointed timeously, the subsequent appointment is deemed to be within the time limit under the former remit[1].

The 'question or difference' for determination specified in the deed of submission or in the application needs to be formulated with care. The Secretary of State's appointment will follow the wording in the application. The arbiter's remit[2] is restricted to the terms of the 'question or difference' referred to him[3].

If the respondent in the arbitration has a counterclaim or requires other questions and differences to be determined, this either has to be included by agreement in the original application or a separate application for appointment has to be made by the respondent. The two arbitrations are then usually co-joined.

Where the Secretary of State is a party to the arbitration, the arbiter is appointed by the Land Court[4]. An arbiter appointed by the Land Court should not be on the panel of arbiters appointed under the 1991 Act, s 63[5].

The panel of arbiters is appointed by the Lord President, after consultation with the Secretary of State. The panel is subject to revision at intervals not exceeding five years[6].

Unless there are good reasons for appointing a private arbiter, parties would be advised to have a statutory arbiter appointed. While a private arbiter requires to follow the statutory procedures[7], he need not give reasons for his decision[8]. In rent arbitrations[9] he cannot state a case for the opinion of the Land Court[10] nor is there an appeal to the Land Court against his decision[11]. A private arbiter requires to be appointed by Deed of Submission in common form. His fees are a matter for agreement[12].

In contrast the statutory arbiter is subject to much more statutory control. He is required to give reasons[13] and to produce his awards in the statutory form[14]. His fee is fixed by the Secretary of State or by the Land Court[15]. In rent arbitrations he may state a case for the opinion of the Land Court[16] and his decision is appealable to the Land Court[17].

---

1 *Dundee Corp v Guthrie* 1969 SLT 93; *Pennington-Ramsden v McWilliam* [1983] CLY 28.
2 Or that of the Land Court if there is a joint remit to it.
3 *Chalmers Property Investment Co v Bowman* 1953 SLT (Sh Ct) 38; *Exven Ltd v Lumsden* 1982 SLT (Sh Ct) 105.
4 1991 Act, s 80.
5 *Secretary of State v Jaffray* 1957 SLCR 27; *Commissioners of Crown Lands v Grant* 1955 SLCR 25.
6 1991 Act, s 63(1) and (2).
7 Ibid, Sch 7.
8 Tribunals and Inquiries Act 1992, s 10 and Sch 1, Pt II; cf the 1991 Act, Sch 7, para 10.
9 Under the 1991 Act, s 13.
10 Ibid, Sch 7, para 22.
11 Ibid, s 61(2).
12 Failing agreement the fees are fixed by the auditor of the sheriff court subject to appeal to the sheriff.
13 Tribunals and Inquiries Act 1992, s 10 and Sch 1, Pt II.
14 Agricultural Holdings (Specification of Forms) (Scotland) Order 1991, SI 1991/2154; in rent arbitrations the 1991 Act, Sch 7, para 10 requires the arbiter to state findings in fact and the reasons for his award.
15 1991 Act, ss 63(3)(a) and 80; *Secretary of State for Scotland v Brown* 1993 SLCR 41.
16 1991 Act, Sch 7, para 22.
17 Ibid, s 61(2).

Where an arbiter is guilty of misconducts the sheriff may remove him[1]. Where a party alleges misconduct he should object at the time, otherwise he might be said to have waived his right to object[2]. Where an arbiter has misconducted himself or the award has been improperly procured[3] the sheriff may set aside the award[4].

While an arbiter is free to appoint a legally-qualified clerk at any time, the clerk's remuneration and expenses can only be claimed if (1) the appointment was made after the submission of the claim and answers, (2) the parties consent or (3) the appointment is sanctioned by the sheriff[5].

The arbiter's appointment can only be revoked by consent of both parties by notice in writing[6].

Particular statutory arbitration provisions

The 1986 and 1991 Acts and the DPQRs 1994 make specific provision for arbitration as follows:

(a) *Leases for less than year to year – s 2.* 'Any question as to the operation of this section in relation to any lease shall be determined by arbitration'[7].

(b) *Terms of the lease – s 4.* Where either party requires the other to enter into a written lease, and if within the period of six months after the giving of notice no lease has been concluded 'the terms of the tenancy shall be referred to arbitration'[8].

Having settled the terms of the lease, the arbiter may vary the rent if it is equitable so to do[9].

Where the arbiter, as part of his award, transfers liability for the maintenance or repair of an item of fixed equipment from the tenant to the landlord, the landlord can require arbitration to determine the compensation to be paid in respect of the previous failure by the tenant to discharge his liability in respect of the fixed equipment[10].

(c) *Fixed equipment – s 5.* 'any question as to the liability of the landlord or tenant under this section shall be determined by arbitration'[11]. It should be noted that this section does not apply to leases entered into before 1 November 1948.

1 1991 Act, Sch 7, para 23. See 2 *Stair Memorial Encyclopaedia* para 483; *Halliday v Semple* 1960 SLT (Sh Ct) 11 (issuing award out of time); *Thomas v Official Solicitor* (1983) 265 EG 601 (failure to postpone hearing to enable tenant to obtain legal representation).
2 *Maclean v Chalmers Property Investment Co Ltd* 1951 SLT (Sh Ct) 71; *Fountain Forestry Holdings Ltd v Sparkes* 1989 SLT 853.
3 2 *Stair Memorial Encyclopaedia* para 484.
4 1991 Act, Sch 7, para 24. NB 'may', so the sheriff has a discretion. Cf *Dunlop v Mundell* 1943 SLT 286, where it was held that the similar wording of the Sheep Stocks Valuation (Scotland) Act 1937, s 1(2) did not oust the Court of Session's concurrent jurisdiction to reduce an award of an arbiter.
5 1991 Act, Sch 7, para 19.
6 Ibid, Sch 7, paras 3 and 4.
7 See p 46 'Arbitration'.
8 1991 Act, s 4(1); see p 26 'Statutory written lease'.
9 Ibid, s 14(a).
10 Ibid, s 46(2). See s 46(3) - the provision for compensation in favour of the tenant is meaningless in the context of the section; cf *Duncan*, note to s 46(3) in *Current Law Statutes*.
11 Ibid, s 5(5); see p 54 'Arbitration on liability under s 5'.

In *Tustian v Johnston*[1], in considering the equivalent English provision[2] it was held[3] that the 'compulsory arbitration provisions apply ... up to the stage of establishing (a) obligation and (b) breach of obligation'. It then fell within the jurisdiction of the courts to order specific performance of the obligation or award damages in lieu.

The arbiter may vary the rent of the holding if it is equitable so to do, having regard to any provision included in the award[4].

Where an arbiter transfers liability for the maintenance or repair of fixed equipment from the tenant to the landlord under the 1991 Act, s 4, an arbiter requires to fix the compensation payable by the tenant to the landlord for past failures by the tenant to discharge the liability[5].

(d) *Freedom of cropping – s 7*[6]. In any proceedings for interdict brought under s 7(3):

'the question of whether a tenant is exercising, or has exercised, his rights under subsection 1 above in such a manner as to injure or deteriorate, or to be likely to injure or deteriorate the holding, shall be determined by arbitration; and the certificate of the arbiter as to his determination of any such question shall ... be conclusive proof of the facts stated in the certificate.'

The remedy of interdict or damages available to the landlord where the tenant exercises his rights 'so as to injure or deteriorate, or to be likely to injure or deteriorate, the holding' should be pursued in the courts[7], although it may be for the arbiter to quantify the loss[8].

(e) *Permanent pasture – s 9*[9]. Where provision is made in the lease for the maintenance of specified land as permanent pasture the tenant may demand;

'a reference to arbitration under this Act of the question whether it is expedient in order to secure the full and efficient farming of the holding that the amount of land required to be maintained as permanent pasture should be reduced.'

(f) *Variation of rent – ss 13 and 14*[10]. Failing agreement 'the question of what rent should be payable in respect of the holding' is to be determined by arbitration[11].

Where, on a reference under ss 4 or 5, it appears equitable that the rent may be varied, the arbiter may vary it.

(g) *Increase of rent for certain improvements – s 15*[12]. Any question arising under this section requires to be determined by arbitration[13].

1 [1993] 2 All ER 673.
2 Agriculture (Maintenance, Repair and Insurance of Fixed Equipment) Regulations 1973, SI 1973/1473.
3 At 681e.
4 1991 Act, s 14(b).
5 Ibid, s 46(2).
6 See p 76.
7 1991 Act, s 7(3).
8 *Hill v Wildfowl Trust (Holdings) Ltd* 1995 SCLR 778.
9 See p 75 'Permanent pasture'.
10 See chap 7.
11 1991 Act, s 13.
12 See p 55 'Increase of rent for landlord's improvements'.
13 1991 Act, s 15(3).

(h) *Payment for implements etc – s 19*[1]. Any question as to the compensation payable by the landlord, where the landlord has failed to pay the agreed price for implements etc sold on quitting the holding, falls to be determined by statutory arbitration.

If it is the incoming tenant that has failed to pay the price, the arbitration will be a common law arbitration[2].

It should be noted that the ascertainment of the price to be paid by valuation required by s 19(1) is a common law arbitration or valuation[3].

(i) *Section 22(2) notices to quit – s 23(2)*[4]. Any question which arises under a s 22(2) notice to quit falls to be determined by arbitration.

If the notice to quit is served under s 22(2)(d) and relates to a demand to remedy breaches of the lease in relation to fixed equipment, then the arbiter also has the powers conferred by s 66 to modify the demand to remedy[5].

(j) *Reduction of rent on dispossession of part of holding – s 31*. Where the tenant is dispossessed of part of the holding either by notice to quit[6] or by a resumption, the amount of the reduction of rent requires to be determined by arbitration.

(k) *Compensation for failure to repair etc fixed equipment – s 46*. Where the arbiter transfers liability for fixed equipment from the tenant to the landlord under s 4, the amount of compensation payable by the tenant for past failures to discharge the liability shall be determined by arbitration[7].

(l) *Compensation where holding divided – s 50*[8]. Where a holding is divided and the rent has not been apportioned between the several persons who constitute the landlords, the arbiter is required to apportion the amount of any compensation awarded to the tenant under the Act between the several persons who constitute the landlord.

(m) *Compensation for game damage – s 52*. The amount of compensation payable to a tenant under a statutory claim for game damage 'shall, in default of agreement made after the damage has been suffered, be determined by arbitration'[9].

Where the right to kill and take game is vested in some person other than the landlord, the landlord is entitled to be indemnified by that person against all claims for compensation under s 52. Any question relating to the landlord's claim for indemnification 'shall be determined by arbitration'[10]. This is a rare instance of the statutory arbitration procedure being extended to a third party.

(n) *Claims on termination of tenancy – s 62*. Any claim by an agricultural tenant against the landlord or by a landlord against the tenant 'being a claim which arises, under this Act or under any custom or agreement, on or out of the

---

1 See p 221 'Payment for implements etc'.
2 *Duncan*, note to the 1991 Act, s 19(3) in *Current Law Statutes*.
3 1991 Act, s 61(7).
4 See chap 17.
5 *Fane v Murray* 1995 SLT 567; see p 194 'Demands in relation to repair . . .'; s 32 also applies.
6 1991 Act, s 29.
7 Ibid, s 46(2).
8 See p 209 'Landlord's interest divided'.
9 Ibid, s 52(3).
10 Ibid, s 51(4).

termination of the tenancy (or part thereof) shall ... be determined by arbitration'[1].

This provision is 'Without prejudice to any other provision of this Act... '[2].

The most important exception is that any:

'valuations of sheep stock, dung, fallow, straw, crops, fences and other specific things the property of the outgoing tenant, agreed under a lease to be taken over from him at the termination of a tenancy by the landlord or the incoming tenant, or to any questions which it may be necessary to determine in order to ascertain the sum to be paid in pursuance of such an agreement, whether such valuations and questions are referred to arbitration under the lease or not.'[3]

have to be referred to a common law arbitration. This subsection only applies where the lease provides for the items to be taken over. This means that in regard to s 17 items which the tenant has to give the landlord or an incoming tenant the opportunity to buy at valuation, it is a common law arbitration if the lease so provides, but that if the lease does not so provide, then the valuation is under s 60 as between landlord and outgoing tenant or at common law as between incoming and outgoing tenants.

(o) *Milk quota compensation – 1986 Act, Sch 2, para 11.* Claims in respect of milk quota compensation are referred to arbitration under s 60[4] by the 1986 Act[5].

(p) *Apportionment of milk quota – DPQRs 1994, Sch 3.* These are carried out by arbitration unless (1) the holding is a landholding or a statutory small tenancy or (2) all parties interested in the apportionment request the Land Court to carry out the apportionment.

The procedures set out in the schedule are very similar to those specified for a statutory arbitration[6].

'Any question or difference'

The catch all provision of s 60(1) provides:

'except where this Act makes express provision to the contrary, any question or difference between the landlord and the tenant of an agricultural holding arising out of the tenancy or in connection with the holding (not being a question or difference as to liability for rent) shall, whether such question or difference arises during the currency or on the termination of the tenancy, be determined by arbitration.'

The essential prerequisite to invoking arbitration under this provision is that there existed, at the time the question or difference arose, or after the

---

1 1991 Act, s 62(1). This provision may be of significance if the decision in *Hill v Wildfowl Trust (Holdings) Ltd* 1995 SCLR 778 was wrongly decided. It empowers the arbiter to deal with 'any claim' including monetary claims as an arbiter does not, in general, have power to assess or award damages or compensation without express power; 2 *Stair Memorial Encyclopaedia* para 448; see p 275, n 8.
2 1991 Act, s 62(1).
3 Ibid, s 61(7).
4 Parties may agree a joint reference to the Land Court under s 60(2).
5 1986 Act, Sch 2, para 11.
6 See p 273, para (h).

termination of a tenancy where there had existed, a relationship between land-lord and tenant[1] of an agricultural holding and that the question of difference is not 'as to liability for rent'[2].

The relationship of landlord and tenant must have existed at the time the question or difference arose or have existed at the time of the termination of the tenancy[3]. It does not matter that the relationship had existed in the past, if it does not exist at the date of the dispute[4]. An agreement between the pur-chaser of a farm prior to entry and the existing tenant, that on entry the tenant would vacate and the new proprietor would take over certain crops, stock and implements at valuation was not a question or difference under the Act, as it did not arise out of the termination of the tenancy, but out of an agreement made between the parties before the relationship of landlord and tenant came into existance.

A 'question or difference as to liability for rent' is

'confined to cases in which liability to pay rent sued for is disputed upon grounds which if sustained, in law extinguish "liability" eg where it is asserted that the rent has in fact been paid in whole or in part ... or where it is asserted that liability for the sums sued for has been discharged by some other transaction personal to the parties and wholly extraneous to the lease and to the relationship of landlord and tenant. But where a ten-ant defends an action of payment of rent by asserting the right of retention, "his liabil-ity" for payment of the rent is not in issue, but is on the contrary admitted.[5]'

Rent arbitrations under s 13 or questions relating to variations of rent arising under the statute or lease are not a question or difference as to *liability for rent*, but are questions relating to the amount of rent to be paid[6].

There is no distinction in 'question or difference' arising between landlord and tenant as between matters of arbitration and valuation, both of which fall to be determined by one arbiter in terms of the statute, unless express provi-sion is made by the 1991 Act to the contrary[7].

An arbiter would appear to have the power to assess damages for breach of obligations under a lease as a 'question or difference', notwithstanding the general rule that an arbiter cannot assess or award damages without express authority[8]. An arbiter cannot assess damages if they arise *ex delicto*[9].

---

1 Cf the 1991 Act, s 85(1) for the extended definitions of 'landlord' and 'tenant'.
2 *Brodie v Ker, McCallum v McNair* 1952 SC 216.
3 *Waddell v Howat* 1925 SC 484.
4 *Hendry v Walker* 1926 SLT 679.
5 *Brodie v Ker, McCallum v McNair* 1952 SC 216 at 226. Cf *Galbraith v Ardnacross Farming Co Ltd* 1953 SLT (Notes) 30, where an action for payment of sheep stock valuation was sisted because the landlord claimed to be entitled to set off an illiquid claim for dilapidations etc on termination of tenancy; *Thorburn's Trs v Ormiston* 1937 SLT (Sh Ct) 26.
6 *Boyd v Macdonald* 1958 SLCR 10.
7 *Stewart v Williamson* 1910 SC (HL) 47.
8 *Hill v Wildfowl Trust (Holdings) Ltd* 1995 SCLR 778; *Thorburn's Tr v Ormston* 1937 SLT (Sh Ct) 26 under the slightly different wording of the arbitration provision in the 1923 Act, s 15(1) (amended by the 1931 Act) which also refers to arbitration as 'any claim'. Cf *McDiarmid v Secretary of State for Scotland* 1970 SLT (Land Ct) 17 at 20L; *Tustian v Johnston* [1993] 2 All ER 673 which appear to be to the contrary. In *Aberdeen Rly Co v Blaikie* (1852) 15 D (HL) 20 the phrase 'all disputes and differences' (which is similar to 'question or difference') was con-strued not to include a power to assess or award damages. 2 *Stair Memorial Encyclopaedia* para 448.
9 *McDiarmid v Secretary of State for Scotland* 1970 SLT (Land Ct) 17.

In defended court proceedings for, eg, interdict[1], removings[2], resumption[3] or decree *ad factum praestandum*[4], where the arbiter does not have jurisdiction, incidental questions such as those relating to whether or not one of the parties is in breach of his obligations under the tenancy, or whether or not the defence to the removing is relevant, will have to be determined by arbitration, while the court action is sisted.

Statutory arbitration procedures[5]

A statutory arbitration is subject to the same principles and procedures as those which apply to a common law submission, except where the principles and procedures are specifically modified by statute[6].

Where statute provides for particular times limits within which the arbitration has to be commenced, those time limits are mandatory and failure to demand arbitration within the time limit means that the right to have the claim or question determined is lost. Where there is no time limit, delay in demanding arbitration may bar the claim on the grounds of waiver or personal bar.

(a) Date of appointment

The date of the appointment of the arbiter is crucial, because time limits relate to that date.

Where the arbiter is appointed by agreement, the date of appointment is the date the submission is executed and not the date of the arbiter's acceptance of the submission[7]. In England it has been held that, where the arbiter has informed the parties that he will accept an appointment, the date of appointment is the date he receives the submission and not the date at which he informs parties that he received it[8].

Where the arbiter is appointed by the Secretary of State or Land Court the date of appointment is the date of the formal appointment and not the date the parties receive copies of it[9].

---

1 Eg 1991 Act, s 7(3) and (4).
2 *McCallum v MacNair* 1952 SC 216 - validity of notice to quit.
3 *Houison-Craufurd's Trs v Davies* 1951 SC 1.
4 *Tustian v Johnston* [1993] 2 All ER 673.
5 See the 1991 Act, Sch 7.
6 Eg ibid, Sch 7; DPQRs 1994, Sch 3; *Mitchell-Gill v Buchan* 1921 SC 390 at 395 per the Lord Principal.
7 *Sheriff v Christie* (1953) 69 Sh Ct Rep 88 at 92.
8 *Hannaford v Smallacombe* [1994] 1 EGLR 9. It was said that if the parties had provided in the submission that it would not be enforceable until they received formal notification from the arbiter of his receipt of the submission, then that might be the date of appointment.
9 *Suggett v Shaw* 1987 SLT (Land Ct) 5. Cf *University College, Oxford v Durdy* [1982] Ch 413, not followed in *Suggett*, where the Court of Appeal suggested obiter that 'date of appointment' should be construed to mean the date at which the parties received notification of the appointment.

(b) Statement of case

The parties require to deliver to the arbiter 'within 28 days[1] from the appointment of the arbiter'[2] their statement of case with all necessary particulars[3].

The arbiter has no power to extend the statutory time limit[4], unless the other party consents or waives its right to object[5]. Where no case has been lodged timeously the party is restricted to challenging the other party's case, but cannot attempt to set up its own case[6].

If the party that has failed to lodge a case is the party on which the onus lies to establish its case, then the respondent has no case to answer[7].

The statement of case, which now normally incorporates the particulars of the claim[8], must be sufficient to give fair notice of the basis and nature of the claim and the particular contractual or statutory provision upon which it is founded[9].

No amendment or addition to the statement of case or particulars is allowed after the 28 days without the consent of the arbiter. If the documents lodged do not amount to a 'statement of case', then that defect cannot be cured by amendment[10].

Normally the arbiter should allow a period of adjustment after the statements of case have been lodged or answers to the other party's case, because neither party can fully anticipate what will be in the other party's case. Such adjustment or answers may include new material, not included within the original statement of case[11].

When the adjustment period is over, amendment of the case can only be made with the arbiter's consent, which is a matter of discretion[12]. In the exercise of his discretion the arbiter should be guided by the practice of the courts and take into account such matters as whether the amendment focuses the real issues between the parties, the time at which amendment is sought to be made and prejudice to the other party[13].

Parties are 'confined at the hearing to matters alleged in the statement and particulars so delivered and any amendment thereof or addition thereto duly

1 See p 6 'Within [21] days after'. In England the time limit is now 35 days.
2 1991 Act, Sch 7, para 5.
3 Cf *Duke of Montrose v Hart* 1925 SC 160 for the meaning of 'particulars' under the 1923 Act.
4 *Jamieson v Clark* (1951) 67 Sh Ct Rep 17; *Stewart v Brims* 1969 SLT (Sh Ct) 2; *Collett v Deeley* (1949) 100 LJ 108; *Hannaford v Smallacombe* [1994] 1 EGLR 9.
5 *Suggett v Shaw* 1987 SLT (Land Ct) 5.
6 *Jamieson v Clark* (1951) 67 Sh Ct Rep 17; *Collett v Deeley* (1949) 100 LJ 108.
7 *Gill* para 628. In rent arbitrations the arbiter has a duty to 'determine, in accordance with subsections (3) to (7) below the rent properly payable' – 1991 Act, s 13(2). He has to carry out this duty whether or not any cases have been lodged; cf *Suggett v Shaw* 1987 SLT (Land Ct) 5 at 6.
8 See *Gill* para 628.
9 *Simpson v Henderson* 1944 SC 365; *Adam v Smythe* 1948 SC 445; cf *Duke of Montrose v Hart* 1925 SC 160 for a consideration of what 'particulars' needed to be given under the 1923 Act prior to arbitration.
10 *Robertson's Trs v Cunningham* 1951 SLT (Sh Ct) 89 at 92.
11 *Strang v Abercairney Estates* 1992 SLT (Land Ct) 32.
12 1991 Act, Sch 7, para 5(a).
13 *Strang v Abercairney Estates* above; *E D & A D Cooke Bourne (Farms) Ltd v Mellows* [1983] QB 104; I D Macphail *Sheriff Court Practice* (W Green, 1988) paras 10–14 to 10–22.

made'[1]. The provision is mandatory and cannot be waived by consent of the parties[2].

(c) Evidence and procedure

The parties to the arbitration 'and all persons claiming through them' require to submit to be examined by the arbiter[3]. Witnesses, subject to legal objection, are examined on oath or affirmation[3], if the arbiter thinks fit[4]. The arbiter is empowered to administer oaths or affirmations[5].

Parties are required to produce before the arbiter all 'samples, books, deeds, accounts, writings and documents, within their possession or power respectfully which may be required or called for'[6]. The normal procedure, if productions are not produced, is to enrol a specification which the arbiter authorises and then recommends to the courts to grant commission and diligence, if the productions are not produced.

The arbiter is engaged as an expert in agricultural matters and valuations. He may and should use his general knowledge of values and his experience in reaching his decision. In rent arbitrations the arbiter may have to use his investigatory powers to obtain proper evidence or to introduce comparables, 'because he has to fulfil his statutory duty and arrive at the rent properly payable'[7].

The arbiter may not take into account some matter of specific knowledge, without giving the parties an opportunity to comment on it[8]. If he forms an opinion from his own expertise he should allow the parties' experts to comment on it[9].

The arbiter has a wide discretion as to procedure, unless the procedure is prescribed by statute[10]. The parties are required to 'do all other things which during the proceedings the arbiter may require'[11].

The arbiter is required to conduct the proceedings in accordance with the rules of natural justice[12]. He may conduct the arbitration informally in a manner appropriate to the remit and the value of the claim[13]. He may order a

1 1991 Act, Sch 7, para 5(b). See *Murray v Fane* (22 April 1996, unreported) Perth Sh Ct.
2 *Stewart v Brims* 1969 SLT (Sh Ct) 2 at 6 and 7. Cf *Jamieson v Clark* (1951) 67 Sh Ct Rep 17. It would be open to a party to seek leave to amend during the course of the hearing.
3 1991 Act, Sch 7, para 6(a).
4 Ibid, Sch 7, para 7; *Maclean v Chalmers Property Investment Co Ltd* 1951 SLT (Sh Ct) 71.
5 Ibid, Sch 7, para 7.
6 Ibid, Sch 7, para 6.
7 *Earl of Seafield v Stewart* 1985 SLT (Land Ct) 35 at 40.
8 *Earl of Seafield v Stewart* above at 40; *Fox v P G Wellfair Ltd* (1982) 263 EG 589, 657. Cf *Fountain Forestry Holdings Ltd v Sparkes* 1989 SLT 853; *Towns v Anderson* 1989 SLT (Land Ct) 17. In both cases the arbiter relied on his own rental comparable, upon which the parties were not given an opportunity to comment.
9 *Fox v P G Wellfair Ltd* above per Dunn LJ.
10 *Strang v Abercairney Estates* 1992 SLT (Land Ct) 32; 2 *Stair Memorial Encyclopaedia* para 442.
11 1991 Act, Sch 7, para 6.
12 *Strang v Abercairney Estates* above; *McNair v Roxburgh* (1855) 17 D 445. Cf *Barrs v British Wool Marketing Board* 1957 SC 72.
13 *Christison's Tr v Callender-Brodie* (1906) 8 F 928; *Paterson v Glasgow Corp* (1901) 3 F (HL) 34; *Gibson v Fotheringham* 1914 SC 987; *Davidson v Logan* 1908 SC 350.

debate. Questions of relevancy or competency are for the arbiter to determine[1]. He may dispense with a proof or hearing if he considers that he has sufficient material before him to dispose of the reference, particularly where the issue is a matter of valuation, which can be determined by inspection[2].

Normally an arbiter will require to carry out an inspection personally[3]. An inspection should be carried out in the presence of both parties. Where the inspection relates to valuation of, eg, growing crops or the present state of the fixed equipment, the arbiter may be able to determine the reference on the inspection alone, with a minimum of further procedure[4]. Where the inspection brings new material to light, the arbiter must allow both parties to comment[5].

(d) Stated case (except in rental arbitrations)

The arbiter may 'at any stage of the proceedings'[6] state a case for the opinion of the sheriff on any question of law arising in the course of the proceedings[7].

Although no provision is made for the issue of proposed findings or draft award, it is common practice to ask an arbiter to issue a draft award, to enable parties (1) to make representations regarding the terms of the draft[8] and (2) to consider whether or not to ask for a stated case[9].

The case can be stated at the request of either party, or on the arbiter's own initiative, or if so directed by the sheriff[10]. The arbiter is under no obligation to state a case if requested by either party[11], but if he intends to refuse the request this should be intimated to the parties so that they can consider making application to the sheriff for a direction before the final award is issued[12].

If a party applies for a stated case and the arbiter does not immediately accede to it, then the party, to safeguard his position, should immediately apply to the sheriff for a direction to the arbiter to state a case[13].

When the arbiter comes to state the case neither the parties, nor the sheriff in directing the arbiter to state a case, can dictate or compel either the form of the findings in fact or the form of the questions in law. The parties can try to

---

1 *Brown v Associated Fireclay Companies* 1937 SC (HL) 42.
2 *Gibson v Fotheringham* 1914 SC 987; *Paterson v Glasgow Corp* (1901) 3 F (HL) 34; *Dundas v Hogg* (1936) 52 Sh Ct Rep 329; *Ledingham v Elphinstone* (1859) 22 D 245.
3 *Fox v P G Wellfair Ltd* (1982) 263 EG 589, 657.
4 *Davidson v Logan* 1908 SC 350 at 367 per Lord Low; *Macnabb v A & J Anderson* 1955 SC 38 at 43 per Lord Russell; *Dundas v Hogg* (1936) 52 Sh Ct Rep 329.
5 *Earl of Seafield v Stewart* 1985 SLT (Land Ct) 35 at 40.
6 Ie at any stage before the final award is delivered to the parties; *Johnson v Gill* 1978 SC 74; *Johnston v Glasgow Corp* 1912 SC 300; *Hendry v Fordyce* (1953) 69 Sh Ct Rep 191; *Fairlie Yacht Slip Ltd v Lumsden* 1977 SLT (Notes) 41.
7 1991 Act, Sch 7, para 20.
8 *M'Laren v Aikman* 1939 SC 222 at 229 where the LJC observed that there is an implied right in every arbitration for parties to make representations to an arbiter about his proposed findings, unless the right was excluded by the reference.
9 A stated case cannot be requested after the formal award is issued.
10 1991 Act, Sch 7, para 20.
11 The arbiter is obliged to state a case if so directed by the sheriff; *Broxburn Oil Co Ltd v Earl of Buchan* (1926) 42 Sh Ct Rep 300.
12 Issue of the final award without intimating that he does not intend to state a case, having been requested to do so, may amount to misconduct; *Johnson v Gill* 1978 SC 74; *Hendry v Forsyth* (1953) 69 Sh Ct Rep 191.
13 *Hendry v Forsyth* above.

influence the arbiter during the adjustment stage of the stated case or, during the course of proceedings before the sheriff, persuade the sheriff to ask the arbiter to amplify or explicate the original case[1].

An arbiter is bound to apply the law as set out in the opinion of the sheriff or, following an appeal, that of the Court of Session, in his award[2]. The opinion of the sheriff is final unless appealed to the Court of Session[3]. It is for the sheriff to deal with the expenses of the stated case in the sheriff court and not the arbiter[4].

It is competent to appeal the sheriff's decision on a stated case to the Court of Session from whose decision no appeal lies[5].

(e) Stated case in rental arbitrations

In rental arbitrations[6] a statutory arbiter[7] may not state a case for the opinion of the sheriff but 'instead the arbiter may at any stage of the proceedings state a case (whether at the request of either party or on his own initiative) on any question of law arising in the course of the arbitration' for the opinion of the Land Court[8].

The comments made in relation to a stated case to the sheriff apply equally in regard to the statement of a case to the Land Court. The Land Court has no power to direct that a case should be stated[9].

The Land Court's decision is final[10]. It is probably competent to ask the Land Court to state a special case for the opinion of the Court of Session during the course of the proceedings before it, but not after the decision is issued[11].

1 *Forsyth-Grant v Salmon* 1961 SC 54; *Gill* para 649(a); *Chalmers Property Co v Bowman* 1953 SLT (Sh Ct) 38.
2 *Mitchell-Gill v Buchan* 1921 SC 390. Failure to apply the law as so stated is misconduct.
3 1991 Act, Sch 7, para 21.
4 *Thomson v Earl of Galloway* 1919 SC 611; *McQuater v Fergusson* 1911 SC 640; *Jamieson v Clark* (1951) 67 Sh Ct Rep 17; *Jack v King* (1932) 48 Sh Ct Rep 242.
5 1991 Act, Sch 7, para 21.
6 Ibid, s 13(1).
7 Ibid, Sch 7, para 22. A privately agreed arbiter cannot state a case for the opinion of the Land Court, but may state a case for the opinion of the sheriff.
8 Ibid, Sch 7, para 22.
9 This is because there is a right of appeal to the Land Court against the decision of the arbiter in a rental arbitration; ibid, s 61(2).
10 Ibid, Sch 7, para 22.
11 Scottish Land Court Act 1993, s 1(7) – 'The Land Court may, if it thinks fit, and shall on the request of any party, state a special case on any question of law arising in any proceedings pending before it . . . for the opinion of the Inner House of the Court of Session'. Scottish Land Court Rules 1992, SI 992/2656, r 88 provides for the request of a special case within one month of the 'intimation to the parties of the decision complained of', but this rule does not appear to exclude the right given by s 1(7) to request a special case during the course of the proceedings. The fact that the 1991 Act, Sch 7, para 22 provides that the decision of the Land Court is final means that a special case cannot be requested after the decision is issued in accordance with the time limits in r 88.

(f) The award

The arbiter requires to make and sign his award[1] within three months of his appointment or within such longer period as may be agreed to in writing by the parties or fixed by the Secretary of State[2]. The extension of the time limit may be agreed before or after the expiry of the three-month period or any extension thereof[3]. Where an arbiter fails to issue his award within the time limit, this amounts to misconduct, and the award may be set aside[4].

The award requires to be in the statutory form[5]. A statutory arbiter, but not one appointed by agreement, is obliged to state his reasons, if requested to do so by either party[6]. In a rent arbitration the statutory arbiter requires, whether or not requested to do so, in making his award, to 'state in writing his findings in fact and the reasons for his decision' and to make the statement available to the Secretary of State[7].

In giving reasons the arbiter should bear in mind that:

'The statutory obligation to give reasons is designed not merely to inform the parties of the result of the committee's deliberations but to make it clear to them and to this Court the basis on which their decision was reached, and that they have reached their result in conformity with the requirements of the statutory provisions and the principles of natural justice. In order to make clear the basis of their decision a committee must state (i) what facts they found admitted or proved; (ii) whether and to what extent the submissions of the parties were accepted as convincing or not; and (iii) by what method or methods of valuation applied to the facts found their determination was arrived at.[8]'

The arbiter is required to state separately in his award the amounts awarded in respect of the several claims referred to him[9]. On the application of either party, he is required to specify the amount 'awarded in respect of any particular improvement or any particular matter which is the subject of the award'[10].

Where by virtue of the 1991 Act compensation under an agreement is to be substituted for compensation under the Act for improvements, the arbiter is required to award compensation in accordance with the agreement instead of in accordance with the Act[11].

---

1 See the Agricultural Holdings (Specification of Forms) (Scotland) Order 1991, SI 1991/2154 which requires the arbiter to sign before two witnesses. This requirement has probably not been altered by the Requirements of Writing (Scotland) Act 1995 – see s 7(1) 'Except where an enactment expressly provides otherwise'.
2 1991 Act, Sch 7, para 8.
3 Ibid, Sch 7, para 8; *Dundee Corp v Guthrie* 1969 SLT 93 at 98 per the LJC. The application is made by the arbiter in the Agricultural Holdings (Specification of Forms) (Scotland) Order 1991, Sch 2, Form C.
4 *Halliday v Semple* 1960 SLT (Sh Ct) 11; 1991 Act, Sch 7, para 24.
5 Ibid, Sch 7, para 11; Agricultural Holdings (Specification of Forms) (Scotland) Order 1991.
6 Tribunals and Inquiries Act 1992, s 10(1) and Sch 1, Pt II.
7 1991 Act, Sch 7, para 10. This is best done when the arbiter applies to have his remuneration fixed by the Secretary of State; ibid, s 63(3)(a).
8 *Albyn Properties Ltd v Knox* 1977 SC 108 at 112 per the Lord Principal. See *Wordie Property Co Ltd v Secretary of State for Scotland* 1984 SLT 345 at 348 per the Lord Principal (a planning case); *Earl of Seafield v Stewart* 1985 SLT (Land Ct) 35 at 40 (an agricultural rent arbitration).
9 1991 Act, Sch 7, para 12(a).
10 Ibid, Sch 7, para 12(b).
11 Ibid, Sch 7, para 13.

The award requires to fix a day not later than one month after delivery of the award for payment of the money awarded as compensation, expenses or otherwise[1]. In arbitrations regarding milk quota compensation the period is three months[2].

An arbiter probably has no power to award interest on his award from a date earlier than the date of the award, although he has power to award interest at the judicial rate from that date[3]. It is therefore in the interest of the person claiming the payment of a sum to ensure that the arbitration proceeds as fast as possible.

An arbiter may, if he thinks fit, make an interim award 'for payment of any sum on account of the sum to be finally awarded'[4].

Any award by an arbiter 'as to compensation, expenses or otherwise' can be enforced, if payment is not made within one month after the date on which it becomes payable, by recording for execution in the Books of Council and Session or in the sheriff court books and is enforceable 'in like manner as a recorded decree arbitral'[5].

An award, except in a rent arbitration[6], is final and binding on the parties and the persons claiming under them[7]. In order to be binding, the award requires to be delivered to the parties[8].

The arbiter may, whether before or after delivery, correct in any award any clerical mistake or error arising from any accidental slip or omission[9]. The correction will have to be made *de recenti*. The arbiter cannot alter the substance of the award. He is not entitled to correct errors of judgment, whether of fact or law, or to have second thoughts[10].

(g) Expenses[11]

The expenses of the arbitration are a matter for the arbiter's discretion. He may direct to and by whom and in what manner those expenses or any part thereof are to be paid[12]. In particular the arbiter is required to take into consideration:

---

1 Ibid, Sch 7, para 14.
2 1986 Act, Sch 2, para 11(4).
3 *Farrans (Construction) Ltd v Dunfermline DC* 1988 SC 120; *John G McGregor (Contractors) Ltd v Grampian RC* 1991 SLT 136.
4 1991 Act, Sch 7, para 9.
5 Ibid, s 65; 1986 Act, Sch 2, para 2 (milk quota compensation awards).
6 Against which provision is made for appeal to the Land Court; 1991 Act, s 61(2).
7 Ibid, Sch 7, para 15.
8 Whether or not delivery has taken place, particularly in circumstances where the signed award has been given to the clerk, is a question of fact and circumstances; ie did the clerk hold the award for the arbiter or the parties?; *Johnson v Gill* 1978 SC 74.
9 1991 Act, Sch 7, para 16.
10 Cf *Macphail* paras 5-99 to 5-102 and *Parliament House Book* vol 2, note to Rules of the Court of Session 1994, r 4.15.6.
11 These are the expenses of the arbitration including the expenses of the preparation of a stated case, but not the expenses of the hearing on a stated case; *Thomson v Earl of Galloway* 1919 SC 611.
12 1991 Act, Sch 7, para 17. Cf *MacGregor v Glencruitten Trs* 1985 SLCR 77.

'the reasonableness or unreasonableness of the claim of either party, whether in respect of amount or otherwise, and any unreasonable demand for particulars or refusal to supply particulars, and generally all the circumstances of the case . . .'

The arbiter may disallow the expenses of any witness whom he considers to have been called unnecessarily and any other expenses he considers to have been unnecessarily incurred[1].

The arbiter should normally award expenses on the basis of expenses follow success. In rental arbitrations, an award of no expenses due to or by is probably appropriate if the arbiter's award falls between each party's contentions or tender. An arbiter should take account of any tender that may have been lodged with his clerk.

The clerk's remuneration or expenses can only be included if the clerk was appointed after the submission of the claim and answers either with the consent of the parties or the sanction of the sheriff[2].

The expenses are subject to taxation by the auditor of the sheriff court on the application of either party, but the taxation is subject to review by the sheriff[3].

(h) Appeals from an arbiter's decision

With the exception of an appeal against an award by a statutory arbiter in a rental arbitration[4], there is no appeal against the final award of an arbiter[5].

In a rental arbitration, any party may appeal to the Land Court against the award of the statutory arbiter[6] 'on any question of law or fact (including the amount of the award)'[7]. The appeal must be brought within two months of the date of the issue of the award[8].

The Land Court's decision, on appeal, may be reviewed by way of a special case on a question of law[9]. A special case has to be requested within one month after the date of the intimation of the decision complained of[10].

The question of appeals to the Land Court in rental arbitrations are more fully dealt with in chapter 7.

(i) Setting aside an award

Where an arbiter has misconducted himself, or an arbitration or award has been improperly procured, the sheriff may set the award aside[11].

1 1991 Act, Sch 7, para 18.
2 Ibid, Sch 7, para 19.
3 Ibid, Sch 7, para 17.
4 Ibid, ss 13(1) and 61(2).
5 Ibid, Sch 7, para 15.
6 There is no appeal to the Land Court from the award of a private arbiter.
7 1991 Act, s 61(2).
8 Ibid, s 61(3).
9 Scottish Land Court Rules 1992, SI 1992/2656, rr 88-94.
10 Ibid, r 88.
11 1991 Act, Sch 7, para 24. The Court of Session has a concurrent jurisdiction to set aside the award; *Dunlop v Mundell* 1943 SLT 286.

'Misconduct' has come to mean 'any mistake committed by the arbiter in the mode of carrying out the arbitration' and does not necessarily imply any improper conduct on the part of the arbiter[1]. It includes acting contrary to natural justice[2]; failure to comply with the mandatory requirements of the 1991 Act, Sch 7, such as failure to issue the award within the time limits[3]; a failure by the arbiter to ensure that his award complies with the statutory requirements[4]; reliance on a comparable or other specific matter within the arbiter's knowledge without giving the parties an opportunity to comment[5]; failure to give a party time to obtain legal representation[6]; and a failure to observe his own procedural directions[7].

Where there is apparent misconduct the parties are advised to complain at the time, although failure to complain will not in general validate the proceedings[8].

Before an award can be set aside on the grounds of misconduct 'very precise averments must be made'[8].

Where an arbiter is removed, then another arbiter can be appointed in his place[9].

COMMON LAW ARBITRATIONS

As noted above[10] waygoing valuations of the outgoing tenant agreed under the lease to be taken over by the landlord or incoming tenant are common law arbitrations[11].

This also applies to the ascertainment of the price of goods by valuation in respect of the payment to be made by the landlord or incoming tenant for implements etc to be sold in terms of the 1991 Act, s 19(1).

With regard to compensation to be paid, where the landlord fails to pay the price determined under s 19(1) within the time limits, then this is a statutory arbitration. If it is the incoming tenant who fails to pay the price, then any question of compensation falls to be determined by a common law arbitration[12].

1 *Paterson v Glasgow Corp* (1901) 3 F (HL) 34 at 38; cf 2 *Stair Memorial Encyclopaedia* para 438 'Misconduct by arbiter'.
2 *Barrs v British Wool Marketing Board* 1957 SC 72 - 'The question is whether the tribunal dealt fairly and equally with the parties before arriving at its result'. It is not necessarily contrary to natural justice for an arbiter neither to hear parties nor hear evidence; *Dundas v Hogg* (1936) 52 Sh Ct Rep 329; cf *JAE (Glasgow) Ltd v City of Glasgow* 1994 SLT 1164.
3 *Halliday v Semple* 1960 SLT (Sh Ct) 11.
4 *Paynter v Rutherford* 1940 SLT (Sh Ct) 18; *Dunlop v Mundell* 1943 SLT 286.
5 *Towns v Anderson* 1989 SLT (Land Ct) 17; cf *Fountain Forestry Holdings v Sparkes* 1989 SLT 853.
6 *Thomas v Official Solicitor* (1983) 265 EG 601.
7 *Control Securities plc v Spencer* (1989) 07 EG 82.
8 *Maclean v Chalmers Property Investment Co Ltd* 1951 SLT (Sh Ct) 71.
9 *Dundee Corp v Guthrie* 1969 SLT 93.
10 See p 215 'Waygoing valuations'.
11 1991 Act, s 61(7).
12 Ibid, s 19(3); *Duncan*, note to s 19(3) in *Current Law Statutes*.

All arbitrations which are not between landlord and tenant when the issue arose, such as arbitrations between outgoing and incoming tenant or circumstances arising before the parties acquired the relationship of landlord and tenant, are common law arbitrations outwith the provisions of the 1991 Act, Sch 7.

A common law arbitration can be conducted by one or two arbiters with an oversman[1], although the provisions as to arbitration will probably be set out in the lease[2].

A single arbiter or oversman in a common law arbitration, unless there is an express provision to the contrary in the submission, may on the application of a party to the arbitration and shall if the Court of Session on such an application directs, state a case for the opinion of the Court of Session on any question of law arising in the arbitration[3].

The application for the stated case has to be made before the award is issued. An application made after the issue of the award is too late[4].

SHEEP STOCK VALUATION ARBITRATIONS

Sheep stock valuation arbitrations[5] are not conducted in terms of Sch 7. They may be conducted by a single arbiter or by two arbiters and an oversman[6]. It will be for the arbiter to settle his own procedure, unless the procedure is laid down in the lease.

In leases entered into after 10 June 1937 the arbiter may at any stage of the proceedings, and if directed by the sheriff must, state a case for the opinion of the sheriff on any question of law arising in the course of the arbitration[7]. The sheriff's decision is final unless either party appeals against it to the Court of Session[8].

Where a stated case is submitted to the sheriff, if the arbiter is satisfied that whatever the decision the amount ultimately to be found due will not be less than a particular sum, then the arbiter may make an interim award for a sum not exceeding that amount, as he may think fit[9].

Where a lease was entered into before or on 6 November 1946 the sheep stock may be valued by the Land Court on the joint application of the parties,

1 With regard to the position of an oversman see *Gibson v Fotheringham* 1914 SC 987 (the arbiters must have differed before an oversman can be consulted; he can act at the request of one of the arbiters. It is competent to devolve some questions to an oversman, while reserving others for consideration by the arbiters); *Davidson v Logan* 1908 SC 350 (circumstances in which oversman's signature does not invalidate award, oversman and arbiters may inspect subjects together); *Cameron v Nicol* 1930 SC 1 (oversman's signature).
2 As to procedures in common law arbitrations see 2 *Stair Memorial Encyclopaedia* para 448 'Arbitration'; *Halliday* vol 1, chap 14.
3 Administration of Justice (Scotland) Act 1972, s 3. Note that there has to be 'express provision to the contrary' to exclude the right to state a case.
4 *Fairlie Yacht Slip Ltd v Lumsden* 1977 SLT (Notes) 41.
5 1991 Act, ss 68-72.
6 Ibid, s 72(b).
7 Ibid, s 69(1).
8 Ibid, s 69(2).
9 Ibid, s 69(3).

in lieu of being determined in the manner provided in the lease. Where the lease was entered into after that date either party may apply to the Land Court to carry out the valuation[1].

Sheep stock valuations by arbitration or by the Land Court are considered more fully at p 217 ('Sheep stock valuations').

ARBITRATIONS TO APPORTION MILK QUOTA

Arbitrations for the apportionment or prospective apportionment of milk quota are carried out in terms of the DPQRs 1994, Sch 3.

The arbiter has to be appointed by agreement[2] or upon an application to the Secretary of State[3] within 28 days of the change of occupation, otherwise thereafter the arbiter is appointed by the Secretary of State[4]. Where the appointment is by agreement the transferee and the occupier in respect of a prospective apportionment, have to intimate the appointment to the Secretary of State within 14 days of the appointment[5].

The procedural provisions of the arbitration, time limits for lodging statements of case and signing the award, provision for expenses and for a stated case are in most respects similar to those for Sch 7 arbitrations[6].

Any person having an interest in the holding[7] may make representations to the arbiter and so may the Intervention Board if it has initiated the arbitration[8].

THE LAND COURT

The jurisdiction of the Land Court and its procedure are laid down by statute[9]. It has been said[10]:

'The Land Court is a creature of statute and its powers and jurisdiction lie solely within the narrow limits which statute lays down.'

The Land Court often raises the question of jurisdiction as a preliminary, if there is doubt as to its jurisdiction[11]. In some applications to the court, which depend on the competency or validity of, eg,. the notice to quit, the court will explicate its own jurisdiction by determining whether or not the notice to quit is valid[12].

The 1991 Act (and in relation to milk quota the 1986 Act) makes specific provision for reference to the Land Court as follows:

---

1 1991 Act, s 70(1).
2 DPQRs 1994, Sch 3, para 3(1).
3 Ibid, Sch 3, para 3(2).
4 Ibid, Sch 3, para 3(3).
5 Ibid, Sch 3, paras 3(1) and 4(2).
6 See above for the meaning and effect of many of the provisions.
7 Eg owner, landlord, mortgagee, fiar (where there is a liferent).
8 DPQRs 1994, reg 12 and Sch 3, para 12.
9 Scottish Land Court Act 1993; Scottish Land Court Rules 1992, SI 1992/2656.
10 *Garvie's Trs v Still* 1972 SLT 29 at 36 per the LJC.
11 Cf *Craig, Applicants* 1981 SLT (Land Ct) 12.
12 Eg *Eagle Star Insurance Co Ltd v Simpson* 1984 SLT (Land Ct) 37; *O' Donnell v Heath* 1995 SLT (Land Ct) 15.

(a) *Record – s 8(6)*. Any question or difference between the landlord and tenant arising out of the making of the record requires to be referred to the Land Court.

(b) *Objection to legatee or acquirer of lease – ss 11(5) and 12(2)*. Where a landlord objects to the legatee, the legatee may apply to the Land Court for an order declaring him to be the tenant of the lease. Where the landlord objects to an acquirer of a lease, he may apply to the Land Court for an order terminating the lease.

(c) *Consent to the operation of a notice to quit – ss 23, 24, 25 and 32*. Where a notice to quit requires the consent of the Land Court, and the tenant has served a counternotice, the landlord may apply to the court for consent to the operation of the notice to quit.

(d) *Certification of bad husbandry – s 26*. The landlord may apply to the Land Court for a certificate of bad husbandry[1] as a prelude to serving an incontestable notice to quit[2].

(e) *Consent to improvement – s 39*. Where a tenant has given notice of intention to carry out an improvement[3] to which the landlord has given written notice of objection, the tenant may apply to the Land Court for consent to carry out the improvement. The Land Court may approve the carrying out of the improvement conditionally or unconditionally[4].

(f) *Consent to treat holding as market garden – s 41*. On the application of the tenant, the Land Court may direct that all or part of a holding may be treated as a market garden.

(g) *Question as to purpose for which tenancy is being terminated – s 55(7)*. Such questions require to be referred to the Land Court.

(h) *Joint application in lieu of arbitration – s 60(2)[5]*. On a joint application the landlord and the tenant may agree that 'Any question or difference which by or under this Act or under the lease is required to be determined by arbitration' may be determined by the Land Court. The procedure is that of the Land Court and not arbitration under Sch 7.

(i) *Rental appeals – s 61(2) and Sch 7, para 22*. The award of a statutory arbiter appointed by the Secretary of State or Land Court[6] under s 13 determining the rent payable for a holding may be appealed to the Land Court on any question of law or fact, including the amount of the award[7].

---

1 See eg *Austin v Gibson* 1979 SLT (Land Ct) 12; *McGill v Bichan* 1982 SLCR 33; *Buchanan v Buchanan* 1983 SLT (Land Ct) 31; *Cambusmore Estate Trust v Little* 1991 SLT (Land Ct) 33.
2 1991 Act, s 22(2)(c).
3 Under ibid, s 38.
4 See eg *Taylor v Burnett's Trs* 1966 SLCR 139; *Fothringham v Fotheringham* 1978 SLCR 144; *Hutchison v Wolfe Murray* 1980 SLCR 112; *Renwick v Rodger* 1988 SLT (Land Ct) 23; *MacKinnon v Arran Estate Trust* 1988 SLCR 32.
5 It is understood that consideration is being given to amend this section, to allow a reference to the Land Court in lieu of arbitration to be made at the instance of either party.
6 But not of an arbiter agreed between the parties.
7 See eg *Aberdeen Endowments Trust v Will* 1985 SLCR 38; *Towns v Anderson* 1989 SLT (Land Ct) 17.

During the course of a rental arbitration, a statutory arbiter may state a case for the opinion of the Land Court whose decision is final[1].

(j) *Appointment of arbiter, where Secretary of State is a party to the tenancy – s 64*[2]. Where the Secretary of State is a party to the tenancy, an arbiter falls to be appointed by the Land Court[3]. The arbiter's remuneration is also fixed by the court[4].

(k) *Termination of tenancy in event of tenant's failure to remedy breach of tenancy in relation to fixed equipment by set date – s 66.* Where an arbiter specifies a period within which a breach of a repair and maintenance obligation in relation to buildings and fixed equipment should be remedied, either the arbiter or the landlord may apply to the Land Court for a date for the termination of the tenancy by notice to quit in the event that the tenant fails to remedy the breach.

(l) *Valuation of sheep stock – s 70.* Where a lease was entered into before or on 6 November 1946, on a joint application or in respect of a lease entered into after that date, on the application of either party, the Land Court may carry out a valuation of sheep stock, in lieu of it being determined in the manner provided in the lease.

(m) *Land Court to act in lieu of Secretary of State – s 80.* Where the Secretary of State is landlord or tenant of a holding, the Land Court acts in place of the Secretary of State in any matter which under the Act is to be referred to the Secretary of State for a decision or which relates to an arbitration concerning the holding. This also applies in relation to milk quota arbitrations[5].

(n) *Milk quota compensation – 1986 Act, Sch 2, para 11(1).* Claims for milk quota compensation in respect of a landholding under the 1911 Act or a statutory small tenancy have to be determined by the Land Court.

In respect of an agricultural holding, on the joint application by the landlord and tenant, the Land Court may determine the compensation in place of an arbiter.

(o) *Apportionment of milk quota – DPQRs 1994, Sch 3, para 1(2) and (3).* The Land Court, in respect of an agricultural holding, may carry out an apportionment of milk quota if all parties interested in the apportionment request the Land Court to do so, within 28 days after the change of occupation.

Where the holding is a landholding under the 1911 Act or a statutory smallholding, the apportionment is to be carried out by the Land Court.

---

1 Eg *Moll v McGregor* 1990 SLT (Land Ct) 59; see p 280. It may be competent to ask the Land Court to state a special case for the opinion of the Court of Session during the proceedings, but not after the decision is issued; see p 280, n 11.
2 Cf the 1991 Act, s 80.
3 *McDiarmid v Secretary of State for Scotland* 1970 SLT (Land Ct) 17. The arbiter so appointed should not be on the panel of arbiters; *Secretary of State v John Jaffray* 1957 SLCR 27; *Commissioners of Crown Lands v Grant* 1955 SLCR 25.
4 *Secretary of State for Scotland v Brown* 1993 SLCR 41.
5 1986 Act, Sch 2, paras 10 and 11.

Where an apportionment is carried out by the Land Court the party applying to the court has to intimate the application to the Secretary of State for Scotland within 14 days of its lodgement in court.

Any person having an interest in the holding[1] is entitled to be a party to the proceedings, as may the Intervention Board where it has initiated the arbitration[2].

Statutory reference is made to the Land Court under the following Acts:

(1) *The Landholders Acts 1886 to 1931.* Issues arising between landlord and tenant of a landholding, and in some cases for a statutory small tenancy, are referred to the Land Court by statute[3].

The particular jurisdictions of the Land Court under these Acts are dealt with in chapter 22.

A claim for compensation for improvements for a statutory small tenancy are referred to arbitration under the 1931 Act unless both parties agree on a reference to the Land Court[4].

(2) *Agriculture (Scotland) Act 1948 – ss 41 and 57.* The Land Court has a jurisdiction under the Act (i) to determine whether s 41 expenses incurred by the Secretary of State are reasonable and (ii) to determine whether a person who incurs expenses in connection with obligations under ss 39 and 40 should be indemnified by another party.

The Land Court can be required by the Secretary of State to conduct an inquiry in connection with a proposal to exercise powers to acquire land compulsorily to ensure its efficient use for agriculture.

(3) *Land Drainage (Scotland) Act 1941 – s 1.* There is a right of appeal to the Land Court against a decision of the Secretary of State concerning the allocation of the expense of and damage incurred by public drainage operations.

(4) *Deer (Scotland) Act 1959 – s 10.* Where the Red Deer Commission is enforcing a control scheme and the expenses incurred exceed the proceeds of the sale of the carcasses, the excess is recoverable from the owner or occupier. The owner or occupier may appeal to the Land Court against the amount sought to be recovered from him and the Land Court has power to vary the amount recoverable, if it appears equitable to do so.

THE ORDINARY COURTS

The following relevant specific jurisdictions are conferred on the ordinary courts.

---

1 Eg owner, landlord, mortgagee, fiar (where there is a liferent).
2 DPQRs 1994,, reg 12 and Sch 3, para 28.
3 See *Scott* for a more detailed consideration of the jurisdiction of the Land Court.
4 1931 Act, s 34.

Sheriff court[1]

(a) Removal of tenant for non-payment of rent for six months[2].

(b) Stated cases by arbiter (except in rent arbitrations where there is a statutory arbiter) on any question of law[3].

(c) Removal of arbiter for misconduct and the setting aside of an award[4].

(d) Sanctioning the appointment of an arbiter's clerk[5].

(e) Review of decision of the auditor of the sheriff court on the question of the arbiter's remuneration where the arbiter has been appointed by agreement[6] and in respect of expenses[7].

(f) In sheep stock valuations, the sheriff may set aside the award of an arbiter, where it fails to comply with the statutory provisions[8].

(g) In sheep stock valuations, where the lease was entered into after 10 June 1937, the arbiter may at any stage of the proceedings, and shall if directed by the sheriff, state a case for the opinion of the sheriff on any question of law.

(h) Where the landlord or the tenant of an agricultural holding is a minor or of unsound mind, not having a tutor, curator or other guardian, the sheriff may, on the application of any person interested, appoint to him for the purposes of the 1991 Act, a tutor or curator and may recall the appointment, if and as the occasion requires[9].

(i) On a summary application, made within one year of the tenant's death, the sheriff is given power to extend the one year in which the executors may transfer the deceased's interest under the lease[10].

Court of Session

(a) There is an appeal to the Court of Session from the decision of a sheriff on a stated case from an arbiter both under Sch 7[11] and in respect of a sheep stock valuation[12].

---

1 Where the matter arises from the 1991 Act, the jurisdiction can be exercised by the sheriff or Sheriff Principal and there is no appeal from the sheriff to the Sheriff Principal. Cf s 67 of the 1991 Act. Other issues arising under the normal jurisdiction are dealt with by the Ordinary Cause Rules 1993 or as otherwise provided in the particular statutes.
2 1991 Act, s 20.
3 Ibid, Sch 7, paras 20 and 21.
4 Ibid, Sch 7, paras 23 and 24.
5 Ibid, Sch 7, para 19.
6 Ibid, s 63(3)(b).
7 Ibid, Sch 7, para 17.
8 Ibid, s 68(4).
9 Ibid, s 77.
10 Succession (Scotland) Act 1964, s 16.
11 1991 Act, Sch 7, para 21.
12 Ibid, s 69(2).

(b) The Court of Session has jurisdiction to deal with special cases stated by the Land Court in any proceedings pending before it[1].

The ordinary courts retain their jurisdiction, except in so far as it is excluded by statute. The most common jurisdictions that the courts are called upon to exercise in relation to agricultural tenancies include:

(a) *Determination of whether or not the relationship of landlord and tenant of an agricultural holding exists.* The courts retain the prime jurisdiction to determine whether or not the relationship of landlord and tenant of an agricultural holding exists[2].

(b) *Removings.* The courts retain the right to remove an occupier of an agricultural holding, after the lease has terminated[3].

Incidental questions, such as the validity of a notice to quit or whether or not the tenant has incurred an irritancy, may have to be determined by arbitration[4].

(c) *Claims arising out of unlawful occupation of a holding.* If a tenant remains on in a holding after the termination of a tenancy, claims such as damages for the unlawful occupation are for the courts to determine[5].

(d) *Interdict.* Any action of interdict between the landlord or tenant will have to be initiated in the courts, although any question or difference arising out of the tenancy or in connection with the holding which arise in the proceedings will have to be determined by arbitration[6].

(e) *Specific performance.* Only the courts can order specific performance of contractual obligations under the lease and, failing implement, authorise the party concerned to carry out the works at the sight of the court and recover the cost from the other party[7].

The application will have to be sisted so that an arbiter may determine whether or not there has been a breach of lease and the extent of the breach[8].

(f) *Succession to a lease.* It is for the courts to determine questions such as whether or not a bequest of a lease is valid, whether the notice given to the landlord by the legatee or acquirer of the lease was valid, and whether the transfer by the executors was effective[9].

1 Scottish Land Court Act 1993, s 1(7).
2 *Brodie v Kerr* 1952 SC 216; *Donaldson's Hospital v Esslemont* 1925 SC 199, 1926 SC (HL) 68; *Exven Ltd v Lumsden* 1982 SLT (Sh Ct) 105; cf *Craig, Applicants* 1981 SLT (Land Ct) 12. Consideration is being given to transferring the jurisdiction to the Land Court.
3 *Hendry v Walker* 1924 SC 757 (declarator and removing where there are different dates for removal for the arable lands and for the house and grass lands); *Rotherwick's Trs v Hope* 1975 SLT 187 (declarator and removing).
4 Cf *Dept of Agriculture for Scotland v Fitzsimmons* 1940 SLT (Sh Ct) 37, where a preliminary plea seeking a reference to arbitration in an action for declarator that a tenancy had incurred an irritancy and for removing was refused where the tenant admitted the sequestration, because there could be no dispute for the arbiter to determine.
5 *Hendry v Walker* 1926 SLT 679, 1927 SLT 333.
6 Cf the 1991 Act, s 7(3) and (4); interdict to restrain the tenant's right of freedom of cropping, where the certificate of the arbiter is stated to be conclusive proof of the facts stated therein.
7 *Davidson v Macpherson* (1899) 30 SLR 2; *Commissioners of Northern Lighthouses v Edmonston* (1908) 16 SLT 439; cf *Tustian v Johnston* [1993] 2 All ER 673.
8 *Tustian v Johnston* above.
9 *Garvie's Trs v Still* 1972 SLT (Land Ct) 29; *Garvie's Trs v Garvie's Tutors* 1975 SLT 94.

It will be for the courts to determine whether or not an an executor acted *auctor in rem suam*[1].

(g) *Supervisory jurisdiction over arbiters.* The court retains the power to interdict an arbiter to prevent him acting, where he has no jurisdiction[2].

The Court of Session retains a concurrent jurisdiction with the sheriff court[3] to remove an arbiter who has misconducted himself or set aside an award obtained in circumstances where the arbiter has misconducted himself or where the arbitration or award has been improperly procured[4].

LANDS TRIBUNAL FOR SCOTLAND

Any dispute as to the additional payments due in consequence of compulsory acquisition[5] have to be referred to and determined by the Lands Tribunal for Scotland[6].

Where the landlord, farmer or tenant of an agricultural holding has a claim for compensation in respect of his holding under the Land Compensation (Scotland) Act 1973, the compensation falls to be assessed by the Lands Tribunal for Scotland.

Where a counternotice is served claiming that the remainder of an agricultural unit or holding cannot be farmed on its own or in conjunction with other relevant land as an agricultural unit, it is for the Lands Tribunal to determine the validity or otherwise of the notice[7].

Questions of disputed compensation arising from the confirmation of a nature conservation order fall to be determined by the tribunal[8].

The tribunal 'may also act as arbitrator under a reference by consent'[9]. Parties should consider using this little known provision in agreed arbitrations on disputed compensation involving compulsory purchase claims, where perhaps statute refers the matter to arbitration or the parties privately agree arbitration under threat of a CPO.

An agreed reference has been used to assess disputed compensation (1) due for an SSSI management agreement[10], (2) in respect of minerals sterilized under a wayleave agreement for a gas distribution pipeline with British Gas[11].

1 *Inglis v Inglis* 1983 SC 8; *Sarris v Clark* 1995 SLT 44.
2 *Cormack v McIldowie's Exrs* 1974 SLT 178; cf *Donaldson's Hospital v Esslemont* 1925 SC 199, 1926 SC (HL) 68.
3 Conferred on the sheriff court by the 1991 Act, Sch 7, paras 23 and 24.
4 *Dunlop v Mundell* 1943 SLT 286.
5 1991 Act, s 56.
6 Ibid, Sch 8, para 1.
7 1973 Act, ss 50(1), 52(1), 55(2) and 58(2).
8 Wildlife and Countryside Act 1981, s 30(8).
9 Lands Tribunal Act 1949, s 1(5).
10 Wildlife and Countryside Act 1981, s 50(3); *Cameron v NCC* 1991 SLT (Lands Tr) 85 and 101.
11 *Jackson v British Gas plc* (26 Aug 1994, unreported) Lands Tribunal.

EUROPEAN COURT OF JUSTICE

Although there are some limited rights of direct action in the European Court of Justice or the Court of First Instance by an individual, they will not be considered here[1].

As EU law impinges to a significant degree on agriculture either through the CAP or through environmental legislation, EU law questions are likely to arise more often in agricultural arbitrations and in cases before the Land Court.

One of the principal jurisdictions of the European Court of Justice is to deal with references for a preliminary ruling on the interpretation of EU legislation, which can then be applied by the national court or tribunal[2]. The Scottish courts provide by their rules of court for the application for and making of such a reference[3].

A statutory arbiter appointed by the Secretary of State[4] may be entitled to make a reference under the treaty, although the matter is not entirely clear[5]. A statutory arbiter would be advised, rather than to make a reference to the European Court of Justice himself, to state a case for the opinion of the sheriff court or Land Court[6] and leave it to those courts to make the reference.

A private arbiter or common law arbiter appointed by the parties probably cannot competently make a reference to the European Court of Justice, although again this is not entirely clear in respect of a private arbiter acting under Sch 7[7]. A private arbiter should state a case for the opinion of the sheriff court, or in the case of a common law arbiter, for the opinion of the Court of Session[8], and leave it to those courts to make the reference.

1 See 10 *Stair Memorial Encyclopaedia* paras 56–70.
2 See 10 *Stair Memorial Encyclopaedia* para 68.
3 Court of Session Rules 1994, chap 55; Ordinary Cause Rules 1993, chap 38. The Land Court has no specific rules on a reference, although no doubt it can competently make a reference.
4 Eg under the 1991 Act or the DPQRs.
5 See 10 *Stair Memorial Encyclopaedia* para 241.
6 In a rental arbitration under the 1991 Act, s 13 and Sch 7, para 22.
7 *Nordsee Deutsche Hochseefischerei Gmbh v Reederei Mond Hochseefischerei Nordstern AG & Co KG* (Case 102/81) [1982] ECR 1095.
8 Administration of Justice (Scotland) Act 1972, s 3.

CHAPTER 24

Milk quotas

Key points

- Milk quota is probably secured by a standard security, although any creditor would be wise to stipulate for specific mention of it in the standard security.

- Any prospective transferee of milk quota by way of a grazing lease should ensure that the lessor is not prohibited from granting a lease under a standard security. Any lease taken in contravention of a standard security and any consequential transfer of milk quota is probably reducible at the instance of the security holder.

- Following upon a transfer of milk quota it is essential that a Form MQ1 is submitted to the Intervention Board (IB) within the time limits otherwise the transferee is not entitled to treat the transferred quota as his in that quota year: DPQRs 1994, reg 7(3).

- Where a producer has special quota of the kind referred to in EC Council Regulation 2055/93, art 1(1) registered as his, then a sale or lease of all or part of the holding before 1 October 1996 will have the effect that the special quota shall be transferred to the national reserve: DPQRs 1994, reg 8(1) and (2).

- A producer who has not made any deliveries or direct sales or temporarily leased quota during the previous quota year will find that his quota has been taken into the national reserve: DPQRs 1994, reg 32(2).

- A producer whose quota has been confiscated has to inform the IB that he wishes to retain the right to seek a restoration of the milk quota.

- Milk quota is transferred by any lease of land for more than eight months. Any milk producer who leases grazing land is advised to take a grazing lease for less than eight months, otherwise some of his quota may be transferred inadvertently to the lessor of the grazings: DPQRs 1994, reg 7(6)(iii).

Time limits

- Before 1 March annually any producer who wishes to benefit from the provisions regarding higher than norm fat content of milk produced from a breed of cows that produces a higher than norm fat content has to confirm to the IB that he has and will maintain a breed of cows with characteristics similar to those in the herd for the first 12 months of production: DPQRs 1994, reg 19.

- No later than 28 days after a change of occupation and in any event no later than seven working days after the end of the quota year in which the transfer takes places, the transferee requires to submit a Form MQ1 to the IB: DPQRs 1994, reg 7(1).

- Where a producer has special quota registered as his and intends to transfer the whole or part of the holding before 1 October 1996, then he requires to submit the prescribed form to the IB at least 28 days before the transfer takes place: DPQRs 1994, reg 9(1).

- Within 45 days of the end of the quota year each purchaser has to give the IB a list of producers who have not made deliveries to him during the year: DPQRs 1994, reg 33(1).

- Any direct selling producer who fails to submit a declaration under EC Commission Regulation 536/93, art 4(2) within 30 days of a notice sent to him by the IB will have his quota taken into the national reserve: DPQRs 1994, reg 33(2)(b).

- Any producer who receives a notification of confiscation requires within 28 days of receipt of the notification to notify any person with an interest in the holding of the content of the notification: DPQRs 1994, reg 33(5)(a).

- Any producer who receives a notification of confiscation requires within six months of the notification to submit a notice to the IB indicating whether he wishes to retain the right to request restoration of the milk quota: DPQRs 1994, reg 33(5)(b).

- A producer who has notified the IB that he wishes to retain the right to restoration of quota may request the IB to restore the quota to him, provided the notice is received by the IB at least six months before the end of the six-year period from the beginning of the quota year in which it was withdrawn: DPQRs 1994, reg 33(6).

- Where a producer has notified the IB that he wishes to retain the right to restoration of quota, and there is a change of occupation of all or part of the holding, the new occupier may request the IB to restore to him the quota relating to that holding or that part, provided the notice is received by the IB at least six months before the end of the six-year period from the beginning of the quota year in which it was withdrawn or within six months of the change of occupation, whichever is the earlier: DPQRs 1994, reg 33(7).

- A producer who has quota restored to him has to make deliveries or direct sales of quota within six months of the restoration or before the end of the six-year period from the beginning of the year in which the quota was withdrawn, whichever is the earlier, or the quota reverts to the national reserve: DPQRs 1994, reg 33(9)(d).

- A producer who has quota restored after a change of occupation has to make deliveries or direct sales of quota within 18 months of the change of occupation or before the end of the six-year period from the beginning of the year in which the quota was withdrawn, whichever is the earlier, or the quota reverts to the national reserve: DPQRs 1994, reg 33(9)(e).

- Deadlines for the conversion of quota from direct sales to wholesale and vice versa are (i) 31 December for permanent conversion and (ii) 28 April following the end of the quota year in which a temporary conversion takes place: DPQRs 1994, reg 18(3).

- The IB has to be notified of a lease of milk quota no later than 15 December: DPQRs 1994, reg 15(3). A producer has to inform his purchaser of any change to his quota within seven working days of the change[1].

- A direct seller or a purchaser requires to inform the IB of the total direct milk sales or actual milk products delivered to the purchaser within 45 days of the end of the quota year, otherwise the direct seller can be charged with levy at a penalty rate or the purchaser lose the benefit of any reallocation of quota: DPQRs 1994, Sch 5, paras 8 and 17.

1 Note to Forms MQ3 and 4.

GENERAL

This chapter aims to deal with aspects of milk quota other than compensation for milk quota paid to a tenant on the termination of a tenancy[1].

Milk quotas were introduced in the United Kingdom with effect from 1 April 1984 by the Dairy Produce Quotas Regulations (DPQRs) 1984[2] in implementation of EC Council Regulations 856/84[3] and 857/84 which set out the levy provisions. EC Commission Regulation 137/84 implemented the community rules in relation to quota and levy.

Quota was allocated to producers on 1 April 1984 based on their levels of production in the calendar year 1983. Additional 'hardship' quota could be allocated after application to the Dairy Produce Quotas Tribunal. Additional special quota called SLOM 1 or SLOM 2 has been allocated under special rules to producers who had gone out of milk production under one of the milk outgoer's schemes and thus did not have a reference quantity in 1983[4]. Further quota was allocated in respect of additional milk products[5] in 1991[6].

Essentially the milk quota regime introduced a levy imposed in each region of the United Kingdom for over-production under either Formula A, payable by the producer on milk delivered to the purchaser in excess of the reference quantity, or Formula B, which is payable by the purchaser on quantities delivered by the producers which exceed the reference quantity in the reference period.

The quota regime is now governed by DPQRs 1994[7], which give effect to the consolidation in EC Council Regulation 3950/92[8] and EC Commission Regulation 536/93[9].

THE INTERVENTION BOARD

The milk quota regime is administered by the Intervention Board for Agricultural Produce (the IB)[10].

The IB is required to maintain milk quota registers[11].

The IB may recover reasonable charges in respect of any visit to a direct seller or any purchaser to obtain information that should have been submitted by that person[12].

---

1 See chap 19.
2 SI 1984/1047.
3 Inserting the quota and levy provisions into EC Council Regulation 804/68.
4 Case 120/86 *Mulder v Minister van Landbouw en Visserij* [1988] ECR 2321 established that farmers who had participated in a milk outgoer's scheme were entitled to an allocation of quota on their outgoer's obligation not to produce milk which had expired.
5 'Dairy produce other than milk, butter, cream or cheese' – DPQRs 1991, reg 2; eg ice cream, yoghurt.
6 DPQRs 1991, reg 24 and Sch 9.
7 SI 1994/672 (amended by SIs 1994/2448, 1994/2919 and 1995/254).
8 Amended by EC Council Regulations 748/93 and 1560/93.
9 Amended by EC Commission Regulations 1765/93 and 470/94.
10 Established under the European Communities Act 1972, s 6(1). Cf the DPQRs 1994, reg 23.
11 See p 299.
12 DPQRs 1994, reg 23(2).

The IB has power to require producers to provide such information and statistics as it requires to perform its functions[1].

MILK QUOTA DEFINITIONS

In order to understand the milk quota regime, it is necessary to have a grasp of certain important definitions:

'*Holding*' – 'means all productions units operated by the single producer and located within the geographical territory of the Community'[2].

The definition of holding includes a mixed holding, provided milk production is actually carried out on the holding[3]. Further, a holding covers all agricultural production units even where the units, as leased, had neither dairy cattle nor the necessary technical facilities for milk production and the lease provides no obligation on the part of the lessee to engage in milk production[4].

A milk quota holding, often called a Euro holding, may well be different from an agricultural holding, if the tenant or farmer occupies a number of agricultural holdings under different titles, but operates them as one dairy unit.

'*Occupier*' – 'includes in relation to land in respect of which there is no occupier, the person entitled to grant occupation of that land to another person'[5].

This definition is of importance, because it is the producer in occupation of land who is entitled to have the milk quota registered as his.

'*Person having an interest in the holding*' – means a person 'having an interest in the holding or part of the holding the value of which interest might be reduced by the apportionment. . . '. Interest includes 'the interest of a mortgagee or heritable creditor and a trustee, but does not include the interest of a beneficiary under a trust or settlement or, in Scotland, the estate of a superior'[6].

'*Producer*'[7] – 'means a natural or legal person or group of natural or legal persons farming a holding within the geographical territory of the Community: – selling milk or other milk products directly to the consumer, – and/or supplying the purchaser'[8].

1 DPQRs 1994, reg 29.
2 EC Council Regulation 3950/92, art 9(d). Cf DPQRs 1994, reg 2.
3 *Wachauf v Bundesamt fur Ernahrung und Forstwirtschaft* [1991] 1 CMLR 328 at 339.
4 *Re the Kuchenhof Farm* [1990] 2 CMLR 289.
5 DPQRs 1994, reg 2.
6 DPQRs 1994, reg 2(1).
7 Cf *R v Dairy Produce Quota Tribunal for England & Wales* [1985] 2 EGLR 10 where it was said that each member of a partnership might be a 'producer' within the DPQRs 1984 and *Stubbs v Hunt & Wrigley* [1992] 1 EGLR 17, which considered EC Regulations in detail and on the facts held that a partnership was the producer and not the individual. These cases dealt with the definition of 'producer' in the context of the DPQRs 1984 where a 'producer' could make an 'exceptional hardship' claim for an additional quota in certain circumstances.
8 EC Council Regulation 3950/92, art 9(c).

'*Purchaser*' – 'means an undertaking or grouping which purchases milk or other milk products from the producer: – to treat or process them, – to sell them to one or more undertakings treating or processing milk or other milk products'[1].

'*Quota year*' – means the 12 months' period in respect of which levy is fixed and payable. This is now 1 April to 31 March in the following year[2].

'*Direct sales quota*' – 'means the quantity of dairy produce which may be sold by direct sale from a holding in a quota year without the direct seller in occupation of that holding being liable to pay levy'[2].
    This is the quota allocated to a milk producing farmer who sells direct to his own customers.

'*Special quota*' – Special quota means quota allocated under Council Regulation 3950/92, art 4(3) and Council Regulation 2055/93, art 1(1) to producers who either had adopted milk production development schemes before the introduction of quota or who had gone out of milk production under an outgoer's scheme which expired after 31 December 1982[3]. Only special quota allocated under Council Regulation 2055/93 is still under transfer restrictions.
    There is a power of entry and inspection in respect of any producer who has special quota registered as his or who has applied for special quota to verify the accuracy of the information provided, and that the requirements or any undertaking given are being complied with by the producer[4].

'*Wholesale quota*' – 'means the quantity of dairy produce which may be delivered by wholesale delivery to the purchaser (to the extent specified in relation to that purchaser under these Regulations) from a holding in a quota year without the producer in occupation of that holding being liable to pay levy'[5].
    This is the wholesale quota allocated to the milk producing farmer, who sells his milk to a purchaser, and who organises collection from the farm for delivery to the commercial dairies.

MILK QUOTA REGISTERS

The IB is required to maintain a direct sales register and a wholesale quota register[6]. The entry for each direct seller and producer[7] is required to contain his name, address, reference number, details of the direct sales of wholesale quota available to the producer, and details of any special quota issues[8]. In respect of

1 EC Council Regulation 3950/92, art 9(e). NB 'any group of purchasers in the same geographical area which carries out administrative and accounting operations necessary for the payment of levy on behalf of its members shall be regarded as a purchaser'.
2 DPQRs 1994, reg 2.
3 DPQRs 1994, reg 2(1).
4 DPQRs 1994, reg 31.
5 DPQRs 1994, reg 2.
6 DPQRs 1994, reg 25.
7 Direct seller and producer includes a person who occupies land with quota whether or not that person is engaged in the sale or delivery of produce; DPQRs 1994, reg 25(6).
8 DPQRs 1994, reg 25(1) and (2).

wholesale quota the register must list the names and addresses of each purchaser, whose quota will be calculated to take into account all or part of that producer's total wholesale quota, and showing the butterfat base of the quota[1].

The IB is required to keep a register of purchaser notices[2]. The entry is required to include the purchaser's name, quota and special quota[3].

Where a holding comprises more than one dairy enterprise, a direct seller or producer may agree with the IB, on presentation of a consent or sole interest notice, a partition of that holding between separate entries in the appropriate register[4].

The IB may amend the registers to record any allocations or adjustments made under the Regulations or to make any correction it reasonably considers necessary[5].

The registers may be inspected during reasonable working hours and, on payment of a reasonable charge, a copy of the entry can be obtained by any person who is the direct seller, the producer or specific purchaser or by a person who in a written statement informs the IB that he has an interest in the holding[6].

A direct seller is required to register his quota with the IB[7].

Each producer is required to register his wholesale quota with a purchaser[8].

Each purchaser is required to maintain, in respect of all producers on his list: a register similar to that maintained by the IB in respect of his producers, a register of particulars of wholesale deliveries from each of those producers, the information in connection with levy assessment required by EC Commission Regulation 536/93, art 7, and a system approved by the IB for sampling the milk of each producer and determining its fat content[9].

Any entry in a register or notice required by the Regulations is evidence in any proceedings of the matters stated therein[10].

NATURE OF MILK QUOTA

The essential characteristic of milk quota is that it attaches to the land of the holding in respect of which it is registered. It is not separate from the land and has the characteristic of a heritable right as 'part and pertinent' of the land[11].

While it has been held that the milk quota is not a separate asset from the land[12] the Inland Revenue still regards milk quota as a separate asset subject to capital gains tax and Customs and Excise considers it liable to VAT, if trans-

---

1 DPQRs 1994, reg 25(2)(a)(vi).
2 DPQRs 1994, reg 25(3).
3 DPQRs 1994, reg 25(3)(a).
4 DPQRs 1994, reg 25(4).
5 DPQRs 1994, reg 25(5); cf *R v MAFF ex p Cox* [1993] 1 EGLR 17.
6 DPQRs 1994, reg 26.
7 DPQRs 1994, reg 27(1).
8 DPQRs 1994, reg 27(2).
9 DPQRs 1994, reg 27(3).
10 DPQRs 1994, reg 28.
11 *Faulks v Faulks* [1992] 1 EGLR 9; *Wachauf v Bundesamt fur Ernehrung und Forstwirtschaft* [1991] 1 CMLR 328.
12 *Faulks v Faulks* above.

ferred without the land[1]. Allocated milk quota has a nil base value, and it is not competent to set off part of the cost of acquisition of the land against the value of the milk quota for CGT purposes[2].

If a dairy farmer disposes of his dairy herd and later disposes of the milk quota, the sale of quota may not attract retirement relief from CGT[3].

The nature of quota was considered in the *Wachauf* case[4] where the Advocate General characterised it on the basis that:

'If one considers the nature of quota from the point of view of the producer then it is plain that what the quota amounts to is a form of licence to produce a given quantity of a commodity (milk) at more or less guaranteed price without incurring a penalty (the additional levy).'

This approach was not followed in *Faulks*[5] where the judge held that the quota was not a separate asset from the land, but this takes no account of the fact that the quota can be divorced from the land on an apportionment following division of the holding, where the location of the quota will depend on the areas used for milk production in the past period. Quota is in the nature of a floating asset in that it attaches only to those areas of land which have been used for milk production and can be moved around a mixed holding depending on which parts have been used for such production.

In *Stubbs v Hunt & Wrigley*[6] it was said that:

'The scheme of milk quotas is territorial. The quota attaches to and enhances the value of the land, to the benefit of the landowner.'

The Advocate General of the European Court of Justice has recognised 'that it is possible for either a landlord or a tenant to have a proprietary interest in a quota'[7].

In *Re the Kuchenhof Farm*[8] a German court, following reference to the European Court of Justice, went further and held that although milk quota for the most part moves with the land it did not constitute a right of the landowner, but of the farmer. It was consequently no part of the landowner's property. In that case it was decided that a tenant could surrender the milk quota under an outgoer's scheme during the currency of the tenancy, without the consent of the landowner, because the requirement to obtain consent led to unequal treatment of milk producers, which could not be objectively justified. An owner-occupier did not need such consent. Accordingly the requirement that the tenant should have obtained the landowner's consent was void.

These conflicting strands of opinion regarding the nature and ownership of milk quota are difficult to reconcile. Perhaps the correct analysis is to hold that milk quota is a part and pertinent of the land to which it is then attached, subject to the right of the occupier to move the quota by reference to areas used for milk production or to transfer the quota to another holding or part of the

---

1 *Cottle v Coldicott* Special Commissioners (27 July 1995, unreported); *Agricultural Law, Tax and Finance* (Longmans, ed Allan A Lennon) para F2.6.
2 *Cottle v Coldicott* above.
3 *Wase (Inspector of Taxes) v Bourke* [1996] STC 18.
4 [1991] 1 CMLR 328 at 342.
5 [1992] 1 EGLR 9.
6 [1992] 1 EGLR 17 at 22.
7 *Wachauf v Bundesamt fur Ernehrung und Forstwirtschaft* [1991] 1 CMLR 328, AG at 343.
8 Case 62/85, [1990] 2 CMLR 289.

existing holding in accordance with EC and UK regulations, subject to any contractual agreement preventing such a transfer. More than one party can have a proprietorial interest in the quota at any one time[1].

An important consequence of the fact that milk quota has been held to be in the nature of a part and pertinent of the land is that it is probably secured by any standard security held over the land[2]. As a transfer of milk quota to another producer can only be achieved by a lease of land[3], the taking of such a lease without the consent of a standard security holder would be in bad faith. The lease and transfer would probably be reducible at the instance of the standard security holder. The onus is on the prospective transferee to establish by a search of the sasine or land registers that the transferor has an unencumbered title permitting him to grant the lease by which transfer of the milk quota is effected, without the consent of a standard security holder[4].

A further consequence of the fact that the milk quota is a part and pertinent of the land is that a person who has no real right in the land will have no real right in the milk quota. The situation where this is most likely to arise is where a farm is farmed by a partnership, but either the land or the tenancy of the land is not a partnership asset. In such circumstances on a termination of the partnership or the retiral from the partnership of a partner, the value of the milk quota will not be a partnership asset available for distribution to the partners or retiring partner. The value of the milk quota will remain with the person entitled to the land or to the tenancy[5]. Conversely if the land or the tenancy was a partnership asset, then the value of the milk quota will also be a partnership asset.

As noted in chapter 19 a tenant, provided he has quota registered as his and fulfils certain other criteria, is entitled to compensation for milk quota on the termination of the tenancy. The compensation provision[6] reinforces the concept that it is only the person with a real right, which includes a tenancy, in the land, and who is also a producer, who has any real right in the milk quota.

ESTABLISHMENT OF MILK QUOTA

A producer's allocated quota is established under the DPQRs 1994 and community legislation[7]. The amount of quota to which a producer is entitled will be that quota which is registered as his by the IB in the wholesale or direct sales quota register, subject to any temporary suspensions of quota.

This quota will be made up of quota originally allocated to the producer in 1984, including any hardship allocations, any allocations of special quota or

1 Cf 'Person having an interest in the holding'; DPQRs 1994, reg 2(1).
2 J Murray QC 'Dairy Quotas' 1986 SLT (News) 153; *Agricultural Law, Tax and Finance* (ed Allan A Lennon) p F45. Lennon recommends that for the avoidance of doubt a standard security should also provide that it secures a particular quantum of milk quota and that the debtor should not do anything to transfer or 'milk' the quota off the holding.
3 See p 305.
4 *Trade Development Bank v Crittall Windows Ltd* 1983 SLT 510; *Rodger (Builders) Ltd v Fawdry* 1950 SC 483.
5 *Faulks v Faulks* [1992] 1 EGLR 9.
6 Agriculture Act 1986, Sch 2.
7 DPQRs 1994, reg 3.

quota for additional milk products and quota obtained by a permanent or temporary transfer of quota to the holding.

The milk equivalence of dairy produce:

'(1) ... shall be calculated on the basis that each kilogram of dairy produce shall equal such quantity of milk referred to in paragraph (2) as is required to make that kilogram of dairy produce.

(2) The milk to which paragraph (1) relates is milk the fat content of which has not been altered since milking.'[1]

Where the fat content of the milk is greater than the EC norm the fat content is translated into additional litres of quota.

A producer who produces milk from breeds of cow that have a higher than average fat content to their milk is required before 1 March annually to confirm to the IB that in that quota year he has maintained in the dairy herd breeds of cow with characteristics similar to those in the herd in the first 12 months of production, and to undertake to maintain such breeds in his dairy herd for the remainder of the quota year.

Where by mistake a person is not allocated quota or is allocated a smaller quantity of quota than he should have been, the Secretary of State may allocate that person such quota as will compensate in whole or in part for the mistake[2].

A producer may apply to convert direct sales quota to wholesale quota and vice versa either temporarily or permanently. An application for a permanent conversion has to be made to the IB by 31 December and for temporary conversion by 28 April in any year following the end of the quota year in which the temporary conversion takes place. A producer who permanently converts quota in any year cannot in that year transfer quota temporarily or otherwise. Special quota cannot be converted[3].

NATIONAL RESERVE

A national reserve of milk quota has been established[4]. It comprises wholesale and direct quota not allocated to any person and includes quota withdrawn pursuant to regs 33 or 34[5].

The Secretary of State for Scotland may make allocations from the national reserve in accordance with community legislation or the DPQRs 1994[6].

Where there is a non-exempt transfer of land before 1 October 1996 with special quota allocated under EC Council Regulation 2055/93, art 1(1) attached to it, then special quota is returned to the national reserve[7].

Within 45 days of the end of the quota year each purchaser is required to provide the IB with a list of producers registered with him who have not made deliveries to him during that year[8].

1 DPQRs 1994, reg 5.
2 DPQRs 1994, reg 17.
3 DPQRs 1994, reg 18. IB Forms MQ6 (permanent conversion) and MQ7 (temporary conversion).
4 EC Council Regulation 3950/94 and DPQRs 1994.
5 DPQRs 1994.
6 DPQRs 1994, reg 14(2).
7 DPQRs 1994, reg 8; see below.
8 DPQRs 1994, reg 33(1).

Any direct selling producer who fails to submit a declaration under EC Commission Regulation 536/93, art 4(2) within 30 days of a notice sent to him by the IB will have his quota taken into the national reserve[1]. If a producer indicates in a notice submitted under those Regulations that he does not wish restoration of quota then he will have the quota taken into the national reserve[2].

Where it appears to the IB that a producer has not made any deliveries or direct sales or temporary transfer of quota during the previous year, then the producer is notified that his quota has been taken into the national reserve[3].

Any quota withdrawn may be restored to the producer in respect of the holding from which it was withdrawn within a period of six years from the beginning of the year in which it was withdrawn[4].

Any producer who receives a notification of confiscation is required (a) within 28 days of receipt of the notification to notify any person with an interest in the holding of the content of the notification and (b) within six months of the confiscation notification submit a notice to the IB indicating whether he wishes to retain the right to request restoration of the milk quota[5]. The notice to the IB is required to include a consent or a sole interest notice in respect of the whole holding or a statement of agreed apportionment signed by every person with an interest in the holding or a statement of apportionment in accordance with an arbitration[6].

A producer who has notified the IB that he wishes to retain the right to restoration of quota may request the IB to restore the quota to him, provided the notice is received by the IB at least six months before the end of the six-year period from the beginning of the quota year in which it was withdrawn[7].

Where a producer has notified the IB that he wishes to retain the right to restoration of quota and there is a change of occupation of all or part of the holding, then the new occupier may request the IB to restore to him the quota relating to that holding or that part provided the notice is received by the IB at least six months before the end of the six-year period from the beginning of the quota year in which it was withdrawn or within six months of the change of occupation, whichever is the earlier[8].

Where quota is restored to part of a holding, the restoration is carried out in the same proportion which the agricultural area concerned bears to the total agricultural area of the holding or apportioned holding from which quota was confiscated[9].

If a producer has quota restored to him and then fails to make deliveries or direct sales of dairy produce from the holding within six months of the restoration of quota or the end of the six-year period from the year in which the quota was confiscated, whichever is the earlier, then the quota is taken

---

1 DPQRs 1994, reg 33(2)(b), (3) and (9)(a).
2 DPQRs 1994, reg 33(9)(b).
3 DPQRs 1994, reg 33(2)(a) and (3); EC Council Regulation 3950/92, art 5. The quota does not go into the national reserve if the producer requests a restoration of the quota; DPQRs 1994, reg 33(9)(c).
4 DPQRs 1994, reg 33(4).
5 DPQRs 1994, reg 33(5).
6 DPQRs 1994, reg 33(5)(b).
7 DPQRs 1994, reg 33(6).
8 DPQRs 1994, reg 33(7).
9 DPQRs 1994, reg 33(8).

into the national reserve[1]. Similarly, on a change of occupation, if the new occupier who has quota restored to him fails to make deliveries or direct sales of dairy produce from the holding within 18 months of the change of occupation or the end of the six-year period from the year in which the quota was confiscated, whichever is the earlier, then the quota is taken into the national reserve[2].

The Secretary of State for Scotland may withdraw the whole or any part of special quota allocated to a producer where it appears to the Secretary of State that the producer made a false or misleading statement in his application for special quota, has failed to comply with the requirements in relation to special quota or has failed to comply with any undertakings given under reg 8(5), which reverts to the national reserve[3].

PRODUCTION CONTRACTS

With the introduction of special quota, which required the producer to make deliveries or direct sales of milk products within certain time limits, similar to the requirement where milk quota is restored, the market developed production contracts.

The milk quota was allocated to the incoming producer's holding, but that producer had no means of producing milk for delivery or sale. In such circumstances, a contract was entered into whereby the incoming producer contracted with an existing producer to produce and sell or deliver milk against the incoming producer's quota. Part of the contract required that certain areas of the existing producer's holding and dairy stock should be leased to the incoming producer so that they formed part of his holding and stock.

Such an arrangement appears to be acceptable to the IB as a means of establishing milk production, where quota is restored to a holding, or maintaining milk production on a holding, where the producer ceases to have the means of production.

TRANSFER OF QUOTA WITH LAND

Milk quota is transferred with all or part of a holding when there is a change of occupation of all or part of the holding.

This section is concerned with the transfer of quota on a permanent or semi-permanent change of occupation of all or part of the holding[4].

A transfer of milk quota does not take place in respect of the grant of a licence to occupy land or a lease of land for less than eight months or on the termination of a licence or lease for less than eight months[5].

Any milk producer who leases in or leases out grazing land is advised to take

---

1 DPQRs 1994, reg 33(9)(d).
2 DPQRs 1994, reg 33(9)(e).
3 DPQRs 1994, reg 34.
4 For sale of milk quota by means of a temporary change of occupation see p 310 'Sale of milk quota'.
5 DPQRs 1994, reg 7(6).

or give a grazing lease for less than eight months, otherwise some of his quota may be transferred inadvertently to the lessee or lessor of the grazings.

Milk quota[1]:

'... shall be transferred with the holding in the case of sale, lease or transfer by inheritance to the producer taking it over ... The same provision shall apply to other cases of transfers involving comparable legal effects for producers.[2]'

'Comparable legal effects' has been held to include the surrender as well as the grant of a lease[3] and that a change in a partnership effected a change in occupation of the holding[4]. A change in a partnership, where the partnership deed provides for a continuing partnership notwithstanding changes in the members, may not give rise to a change in occupation[5].

Where all or part of a holding is transferred, no later than 28 days after the change of occupation and in any event no later than seven working days after the end of the quota year in which the transfer takes places, the transferee is required to submit Form MQ1 to the IB[6]. The IB requires evidence of the land transfer. A copy of the disposition or lease is usually sufficient.

Form MQ1 is required to include a statement by the transferor and transferee that they have agreed that the quota will be apportioned[7] taking account of areas used for milk production, and in the case of transfer of the whole holding a consent by any other party interested in the holding or sole interest notice by the transferor[8].

Where a transferee fails to submit Form MQ1 no later than seven working days after the end of the quota year in which the transfer takes place then, for the purposes of levy calculation, the unused quota is not treated as part of the transferee's quota entitlement for the quota year in which the transfer takes place[9]. The transferee cannot demand that an amendment be made to the amount of quota, which has been reallocated to him[10].

The purpose of this regulation appears to be administrative, in that it is only 'for the purposes of levy calculation' that the quota is not treated as part of the transferee's quota. This non-treatment would appear to apply in that year only and not in subsequent years. The quota would 'belong' to the transferee[11].

Where there is a transfer of part of a holding (a) there requires to be an apportionment[12] and (b) any dairy produce sold by direct sale or delivered by wholesale from the holding during the quota year prior to the transfer, is deemed for the purpose of any levy calculation to have been sold or delivered

1 There are particular rules for special quota; see p 308.
2 EC Council Regulation 3950/92, art 7.1.
3 *Wachauf v Bundesamt fur Ernehrung und Forstwirtschaft* [1991] 1 CMLR 328; cf 347 – this applies even if the unit as leased had neither dairy cows nor the technical facilities necessary for milk production and the lease provided no obligation on the tenant to produce milk.
4 *Holdcroft v Staffordshire CC* [1994] EGCS 56.
5 Cf *IRC v Graham's Trs* 1971 SC (HL) 1.
6 DPQRs 1994, reg 7(1).
7 Cf p 306 'Apportionment'.
8 DPQRs 1994, reg 7(1), (2).
9 DPQRs 1994, reg 7(3)(a). The IB may reallocate the quota in accordance with ibid, Sch 5, para 7.
10 DPQRs 1994, reg 7(3)(b).
11 *Agricultural Law, Tax and Finance* (ed Allan A Lennon) para F2.4 'Penalties for not giving notice of transfer'.
12 See 'Apportionment' below.

from each part of the holding in proportion to the apportionment, unless the parties agree otherwise[1].

APPORTIONMENT

Where there is a transfer of part of a holding, including a part of a holding to which reg 8(5)[2] applies, then the quota requires to be apportioned either by agreement or by arbitration[3].

Apportionment, whether it is an agreed apportionment or an apportionment by arbitration, is carried out by reference to 'areas used for milk production'[4].

The phrase 'areas used for milk production' was defined in *Puncknowle Farms Ltd v Kane*[5] to mean:

'not merely the farm area used for current production but includes land which, taking the annual cycle of agriculture, is used for the support of the dairy herd and to provide for future milk production, and therefore includes land used for dry cows and heifers, land used for buildings and yards of a dairy unit and land used for dairy or dual-purpose bulls, which have been bred to enter the production herd rather than for sale.'

The arbiter's definition, which was approved with the additions noted above was that areas used for milk production:

'are the forage areas used by the dairy herd and to support the dairy herd by the growing of grass and any fodder crop for the milking dairy herd, dry cows and all dairy following female youngstock (and home bred dairy and dual purpose bulls for use on the premises if applicable) if bred to enter the production herd and not for sale. In this case maize, silage, hay and grass were fodder crops, but consideration would have been given to corn crops or part of corn crops grown for consumption by the dairy herd or youngstock, including the use of straw.'

In *Posthumus v Oosterward*[6] the Advocate expressed the view that areas used for milk production should include the yards, buildings and roads provided they made a significant contribution to the production of milk.

The *Posthumus* case determined that 'areas used for milk production' are to be established by an arithmetical approach, without regard to the productivity or otherwise of the holding, even if this arrived at an inequitable result. This was followed in *Knufer v Buchmann*[7] where it was held that the milk quota had to be divided strictly in proportion to the size of the areas without it being possible to make any distinction according to the nature of the use of the areas.

Where land is used 'concomitantly'[8], such as for summer grazing of dairy

---

1 DPQRs 1994, reg 7(4).
2 DPQRs 1994. Special quota, where part of a holding is transferred by inheritance, gift or grant of a tenancy.
3 DPQRs 1994, reg 10.
4 NB 'Areas used for milk production' should not be confused with the definition of 'relevant number of hectares' (Agriculture Act 1986, Sch 2, para 6(5)(a)) used for the purpose of determining a tenant's compensation for milk quota. As a result of the different definitions it is possible to have milk quota apportioned to a leased farm, in respect of which the tenant is entitled to no compensation, because of the different definitions.
5 [1985] 3 All ER 790, (1985) 275 EG 1283.
6 Case C-121/90 ECJ 5833, [1992] 2 CMLR 336.
7 Case C-79/91, [1993] 1 CMLR 692.
8 *Puncknowle Farms Ltd v Kane* [1985] 3 All ER 790.

cows and over-wintering of sheep, the dairy and non-dairy use requires to be apportioned out.

An agreed apportionment is carried into effect by the statement in Form MQ1 in which the transferor and the transferee agree the apportionment 'taking account of areas used for milk production'[1]. Other parties interested in the holding also have to agree the apportionment.

Where the apportionment is agreed, there is no requirement when considering 'areas used for milk production' to have regard to any period of time to which reference must be made in determining what areas have been used for milk production[2].

An occupier of a holding may apply to the IB for a prospective apportionment to be made, either by taking account of areas used for milk production, as set out in the application Form MQ8, or by arbitration[3]. A request for a prospective apportionment may be revoked by notice to the IB in writing[4]. Where an occupier requests, or revokes the request, for a prospective apportionment, the notice requires to be accompanied by a consent or sole interest notice in respect of the entire holding[5].

A prospective apportionment is only valid in respect of a change of occupation which takes place in the six months preceding the change[6].

Where the IB has reasonable grounds for believing that the areas used for milk production are not as specified in Forms MQ1 or MQ8 or were not as agreed, even if no form was submitted, then on giving notice to the person who submitted the form or where no form was submitted the transferee, the apportionment or prospective apportionment falls to be carried out by arbitration[7].

Where parties cannot agree an apportionment, the apportionment will be carried out by arbitration or by the Land Court[8]. The procedures for the appointment of the arbiter or reference to the Land Court and the conduct of the arbitration, which are set out in Sch 3[9], have already been considered[10].

In carrying out an apportionment under Sch 3 the arbiter or the Land Court is required to:

'decide the apportionment on the basis of findings made as to areas used for milk production in the last five-year period during which production took place before the change of occupation or, in the case of a prospective apportionment, in the last five-year period during which production took place before the appointment of the arbiter or the application to the Scottish Land Court.[11]'

There is no difficulty if the whole holding has been used for milk production in the preceding five years.

1 DPQRs 1994, regs 7(2)(a) and 10 (a).
2 Derek Wood et al *Milk Quotas, Law and Practice* (Farmgate Communications Ltd, 1986) para 4.07.
3 DPQRs 1994, reg 11(1).
4 DPQRs 1994, reg 11(2).
5 DPQRs 1994, reg 11(3).
6 DPQRs 1994, reg 11(4).
7 DPQRs 1994, reg 12.
8 DPQRs 1994, reg 10(b)(ii) and Sch 3.
9 DPQRs 1994.
10 See chap 23.
11 DPQRs 1994, Sch 3, para 2(1). Prior to the coming into force of the DPQRs 1994 the schedule referred to the five-year period prior to the change of occupation. This led to difficulties if milk production had ceased some years before there was a change of occupation, which is why the regulation was amended.

The arbiter has to determine what areas of that holding or each part of that holding have been used for milk production and to apportion dairy usage where there has been a mixed use of parts of the holding. Milk quota cannot be apportioned to an area of the holding which has not been and cannot be used for milk production.

Where part of a holding is an agricultural tenancy, it should be noted the definitions of 'areas used for used for milk production' used in an apportionment and 'relevant number of hectares'[1] used for milk quota compensation, may have the effect of apportioning milk quota to the tenancy for which no compensation may be payable. Such a situation arises where the tenancy is used for feeding young stock, while the rest of the holding is used for 'feeding of dairy cows'.

EFFECT OF LAND TRANSFERS ON SPECIAL QUOTA

Where a producer has special quota of the kind referred to in EC Council Regulation 2055/93, art 1(1)[2] registered as his, then a sale or lease of all or part of the holding before 1 October 1996 will have the effect that the special quota shall be transferred to the national reserve[3].

The proportion of special quota to be returned to the national reserve is in the same proportion as the agricultural area of the holding transferred bears to the total agricultural area[4] farmed by the producer[5].

Special quota is not transferred in circumstances where there is a transfer of a holding or part of the holding (a) by a license to occupy or (b) the grant of a lease for a period of less than eight months[6].

No special quota requires to be transferred to the national reserve where the holding or part of the holding is transferred (a) by inheritance (b) by gift for which no consideration is given or (c) by the granting of a tenancy by the landlord to a successor tenant who had died or retired, provided the person to whom the holding or part of the holding is transferred undertakes to comply with the special quota undertakings given by his predecessor[7].

It should be noted that a change in a partnership may trigger return of all or part of the quota to the national reserve, unless a new lease is granted by the landlord to the newly constituted partnership[8].

Where a producer has special quota registered as his and intends to transfer

---

1 Agriculture Act 1986, Sch 2, para 6.
2 NB special quota referred to in EC Council Regulation 3950/92, art 4(3) is no longer subject to the return to the national reserve rules – DPQRs 1994, reg 8(2)(a).
3 DPQRs 1994, reg 8(1) and (2).
4 Defined by DPQRs 1994, reg 2(1) to include 'areas used for horticulture, fruit growing, seed growing, dairy farming and livestock breedings and keeping, areas of land used as grazing land, meadow land, osier land, market gardens and nursery grounds and areas of land used for woodlands where that use is ancillary to the farming of the land for agricultural purposes'.
5 DPQRs 1994, reg 8(3). NB the apportionment is **not** in accordance with areas used for milk production.
6 DPQRs 1994, reg 8(4).
7 DPQRs 1994, reg 8(5).
8 *Holdcroft v Staffordshire CC* [1994] EGCS 56.

the whole or part of the holding before 1 October 1996, then he requires to submit the prescribed form to the IB at least 28 days before the transfer takes place, together with such other evidence of the proposed transfer as the IB may require[1].

Where the transfer is an exempt transfer under reg 8(5) then the notification provisions of reg 7 apply[2].

TRANSFER OF QUOTA WITHOUT TRANSFER OF LAND

EC Council Regulation 3950/92, art 8, 5th indent permits the transfer of milk quota without transfer of the corresponding land 'with the aim of improving the structure of milk production at the level of the holding or to allow for extensification of production'[3].

An application[4] for a transfer of quota requires to be made to the IB by the transferee not later than ten working days before the intended date of transfer[5].

The application requires a statement by both the transferor and transferee explaining how the transfer is necessary to improve the structure of the business of **both** the transferor and transferee[6].

The application requires an undertaking by the transferor that he has not transferred quota onto the holding 'in accordance with the provisions of this regulation'[7] in the course of the quota year in which the application is made, or in the preceding year, and that he will not transfer quota onto his holding in the period between the date of submission of the application and the end of the quota year following the quota year in which the transfer of quota without land takes place. The transferee has to undertake not to transfer quota from his holding in the period between the date of submission of the application and the end of the quota year following the year in which the transfer takes place. Further, both parties have to undertake that they will not try to circumvent the former provisions through their involvement in any other business[8].

Where the IB approves the application, then the transferee has to submit Form MQ1 to the IB within the time limits that apply to a transfer of quota with land. Where the transferee fails to submit the form the IB revokes its approval[9].

---

1 DPQRs 1994, reg 9(1).
2 DPQRs 1994, reg 9(2).
3 See DPQRs 1994, reg 13.
4 Form MQ2.
5 DPQRs 1994, reg 13(1).
6 DPQRs 1994, reg 13(2)(a). *Quaere* whether the requirement that both the transferor and transferee have to improve their business conforms to EC Council Regulation 3950/92, art 5, 5th indent, which refers only to 'the producer' who applies and improving the structure of milk production 'at the level of the holding'. This suggests the article is dealing with only the improvement of production on the producer's holding.
7 DPQRs 1994, art 13(2)(c)(i). It is queried in *Agricultural Law, Tax and Finance* (ed Allan A Lennon) at p F28 whether 'regulation' should be 'this SI', but the reference appears to mean that in the year there have been no other transfers without land.
8 DPQRs 1994, reg 13(2)(c). The undertaking applies to temporary transfers as well, but not to transfer on inheritance; reg 13(3).
9 DPQRs 1994, reg 13(5) and (6); not later than 28 days after the transfer and in any event not later than 7 working days after the end of the quota year in which the transfer takes place.

A producer may be released from his undertaking:

'where the Intervention Board is satisfied that exceptional circumstances, resulting in a significant fall in milk production which could not have been avoided or foreseen by the transferee at the time of the submission of the application ... justify that release.'[1]

A transfer without land cannot be authorised if it affects the total quota available within a Scottish islands area[2].

SALE OF MILK QUOTA

The sale of quota can be effected by the temporary lease of land by the transferor to the transferee for a period of more than eight months[3]. Such a lease and transfer of quota requires to be consented to by anyone having an interest in the land or quota and in particular by a standard security holder[4].

The procedure usually adopted is for the transferor to grant a grazing lease for more than eight months, but less than a year[5], to the transferee of part of the holding. The leased area accordingly becomes part of the transferee's milk quota holding and the quota attached to that area becomes part of the transferee's quota.

The transferee then arranges to graze the area with non-dairy stock for the duration of the lease. At the end of the grazing lease, the milk quota formerly adhering to the area of the grazing lease has moved to the main holding. The theory is that the area was not an area used for milk production during the lease period, so on a transfer back of that part of the holding no quota transfers, because none was attached to that area[6].

The transferee must produce milk from his own holding during this arrangement[7].

On a sale or transfer of quota by this method at the time of the lease, the milk quota relating to the transferor's holding needs to be apportioned between the leased area and the remaining area. In general the IB accepts as a reasonable apportionment a transfer which involves no more than 25,000 litres per hectare.

The notification of a transfer to the IB requires to be made in the same manner and within the same time limits as noted above, because a sale is a transfer of quota with land[8].

1   DPQRs 1994, reg 13(7). Reg 13(8) defines the exceptional circumstances; eg ill health, natural disaster affecting the holding, accidental destruction of buildings, illness or disease affecting the herd, a notice or declaration under the Animal Health Act 1981 etc, loss of forage due to compulsory purchase, service of an incontestible notice to quit.
2   DPQRs 1994, reg 13(9).
3   Eg DPQRs 1994, reg 7(6)(iii).
4   See p 300 'Nature of milk quota'.
5   Cf the Agricultural Holdings (Scotland) Act 1991, s 2(2), to avoid granting an agricultural tenancy.
6   See *R v MAFF ex p Cox* [1993] 1 EGLR 17. In that case the court refused to cancel a transfer where the lease had given an exclusive right of occupation without there having been actual occupation, because MAFF had accepted such arrangements in the past. The court did suggest that if the lease arrangement was a sham that it could be set aside. The safe course is for the transferee to take actual occupation under the grazing lease with non-dairy stock.
7   IB 'Milk Quota Transfers with Land, Explanatory Notes'; *R v MAFF ex p Cox* [1993] 1 EGLR 17.
8   IB 'Transfer of Quota with Land' above.

As noted above the transferee who is taking a lease of the land should confirm that the lessor is entitled to grant a lease and so to transfer the milk quota[1].

It has been opinioned that the UK provisions allowing transfer of milk quota by short-term grazing leases is contrary to EC law[2].

CHANGE OF SUPPLY ADDRESS AND MOVING QUOTA

It is possible to shift quota from one area of land[3] to another by the use of the transfer of quota with land. For example if the farmer dairies on farm A, but wishes to move the quota to farm B, he can occupy farm B as part of his quota holding and move the supply address to farm B by notifying the IB. He then ceases milk production on farm A.

In due course[4] he apportions all the quota to farm B on the basis of areas used for milk production. The farmer is then free to dispose of farm A without any milk quota attached thereto.

LEASING QUOTA

Leasing quota or temporary transfer of quota was first authorised in 1991 and is now governed by reg 15[5].

Any producer, who is the registered holder of quota[6], may agree with another producer[7] to make a temporary transfer of milk quota of all or part of any unused quota registered as his for the period of one quota year[8].

The IB has to be notified, no later than 15 December in the year in which the agreement is made, of the temporary transfer[9]. A producer has to inform his purchaser of any change to his quota within seven working days of the change[10].

Where milk quota has been permanently transferred off the holding, quota cannot be leased in during the same or the following quota year[11].

Wholesale quota which is leased out will have the lessor's permanent butterfat base attached to it. The IB then adjusts the lessee's ongoing butterfat base, for the purposes of levy calculation[12].

Leasing out of quota is not farming the land in accordance with the rules of good husbandry or the maintenance of a reasonable standard of efficient production from an agricultural holding[13].

1 Eg the land is not subject to a standard security which prohibits leases.
2 Opinion for arbiter by Professor John Usher in *Carson v Cornwall CC* [1993] 1 EGLR 21, confirmed in a lecture to the Faculty of Advocates – March 1996.
3 Excluding transfers in or out of a Scottish islands area; DPQRs 1994, reg 7(7).
4 The IB is sometimes prepared to accept quite short time spans for such a move of quota.
5 DPQRs 1994 and EC Council Regulation 3950/92, reg 6.
6 Note to forms MQ3 and 4.
7 Except in respect of a Scottish islands area if the lease results in an increase or reduction of the milk quota available in the area; DPQRs 1994, reg 15(4).
8 DPQRs 1994, reg 15(1).
9 DPQRs 1994, reg 15(3) – Forms MQ3 (wholesale quota) and MQ4 (direct sales quota).
10 Note to Forms MQ3 and 4.
11 DPQRs 1994, reg 13(2)(c).
12 Note to Forms MQ3 and 4.
13 *Cambusmore Estate Trust v Little* 1991 SLT (Land Ct) 33 at 41H.

TEMPORARY REALLOCATION OF QUOTA

Regulation 16[1] makes provision for the temporary reallocation of quota in any quota year in specific circumstances. The temporary reallocation is of an amount of quota corresponding to a proportion of any levy collected in excess of the levy actually due in that year.

Temporary reallocation of quota may be made (i) where the producer has received a formal acknowledgement of an error in quota calculation or (ii) has quota registered in respect of a holding subject to an order prohibiting or regulating the movement of dairy cows under the Animal Health Act 1981 or (iii) is situated in an area designated by an emergency order under the Food and Environment Protection Act 1985, s1.

The rules as to reallocation are complex and detailed reference should be made to reg 16 by anyone who may be eligible.

SPECIAL ALLOCATION OF QUOTA

Regulation 17[2] provides that where, by reason of a mistake by the minister or anyone acting on his behalf, a person has not been allocated quota or has been allocated a smaller quantity of quota than he should have been, then quota may be allocated from the national reserve to compensate in whole or in part for that mistake.

CALCULATION OF LEVY

Schedule 5[2] provides for the reallocation of quota and the calculation of levy applicable both to direct and wholesale quotas.

If the direct seller or the purchaser fails to notify the IB within 45 days of the end of the quota year of the actual milk quantity sold direct or delivered to the purchaser, then the IB may decide that the direct seller shall pay the rate of levy calculated under EC Council Regulation 3950/92, art 1 or that producer shall not benefit from any reallocation of quota[3].

Where a direct seller or a producer cannot provide the IB with such proof as the IB may require of the quantities of dairy produce sold direct or delivered to the purchaser, then the IB may make its own determination of those quantities for levy purposes[4].

Levy has to be paid by direct sellers or purchasers to the IB by 1 September. Interest runs at 1 per cent above the sterling three-month London interbank offered rate on unpaid levy[5]. Where a producer exceeds his quota in deliveries to a purchaser, the purchaser may retain against monies due by him to the producer the potential sum due in respect of levy[6].

1 DPQRs 1994.
2 DPQRs 1994.
3 DPQRs 1994, Sch 5, paras 8 and 17.
4 DPQRs 1994, Sch 5, paras 9 and 18.
5 DPQRs 1994, reg 22(1) and (2).
6 DPQRs 1994, reg 22(3).

To prevent avoidance of levy, where a producer makes sales or deliveries of milk or milk products and subsequently another producer makes sales or deliveries of milk or milk products from any or all of the same cows, then the second producer is deemed to have made those sales or deliveries as the agent for the first producer[1].

PENALTIES

Any person who without reasonable excuse fails to comply with any requirement imposed on him by or under the DPQRs 1994, or in connection with the Regulations or any community legislation 'makes a statement or uses a document which he knows to be false in a material particular or recklessly makes a statement or uses a document which is false', is guilty of an offence and liable, on summary conviction, to a fine not exceeding level 5 or, on indictment, to a fine[2].

Following upon a conviction against which there is no successful appeal, the Secretary of State for Scotland may by notice[3] served on the person to whose quota the conviction relates, reduce the quota by the amount he considers as being attributable to the falsehood on which the conviction was founded[4].

THE DAIRY PRODUCE QUOTA TRIBUNAL FOR SCOTLAND

The Dairy Produce Quota Tribunals were established under the DPQRs 1984, reg 6 to deal with claims for the additional allocation quota in 1984. Their existence has been continued by reg 35[5]. Rules regarding the conduct and sittings of the Tribunals are governed by Part II of Schedule 6[5].

There is no appeal from a decision of the Tribunal. Procedural irregularities, errors in law or any unreasonableness in the decision-making are subject to judicial review, although an application will have to be made promptly[6].

CONVEYANCING CONSIDERATIONS[7]

In offering for a dairy farm consideration should be given to the following:

(a) Obtaining confirmation from the seller as to whether or not the farm being sold is the whole quota holding. If the sale is only part of the quota holding, then consideration will have to be given to an apportionment of milk quota

---

1 DPQRs 1994, reg 21.
2 DPQRs 1994, reg 32(1).
3 Served within 12 months of the expiry of the appeal period after a conviction or further appeal or the date of appeal from which there is no further appeal; DPQRs 1994, reg 32(3).
4 DPQRs 1994, reg 32(2).
5 DPQRs 1994.
6 Cf R v Dairy Produce Quota Tribunal ex p Caswell [1989] 3 All ER 205; Miller and Budge v Dairy Produce Quota Tribunal for Scotland 1992 GWD 39-2280.
7 Cf 'Conveyancing considerations' in chap 25 regarding arable and stock farms.

between the farm being sold and the remainder of the quota holding. Apportionment may involve other parties, if for example there is a standard security over the farm being sold or if the remainder of the quota holding is a tenanted farm. These other parties will have to concur in any agreed apportionment.

(b) It is not uncommon for a farm to be owned by one person, but for the farming to be carried on by a partnership in whose name the quota is registered. The partnership may have an interest in the quota and its consent will have to be obtained to any transfer.

(c) Obtaining from the seller a printout from (i) the IB milk quota register to show the amount and type of quota registered in his name and (ii) if wholesale quota, a printout of the purchaser's register to show how much milk has been delivered in the current quota year.

(d) The seller should be taken bound to warrant (i) that he is entitled to transfer the quota to the purchaser (ii) that there are no other parties with an interest in the quota and (iii) that the quota has been used to confirm that the quota is not vulnerable to being withdrawn.

(e) Confirmation should be obtained that no special quota is involved.

(f) Confirmation should be obtained that no milk quota is currently being leased out or if it is details of the leasing situation obtained.

(g) Obtaining from the seller details of the butterfat content of the milk quota, the amount of quota used to the date of missives and an estimate of the amount of quota that will have been used by the date of entry, to confirm that there is sufficient unused quota to cover the purchaser's needs. The seller should be taken to warrant these figures.

(h) There should be an apportionment between the price of the farm and the price for the milk quota. Unless the farm, with its stock etc, is being sold as a going concern VAT will be payable on the price of the milk quota. The missives should make clear whether the price is inclusive or exclusive of VAT.

(i) Provision should be made for retention on joint deposit receipt of the price allocated to the milk quota until confirmation is received from the IB (i) that the transfer has been effected and (ii) that if there is an agreed apportionment to the farm, that this apportionment is accepted.

(j) In the event that super levy might be payable at the end of the year if there appears a risk of over-production, provision should be made as to who is to pay super levy or how the levy is to be apportioned between the parties. If the seller is to bear part of the super levy, provision should be made for a sum to cover the potential liability.

CHAPTER 25

IACS, subsidies and quotas

Key points

- The rules relating to the Integrated Administration and Control System (IACS) forms and the various support and quota schemes are changed regularly so it is important to obtain the most up-to-date information booklets and leaflets from SOAEFD.

- IACS forms are completed by each separate business. Where farmers are engaged in more than one farming business it is vital to ensure that the businesses are kept separate within the requirements of the IACS Scheme.

- The person completing the IACS form in any year is the person entitled to the Arable Area Payments Scheme (AAP) payments for that year, irrespective of whether or not the farm is transferred to a new occupier after that date.

- Sheep annual premium and suckler cow premium quota does not attach to the land. It is not secured by a standard security. It does not belong to the landlord of an agricultural holding if it was allocated to the tenant. It is moveable in succession.

- Quota is ring-fenced within sensitive zones consisting of (i) the Highlands and Islands less favoured areas (LFA), (ii) the rest of Scotland LFA and (iii) the GB non-LFA area: Sheep Annual Premium and Suckler Cow Premium (Quotas) Regulations 1993 (1993 Regs), reg 3 and Sch 1.

- With limited exceptions, quota may only be transferred within its own sensitive zone: 1993 Regs, regs 8 and 9.

- To avoid the 15 per cent siphon to the national reserve, quota must be transferred with the whole holding: 1993 Regs, regs 6 and 7.

- Where quota is transferred or leased the transferee or lessee would be well advised to make provision for the deposit of any sum due to be paid, until such time as SOAEFD confirms the transfer or lease.

- Where only part of a producer's quota is transferred or leased there are minimum numbers that may be leased out: cf 1993 Regs, reg 6(2).

- There are strict rules regarding the use of quota, which must be complied with otherwise quota is surrendered to the national reserve: see 'Usage rules' below.

- In any conveyancing transaction for the transfer of occupation of a holding, detailed consideration needs to be given to rules and dates relating to IACS forms, AAP payments, premium claim dates, retention periods, etc: see 'Conveyancing considerations' below.

Time limits

- IACS forms have to be submitted to SOAEFD by 30 March annually, although the UK may obtain an extension from the EC to 15 May.

- Under the Beef Special Premium Scheme (BSPS) a Cattle Control Document (CCD) has to be applied for within ten weeks of the birth of the male bovine animal.

- Under the BSPS deaths, thefts, losses through natural circumstances or *force majeure* preventing completion of the retention period have to be notified in writing to SOAEFD within ten days: BSPS 'Notes for Guidance', SOAEFD Beef 4.

- The application period for sheep annual premium quota is notified in each year for a period from early December to early February (1996 dates were 4 December 1995 to 4 February 1996). Late applications, subject to penalty, can be received in the following 20 calendar days, but no later applications can be received after that date: Sheep Annual Premium Regulations 1992, reg 3(2) amended by SI 1995/2779.

- Applications for suckler cow premium may be submitted during the period from and including 1 July to and including 8 December in any calendar year. Late applications, subject to penalty, can be received in the following 20 calendar days, but no later applications can be received after that date: Suckler Cow Premium Regulations 1993, reg 3.

- The retention period for sheep is 100 consecutive days starting the day after the last day of the application period.

- The retention period for suckler cows is six months from the day on which SOAEFD stamps the lodged application form.

- Hill Livestock Compensatory Allowance (HLCA) application dates and retention periods are now the same as those for the Suckler Cow Premium Scheme (SCP) or Sheep Annual Premium Scheme (SAPS).

- The transfer or lease of sheep annual premium quota has to be notified to SOAEFD by the producer receiving the quota either by the date at which he lodges his SAP claim form in any marketing year or, where no SAPS claim is lodged, by the end of the period for lodging such claims: 1993 Regs, reg 5(2).

- The transfer or lease of suckler cow premium quota has to be notified to SOAEFD by the producer receiving the quota either by the date at which he lodges his application form in respect of any calendar year or, where no application is lodged, by the end of the period for lodging such applications: 1993 Regs, reg 5(3).

GENERAL

The Integrated Administration and Control System (IACS) was introduced in 1992[1] as part of the reform of the CAP, as an anti-fraud measure. The system is used throughout the EC to allow monitoring of what is being grown and where so that consideration can be given to the different subsidy schemes.

The IACS scheme requires completion of two forms each year, namely (i) the Area Aid Application Base Form[1] and (ii) the Field Data Sheets[2].

The following subsidies are dependent on the submission of an AAP form:

(a) Arable Area Payments Scheme, including set-aside (AAP);

(b) Beef Special Premium Scheme (BSPS);

(c) Suckler Cow Premium Scheme (SCP);

(d) Hill Livestock Compensatory Allowance (HLCA);

(e) Sheep Annual Premium (SAP) if claimed with any of the above subsidies[3].

This chapter does not seek to explain the details of the CAP subsidies or the quotas linked to them, because the schemes are subject to frequent alteration by the EC[4]. The aim is to highlight areas of the schemes which affect the advice that a lawyer might have to give to a farming client. Practitioners are advised to obtain from SOAEFD the most up-to-date guidance leaflets and explanatory booklets issued regarding any scheme on which they have to give advice[5].

IACS FORMS AND THE AAP SCHEME

The 1992 CAP reforms included a substantial reduction in the support of prices for cereals over three years. The AAP was established[6] with the dual purpose of offsetting these reductions in income, while attempting to reduce production by extending set-aside through making it a pre-condition of subsidy.

Full details of IACS and the AAP scheme can be found in (i) the IACS Regulations 1993[7] and IACS 1996 explanatory booklet[8] (ii) the Arable Area Payment Regulations 1995[9] and (iii) the Arable Area Payments Scheme 1996 explanatory leaflets[10].

The IACS system for AAP is based on the use to which the fields were put on 31 December 1991 and in the preceding five years. The system is map-

---

1 EC Council Regulation 3508/92.
2 Form IACS 2 (1996).
3 Forms IACS 3 and 4 (1996).
4 An Area Aid Application Base Form is not required if SAP is the only subsidy claimed.
5 Eg up to December 1995 there were about 120 EC Council and Commission Regulations affecting EC Council Regulation 1765/92, which introduced the AAP scheme.
6 Where dates, details of a scheme or particular figures are given, those are for the 1996 schemes and are subject to variation in subsequent years. NB in a number of instances there are differences between schemes in Scotland and England, so MAFF booklets should be treated with caution.
7 EC Council Regulation 1765/92 and EC Commission Regulation 2293/92.
8 SI 1993/1317 as amended by SI 1994/1134.
9 SOAEFD, IACS 1 (1996).
10 SI 1995/1738 as amended by SI 1995/2780.
11 SOAEFD, AAP 1 (1996) 'General Rules'; AAP 2 (1996) 'Set Aside Rules'.

based on the Ordnance Survey maps at 1:25,000 or 1:10,000 scales showing farm boundaries and Ordnance Survey field numbers.

Provided the fields were in arable rotation at that date, claims can thereafter be made in respect of cereal, linseed, oilseed rape, peas for harvesting dry, and field beans grown in those fields. Land use on 31 December 1991 and in the preceding five years governs the eligibility of that land to be included in the AAP scheme[1]. In general, permanent grass[2], rough grazing, permanent crops, woodland and non-agricultural land on 31 December 1991 are not eligible for inclusion in the scheme[3].

Cereal and other crops may be grown on non-eligible land, but payments under the AAP cannot be claimed.

One to one switches of eligible land for ineligible land may be possible for agronomic, plant health or environmental reasons[4].

Eligibility of a field to participate in the AAP scheme does not necessarily mean that the land is also eligible to participate in set-aside.

IACS forms require to be submitted by each separately managed business in respect of the holding managed by that business[5].

The criteria applied by SOAEFD in determining what constitutes a separately managed business has changed over the years. For 1996 the criteria[6] is that (i) the business has a separate legal status, (ii) the economic structure of the business is separate from any other business in which the applicant participates, (iii) the operational management is separate and (iv) the commercial management is separate[7]. Other farming businesses in which an applicant has an interest, including sheep producer groups, have to be notified in the IACS forms. Where farming businesses are deemed not to be sufficiently separate, subsidies will be lost and claims limited to one IACS form only.

It is therefore of prime importance that farmers, who are participating in a number of separate businesses[8], are advised on the necessary steps to take to try to ensure that their businesses will be treated as separate businesses for the purpose of IACS.

In the IACS form producers may register land growing crops eligible for AAP payments as a forage area[9] for the purpose of claiming livestock subsidies, which are based on stocking density levels per 'forage' hectare. AAPS payment

---

1 SOAEFD should be able to provide a letter confirming what land is registered as eligible for a farm under IACS for AAP.
2 Land in grass at 31 December 1991 and during the preceding five years, unless it can be proved that the older grass leys were part of a normal arable rotation.
3 NB there are some exceptions – permanent grass, which was in a long but normal rotation, and woodland and non-agricultural land.
4 Cf the Explanatory Booklet AAP 1 (1996), paras 23-28; Form IACS 21. For 1996, applications to switch had to be made by 30 November 1995.
5 Cf EC Council Regulation 3508/92, art 1(4) where the definition of 'Farmer shall mean an individual agricultural producer, whether a natural or legal person or a group of natural or legal persons, whatever legal status is granted the group and its members by national law' and 'holding shall mean all the production units managed by a farmer situated within the same member state's territory'.
6 See the Explanatory Booklet IACS 1 (1996), para 15.
7 Ie separately taxed; the management is separate and has its own discretion in regard to management decisions, and any transactions between businesses are commercial and invoiced at the time.
8 Eg dairy farming with a separate suckler cow business.
9 Defined by EC Council Regulation 805/68, art 4g(3).

may then not be claimed on that area. The forage area is the area of the holding available throughout the calendar year for feeding or grazing of livestock[1].

The IACS forms require to be lodged with SOAEFD by 30 March annually, although up to 1996 the UK obtained from the EC an extension to 15 May[2]. The person submitting the form in any year is the person eligible to receive the AAP payments for that year, even if the farm is transferred to another occupier after that date.

There are two options under the AAP scheme, namely[3]:

(a) *The simplified scheme*. Open to small producers claiming in respect of crops grown on 17.66 hectares of less favoured areas (LFA) land or 16.23 hectares of non-LFA land. The claimant is not required to set aside land. Payment is at the rate for cereals even if other crops are grown.

(b) *The general scheme*. Open to all producers.

The 1996 AAP payments in £/ha for cereals[4] were LFA £242.49 and non-LFA £263.90 and for set-aside were LFA £307.16 and non-LFA £334.28.

Eligible crops are specified, which have to be planted by 15 May[5]. Crops have to be maintained until the beginning of flowering[6] or 30 June[7].

Claims can only be made in respect of a minimum area of 0.3 hectares with a corresponding area for set-aside meeting the minimum plot size. Payment is made based upon EC rates per tonne calculated at a rate per hectare using historic average yields set in ECUs and then converted per the green ECU exchange rate.

In order to qualify for AAP payments the applicant under the general scheme has to set aside land in accordance with one of the options. The options include:

(a) *six-year rotational set-aside* (10 per cent) which is land that has not been set aside under AAP at any time during the five years, unless no other eligible land is available;

(b) *flexible set-aside* (10 per cent) which is set-aside that may be left in one place or moved from year to year. If some of the land is rotated, then all land must be in flexible set-aside;

(c) *guaranteed set-aside* (10 per cent) is land which the farmer undertakes to keep set aside for five years. From 1996 onwards applications are restricted to narrowly defined categories;

(d) *voluntary set-aside* is land set aside in excess of the basic 10 per cent, provided that the total of land set aside, excluding penalty set-aside, does not exceed the cropped area (plus penalty set-aside) for which AAP payments are claimed;

---

1 It excludes buildings, areas used for other crops benefiting from EC aid, set-aside areas, areas used for permanent crops (eg fruit and vegetables) and land on which potatoes, sugar-beet or vining peas are grown: Beef Special Premium Scheme 'Notes for Guidance' SOAEFD Beef 4.
2 Whether or not this extension will be allowed in future years is a matter of conjecture.
3 EC Council Regulation 1765/92, art 5.
4 Support for other crops is different.
5 With the exception of maize which has to be planted by 31 May. Where *force majeure* prevents sowing a written exemption may be applied for.
6 Cereals and linseed.
7 Oilseeds and protein crops.

(e) *additional voluntary set-aside* is a special option available to farmers who had land in the old five year set-aside scheme[1];

(f) *penalty set-aside* is additional uncompensated set-aside imposed if the regional base is exceeded[2];

(g) *Farm Woodland or Habitats Scheme* land included as set-aside. Eligible land withdrawn from the FWPS/WGS or the Habitat scheme can count towards set-aside, provided the land was entered into the scheme on or after 1 July 1995.

There is an option to put set-aside land into the Set-Aside Access Scheme[3], but such land has to be put into guaranteed set-aside.

Producers have an option to transfer their set-aside obligations to another producer. The rules are complex and in the majority of situations the financial benefits are marginal. The farmer transferring set-aside is the 'exporter' and the receiving farmer is the 'importer'. All or part of the set-aside obligation may be transferred to one or two other producers. Transferred set-aside is subject to additional set-aside requirements. Set-aside cannot be imported into the five year guaranteed set-aside or into the set-aside access scheme. Set-aside may be transferred within a 20 km radius from a fixed point on the holding[4] to the border of the plot to which it is transferred. There are additional rules for transfers into environmental target areas[5].

Land set aside must be of at least 0.3 hectares, with smaller areas only acceptable if they are completely bounded by permanent, fixed boundaries such as walls, hedges or water courses. Set-aside land can only count as a single block if one can walk from any point on the set-aside area to every other point without leaving the set-aside. Each set-aside block, including field margins, must be at least 20 meters wide throughout.

Where land is in different yield regions[6], generally set-aside requirements require to be calculated separately in each region, although there are exceptions[7].

The rules governing the use of set-aside land, the crops that may be grown and the green cover that may be established, what may and may not be grazed etc are complex. Reference should be made to the current Regulations and to the current explanatory booklets.

The set-aside period lasts from 15 January to 31 August on most set-asides, but for the whole year under guaranteed set-aside. With rotational set-aside, after 15 July the ground may be prepared and crops (but not a horticultural crop) sown for harvest or grazing after 15 January 1997. There are restrictions

---

1 Set-Aside Regulations 1988, SI 1988/1352 amended by SIs 1989/1042, 1990/1716, 1991/1993.
2 Applied in Scottish less favoured areas (LFA) in 1996.
3 Set-Aside Access (Scotland) Regulations 1994, SI 1994/3085.
4 Generally measured from the farmhouse or farm headquarters or the covered area where the main agricultural machinery is stored. Where the holding consists of individual farms the rules are more complex. The measurement is to any point on the border of the plot to which the set-aside is transferred.
5 SOAEFD, 'Transfers of Set-Aside Obligations Between Producers' AAP 3 (REV) (1995).
6 Eg LFA and non-LFA.
7 AAP 1 (1996), para 129 and App 2.

on the use to which the green cover may be put in the period 1 September to 15 January[1].

In summary set-aside land cannot be used (i) for a non-agricultural use, which brings a commercial return or a direct benefit[2], (ii) for an activity which is incompatible with the set-aside management rules[3], (iii) for any form of non-agricultural production (including horticulture and grazing)[4] other than the growing of non-food crops[5].

Where there are land transfers on a sale or other change of occupancy, the parties are advised to have a written agreement confirming that the outgoer has observed all the conditions of the AAP and that the incomer will continue to do so. The incomer should confirm that he is able to comply with all the AAP and set-aside requirements, including the exemptions from the requirement to have farmed the set-aside land for two years.

In particular a check should be made, by requiring a sight of the previous IACS forms, the crop rotations submitted to SOAEFD and the letter from SOAEFD confirming the registration of the land as eligible.

While the AAP payment is due to the person submitting the IACS form, it is possible to transfer the land for the purposes of making a claim under the livestock schemes. Transfer of the Area Aid Application is only possible where all of the land covered by the AAA form is transferred.

With regard to tenants, it is important that the tenant agrees with the landlord any guaranteed set-aside schemes, otherwise there may be difficulties on the termination of a tenancy.

BEEF SPECIAL PREMIUM SCHEME[6]

The BSPS is designed to make payment of premium for producers of male bovine animals under EC Council Regulation 805/68, art 4b and h[7].

To qualify for premium the animal must be registered with SOAEFD, by applying for a Cattle Control Document (CCD)[8] within ten weeks of birth giving the eartag[9], breed and date of birth. A white CCD is then issued to the

---

1 AAP 1 (1996), paras 123-124. Green cover must not be put to any commercial use, although the farmer's own animals may be grazed or silage cut for own use. Letting the grazing or selling, bartering or exchanging the hay or silage harvested, putting the land to any non-agricultural use dependent on there being a green cover or wild bird cover or use for any other commercial purpose during this period is contrary to the management rules.
2 Eg free car parking in connection with a fund-raising event. Any non-agricultural use, except grazing the farmer's own non-agricultural animals (eg ponies) kept for own use, requires the consent in writing of the SOAEFD.
3 AAP 2 (1996), s 2.
4 Or sowing an agricultural crop, other than an approved green cover, even if crop is not taken to harvest or is subsequently destroyed.
5 AAP 2 (1996), s 4, paras 132-140.
6 Beef Special Premium Regulations 1993, SI 1993/1734 amended by SIs 1994/3131 and 1995/14. See Beef Special Premium Scheme 'Notes for Guidance' SOAEFD Beef 4.
7 Support in 1996 was £93.09 per claim, up to 90 claims. 75 per cent of payment is made by 30 October or within six weeks of the end of the retention period, whichever is the later, and the remainder of the subsidy, with any extensification payment, is made by the following 30 June.
8 Beef Special Premium Regulations 1993, reg 3; Form Beef 2. In England a CCD is known as a Cattle Identification Document (CID).
9 Issued under the Eartag Allocation System; eg 'UK AB12345/00023'.

producer. A blue CCD is issued when the first premium is claimed and a pink CCD when the second premium is claimed at the end of each retention period. The CCD must accompany any animal over three months in age when traded or sent for slaughter[1]. CCDs require to be surrendered within three months of the death, loss, theft or export of the animal[2].

Premium may be claimed twice during the life of the male bovine. A first premium can be claimed when the animal is eight months old and less than 21 months of age at the start of the retention period[3]. The second premium can be claimed when the animal is 21 months old or older, at the start of the retention period[4]. Claims are restricted to 90 animals in each age group in each calendar year[5].

The retention period is a minimum of two months starting with the date the form is received by SOAEFD, but a producer may submit his application up to two months before the start of the retention period, stating a date for the start of the period[6].

If the animal is lost due to natural circumstances and this is accepted by SOAEFD upon notification within ten days, then the lost animal will not count against the headage limit. If through *force majeure* the animal cannot be kept for the retention period and this is accepted by SOAEFD upon written notification within ten days of the event, then premium can still be claimed[7].

The stocking density limits for claiming BSPS in 1996 is two livestock units per forage hectare[8]. If SOAEFD considers that the land is being over-grazed or that supplementary feeding is unsuitable, and after notification this is not corrected, premium may be reduced[9].

An extensification payment[10] may be claimed under the BSPS where the stocking density is below 1.4 livestock units per forage area.

Records in relation to an animal for which premium has been claimed must be retained for a period of four years from the relevant date of any bill, account, receipt, voucher or other record including records relating to the animal movements etc regulations[11].

---

1 Beef Special Premium Regulations 1993, reg 4. Cf exemptions in respect of animals imported into the United Kingdom in the preceding three months.
2 Beef Special Premium Regulations 1993, reg 5.
3 Beef Special Premium Regulations 1993, reg 7; BSPS 'Notes for Guidance' SOAEFD Beef 4, para 4.1.
4 Beef Special Premium Regulations 1993, reg 8; BSPS 'Notes for Guidance' SOAEFD Beef 4, para 4.1.
5 Beef Special Premium Regulations 1993, regs 7(4)(a) and 8(4)(a); BSPS 'Notes for Guidance' SOAEFD Beef 4, para 4.1.
6 Beef Special Premium Regulations 1993, reg 9; BSPS 'Notes for Guidance' SOAEFD Beef 4, para 4.1.
7 BSPS 'Notes for Guidance' SOAEFD Beef 4, para 6.1.4
8 Forage area defined by EC Council Regulation 805/68, art 4g(3).
9 Beef Special Premium Regulations 1993, regs 9A and 9B amended by SI 1994/3131; BSPS 'Notes for Guidance' SOAEFD Beef 4, para 6.9.
10 £31.03 in 1996.
11 Beef Special Premium Regulations 1993, reg 10 as amended by SI 1995/14, reg 2(3).

SUCKLER COW ANNUAL PREMIUM SCHEME[1]

From 1996, claims for SCP are now combined with the claim for HLCA (Cattle), so the time limits for lodging applications etc is the same for both subsidies. The crucial difference is that the SCP claim is related to the forage areas in the current year's IACS form, whereas the HLCA claim is related to the previous year's IACS form[2].

SCP is an EC-funded scheme to help maintain the incomes of specialist beef producers at a satisfactory level. Premium is paid in respect of suckler cows forming part of a regular breeding herd used for rearing calves for meat production. A producer of suckler cows in respect of which premium is claimed, with limited exceptions[3], may not sell milk or milk products from the holding for 12 months after making the claim and may not manufacture milk products on the holding for sale after that date.

This section will only provide sufficient information about the scheme to allow for a better understanding of the regime and its quotas[4].

Application forms[5] for both SCP and HLCA (Cattle) require to be submitted to SOAEFD in the period from and including 1 July to and including 8 December (approximately) in any calendar year. Late applications can be received for up to 20 calendar days after the last date for lodging subject to a penalty of 1 per cent of premium for each day late. No applications can be received after that date unless *force majeure* can be proved[6].

The suckler cows (or replacements which may be suckler cows or in-calf heifers) at least equal to the number on which premium has to be claimed have to be retained on the holding for six months (the retention period). The six months run from the starting date, which is the day after the day the department receives and stamps the form[7].

An applicant requires to retain for a period of four years from the date an application was submitted any bill, account, receipt, voucher or other record in relation to:

(i) the number of cattle kept on the holding during the period of six months following that date; and

(ii) any transaction concerning cattle, milk or milk products carried out by him on that date and during the period of twelve months following that date; and

---

1 Suckler Cow Premium Regulations 1993, SI 1993/1441 as amended by SI 1994/1528.
2 Eg 1996 SCP claims relate to 1996 IACS forms; 1996 HLCA claims relate to 1995 IACS forms.
3 Eg milk producers with no more than 120,000 litres of quota may be eligible for reduced SCP; see Explanatory Leaflet SCP 2 (1994), para 34.
4 For fuller details of the scheme in any year reference should be made to the current SOAEFD explanatory leaflet obtainable from the department: 'Suckler Cow Premium Scheme 1996 (including Small Milk Producers) and Hill Livestock Compensatory Allowance (Cattle) 1996' – SCP 1995/HLCA 1996 (Cattle) 2.
5 SCP 1995/HLCA (Cattle) 1.
6 Suckler Cow Premium Regulations 1993, reg 3 (as amended); see Explanatory Leaflet SCP 1995/HLCA 1996 (Cattle) 2 above.
7 The retention period ends on the same numerical day six months hence, or if there is no same numerical day on the last day of the month; eg if form stamped 12 October 1996 the retention period would end at midnight 12 April 1997, or if form stamped 30 August 1996 the retention period would end on 28 February 1997. See Explanatory Leaflet SCP 1995/HLCA 1996 (Cattle) 2, para 13.

(iii) the register referred to in EC Council Regulation 805/68, art 4g(4); and

(iv) records kept under the Movements of Animals (Records) Order 1960, art 3(1)[1]; the Bovine Animals (Identification, Marketing and Breeding Records) Order 1990, art 9(1)[2] and the Bovine Animals (Records, Identification and Movement) Order 1995, art 5(1)[3].

With effect from 1 July 1994 the amount of SCP which can be paid to a producer in any calendar year shall not exceed the number of suckler cows which the Secretary of State for Scotland notifies[4] the producer as being the number that can be carried without over-grazing the whole or part of the holding. Where the number of suckler cows being grazed exceeds the notified number, then all SCP otherwise due may be withheld[5]. These figures are related to the livestock units carried on the forage area as notified in the IACS form[6].

The amount of premium that can be claimed is restricted to the quantum of the relevant quota owned or leased in for that year.

The CAP allows an additional extensification payment to be made in respect of SCP where the stocking density is below 1.4 livestock units per hectare[7].

SHEEP ANNUAL PREMIUM SCHEME[8]

From 1996, claims for SAP are now combined with the claim for HLCA (Sheep), so the time limits for lodging applications etc is the same for both subsidies. The crucial difference is that the SAP claim is related to the forage areas in the current year's IACS form, whereas the HLCA claim is related to the previous year's IACS form[9].

SAP is an EC-funded scheme to compensate a sheep producer if the average market price of sheep falls below the basic price agreed by the EC. This section will only provide sufficient information about the scheme to allow for a better understanding of the quota regime[10].

The producer is required to keep a register containing information on (i) the dates on which female sheep put to the ram for the first time gave birth and the

---

1 SI 1960/105 amended by SIs 1961/1493 and 1989/879 (no longer applicable to bovine animals – SI 1995/12).
2 SI 1990/1867 amended by SI 1993/503 (no longer applicable to bovine animals – SI 1995/12).
3 SI 1995/12.
4 The notified numbers were: in 1994, 3 livestock units per hectare of forage area; in 1995, 2.5 livestock units per hectare of forage area; and from 1995 onwards 2 livestock units per hectare of forage area. For the complicated definition of 'forage area' as applicable to this regulation see Explanatory Leaflet SCP 2 (1994), para 44.
5 Suckler Cow Premium Regulations 1993, reg 3A amended by SI 1994/1528.
6 Stocking density for 1996 was 2 livestock units per hectare of forage area.
7 Extensification premium was fixed at 36.2 ECUs (about £31.03) in 1996.
8 Sheep Annual Premium Regulations 1992, SI 1992/2677 amended by SI 1994/2741.
9 Eg 1996 SAP claims relate to 1996 IACS forms; 1996 HLCA claims relate to 1995 IACS forms.
10 For fuller details of the scheme in any year reference should be made to the current SOAEFD explanatory leaflet obtainable from the department: 'Sheep Annual Premium Scheme 1996 and Hill Livestock Compensatory Allowance (Sheep) 1996' – SAP 1996/HLCA 1996 (Sheep) 2.

number of lambs produced, (ii) the number and dates of any sheep purchased, sold or otherwise disposed of: the name and address of the seller, buyer or other recipient, or, in the case of sheep purchased or sold at livestock market, the name and address of that market, and (iii) in the case of losses of sheep the date the producer discovered the loss, the number lost and the circumstances of the loss. The register has to be retained for three years from the end of the marketing year to which the last entry relates[1].

The scheme requires the producer to submit a claim form[2] for both SAP and HLCA by the closing date of the application periods and then to keep on the holding (or in agistment) the number of eligible sheep claimed on the form for 100 consecutive days (the retention period) after the application period[3].

The producer is required to lodge his claim on or after 4 December in the preceding marketing year, but not later than 4 February in that marketing year[4]. The retention period runs from midnight on 4 February to midnight on 14 May. Claims can be received in the 20 calendar days following the closing date, but premium is reduced by 1 per cent for each day late. Claims cannot be received after the 20 days unless the claimant can demonstrate *force majeure*.

From 15 November 1994, SAP otherwise payable may be reduced or withheld or payments already made recovered, where the Secretary of State for Scotland forms the opinion that a particular parcel of land is being overgrazed, or if the applicant uses unsuitable supplementary feeding methods[5]. These figures are related to the livestock units carried on the forage area as notified in the IACS form[6].

The amount of SAP that may now be claimed is limited by the amount of quota owned or leased by the producer to cover the particular SAP claim.

SHEEP AND SUCKLER COW PREMIUM QUOTA

Sheep annual premium and suckler cow premium quotas were introduced in 1993[7] to provide a cap on the quantum of sheep annual premium (SAP) and suckler cow premium (SCP)[8] payable in 1993 under CAP. The EC legislation has been modified a number of times since the first introduction of quotas[9].

---

1 Sheep Annual Premium Regulations 1992.
2 SAP 1996/HLCA 1996 (Sheep) 1.
3 The single claim and retention period introduced in 1996 replaces the two claim and retention periods which existed before that.
4 Sheep Annual Premium Regulations 1992, reg 3(2) as amended by SI 1995/2779.
5 Sheep Annual Premium Regulations 1992, regs 3A and 3B as amended by SI 1994/2741.
6 Stocking density for 1996 was 2 livestock units per hectare of forage area.
7 Sheep Annual Premium and Suckler Cow Premium (Quotas) Regulations 1993, SI 1993/1626 (the 1993 Regs) amended by SIs 1993/3036 and 1994/2894. Cf SOAFDS 'Sheep Annual Premium – Quotas, Explanatory Guide' and 'Suckler Cow Premium Scheme – Quotas, Explanatory Guide'.
8 Suckler Cow Premium Regulations 1993 amended by SIs 1994/1528 and 1995/15.
9 SAP – EC Commission Regulation 3567/92 modified by EC Commission Regulations 1845/93, 3534/93, 0826/94, 1720/94, 2527/94. SCP – EC Commission Regulation 3886/92 modified by EC Commission Regulations 0538/93, 1433/93, 1909/93, 2889/93, 3429/93, 3484/93, 00129/94, 0489/94, 1034/94, 1719/94, 2526/94, 3269/94. Reference in the following footnotes to the principal regulations means the principal regulation as modified, where applicable.

Nature of sheep annual premium and suckler cow premium quota

Unlike milk quota, which attaches to the land, sheep annual premium and suckler cow premium quota are personal to the producer to whom they are allocated. There is no restriction in the regulations on the transfer or lease of these quotas, which make it necessary to link a transfer or lease to any change of occupation of the land.

These quotas are therefore moveable in character. They will not be secured by a standard security. They will be moveable in any succession, potentially liable to claims for prior and legal rights[1].

Any producer holding quota should be advised to make provision in his will for an appropriate transfer of these quotas to avoid any difficulty of the person succeeding to the holding not succeeding to the quota.

Tenants and sheep annual premium and suckler cow premium quota

In general, as these quotas are personal to the producer, they will not belong to the landlord of an agricultural holding. The producer will be entitled to transfer the quota from the holding on the termination of the tenancy, unless there is an appropriate provision in the lease binding the tenant to make the quota over to the landlord on the termination of the tenancy.

Such a provision is unlikely to exist in leases granted prior to the introduction of quotas. In leases drafted after the introduction of milk quota a general provision was often included requiring tenants to make over future allocations of quota to the landlord. Many such clauses were drafted with milk quota in mind and were phrased in a way that required quotas 'allocated to the holding' to be made over to the landlord on the termination of the tenancy. A clause so worded may not be effective to require a tenant to transfer these quotas to the landlord on the termination of the tenancy.

As these quotas are personal to the tenant, the landlord will not be entitled to have the value of the quotas included in the rental value of the holding.

Where the landlord transfers quota to the tenant at the commencement of a lease, detailed contractual provisions will have to be made to prevent the tenant transferring quota off the holding during the currency of the lease and to ensure that the tenant does not prejudice the quota by failing to use it as required by the regulations. Such provisions will be difficult to police.

The existence of sheep annual premium quota may well have an impact on the valuation of bound sheep stock on the termination of a tenancy[2].

Quota zoning

In Scotland, sheep and suckler cow quota is divided into three categories, which are ring fenced within their designated areas called 'sensitive zones'[3]. The quota is divided into (i) the Scottish Highlands and Islands[4] less favoured

1 Cf the Succession (Scotland) Act 1964, ss 8-13.
2 See 'Sheep stock valuations' in chap 19.
3 1993 Regs, Sch 1, para 2.
4 The areas of operation of Highlands and Islands Enterprise; 1993 Regs, Sch 1, para 3.

areas (LFAs), (ii) Scottish LFAs[1], other than those in the Scottish Highlands and Islands, and (iii) non-LFAs of Great Britain[2].

The importance of the categorisation of sensitive areas is that (i) a producer can only use quota applicable to that zone[3], and (ii) transfers or leases of quota, with limited exceptions, cannot be made outwith the sensitive zone to which the allocation has been made[4].

Where a producer's holding is not situated entirely within a single sensitive zone, then the holding is treated as being in the sensitive zone in which the greater part of the 'agricultural area utilised for farming'[5] on the holding is situated[6].

Quota allocated from the national reserve to any producer is regarded as belonging to the sensitive zone in which the producer's holding is situated[7].

Allocation of sheep annual premium quota

An initial allocation of sheep quota was made on 3 February 1993 to an individual producer or member of a producer group[8] who had received SAP 1991 and submitted a claim for SAP 1992. An initial allocation was also made to a producer or member of a producer group which had inherited or taken over all the holding of a producer who received SAP 1991, but did not claim SAP 1992.

Normally the allocation of quota was based on the number of eligible ewes in respect of which premium was paid under SAP 1991, although a special calculation was applied where the producer was a member of a producer group for SAP 1991, the producer or producer group was subject to penalties under SAP 1991 or had suffered a loss as a consequence of 'natural circumstances' under SAP 1991.

Sheep quota was issued as either full-rate quota or a combination of full-rate and half-rate quota. There was a headage limit of 1,000 ewes in LFAs and 500 ewes in non-LFAs. Full-rate quota was issued up to the headage limit and half-rate quota for numbers above the headage limit claimed in 1991.

Headage limits were abolished by the EC for SAP 1995 with the effect that two half-rate units were converted into a full-rate unit for SAP 1995[9].

Where the producer was a producer group, the apportionment calculation was applied to the total eligibility figure, to apportion the quota between individual members of the group, based on either the 1991 flock apportionment declaration or on the breakdown of flock ownership or apportionment on the 1992 SAPS application. The quota was then issued to the individual.

---

1 1993 Regs, Sch 1, para 4.
2 1993 Regs, Sch 1, para 5.
3 1993 Regs, reg 8(3).
4 1993 Regs, regs 8 and 9. See below.
5 Defined by EC Council Regulation 571/88, art 5(b) to mean 'the total area taken up by arable land, permanent pasture and meadow, land used for permanent crops and kitchen gardens'; cf 1993 Regs, reg 3(3).
6 1993 Regs, reg 3.
7 1993 Regs, reg 8(1).
8 Scottish partnerships were treated as producer groups rather than as a separate legal persona.
9 Cf EC Commission Regulation 826/94.

In a partnership, while the quota may have been issued to the individuals, the quota will probably be held by each partner in trust as a partnership asset.

Provision was made for a supplementary allocation of quota to (i) participants in a community extensification scheme or an environmentally sensitive area scheme or (ii) those who had suffered a serious natural disaster affecting the holding, accidental destruction of fodder or buildings or a disease which led to the destruction of at least half the flock.

Apart from the initial or supplementary allocations all other claimants had to apply for an allocation of quota from the national reserve. The 1993 national reserve was divided initially into five categories but these were subsequently extended to seven with categories two, three and four having sub-categories. Detailed rules for allocation of quota from the national reserve were promulgated by regulation[1]. Subsequently it was held that the rules applied to the allocation of quota from the national reserve failed to implement EC law and were unlawful[2]. In consequence there was a re-adjustment of the initial allocations of quota from the national reserve.

As the initial allocations of quota from the national reserve have now been completed, the issues arising as to eligibility are now only of historic interest.

Allocations of suckler cow premium quota

Suckler cow premium quota was allocated to claimants who received suckler cow premium under the 1992 scheme. Unlike sheep annual premium quota, where the claimant was a producer group or limited company, the quota was allocated to the group or company.

The allocation was based on the number of eligible suckler cows in respect of which premium was paid, less 1 per cent reduction taken into the national reserve. The effect of the 1 per cent reduction in some cases led to an allocation of quota, which included a partial quota right.

Provision was also made for allocations from the national reserve, the majority of which has now been completed.

National reserve

An initial national reserve of SAP and SCP quota was established, from which the initial allocations to hardship cases were made. The national reserve is topped up by the 15 per cent siphon on any transfer of quota without a transfer of the holding and any quota surrendered of withdrawn.

The initial allocations from the national reserve in 1993 and the problems relating thereto are now only of historic interest.

There are six categories of person entitled to apply for an allocation of quota from the national reserve for any given year later than 1993[3].

---

1 Sheep Annual Premium and Suckler Cow Premium (Quotas) (Amendment) Regulations 1993, SI 1993/3036.
2 *R v Ministry of Agriculture, Fisheries and Food ex p National Farmers Union* (1995) Independent, 25 September.
3 Sheep Annual Premium and Suckler Cow Premium (Quotas) (Amendment) Regulations 1993, Sch 2, Pt II.

The categories may be summarised as follows[1]:

1. Any producer who participated in an approved environmental type scheme or management agreement[2] and has ceased or demonstrated that he is committed to ending his participation in the scheme.

2. Any producer who can demonstrate that at the time he makes the application he has taken over or is committed to take over any proportion of land from which the allocated quota[3] has been removed by a departing tenant or share-farmer, or he has committed himself to take over any proportion of such land by the close of the application period in respect of which he intends to use the quota allocated to him.

3. (a) Any person who at the time he makes an application to the national reserve can demonstrate that he has become a producer, or, already being a producer, has increased the size of his existing flock or herd, in consequence of reverting from arable to livestock farming, intending to make an application for premium, and giving certain environmental undertakings or obligations; eg under a nitrate sensitive area scheme, a habitat scheme or a development project under the Natural Heritage (Scotland) Act 1991.

   (b) Any young person who is a newcomer to farming who intends to apply for premium.

4. Any person, other than a young person, who is a newcomer to farming who intends to apply for premium.

5. Any person who has become a producer, or, already being a producer, has increased the size of his flock, in consequence of, or has committed himself to, a plan approved by the EC aid scheme for organic farming.

6. (a) Any producer who has not previously applied for premium, and who can demonstrate that he proposes to apply for premium using the quota allocated to him and does not fall within a higher category.

   (b) Any producer who at the time he makes his application for an allocation of quota from the appropriate national reserve can demonstrate that he has acquired a part of any area or irrevocably committed himself to acquire such an area before the close of the application period in respect of which he intends to use the quota, formerly used for sheep or suckler cow production, and which was temporarily taken out of agricultural use.

A producer who has transferred or temporarily leased out quota rights is not eligible to apply for quota from the national reserve in that year.

Where a producer obtains rights free from the national reserve he may not transfer or temporarily lease his rights in the following three marketing years, except in duly justified cases. If he does not use at least 90 per cent of those rights during the following three marketing years, the average of the rights not

---

1 What follows is only a summary. Any producers considering that they may be eligible will require to have regard to the details of the regulations.
2 Eg under the Countryside Act 1968 or the Wildlife and Countryside Act 1981.
3 But not transferred or leased in quota.

used in the three years are withdrawn and surrendered to the national reserve[1].

Quota registers

The Secretary of State for Scotland is required to maintain a register for each producer who has been allocated sheep annual premium or suckler cow premium quota. The register is required to include an entry for each producer giving his name and address, his holding number and producer identification, an indication of the sensitive zone within which the holding is situated, a statement of the amount of his quota, a statement of the amount of quota that has either been leased out or leased in, with the expiry date of the lease[2].

Any person who has entered into a financial arrangement with the producer in expectation of the producer's continuing right to receive the appropriate premium may request a copy of that producer's entry. The Secretary of State is required to inform the producer of the request, and after having taken into account any representations, has to decide whether or not to meet the request[3].

The copy of the entry must be provided to the producer on request or to any person who has the producer's written consent. A reasonable charge may be made for the supply of the copy entry, to anyone other than the producer to whom it relates[4].

Notification of transfer or lease of quota

The transfer or lease of quota has to be notified to SOAEFD on the appropriate form[5].

Quota may only be transferred or leased out in each year within particularly narrow time limits to be effective for the following year.

The initial regulations[6] provided for a deadline, which was the latest date that left two clear months before the first day (i) for sheep quota – of the first period for submitting applications for premium in respect of the next marketing year and (ii) for suckler cow quota – the next period for submitting applications for premium.

This gave rise to substantial difficulties. Further, the deadlines were such that a person transferring a holding by sale or lease to a new owner or occupier at Martinmas 1993 or 1994, could not effect a transfer of the quota with the

---

1 SAP – EC Commission Regulation 3567/92, art 6; SCP – EC Commission regulation 3886/92, art 32.
2 1993 Regs, reg 14.
3 1993 Regs, reg 15(1), (2) and (3).
4 1993 Regs, reg 15(4).
5 1993 Regs, reg 5(1). See Explanatory Leaflet 'Notification of Permanent Transfer of Quota (Sheep)' – SAP (Q) 2 (1996) and form 'Notification of Permanent Transfer of Quota (Sheep)' – SAP (Q) 1 (1996); 'Notification of Permanent Transfer of Suckler Cow Quota' – SCP (Q) 2 (1995) and form 'Notification of Permanent Transfer of Suckler Cow Quota' – SCP (Q) 1 (1995); 'Notification of Lease of Quota (Sheep)' – SAP (Q) 4 (1996) and form 'Notification of Lease of Quota (Sheep)' – SAP (Q) 3 (1996); 'Notification of Lease of Suckler Cow Quota' – SCP (Q) 4 (1995) and form 'Notification of Lease of Suckler Cow Quota' – SCP (Q) 3 (1995).
6 1993 Regs, reg 5(2) and (3) (prior to amendment by SI 1994/2894).

holding on that date. In consequence such a transfer of quota was a transfer without a holding and 15 per cent of the quota was siphoned to the national reserve[1].

Further, because of difficulties in making allocations from the national reserve, these allocations were often made after the deadline for transfer or leases had expired. In consequence, farmers who were waiting for an allocation of quota from the national reserve, would lease or buy in quota before the deadline expired in order to safeguard their position if no quota was allocated. Where they then received a subsequent allocation from the national reserve for the year in respect of which they had bought in or leased quota, they then found themselves with unused quota, which could neither be leased out or sold for the year.

In consequence, the EC authorised the re-opening of the notification periods for further limited periods and in some cases for limited categories, for lease and transfers for the years 1993 to 1995 to take account of these difficulties[2]. It also authorised sub-leasing of quotas in 1993 and 1994.

Additional deadlines are notified by the Secretary of State by such means as he considers likely to bring the matter to the attention of producers[3]. It is unlikely that the transfer and lease period will be re-opened in the future.

To take account of some of the difficulties encountered the regulations have now been amended to alter the transfer deadlines[4].

The deadline for notification of transfers or leases of sheep annual premium quotas is[5]:

(a) if the producer receiving the quota under the transfer or lease lodges his application for SAP by the end of the period specified for delivering such applications in respect of any marketing year, the date of lodgement of that application; or

(b) if no such application is lodged by that time, by the end of the period.

The closing date for receipt of registration forms for permanent transfer of quota is the date on which the application for SAP submitted by the recipient of the quota is received by the department and in any event by no later than the closing date for the receipt of SAP claims, excluding the days of grace.

The deadline for notification of transfers or leases of suckler cow premium quotas is[6]:

(a) if the producer receiving the quota under the transfer or lease lodges his premium application by the end of the period specified for delivering such applications in respect of any calendar year, the date of lodgement of that application; or

(b) if no such application is lodged by that time, by the end of the period[7].

---

1 Cf p 333 'Siphon'.
2 1993 Regs, reg 5(5) as amended by SI 1994/2894.
3 1993 Regs, reg 5(6).The onus is on the party wanting a transfer to find out about any notice.
4 SI 1994/2894.
5 1993 Regs, reg 5(2).
6 1993 Regs, reg 5(3).
7 The deadline for claims in 1994 was 8 December 1994.

The siphon

Where quota is transferred by a producer without the transfer of the whole holding, then 15 per cent of the quota is surrendered to the national reserve[1].

The 15 per cent siphon also applies where part of the holding is transferred with the quota. It is possible to reduce the size of a holding by transferring land without quota, prior to transferring the balance of the holding with all the quota, thus avoiding siphon.

In some cases, where a holding is being divided and one part does not wish to receive quota, it is more economic to transfer at least the minimum quota to that part, so that all the quota is in fact being transferred with the whole holding. If the occupier of the land with the minimum of quota then loses that minimum, the minimum is often less than 15 per cent of the whole quota.

A practical difficulty arising out of this siphon is that parties to the sale or lease of a holding have to co-ordinate the entry date to the holding with a date that allows for notification of the transfer to SOAEFD before the applicable deadline, if the intention is to avoid siphon. This needs to be related to the relevant retention dates unless a management or agistment agreement is entered into.

Transfer considerations

A producer with suckler cow premium for less than ten animals may not transfer quota for less than one animal without transferring his holding[2].

Where a producer obtains rights free from the national reserve he may not transfer or temporarily lease his rights in the following three marketing years except in duly justified cases[3].

Quota is regularly bought and sold through brokers or at auction. A purchaser would be well advised to make provision for the retention of the price until such time as SOAEFD confirms the transaction.

Where the producer is a tenant, regular leasing out of quota, for the purposes of the obligation to farm in accordance with the rules of good husbandry, probably does not amount to farming in accordance with those rules[4].

Transfer of quota with holding

Where a producer transfers quota to a person taking over his holding, the notification of the transfer requires to be made no later than the next deadline for notification[5].

The notification has to be accompanied by satisfactory evidence of the taking over of the holding. The evidence must demonstrate either that the transferee of the quota has already taken over the holding or is under an obligation to take it over before the next deadline for notification[6].

1 1993 Regs, reg 6(1). Transfer of quota between partners is not now subject to siphon.
2 1993 Regs, reg 6(2)(a).
3 SAPS – EC Commission Regulation 3567/92, art 6; SCP – EC Commission Regulation 3886/92, art 32.
4 *Cambusmore Estate Trust v Little* 1991 SLT (Land Ct) 33 at 41H.
5 1993 Regs, reg 7(1).
6 1993 Regs, reg 7(1) and (2).

If the requirements of reg 7 are not met in respect of the transfer or if the Secretary of State for Scotland is not satisfied that the transferee of the quota has taken over or is under an obligation to take over the holding, then the transfer is treated as a transfer to which siphon applies[1].

As it is possible to have a date of entry different from, and earlier than, the date of transfer of the quota, in any transaction for the purchase of the holding with quota provision should be made for the retention of the quota price until such time as SOAEFD confirms the transfer.

Transfer of quota by executors

Following a death, the executors will be required to make provision to transfer the deceased's quota to the appropriate beneficiary within the notification deadlines.

Where the death occurs after the closing date for transfers, then the claim for SAP should be made by the executors, prior to transferring the quota during the next transfer period[2].

Where the holder of the quota is a tenant under an agricultural lease, the executors will have to take care to co-ordinate transfer of the lease with the timetable applicable for the notification of transfers of the relevant quota.

Particular exemptions were promulgated for SAP 1993 and 1994, but thereafter executors are required to transfer quota to the person who will be claiming quota, using the normal notification forms[3].

Usage rules

A producer is required to use at least 50 per cent of the allocated quota every other year. If less than 50 per cent of the quota is used in each of two consecutive years, then the number of units not used in the second year will be withdrawn from the producer without compensation from 1996[4].

In any five years, starting with the first year in which a producer leases out quota, there must be two consecutive years in which the producer does not lease out the quota. In each of those years the producer must use at least 50 per cent of the quota, to avoid a withdrawal[5].

In 1993 and 1994, where quota was allocated from the national reserve, the producer was required to use all his quota. From 1995 the producer is required to use on average 90 per cent of his quota during the three years following the

1 1993 Regs, reg 7(4).
2 'Notification of Permanent Transfer of Quota' explanatory leaflets.
3 SOAEFD letters – 'Changes to SAP Quota Regulations' and 'Changes to SCP Quota Regulations' – November 1994.
4 SAP – EC Commission Regulation 3567/92, art 6(2); SCP – EC Commission Regulation 3886/92, art 34(3).
5 SAP – EC Commission Regulation 3567/92, art 7(4); SCP – EC Commission Regulation 3886/92, art 34(3).

receipt of the quota from the national reserve, otherwise the average of the number of rights not used is withdrawn to the national reserve[1].

Producers who signed and lodged an application form for an environmentally sensitive area scheme or a pilot sheep extensification scheme before 22 July 1994 may lease out quota throughout the lifetime of the scheme. Where forms were signed and lodged after that date quota may only be leased out for three of the five years, but must be used by the producer in two consecutive years[2].

The EC is reconsidering usage rules, which may be changed in 1996.

Restrictions on numbers of quota units for transfer or lease

Where a producer wishes to transfer or lease out some, but not all, of his sheep annual premium quota there is a minimum number that may be leased. The minimum numbers are: 100 and over – 10; 20 to 99 – 5; fewer than 20 – 1[3].

A producer with suckler cow premium for less than ten animals may not transfer quota for less than one animal without transferring the holding, and may not lease out quota for less than one animal to another producer unless he transfers or leases the whole of his quota[4].

HILL LIVESTOCK COMPENSATORY ALLOWANCE[5]

The HLCA is funded by SOAEFD and not the EC and is designed to help maintain livestock farming in LFAs. There are separate schemes for sheep and cattle.

In order to claim HLCA an IACS form for the preceding year will have had to be completed.

Claims for HLCA are now combined with the SCP and SAP claims forms[6]. The application and retention periods for HLCA are now concurrent with the SCP and SAP schemes, although HLCA is based on the preceding year's AICS form, and SCP and SAP on the current year's IACS form.

Payments are made to the 'person who is on the qualifying day for that year the occupier of eligible land' in respect of cattle kept in regular breeding herds and/or sheep kept in qualified flocks, which are owned or leased[7]. 'Eligible land' means land of not less than three hectares within the LFA which is severely disadvantaged or disadvantaged land[8]. 'Qualifying day' is the day of submission of the claim[8].

1 SAP – EC Commission Regulation 3567/92, art 6(1)(b); SCP – EC Commission Regulation 3886/92, art 32(b).
2 1993 Regs, reg 10; cf SOAEFD letters – 'Changes to SAP Quota Regulations' and 'Changes to SCP Quota Regulations' – November 1994.
3 SOAEFD letters above.
4 1993 Regs, reg 6(2).
5 Hill Livestock (Compensatory Allowances) Regulations 1994, SI 1994/2740 amended by SIs 1995/100, 1995/1481, 1995/2778 and 1996/27. The SOAEFD explanatory leaflets are the same as those issued for claiming SCP or SAP.
6 Qv for dates of lodging etc.
7 Hill Livestock (Compensatory Allowances) Regulations 1994, reg 3(1).
8 Hill Livestock (Compensatory Allowances) Regulations 1994, reg 2.

The animals must have been substantially maintained on LFA land during the grazing season[1].

Claims for HLCA are not restricted by quota. There are restrictions to prevent over-grazing[2].

A claimant is not excluded from claiming HLCA if there is some milk production, although animals kept mainly for milk production are excluded[3]. Under the 1996 scheme, in-calf heifers are no longer eligible.

There is a stocking density limit of 1.4 livestock units per forage hectare. Claims for breeding ewes are restricted to nine ewes per hectare on disadvantaged land, up to a total of £60.85 per hectare, and six ewes per hectare on severely disadvantaged land, up to £81.13 per hectare.

A claimant must undertake to remain in LFA farming on at least three hectares for five years following the first claim, or until receiving the state retirement pension[4]. A partner receiving a state retirement pension does not relieve the partnership of the obligation.

OFFENCES AND PENALTIES

The regulations relating to SAP, SCP, HLCA and quota all provide for powers to authorised officers to enter upon certain premises to count stock or require information etc. The regulations provide for various penalties for failing to comply with requests by authorised officers, or for providing false or misleading information or recklessly providing such information[5].

Where an offence is committed by a body corporate and it is proved that it was committed with the consent or connivance of, or to be attributable to the neglect on the part of, any director, manager, secretary of similar officer, then that person is also deemed guilty of the offence.

CONVEYANCING CONSIDERATIONS[6]

The form completion dates, claim dates, retention periods, transfer dates for quotas and other aspects of CAP subsidy and quota regimes give rise to problems in the conveyance of farms or in the grant and surrender of leases.

The traditional entry dates of Whitsunday and Martinmas are difficult in relation to the subsidy and quota regimes. It may be necessary to avoid those dates, depending on the nature of the farming operations carried out on the holding.

1 Cf SAP Explanatory Leaflet, para 68(b) and SCP Explanatory Leaflet, para 85(b).
2 Hill Livestock (Compensatory Allowances) Regulations 1994, reg 6.
3 Hill Livestock (Compensatory Allowances) Regulations 1994, reg 4(5); Explanatory Leaflet SCP 1995/HLCA (Cattle) 1996.
4 Hill Livestock (Compensatory Allowances) Regulations 1994, reg 3(2); Explanatory Leaflets SAPS 1996/HLCA (Sheep) 1996, para 70 and SCP 1995/HLCA (Cattle) 1996, para 87.
5 Sheep Annual Premium Regulations 1992, SI 1992/2677, regs 5 and 9; Suckler Cow Premium Regulations 1993, SI 1993/1441, regs 5 and 9; Sheep Annual Premium and Suckler Cow Premium (Quotas) Regulations 1993, SI 1993/1626.
6 Cf PQLE course papers 'Conveyancing Aspects of Agricultural Quotas' (Law Society of Scotland, 29 Feb 1996).

Points which should be considered by conveyancers of farms include[1]:

(a) A check should be made as to whether or not the subjects for sale are or are not the seller's whole holding in terms of CAP. This may have important consequences in relation to various subsidies and quotas, which need to be provided for in the missives.

(b) A check should be made as to the classification of fields and other areas of the farm at 31 December 1991 to confirm their eligibility for AAP, set-aside, forage areas etc. A letter from SOAEFD should be obtained confirming the classification.

(c) A check should be made as to what aid or set-aside payments and premiums have been claimed, when claimed, when payment is due and when the relevant retention periods started and finish. These dates are necessary in order to plan an appropriate entry date.

(d) A warranty should be sought that (i) the IACS forms have been properly completed for the past four years, (ii) the requirements for the arable aid/set-aside payments have been and will be complied with to date of entry, (iii) no part of the subjects are subject to penalty set-aside.

(e) A check should be made, and the seller should be obliged to warrant the areas, that there are sufficient areas registered as eligible for arable area payments or as forage areas, and are accepted as such by SOAEFD, to satisfy the incomer's requirements. A letter to this effect should be obtained from SOAEFD.

(f) Confirmation should be obtained of what, if any, land is in guaranteed set-aside, which the seller should warrant.

(g) The seller should be taken bound to indemnify the purchaser against all losses, costs, claims and liability which the purchaser may suffer as a result of any breach by the seller of obligations under AAP and set-aside schemes.

(h) It may be necessary to take the purchaser bound to implement the AAP and set-aside obligations continuing after entry or bound to relieve the seller of any obligations that might arise from the buyer's breach of the continuing obligations.

(i) As the person who completes the IACS form is entitled to the payments due thereunder, contractual provision will either have to be made for a transfer of the holding before the form completion date or for who is to receive the AAP and set-aside payments after entry.

(j) The seller should be taken bound to deliver a completed land transfer form under the IACS scheme at the date of entry.

(k) If the date of entry is to be at a date when there are growing crops in the ground or the ground has been ploughed, provision will have to be made for valuations of the ploughing and/or growing crops.

(l) Depending on the entry date in relation to deadlines for lodging premium claims, if entry is in the middle of a claim period it may be necessary to take the

---

1 Cf 'Conveyancing considerations' in relation to milk quota in chap 24.

seller bound not to put in a claim before entry, but to leave it to the purchaser to put in the claim after entry.

(m) Where the outgoer is claiming SCP, SAP and HLCA and/or BSPS, provision will have to be made for entry after the retention periods are over or for an agistment contract for the incomer to look after the stock for the outgoer. Provision should be made as to who is to retain the SCP, SAP and HLCA payments or if they are to be apportioned.

(n) Where there is sheep and/or suckler cow quota on the holding the transfer date will have to be at a date which will permit the relevant quotas to be transferred to the incomer for use in the following quota year. The seller should be taken bound to do nothing which might have the effect of reducing the quota rights which fall to be transferred.

(o) Where quota is to be transferred as part of the contract, separate provision should be made for an apportionment of the price for the quota. An appropriate joint deposit arrangement should be made to safeguard the quota price, until SOAEFD confirms that the quota has been safely transferred to the incomer[1]. VAT is payable on the price of quota[2]. The missives should make clear whether the price is inclusive or exclusive of VAT.

(p) The seller should be asked to exhibit the relevant entries in the quota registers and the SOAEFD stocking density calculations.

(q) The seller should be asked to confirm and to grant warranty in relation to his usage of quota, to confirm that the usage rules have been complied with and that the quota can be used or leased out by the purchaser in accordance with his requirement.

(r) Where sheep or suckler cow quota is to be transferred, provision should be made to avoid siphon if possible. It may be possible to arrange for the sale of part of the holding without quota first, thus leaving the rest of the holding to be transferred with all the quota. Alternatively, if the holding is to be sold in parts, the transfer of a minimum of quota to one part, with the bulk to the other part, may in fact give rise to a smaller loss of quota than the 15 per cent siphon.

(s) A check should be made that the quota to be transferred with the holding, if it is only part of the producer's farming enterprise, is in the correct ring fence.

(t) The contract should make provision for the seller to hand over to the purchaser all documentation, letters, vouchers, invoices, movement orders etc, that the seller was required to keep in terms of the regulations under the

---

1 NB most quotas are in the name of the business/partnership, but sheep quota is in the name of the individual partners. If the holding belongs to one of the partners then the contract should bind the other members of the parnership to transfer their quota along with the holding. Further provision will have to be made for the transfer of quota by the seller to the members of a purchasing partnership.

2 If the quota is transferred along with the whole 'going concern' VAT is not payable; VAT Special Provisions Order 1995, SI 1995/3129, art 5.

various aid schemes. In general the documentation has to be kept for four years so the contract should provide for the preceding four years' documentation to be handed over.

Although under review, a retiral of a partner from a partnership is currently viewed as constituting a new business. Transfer of the IACS to the new partnership and other quotas is essential.

Appendices

Appendix 1. Statutory materials

Appendix 2. Styles*

* Styles by courtesy of W & J Burness WS

Appendix 1

Statutory materials

Agriculture (Scotland) Act 1948
(c 45)

* * *

Part II

Good Estate Management and Good Husbandry

26. Duties of good estate management and good husbandry

(1) ...

(2) The provisions of the Fifth Schedule and of the Sixth Schedule to this Act shall have effect respectively for the purpose of determining for the purposes of this Act whether the owner of agricultural land is fulfilling his responsibilities to manage it in accordance with the rules of good estate management, and whether the occupier of an agricultural unit is fulfilling his responsibilities to farm it in accordance with the rules of good husbandry.

NOTE to s 26
Sub-s 1: Sub-s repealed by the Agriculture Act 1958 (c 71), s 10(1), Sch 2, pt I.

* * *

Section 26　　　　　　　　　　Fifth Schedule

Rules of Good Estate Management

1. For the purposes of this Act, the owner of agricultural land shall be deemed to fulfil his responsibilities to manage it in accordance with the rules of good estate management in so far as his management of the land and (so far as it affects the management of that land) of other land managed by him is such as to be reasonably adequate, having regard to the character and situation of the land and other relevant circumstances, to enable an occupier of the land reasonably skilled in husbandry to maintain efficient production as respects both the kind of produce and the quality and quantity thereof.

2. In determining whether the management of land is such as aforesaid regard shall be had, but without prejudice to the generality of the provisions of the last foregoing paragraph, to the extent to which the owner is making regular muirburn in the interests of sheep stock, exercising systematic control of vermin on land not in the control of a tenant, and undertaking the eradication of bracken, whins and broom so far as is reasonably practicable, and to the extent to which the owner is fulfilling his responsibilities in relation to the provision, improvement, replacement and renewal of the fixed equipment on the land in so far as is necessary to enable an occupier reasonably skilled in husbandry to maintain efficient production as aforesaid.

Section 26   SIXTH SCHEDULE

RULES OF GOOD HUSBANDRY

1. For the purposes of this Act, the occupier of an agricultural unit shall be deemed to fulfil his responsibilities to farm it in accordance with the rules of good husbandry in so far as the extent to which and the manner in which the unit is being farmed (as respects both the kind of operations carried out and the way in which they are carried out) are such that, having regard to the character and situation of the unit, the standard of management thereof by the owner and other relevant circumstances, the occupier is maintaining a reasonable standard of efficient production, as respects both the kind of produce and the quality and quantity thereof, while keeping the unit in a condition to enable such a standard to be maintained in the future.

2. In determining whether the manner in which a unit is being farmed is such as aforesaid regard shall be had, but without prejudice to the generality of the provisions of the last foregoing paragraph, to the following:—

(a) the maintenance of permanent grassland (whether meadow or pasture) properly mown or grazed and in a good state of cultivation and fertility;

(b) the handling or cropping of the arable land, including the treatment of temporary grass, so as to maintain it clean and in a good state of cultivation and fertility;

(c) where the system of farming practised requires the keeping of livestock, the proper stocking of the holding;

(d) the maintenance of an efficient standard of management of livestock;

(e) as regards hill sheep farming in particular:–

　　(i)   the maintenance of a sheep stock of a suitable breed and type in regular ages (so far as is reasonably possible) and the keeping and management thereof in accordance with the recognised practices of hill sheep farming;

　　(ii) the use of lug, horn or other stock marks for the purpose of determining ownership of stock sheep;

　　(iii) the regular selection and retention of the best female stock for breeding;

　　(iv) the regular selection and use of tups possessing the qualities most suitable and desirable for the flock;

　　(v)   the extent to which regular muirburn is made;

(f) the extent to which the necessary steps are being taken—

　　(i)   to secure and maintain the freedom of crops and livestock from disease and from infestation by insects and other pests;

　　(ii) to exercise systematic control of vermin and of bracken, whins, broom and injurious weeds;

　　(iii) to protect and preserve crops harvested or in course of being harvested;

　　(iv) to carry out necessary work of maintenance and repair of the fixed and other equipment.

Succession (Scotland) Act 1964
(c 41)

* * *

PART III

ADMINISTRATION AND WINDING UP OF ESTATES

14. Assimilation for purposes of administration etc, of heritage to moveables

(1) Subject to subsection (3) of this section the enactments and rules of law in force immediately before the commencement of this Act with respect to the administration and winding up of the estate of a deceased person so far as consisting of moveable property shall have effect (as modified by the provisions of this Act) in relation to the whole of the estate without distinction between moveable property and heritable property; and accordingly on the death of any person (whether testate or intestate) every part of his estate (whether consisting of moveable property or heritable property) falling to be administered under the law of Scotland shall, by virtue of confirmation thereto, vest for the purposes of administration in the executor thereby confirmed and shall be administered and disposed of according to law by such executor.

(2) Provision shall be made by the Court of Session by act of sederunt made under the enactments mentioned in section 22 of this Act (as extended by that section) for the inclusion in the confirmation of an executor, by reference to an appended inventory or otherwise, of a description, in such form as may be so provided, of any heritable property forming part of the estate.

(3) Nothing in this section shall be taken to alter any rule of law whereby any particular debt of a deceased person falls to be paid out of any particular part of his estate.

15. Provisions as to a transfer of heritage

(1) Section 5(2) of the Conveyancing (Scotland) Act, 1924 (which provides that a confirmation which includes a heritable security shall be a valid title to the debt thereby secured), shall have effect as if any reference therein to a heritable security, or to a debt secured by a heritable security, included a reference to any interest in heritable property which has vested in an executor in pursuance of the last foregoing section by virtue of a confirmation:

Provided that a confirmation [(other than an implied confirmation within the meaning of the said section 5(2))] shall not be deemed for the purposes of the said section 5(2) to include any such interest unless a description of the property, in accordance with any act of sederunt such as is mentioned in subsection (2) of the last foregoing section, is included or referred to in the confirmation.

(2) Where in pursuance of the last foregoing section any heritable property has vested in an executor by virtue of a confirmation, and it is necessary for him in distributing the estate to transfer that property–

(a) to any person in satisfaction of a claim to legal rights or the prior rights of a surviving spouse out of the estate, or

(b) to any person entitled to share in the estate by virtue of this Act, or

(c)   to any person entitled to take the said property under any testamentary disposition of the deceased,

the executor may effect such transfer by endorsing on the confirmation (or where a certificate of confirmation relating to the property has been issued in pursuance of any act of sederunt, on the certificate) a docket in favour of that person in the form set out in Schedule 1 to this Act, or in a form as nearly as may be to the like effect, and any such docket may be specified as a midcouple or link in title in any deduction of title; but this section shall not be construed as prejudicing the competence of any other mode of transfer.

NOTE to s 15

s 15(1): Words inserted by the Law Reform (Miscellaneous Provisions) (Scotland) Act 1968 (c 70), s 19.

---

16. Provisions relating to leases

(1) This section applies to any interest, being the interest of a tenant under a lease, which is comprised in the estate of a deceased person and has accordingly vested in the deceased's executor by virtue of section 14 of this Act; and in the following provisions of this section 'interest' means an interest to which this section applies.

(2) Where an interest—

(a)   is not the subject of a valid bequest by the deceased, or

(b)   is the subject of such a bequest, but the bequest is not accepted by the legatee, or

(c)   being an interest under an agricultural lease, is the subject of such a bequest, but the bequest is declared null and void in pursuance of section 16 of the Act of 1886 or [section 11 of the 1991 Act] [or becomes null and void under section 10 of the Act of 1955];

and there is among the conditions of the lease (whether expressly or by implication) a condition prohibiting assignation of the interest, the executor shall be entitled, notwithstanding that condition, to transfer the interest to any one of the persons entitled to succeed to the deceased's intestate estate, or to claim legal rights or the prior rights of a surviving spouse out of the estate, in or towards satisfaction of that person's entitlement or claim; but shall not be entitled to transfer the interest to any other person without the consent

[(i)   in the case of an interest under an agricultural lease, being a lease of a croft within the meaning of section 3(1) of the Act of 1955, of the Crofters Commission;

(ii)   in any other case, of the landlord.]

(3) If in the case of any interest—

(a)   at any time the executor is satisfied that the interest cannot be disposed of according to law and so informs the landlord,
      or

(b)   the interest is not so disposed of within a period of one year or such longer period as may be fixed by agreement between the landlord and the executor or, failing agreement, by the sheriff on summary application by the executor—

      (i)   in the case of an interest under an agricultural lease which is the subject of a petition to the Land Court under section 16 of the Act of 1886 or an application to that Court under [section 11 of the 1991 Act], from the date of the determination or withdrawal of the petition or, as the case may be, the application

      [(ia) in the case of an interest under an agricultural lease which is the subject of an application by the legatee to the Crofters Commission under section 10(1) of the Act of 1955, from the date of any refusal by the Commission to determine that the bequest shall not be null and void,

      (ib) in the case of an interest under an agricultural lease which is the subject of an intimation of objection by the landlord to the legatee and the Crofters Commission under section 10(3) of the Act of 1955, from the date of any decision of the Commission upholding the objection,]

(ii) in any other case, from the date of death of the deceased,
either the landlord or the executor may, on giving notice in accordance with the next following subsection to the other, terminate the lease (in so far as it relates to the interest) notwithstanding any provision therein, or any enactment or rule of law, to the contrary effect.

(4) The period of notice given under the last foregoing subsection shall be—

(a) in the case of an agricultural lease, such period as may be agreed, or, failing agreement, a period of not less than one year and not more than two years ending with such term of Whitsunday or Martinmas as may be specified in the notice; and

(b) in the case of any other lease, a period of six months:

Provided that paragraph (b) of this subsection shall be without prejudice to any enactment prescribing a shorter period of notice in relation to the lease in question.

(5) Subsection (3) of this section shall not prejudice any claim by any party to the lease for compensation or damages in respect of the termination of the lease (or any rights under it) in pursuance of that subsection; but any award of compensation or damages in respect of such termination at the instance of the executor shall be enforceable only against the estate of the deceased and not against the executor personally.

(6) Where an interest is an interest under an agricultural lease, and—

(a) an application is made under section 3 of the Act of 1931 [or section 13 of the Act of 1955] to the Land Court for an order for removal, or

(b) a reference is made under [section 23(2) and (3) of the 1991 Act] to an arbiter to determine any question which has arisen under [section 22(2)(e)] of that Act in connection with a notice to quit,

the Land Court shall not make the order, or, as the case may be, the arbiter shall not make an award in favour of the landlord, unless the court or the arbiter is satisfied that it is reasonable, having regard to the fact that the interest is vested in the executor in his capacity as executor, that it should be made.

(7) Where an interest is not an interest under an agricultural lease, and the landlord brings an action of removing against the executor in respect of a breach of a condition of the lease, the court shall not grant decree in the action unless it is satisfied that the condition alleged to have been breached is one which it is reasonable to expect the executor to have observed, having regard to the fact that the interest is vested in him in his capacity as an executor.

(8) Where an interest is an interest under an agricultural lease and is the subject of a valid bequest by the deceased, the fact that the interest is vested in the executor under the said section 14 shall not prevent the operation, in relation to the legatee, of paragraphs (a) to (h) of section 16 of the Act of 1886, or, as the case may be [section 11(2) to (8) of the 1991 Act] [or, as the case may be, subsections (2) to (7) of section 10 of the Act of 1955].

(9) In this section—

'agricultural lease' means a lease of a holding within the meaning of the Small Landholders (Scotland) Acts, 1886 to 1931, or of [the 1991 Act] [, or a lease of a croft within the meaning of section 3(1) of the Act of 1955];

'the Act of 1886' means the Crofters Holdings (Scotland) Act, 1886;

'the Act of 1931' means the Small Landholders and Agricultural Holdings (Scotland) Act, 1931;

['the 1991 Act' means the Agricultural Holdings (Scotland) Act 1991;]

['the Act of 1955' means the Crofters (Scotland) Act 1955;]

'lease' includes tenancy.

NOTE to s 16

sub-s (2): Words substituted by the Agricultural Holdings (Scotland) Act 1991 (c 55), Sch 11, para 24 and the Law Reform (Miscellaneous Provisions) (Scotland) Act 1968 (c 70), Sch 2, para 22. Words added by the 1968 Act, Sch 2, para 22.

sub-s (3): Words substituted by the 1991 Act, Sch 11, para 24. Words added by the 1968 Act, Sch 2, para 23.

sub-s (6): Words added by the 1968 Act, Sch 2, para 24. Words substituted by the 1991 Act, Sch 11, para 24.

sub-s (8): Words substituted by the 1991 Act, Sch 11, para 24. Words added by the 1968 Act, Sch 2, para 25.

sub-s (9): Words substituted by the 1991 Act, Sch 11, para 24. Words added by the 1968 Act, Sch 2, para 26.

*   *   *

36. Interpretation

*   *   *

(2) Any reference in this Act to the estate of a deceased person shall, unless the context otherwise requires, be construed as a reference to the whole estate, whether heritable or moveable, or partly heritable and partly moveable, belonging to the deceased at the time of his death or over which the deceased had a power of appointment and, where the deceased immediately before his death held the interest of a tenant under a tenancy or lease which was not expressed to expire on his death, includes that interest: Provided that—

(a) where any heritable property belonging to a deceased person at the date of his death is subject to a special destination in favour of any person, the property shall not be treated for the purposes of this Act as part of the estate of the deceased unless the destination is one which could competently be, and has in fact been, evacuated by the deceased by testamentary disposition or otherwise; and in that case the property shall be treated for the purposes of this Act as if it were part of the deceased's estate on which he has tested; and

(b) where any heritable property over which a deceased person had a power of appointment has not been disposed of in exercise of that power and is in those circumstances subject to a power of appointment by some other person, that property shall not be treated for the purposes of this Act as part of the estate of the deceased.

Agriculture Act 1986
(c 41)

* * *

14. Compensation to outgoing tenants for milk quota: Scotland

Schedule 2 to this Act shall have effect in connection with the payment of outgoing tenants who are—

(a) tenants of agricultural holdings within the meaning of [the 1991 Act];

(b) landholders within the meaning of section 2 of the Small Landholders (Scotland) Act 1911;

(c) statutory small tenants within the meaning of section 32(1) of that Act;

(d) crofters within the meaning of section 3(2) of the Crofters (Scotland) Act 1955,

of compensation in respect of milk quotas.

NOTE to s 14

Words substituted by the Agricultural Holdings (Scotland) Act 1991 (c 55), Sch 11, para 43.

---

16. Rent arbitrations: milk quotas, Scotland

(1) Paragraph 1 and the other provisions of Schedule 2 to this Act referred to therein shall have effect for the interpretation of this section, as they do in relation to that Schedule.

(2) This section applies where an arbiter or the Scottish Land Court is dealing with a reference under—

(a) section 6 of the 1886 Act;

(b) section 32(7) of the 1911 Act;

(c) [section 13 of the 1991 Act]; or

(d) section 5(3) of the 1955 Act;

(determination of rent) and the tenant has milk quota, including transferred quota by virtue of a transaction the cost of which was borne wholly or partly by him, registered as his in relation to a holding consisting of or including the tenancy.

(3) Where this section applies, the arbiter or, as the case may be, the Land Court shall disregard any increase in the rental value of the tenancy which is due to—

(a) where the tenancy comprises the holding, the proportion of the transferred quota which reflects the proportion of the cost of the transaction borne by the tenant;

(b) where such transferred quota affects part only of the tenancy, that proportion of so much of the transferred quota as would fall to be apportioned to the tenancy under the 1986 Regulations on a change of occupation of the tenancy.

(4) For the purposes of determining whether transferred quota has been acquired by virtue of a transaction the cost of which was borne wholly or partly by the tenant any payment by a tenant when he was granted a lease, or when a lease was assigned to him, shall be disregarded.

(5) Paragraph 3 of Schedule 2 to this Act (in so far as it relates to transferred quota) shall apply in relation to the operation of this section as it applies in relation to the operation of that Schedule.

(6) This section shall apply where paragraph 4 of Schedule 2 to this Act applies, and in any question between the original landlord and the head tenant, this section shall

apply as if any transferred quota acquired by the sub-tenant by virtue of any transaction during the subsistence of the sub-lease had been acquired by the head tenant by virtue of that transaction.

(7) [Section 79 of the 1991 Act] (Crown land) shall have effect in relation to this section as it does in relation to that Act.

NOTE to s 16

Words substituted by the Agricultural Holdings (Scotland) Act 1991 (c 55), Sch 11, para 44.

★ ★ ★

Section 14

SCHEDULE 2

TENANTS' COMPENSATION FOR MILK QUOTA: SCOTLAND

1. Interpretation

(1) In this Schedule, except where the context otherwise requires or provision is made to the contrary–

'allocated quota' has the meaning given in paragraph 2(1) below;

'holding' has the same meaning as in the 1986 Regulations;

'landlord' means—

(a) in the case of an agricultural holding to which [the 1991 Act] applies, the landlord within the meaning of [section 85(1)] of that Act;

(b) in the case of a croft within the meaning of the 1955 Act, the landlord within the meaning of section 37(1) of that Act;

(c) in the case of a holding within the meaning of the 1911 Act to which [the 1991 Act] does not apply, the same as it means in the 1911 Act;

'milk quota' means—

(a) in the case of a tenant registered in the direct sales register maintained under the 1986 Regulations, a direct sales quota within the meaning of those Regulations; and

(b) in the case of a tenant registered in the wholesale register maintained under those Regulations, a wholesale quota within the meaning of those Regulations;

'registered', in relation to milk quota, means—

(a) in the case of direct sales quota within the meaning of the 1986 Regulations, registered in the direct sales register maintained under those Regulations; and

(b) in the case of a wholesale quota within the meaning of those Regulations, registered in a wholesale register maintained under those Regulations;

'relevant quota' has the meaning given in paragraph 2(2) below;

'standard quota' means standard quota as calculated under paragraph 6 below;

'tenancy' means, as the case may be—

(a) the agricultural holding, within the meaning of section 1 of [the 1991 Act];

(b) the croft within the meaning of section 3(1) of the 1955 Act;

(c) the holding within the meaning of section 2 of the 1911 Act;

(d) the holding of a statutory small tenant under section 32 of the 1911 Act;

(e) any part of a tenancy which is treated as a separate entity for the purposes of succession, assignation or sub-letting,

'tenant' means—

(a) in the case of an agricultural holding to which [the 1991 Act] applies, the tenant within the meaning of [section 85(1)] of that Act;

(b) in the case of a croft within the meaning of the 1955 Act, the crofter within the meaning of section 3(2) of that Act;

(c) in the case of a holding within the meaning of the 1911 Act to which [the 1991 Act] does not apply, the landholder within the meaning of section 2(2) of the 1911 Act;

'tenant's fraction' has the meaning given in paragraph 7 below;
'termination' means the resumption of possession of the whole or part of the tenancy by the landlord by virtue of any enactment, rule of law or term of the lease which makes provision for removal of or renunciation by a tenant, or resumption of possession following—
  (a) vacancy arising under section 11(5) of the 1955 Act;
  (b) termination of a lease in pursuance of section 16(3) of the Succession (Scotland) Act 1964;
'transferred quota' has the meaning given in paragraph 2(2) below;
'the 1886 Act' means the Crofters Holdings (Scotland) Act 1886;
'the 1911 Act' means the Small Landholders (Scotland) Act 1911;
'the 1949 Act' means the Agricultural Holdings (Scotland) Act 1949;
'the 1955 Act' means the Crofters (Scotland) Act 1955;
'the 1986 Regulations' means the Dairy Produce Quotas Regulations 1986.
(2) For the purposes of this Schedule, the designations of landlord and tenant shall continue to apply to the parties to any proceedings taken under or in pursuance of it until the conclusion of those proceedings.

2. Tenant's right to compensation

(1) Subject to this Schedule, where, on the termination of the lease, the tenant has milk quota registered as his in relation to a holding consisting of or including the tenancy, he shall be entitled, on quitting the tenancy, to obtain from his landlord a payment—
  (a) if the tenant had milk quota allocated to him in relation to a holding consisting of or including the tenancy ('allocated quota'), in respect of so much of the relevant quota as consists of allocated quota; and
  (b) if the tenant had quota allocated to him as aforesaid or was in occupation of the tenancy as a tenant on 2nd April 1984 (whether or not under the lease which is terminating), in respect of so much of the relevant quota as consists of transferred quota by virtue of a transaction the cost of which was borne wholly or partly by him.
(2) In sub-paragraph (1) above—
'the relevant quota' means—
  (a) where the holding consists only of the tenancy, the milk quota registered in relation to the holding; and
  (b) otherwise, such part of that milk quota as falls to be apportioned to the tenancy on the termination of the lease;
'transferred quota' means milk quota transferred to the tenant by virtue of the transfer to him of the whole or part of a holding.
(3) A tenant shall not be entitled to more than one payment under this paragraph in respect of the same tenancy.
(4) Nothing in this paragraph shall prejudice the right of a tenant to claim any compensation to which he may be entitled under an agreement in writing, in lieu of any payment provided by this paragraph.

3. Succession to lease of tenancy

(1) This paragraph applies where a person (the successor) has acquired right to the lease of the tenancy after 2nd April 1984—
  (a) under section 16 of the Succession (Scotland) Act 1964;
  (b) as a legatee, under [section 11 of the 1991 Act] or under section 16 of the 1886 Act;
  (c) under a bequest of a croft under section 10 of the 1955 Act, or following nomination under section 11 of that Act;
  (d) under a lawful assignation of the lease,
and the person whom he succeeded or, as the case may be, who assigned the lease to him is described in this paragraph as his 'predecessor'.
  (2) Where this paragraph applies—
  (a) any milk quota allocated or transferred to the predecessor (or treated as having been allocated or transferred to him) in respect of the tenancy shall be treated as if it had been allocated or transferred to his successor;

(b) where under (a) above, milk quota is treated as having been transferred to the successor, he shall be treated as if he had paid so much for the cost of the transaction by virtue of which the milk quota was transferred as his predecessor bore (or is treated as having borne).

4. Sub-tenants

In the case of a tenancy which is sub-let, if the sub-tenant quits the tenancy—

(a) paragraph 2 above shall apply so as to entitle the sub-tenant to obtain payment from the head tenant, and for that purpose, references to the landlord and the tenant in this Schedule shall be respectively construed as references to the head tenant and the sub-tenant; and

(b) for the purposes of the application of paragraph 2 above as between the original landlord and the head tenant—
   (i)   the head tenant shall be deemed to have had the relevant quota allocated to him, and to have been in occupation of the tenancy as a tenant on 2nd April 1984; and
   (ii) if the head tenant does not take up occupation of the tenancy when the sub-tenant quits, the head tenant shall be treated as if he had quitted the tenancy when the sub-tenant quitted it.

5. Calculation of payment

(1) The amount of the payment to which a tenant is entitled under paragraph 2 above on the termination of the lease shall be determined in accordance with this paragraph.

(2) The amount of the payment in respect of allocated quota shall be equal to the value of—

(a) where the allocated quota exceeds the standard quota for the tenancy—
   (i)   the tenant's fraction of so much of the allocated quota as does not exceed the standard quota; together with
   (ii) the amount of the excess;

(b) where the allocated quota is equal to the standard quota, the tenant's fraction of the allocated quota;

(c) where the allocated quota is less than the standard quota, such proportion of the tenant's fraction of the allocated quota as the allocated quota bears to the standard quota.

(3) The amount of the payment in respect of transferred quota shall be equal to the value of—

(a) where the tenant bore the whole of the cost of the transaction by virtue of which the transferred quota was transferred to him, the transferred quota; and

(b) where the tenant bore only part of that cost, the corresponding part of the transferred quota.

6. Standard quota

(1) Subject to this paragraph, the 'standard quota' for any tenancy for the purposes of this Schedule shall be calculated by multiplying the relevant number of hectares by the standard yield per hectare.

(2) Where by virtue of the quality of the land in question or of climatic conditions in the area the amount of milk which could reasonably be expected to have been produced from one hectare of the tenancy during the relevant period ('the reasonable amount') is greater or less than the average yield per hectare then sub-paragraph (1) above shall not apply and the standard quota shall be calculated by multiplying the relevant number of hectares by such proportion of the standard yield per hectare as the reasonable amount bears to the average yield per hectare; and the Secretary of State shall by order prescribe the amount of milk to be taken as the average yield per hectare for the purposes of this sub-paragraph.

(3) Where the relevant quota includes milk quota allocated in pursuance of an award of quota made by the Dairy Produce Quota Tribunal for Scotland which has not been

allocated in full, the standard quota shall be reduced by the amount by which the milk quota allocated in pursuance of the award falls short of the amount awarded (or, in the case where only part of the milk quota allocated in pursuance of the award is included in the relevant quota, by the corresponding proportion of that shortfall).

(4) In sub-paragraph (3) above the references to milk quota allocated in pursuance of an award of quota include references to quota allocated by virtue of the amount awarded not originally having been allocated in full.

(5) For the purposes of this paragraph—

(a) 'the relevant number of hectares' means the average number of hectares of the tenancy used during the relevant period for the feeding of dairy cows kept on the tenancy or, if different, the average number of hectares of the tenancy which could reasonably be expected to have been so used (having regard to the number of grazing animals other than dairy cows kept on the tenancy during that period); and

(b) 'the standard yield per hectare' means such number of litres as the Secretary of State may from time to time by order prescribe for the purposes of this sub-paragraph.

(6) In this and in paragraph 7 below—

(a) references to the area of a tenancy used for the feeding of dairy cows kept on the tenancy do not include references to land used for growing cereal crops for feeding to dairy cows in the form of loose grain; and

(b) 'dairy cows' means milking cows and calved heifers.

(7) An order under this paragraph may make different provision for different cases.

(8) The powers to make an order under this paragraph shall be exercisable by statutory instrument and any statutory instrument containing such an order shall be subject to annulment in pursuance of a resolution of either House of Parliament.

7. Tenant's fraction

(1) For the purposes of this Schedule 'the tenant's fraction' means the fraction of which—

(a) the numerator is the annual rental value at the end of the relevant period of the tenant's dairy improvements and fixed equipment; and

(b) the denominator is the sum of that value and such part of the rent payable by the tenant in respect of the relevant period as is attributable to the land used in that period for the feeding, accommodation or milking of dairy cows kept on the tenancy.

(2) For the purposes of sub-paragraph (1)(a) above, in the case of an agricultural holding within the meaning of [the 1991 Act], the annual rental value of the tenant's dairy improvements and fixed equipment shall be taken to be the amount which would be disregarded, on a reference to arbitration made in respect of the tenancy under [section 13 of the 1991 Act] (variation of rent), as being—

(a) an increase in annual rental value due to dairy improvements at the tenant's expense (in terms of subsection (2)(a) of that section); or

(b) the value of tenant's fixed equipment and therefore not relevant to the fixing of rent under that section,

so far as that amount is attributable to tenant's dairy improvements and fixed equipment which are relevant to the feeding, accommodation or milking of dairy cows kept on the tenancy.

(3) Where—

(a) the relevant period is less than or greater than 12 months; or

(b) rent was payable by the tenant in respect of only part of the relevant period,

the average rent payable in respect of one month in the relevant period or, as the case may be, in that part shall be determined and the rent referred to in sub-paragraph (1)(b) above shall be taken to be the corresponding annual amount.

(4) For the purposes of this paragraph—

(a) 'dairy improvement'—

(i) in the case of an agricultural holding or a statutory small tenancy, means a 'new improvement' or an 'old improvement' within the meaning of [section 85 of the 1991 Act];

(ii) in the case of a croft, means a 'permanent improvement' within the meaning of section 37 of the 1955 Act;

(iii) in the case of a holding under the 1911 Act to which [the 1991 Act] does not apply, means a 'permanent improvement' within the meaning of section 34 of the 1886 Act,

so far as relevant to the feeding, accommodation or milking of dairy cows kept on the tenancy;

(b) 'fixed equipment' means fixed equipment, within the meaning of [section 85 of the 1991 Act], so far as relevant to the feeding, accommodation or milking of dairy cows kept on the tenancy;

(c) all dairy improvements and fixed equipment provided by the tenant shall be taken into account for the purposes of sub-paragraph (1)(a) above, except for such improvements and fixed equipment in respect of which he has, before the end of the relevant period, received full compensation directly related to their value.

(5) For the purposes of this paragraph—

(a) any allowance made or benefit given by the landlord after the end of the relevant period in consideration of the execution of dairy improvements or fixed equipment wholly or partly at the expense of the tenant shall be disregarded;

(b) any compensation received by the tenant after the end of the relevant period in respect of any dairy improvement or fixed equipment shall be disregarded; and

(c) where paragraph 3 above applies, dairy improvements which would be regarded as tenant's dairy improvements or fixed equipment on the termination of a former tenant's lease (if he were entitled to a payment under this Schedule in respect of the land) shall be regarded as the new tenant's dairy improvements or fixed equipment.

8. Relevant period
In this Schedule 'the relevant period' means—

(a) the period in relation to which the allocated quota was determined; or

(b) where it was determined in relation to more than one period, the period in relation to which the majority was determined or, if equal amounts were determined in relation to different periods, the later of those periods.

9. Valuation of milk quota
The value of milk quota to be taken into account for the purposes of paragraph 5 above is the value of the milk quota at the time of the termination of the lease and in determining that value there shall be taken into account such evidence as is available, including evidence as to the sums being paid for interests in land—

(a) in cases where milk quota is registered in relation to land; and

(b) in cases where no milk quota is so registered.

10. Determination of standard quota and tenant's fraction before end of lease
(1) Where it appears that on the termination of a lease, the tenant may be entitled to a payment under paragraph 2 above, the landlord or tenant may at any time before the termination of the lease by notice in writing served on the other demand that the determination of the standard quota for the land or the tenant's fraction shall be referred—

(a) in the case of an agricultural holding within the meaning of [the 1991 Act] to arbitration under that Act or, under [section 60(2)] of that Act to the Scottish Land Court;

(b) in any other case, to the Scottish Land Court, for determination by that court,

and where (a) above applies, [section 60(1) (or, where the circumstances require, sections 64 and 80) of the 1991 Act] shall apply, as if the matters mentioned in sub-paragraph (1) above were required by that Act to be determined by arbitration.

(2) On a reference under this paragraph the arbiter or, as the case may be, the

Scottish Land Court shall determine the standard quota for the land or, as the case may be, the tenant's fraction (as nearly as is practicable at the end of the relevant period).

11. Settlement of tenant's claim on termination of lease

(1) Subject to this paragraph, any claim arising under paragraph 2 above shall be determined—

(a) in the case of an agricultural holding within the meaning of [the 1991 Act] by arbitration under that Act, or, under [section 60(2)] of that Act, by the Scottish Land Court;

(b) in any other case, by the Scottish Land Court,

and no such claim shall be enforceable unless before the expiry of the period of 2 months from the termination of the lease the tenant has served notice in writing on the landlord of his intention to make the claim specifying the nature of the claim.

(2) The landlord and tenant may within the period of 8 months from the termination of the lease by agreement in writing settle the claim but where the claim has not been settled during that period it shall be determined as provided in sub-paragraph (1) above.

(3) Where a tenant lawfully remains in occupation of part of the tenancy after the termination of the lease, the references in sub-paragraphs (1) and (2) above to the termination of the lease shall be construed as references to the termination of the occupation.

(4) In the case of an arbitration under this paragraph [section 60(1) (or, where the circumstances require, sections 64 and 80) of the 1991 Act] (arbitrations) shall apply as if the requirements of this paragraph were requirements of that Act, but paragraph 13 of the Sixth Schedule to that Act (arbitration awards to fix day for payment not later than one month after award) shall have effect for the purposes of this paragraph with the substitution for the words 'one month' of the words 'three months'.

(5) In the case of an arbitration under this paragraph [section 50 of the 1991 Act] (determination of claims for compensation where landlord's interest is divided) shall apply, where the circumstances require, as if compensation payable under paragraph 2 above were compensation payable under that Act.

(6) Where—

(a) before the termination of the lease of any land the landlord and tenant have agreed in writing the amount of the standard quota for the land or the tenant's fraction or the value of milk quota which is to be used for the purpose of calculating the payment to which the tenant will be entitled under this Schedule on the termination of the lease; or

(b) the standard quota or the tenant's fraction has been determined by arbitration in pursuance of paragraph 10 above,

the arbiter or, as the case may be, the Scottish Land Court in determining the claim under this paragraph shall, subject to sub-paragraph (7) below, award payment in accordance with that agreement or determination.

(7) Where it appears to the arbiter or, as the case may be, the Scottish Land Court that any circumstances relevant to the agreement or determination mentioned in sub-paragraph (6) above were materially different at the time of the termination of the lease from those at the time the agreement or determination was made, he shall disregard so much of the agreement or determination as appears to him to be affected by the change in circumstances.

12. Enforcement

[Sections 65 and 75(1), (2), (4) and (6) of the 1991 Act (recovery of sums due and power of tenant to obtain charge on holding) shall apply in relation to any sum payable to the tenant under this Schedule as they apply to sums payable under that section.]

13. Powers of limited owners

Whatever his interest in the tenancy, the landlord may, for the purposes of this Schedule, do or have done to him anything which might be so done if he were absolute owner of the tenancy.

14. Notices

(1) Any notice or other document required or authorised by this Schedule to be served on any person shall be duly served if it is delivered to him, or left at his proper address, or sent to him by post in a recorded delivery or a registered letter.

(2) In the case of an incorporated company or body, any such document shall be duly served if served on the secretary or clerk or the company or body.

(3) Any such document to be served by or on a landlord or tenant shall be duly served if served by or on any agent of the landlord or tenant.

(4) For the purposes of this paragraph and of section 7 of the Interpretation Act 1978, the proper address of a person is—

(a) in the case of a secretary or clerk to a company or body, that of the registered or principal office of the company or body;

(b) in any other case, the person's last known address.

(5) Unless and until the tenant receives notice of a change of landlord, any document served by him on the person previously known to him as landlord shall be deemed to be duly served on the landlord under the tenancy.

15. Crown land

(1) This Schedule shall apply to land belonging to Her Majesty in right of the Crown, subject to such modifications as may be prescribed; and for the purposes of this Schedule the Crown Estates Commissioners or other proper officer or body having charge of the land for the time being or, if there is no such officer or body, such person as Her Majesty may appoint in writing under the Royal Sign Manual, shall represent Her Majesty and shall be deemed to be the landlord.

(2) Without prejudice to sub-paragraph (1) above, subject to such modifications as may be prescribed, section 14 of this Act and this Schedule shall apply to land where the interest of the landlord or of the tenant belongs to a government department or is held on behalf of Her Majesty for the purposes of a government department.

NOTE to Sch 2

Para 3: Words substituted by the Agricultural Holdings (Scotland) Act 1991 (c 55), Sch 11, para 49.

Para 7: Words substituted by the 1991 Act, Sch 11, para 50.

Para 10: Words substituted by the 1991 Act, Sch 11, para 51.

Para 11: Words substituted by the 1991 Act, Sch 11, para 52.

Para 12: Para substituted by the 1991 Act, Sch 11, para 53.

Agricultural Holdings (Scotland) Act 1991
(c 55)

ARRANGEMENT OF SECTIONS

PART I

AGRICULTURAL HOLDINGS

PART II

TERMS OF LEASES AND VARIATIONS THEREOF

Variation of rent

Termination of tenancy

PART III

NOTICE TO QUIT AND NOTICE OF INTENTION TO QUIT

PART IV

COMPENSATION FOR IMPROVEMENTS

PART V

OTHER PROVISIONS REGARDING COMPENSATION

Market gardens

Miscellaneous

PART VI

ADDITIONAL PAYMENTS

Consequential amendments and repeals

88. Consequential amendments and repeals.

Citation, commencement and extent

89. Citation, commencement and extent

An Act to consolidate the Agricultural Holdings (Scotland) Act 1949 and other enactments relating to agricultural holdings in Scotland.

[25th July 1991]

PART I

AGRICULTURAL HOLDINGS

1. Meaning of 'agricultural holding' and 'agricultural land'

(1) In this Act (except sections 68 to 72) 'agricultural holding' means the aggregate of the agricultural land comprised in a lease, not being a lease under which the land is let to the tenant during his continuance in any office, appointment or employment held under the landlord.

(2) In this section and in section 2 of this Act, 'agricultural land' means land used for agriculture for the purposes of a trade or business, and includes any other land which, by virtue of a designation of the Secretary of State under section 86(1) of the Agriculture (Scotland) Act 1948, is agricultural land within the meaning of that Act.

2. Leases for less then year to year

(1) Subject to subsection (2) below, where, under a lease entered into on or after November 1, 1948, land is let for use as agricultural land for a shorter period than from year to year, and the circumstances are such that if the lease were from year to year the land would be an agricultural holding, then, unless the letting was approved by the Secretary of State before the lease was entered into, the lease shall take effect, with the necessary modifications, as if it were a lease of the land from year to year.

(2) Subsection (1) above shall not apply to—
(a) a lease entered into (whether or not the lease expressly so provides) in contemplation of the use of the land only for grazing or mowing during some specified period of the year;
(b) a lease granted by a person whose interest in the land is that of a tenant under a lease for a shorter period than from year to year which has not by virtue of that subsection taken effect as a lease from year to year.

(3) Any question arising as to the operation of this section in relation to any lease shall be determined by arbitration.

3. Leases to be continued by tacit relocation

Notwithstanding any agreement or any provision in the lease to the contrary, the tenancy of an agricultural holding shall not come to an end on the termination of the stipulated endurance of the lease, but shall be continued in force by tacit relocation for another year and thereafter from year to year, unless notice to quit has been given by the landlord or notice of intention to quit has been given by the tenant.

PART II

TERMS OF LEASES AND VARIATIONS THEREOF

4. Written leases and the revision of certain leases

(1) Where in respect of the tenancy of an agricultural holding—
  (a) there is not in force a lease in writing; or
  (b) there is in force a lease in writing, being either—
      (i) a lease entered into on or after November 1, 1948, or
      (ii) a lease entered into before that date, the stipulated period of which has expired and which is being continued in force by tacit relocation,
  but such lease contains no provision for one or more of the matters specified in Schedule 1 to this Act or contains a provision inconsistent with that Schedule or with section 5 of this Act,

either party may give notice in writing to the other requesting him to enter into a lease in writing containing, as the case may be, provision for all of the matters specified in Schedule 1 to this Act, or a provision which is consistent with that Schedule or with section 5 of this Act; and if within the period of 6 months after the giving of such notice no such lease has been concluded, the terms of the tenancy shall be referred to arbitration.

(2) On a reference under subsection (1) above, the arbiter shall by his award specify the terms of the existing tenancy and, in so far as those terms do not make provision for all the matters specified in Schedule 1 to this Act or make provision inconsistent with that Schedule or with section 5 of this Act, make such provision for those matters as appears to the arbiter to be reasonable.

(3) On a reference under subsection (1) above, the arbiter may include in his award any further provisions relating to the tenancy which may be agreed between the landlord and the tenant, and which are not inconsistent with this Act.

(4) The award of an arbiter under this section or section 5 of this Act shall have effect as if the terms and provisions specified and made therein were contained in an agreement in writing between the landlord and the tenant, having effect as from the making of the award or from such later date as the award may specify.

5. Fixed equipment and insurance premiums

(1) When a lease of an agricultural holding to which this section applies is entered into, a record of the condition of the fixed equipment on the holding shall be made forthwith, and on being so made shall be deemed to form part of the lease; and section 8 of this Act shall apply to the making of such a record and to the cost thereof as it applies to a record made under that section.

(2) There shall be deemed to be incorporated in every lease of an agricultural holding to which this section applies—

   (a) an undertaking by the landlord that, at the commencement of the tenancy or as soon as is reasonably practicable thereafter, he will put the fixed equipment on the holding into a thorough state of repair and will provide such buildings and other fixed equipment as will enable an occupier reasonably skilled in husbandry to maintain efficient production as respects both—

      (i) the kind of produce specified in the lease, or (failing such specification) in use to be produced on the holding, and

      (ii) the quality and quantity thereof,

   and that he will during the tenancy effect such replacement or renewal of the buildings or other fixed equipment as may be rendered necessary by natural decay or by fair wear and tear; and

   (b) a provision that the liability of the tenant in relation to the maintenance of fixed equipment shall extend only to a liability to maintain the fixed equipment on the holding in as good a state of repair (natural decay and fair wear and tear excepted) as it was in—

      (i) immediately after it was put in repair as aforesaid, or

      (ii) in the case of equipment provided, improved, replaced or renewed during the tenancy, immediately after it was so provided, improved, replaced or renewed.

(3) Nothing in subsection (2) above shall prohibit any agreement made between the landlord and the tenant after the lease has been entered into whereby one party undertakes to execute on behalf of the other, whether wholly at his own expense or wholly or partly at the expense of the other, any work which the other party is required to execute in order to fulfil his obligations under the lease.

(4) Any provision in a lease to which this section applies requiring the tenant to pay the whole or any part of the premium due under a fire insurance policy over any fixed equipment on the holding shall be null and void.

(5) Any question arising as to the liability of a landlord or tenant under this section shall be determined by arbitration.

(6) This section applies to any lease of an agricultural holding entered into on or after November 1, 1948.

6. Sums recovered under fire insurance policy

Where the tenant of an agricultural holding is responsible for payment of the whole or part of the premium due under a fire insurance policy in the name of the landlord over any buildings or other subjects included in the lease of the holding and the landlord recovers any sum under such policy in respect of the destruction of, or damage to, the buildings or other subjects by fire, the landlord shall be bound, unless the tenant otherwise agrees, to expend such sum on the rebuilding, repair, or restoration of the buildings or subjects so destroyed or damaged in such manner as may be agreed or, failing agreement, as may be determined by the Secretary of State.

7. Freedom of cropping and disposal of produce

(1) Subject to subsections (2) and (5) below, the tenant of an agricultural holding shall, notwithstanding any custom of the country or the provisions of any lease or of any agreement respecting the disposal of crops or the method of cropping of arable lands, have full right, without incurring any penalty, forfeiture or liability,—

(a) to dispose of the produce of the holding, other than manure produced thereon;

(b) to practise any system of cropping of the arable land on the holding.

(2) Subsection (1) above shall not have effect unless, before exercising his rights thereunder or as soon as is practicable after exercising them, the tenant makes suitable and adequate provision—

(a) in the case of an exercise of the right to dispose of crops, to return to the holding the full equivalent manurial value to the holding of all crops sold off or removed from the holding in contravention of any such custom, lease or agreement; and

(b) in the case of an exercise of the right to practise any system of cropping, to protect the holding from injury or deterioration.

(3) If the tenant of an agricultural holding exercises his rights under subsection (1) above so as to injure or deteriorate, or to be likely to injure or deteriorate, the holding, the landlord shall have the following remedies, but no other—

(a) should the case so require, he shall be entitled to obtain an interdict restraining the exercise of the tenant's rights under that subsection in that manner;

(b) in any case, on the tenant quitting the holding on the termination of the tenancy the landlord shall be entitled to recover damages for any injury to or deterioration of the holding attributable to the exercise by the tenant of his rights under that subsection.

(4) For the purposes of any proceedings for an interdict brought under subsection (3)(a) above, the question whether a tenant is exercising, or has exercised, his rights under subsection (1) above in such a manner as to injure or deteriorate, or to be likely to injure or deteriorate the holding, shall be determined by arbitration; and a certificate of the arbiter as to his determination of any such question shall, for the purposes of any proceedings (including an arbitration) brought under this section, be conclusive proof of the facts stated in the certificate.

(5) Subsection (1) above shall not apply—

(a) in the case of a tenancy from year to year, as respects the year before the tenant quits the holding or any period after he has received notice to quit or given notice of intention to quit which results in his quitting the holding; or

(b) in any other case, as respects the year before the expiry of the lease.

(6)—

(a) In this section 'arable land' does not include land in grass which, by the terms of a lease, is to be retained in the same condition throughout the tenancy;

(b) the reference in paragraph (a) above to the terms of a lease shall, where the Secretary of State has directed under section 9 of the 1949 Act or an arbiter has directed under that section or under section 9 of this Act that the lease shall have effect subject to modifications, be construed as a reference to the terms of the lease as so modified.

8. Record of condition, etc, of holding

(1) The landlord or the tenant of an agricultural holding may, at any time during the tenancy, require the making of a record of the condition of the fixed equipment on, and of the cultivation of, the holding.

(2) The tenant may, at any time during the tenancy, require the making of a record of—

(a) existing improvements carried out by him or in respect of the carrying out of which he has, with the consent in writing of his landlord, paid compensation to an outgoing tenant;

(b) any fixtures or buildings which, under section 18 of this Act, he is entitled to remove.

(3) A record under this section shall be made by a person to be appointed by the Secretary of State, and shall be in such form as may be prescribed.

(4) A record made under this section shall show any consideration or allowances which have been given by either party to the other.

(5) Subject to section 5 of this Act, a record may, if the landlord or the tenant so requires, be made under this section relating to a part only of the holding or to the fixed equipment only.

(6) Any question or difference between the landlord and the tenant arising out of the making of a record under this section shall, on the application of the landlord or the tenant, be referred to the Land Court for determination by them.

(7) The cost of making a record under this section shall, in default of agreement between the landlord and the tenant, be borne by them in equal shares.

(8) The remuneration of the person appointed by the Secretary of State to make a record under this section shall be such amount as the Secretary of State may fix, and any other expenses of and incidental to the making of the record shall be subject to taxation by the auditor of the sheriff court, and that taxation shall be subject to review by the sheriff.

(9) The remuneration of the person appointed by the Secretary of State to make a record under this section shall be recoverable by that person from either the landlord or the tenant, but any amount paid by either of those parties in respect of—

(a) that remuneration, or

(b) any other expenses of and incidental to the making of the record, in excess of the share payable by him under subsection (7) above of the cost of making the record, shall be recoverable by him from the other party.

9. Arbitration as to permanent pasture

(1) Where under the lease of an agricultural holding, whether entered into before or after the commencement of this Act, provision is made for the maintenance of specified land, or a specified proportion of the holding, as permanent pasture, the landlord or the tenant may, by notice in writing served on the other party, demand a reference to arbitration under this Act of the question whether it is expedient in order to secure the full and efficient farming of the holding that the amount of land required to be maintained as permanent pasture should be reduced.

(2) On a reference under subsection (1) above the arbiter may by his award direct that the lease shall have effect subject to such modifications of its provisions as to land which is to be maintained as permanent pasture or is to be treated as arable land, and as to cropping, as may be specified in the direction.

(3) If the arbiter gives a direction under subsection (2) above reducing the area of land which is to be maintained as permanent pasture, he may also by his award direct that the lease shall have effect as if it provided that on quitting the holding on the termination of the tenancy the tenant should leave—

(a) as permanent pasture, or

(b) as temporary pasture sown with seeds mixture of such kind as may be specified in that direction,

(in addition to the area of land required by the lease, as modified by the direction, to be maintained as permanent pasture) a specified area of land not exceeding the area by which the land required to be maintained as permanent pasture has been reduced by the direction under subsection (2) above.

10. Power of landlord to enter on holding

The landlord of an agricultural holding or any person authorised by him may at all reasonable times enter on the holding for any of the following purposes—

(a) viewing the state of the holding;

(b) fulfilling the landlord's responsibilities to manage the holding in accordance with the rules of good estate management;

(c) providing, improving, replacing or renewing fixed equipment on the holding otherwise than in fulfilment of such responsibilities.

11. Bequest of lease

(1) Subject to subsections (2) to (8) below, the tenant of an agricultural holding may, by will or other testamentary writing, bequeath his lease of the holding to his son-in-law or daughter-in-law or to any one of the persons who would be, or would in any circumstances have been, entitled to succeed to the estate on intestacy by virtue of the Succession (Scotland) Act 1964.

(2) A person to whom the lease of a holding is so bequeathed (in this section referred to as 'the legatee') shall, if he accepts the bequest, give notice of the bequest to the landlord of the holding within 21 days after the death of the tenant, or, if he is prevented by some unavoidable cause from giving such notice within that period, as soon as practicable thereafter.

(3) The giving of a notice under subsection (2) above shall import acceptance of the lease and, unless the landlord gives a counter-notice under subsection (4) below, the lease shall be binding on the landlord and on the legatee, as landlord and tenant respectively, as from the date of the death of the deceased tenant.

(4) Where notice has been given under subsection (2) above, the landlord may within one month thereafter give to the legatee a counter-notice intimating that he objects to receiving him as tenant under the lease.

(5) If the landlord gives a counter-notice under subsection (4) above, the legatee may make application to the Land Court for an order declaring him to be tenant under the lease as from the date of the death of the deceased tenant.

(6) If, on the hearing of such an application, any reasonable ground of objection stated by the landlord is established to the satisfaction of the Land Court, they shall declare the bequest to be null and void, but in any other case they shall make an order in terms of the application.

(7) Pending any proceedings under this section, the legatee, with the consent of the executor in whom the lease is vested under section 14 of the Succession (Scotland) Act 1964, shall, unless the Land Court on cause shown otherwise direct, have possession of the holding.

(8) If the legatee does not accept the bequest, or if the bequest is declared null and void under subsection (6) above, the right to the lease shall be treated as intestate estate of the deceased tenant in accordance with Part I of the Succession (Scotland) Act 1964.

12. Right of landlord to object to acquirer of lease

(1) A person to whom the lease of an agricultural holding is transferred under section 16 of the Succession (Scotland) Act 1964 (referred to in this section as 'the acquirer') shall give notice of the acquisition to the landlord of the holding within 21 days after the

date of the acquisition, or, if he is prevented by some unavoidable cause from giving such notice within that period, as soon as is practicable thereafter and, unless the landlord gives a counter-notice under subsection (2) below, the lease shall be binding on the landlord and on the acquirer, as landlord and tenant respectively, as from the date of the acquisition.

(2) Within one month after receipt of a notice given under subsection (1) above the landlord may give a counter-notice to the acquirer intimating that the landlord objects to receive him as tenant under the lease; and not before the expiry of one month from the giving of the counter-notice the landlord may make application to the Land Court for an order terminating the lease.

(3) On an application under subsection (2) above, the Land Court shall, if they are satisfied that the landlord has established a reasonable ground of objection, make an order terminating the lease, to take effect as from such term of Whitsunday or Martinmas as they may specify.

(4) Pending any proceedings under this section, the acquirer, with the consent of the executor in whom the lease is vested under section 14 of the Succession (Scotland) Act 1964 shall, unless the Land Court on cause shown otherwise direct, have possession of the holding.

(5) Termination of the lease under this section shall be treated, for the purposes of Parts IV and V of this Act (compensation), as termination of the acquirer's tenancy of the holding; but nothing in this section shall entitle him to compensation for disturbance.

Variation of rent

13. Variation of rent

(1) Subject to subsection (8) below, the landlord or the tenant of an agricultural holding may, whether the tenancy was created before or after the commencement of this Act, by notice in writing served on the other party, demand a reference to arbitration of the question what rent should be payable in respect of the holding as from the next day after the date of the notice on which the tenancy could have been terminated by notice to quit (or notice of intention to quit) given on that date, and the matter shall be referred accordingly.

(2) On a reference under subsection (1) above, the arbiter shall determine, in accordance with subsections (3) to (7) below the rent properly payable in respect of the holding as from the 'next day' mentioned in subsection (1) above.

(3) For the purposes of this section the rent properly payable in respect of a holding shall normally be the rent at which, having regard to the terms of the tenancy (other than those relating to rent), the holding might reasonably be expected to be let in the open market by a willing landlord to a willing tenant, there being disregarded (in addition to the matters referred to in subsection (5) below) any effect on rent of the fact that the tenant is in occupation of the holding.

(4) Where the evidence available to the arbiter is in his opinion insufficient to enable him to determine the rent properly payable or he is of the view that the open market for rents of comparable subjects in the surrounding area is distorted by scarcity of lets or by other factors, the rent properly payable for the purposes of this section shall be the rent which he would expect to be paid, in a market which was not affected by such distortion, having particular regard to the following—

(a) information about open market rents of comparable subjects outside the surrounding area;

(b) the entire range of offers made as regards any lease of subjects which are comparable after regard is had to the terms of that lease;

(c) sitting tenants' rents fixed by agreement for subjects in the surrounding area which are comparable after regard is had to any element attributable to goodwill between landlord and tenant or to similar considerations; and

(d) the current economic conditions in the relevant sector of agriculture.

(5) The arbiter shall not take into account any increase in the rental value of the holding which is due to improvements—

(a) so far as—

(i) they have been executed wholly or partly at the expense of the tenant (whether or not that expense has been or will be reimbursed by a grant out of moneys provided by Parliament) without equivalent allowance or benefit having been made or given by the landlord in consideration of their execution; and

(ii) they have not been executed under an obligation imposed on the tenant by the terms of his lease;

(b) which have been executed by the landlord, in so far as the landlord has received or will receive grants out of moneys provided by Parliament in respect of the execution thereof,

nor fix the rent at a higher amount than would have been properly payable if those improvements had not been so executed.

(6) The continuous adoption by the tenant of a standard of farming or a system of farming more beneficial to the holding than the standard or system required by the lease or, in so far as no system of farming is so required, than the system of farming normally practised on comparable holdings in the district, shall be deemed, for the purposes of subsection (5) above, to be an improvement executed at his expense.

(7) The arbiter shall not fix the rent at a lower amount by reason of any dilapidation or deterioration of, or damage to, fixed equipment or land caused or permitted by the tenant.

(8) Subject to subsection (9) below, a reference to arbitration under subsection (1) above shall not be demanded in circumstances which are such that any increase or reduction of rent made in consequence thereof would take effect as from a date earlier than the expiry of 3 years from the latest in time of the following—

(a) the commencement of the tenancy;

(b) the date as from which there took effect a previous variation of rent (under this section or otherwise);

(c) the date as from which there took effect a previous direction under this section that the rent should continue unchanged.

(9) There shall be disregarded for the purposes of subsection (8) above—

(a) a variation of rent under section 14 of this Act;

(b) an increase of rent under section 15(1) of this Act;

(c) a reduction of rent under section 31 of this Act.

14. Arbitrations under sections 4 and 5

Where it appears to an arbiter—

(a) on a reference under section 4 of this Act that, by reason of any provision which he is required by that section to include in his award, or

(b) on a reference under section 5 of this Act that, by reason of any provision included in his award,

it is equitable that the rent of the holding should be varied, he may vary the rent accordingly.

15. Increase of rent for certain improvements by landlord

(1) Where the landlord of an agricultural holding has, whether before or after the commencement of this Act, carried out on the holding an improvement (whether or not one for the carrying out of which compensation is provided for under Part IV of this Act)—

(a) at the request of, or in agreement with, the tenant,

(b) in pursuance of an undertaking given by the landlord under section 39(3) of this Act, or

(c) in compliance with a direction given by the Secretary of State under powers conferred on him by or under any enactment,

subject to subsections (2) and (3) below, the rent of the holding shall, if the landlord by notice in writing served on the tenant within 6 months from the completion of the improvement so requires, be increased as from the completion of the improvement by an amount equal to the increase in the rental value of the holding attributable to the carrying out of the improvement.

(2) Where any grant has been made to the landlord out of moneys provided by Parliament, in respect of an improvement to which subsection (1) above applies, the increase in rent provided for by that subsection shall be reduced proportionately.

(3) Any question arising between the landlord and the tenant in the application of this section shall be determined by arbitration.

Termination of tenancy

16. Leases not terminated by variation of terms, etc

The lease of an agricultural holding shall not be brought to an end, and accordingly neither party shall be entitled to bring proceedings to terminate the lease or, except with the consent of the other party, to treat it as at an end, by reason only that any new term has been added to the lease or that any terms of the lease (including the rent payable) have been varied or revised in pursuance of this Act.

17. Prohibition of removal of manure, etc., after notice to quit, etc

Where, in respect of an agricultural holding, notice to quit is given by the landlord or notice of intention to quit is given by the tenant, the tenant shall not, subject to any agreement to the contrary, at any time after the date of the notice, sell or remove from the holding any manure or compost, or any hay, straw or roots grown in the last year of the tenancy, unless and until he has given the landlord or the incoming tenant a reasonable opportunity of agreeing to purchase them on the termination of the tenancy at their fair market value, or at such other value as is provided by the lease.

18. Tenant's right to remove fixtures and buildings

(1) Subject to subsections (2) to (4) below, and to section 40(4)(a) of this Act—
  (a) any engine, machinery, fencing or other fixture affixed to an agricultural holding by the tenant thereof; and
  (b) any building (other than one in respect of which the tenant is entitled to compensation under this Act or otherwise) erected by him on the holding,

not being a fixture affixed or a building erected in pursuance of some obligation in that behalf, or instead of some fixture or building belonging to the landlord, shall be removable by the tenant at any time during the continuance of the tenancy or before the expiry of 6 months, or such longer period as may be agreed, after the termination of the tenancy and shall remain his property so long as he may remove it by virtue of this subsection.

(2) The right conferred by subsection (1) above shall not be exercisable in relation to a fixture or building unless the tenant—
  (a) has paid all rent owing by him and has performed or satisfied all his other obligations to the landlord in respect of the holding; and
  (b) has, at least one month before whichever is the earlier of the exercise of the right and the termination of the tenancy, given to the landlord notice in writing of his intention to remove the fixture or building.

(3) If, before the expiry of the period of notice specified in subsection (2)(b) above, the landlord gives to the tenant a counter-notice in writing electing to purchase a fixture or building comprised in the notice, subsection (1) above shall cease to apply to that fixture or building, but the landlord shall be liable to pay to the tenant the fair value thereof to an incoming tenant of the holding.

(4) In the removal of a fixture or building by virtue of subsection (1) above, the tenant shall not do to any other building or other part of the holding any avoidable damage, and immediately after the removal shall make good all damage so occasioned.

19. Payment for implements, etc., sold on quitting holding

(1) Where a tenant of an agricultural holding has entered into an agreement or it is a term of the lease of the holding that the tenant will on quitting the holding, sell to the landlord or to the incoming tenant any implements of husbandry, fixtures, farm produce or farm stock on or used in connection with the holding, notwithstanding anything in the agreement or lease to the contrary, it shall be deemed to be a term of the agreement or of the lease, as the case may be, that the property in the goods shall not pass to the buyer until the price is paid and that payment of the price shall be made within one month after the tenant has quitted the holding or, if the price of the goods is to be ascertained by a valuation, within one month after the delivery of the award in the valuation.

(2) Where payment of the price is not made within one month as aforesaid the outgoing tenant shall be entitled to sell or remove the goods and to receive from the landlord or the incoming tenant, as the case may be, by whom the price was payable, compensation of an amount equal to any loss or expense unavoidably incurred by the outgoing tenant upon or in connection with such sale or removal, together with any expenses reasonably incurred by him in the preparation of his claim for compensation.

(3) Any question arising as to the amount of compensation payable under subsection (2) above shall be determined by arbitration.

20. Removal, of tenant for non-payment of rent

(1) When 6 months' rent of an agricultural holding is due and unpaid, the landlord shall be entitled to raise an action of removing in the sheriff court against the tenant, concluding for his removal from the holding at the term of Whitsunday or Martinmas next ensuing after the action is raised.

(2) In an action raised under subsection (1) above, the sheriff may, unless the arrears of rent then due are paid or caution is found to his satisfaction for them, and for one year's rent further, decern the tenant to remove, and may eject him at the said term in like manner as if the lease were determined and the tenant had been legally warned to remove.

(3) A tenant of a holding removed under this section shall have the rights of an outgoing tenant to which he would have been entitled if his tenancy had terminated by operation of notice to quit or notice of intention to quit at the term when he is removed.

(4) Section 5 of chapter XV of Book L of the Codifying Act of Sederunt of June 14, 1913, anent removings, shall not apply in any case where the procedure under this section is competent.

PART III

NOTICE TO QUIT AND NOTICE OF INTENTION TO QUIT

21. Notice to quit and notice of intention to quit

(1) Subject to section 20 of this Act and to subsections (6) and (7) below a tenancy of an agricultural holding shall not come to an end except by operation of a notice which complies with this subsection notwithstanding any agreement or any provision in the lease to the contrary.

(2) In this Act, a notice which complies with subsection (1) above is referred to as a 'notice to quit' if it is given by the landlord to the tenant and as a 'notice of intention to quit' if it is given by the tenant to the landlord.

(3) A notice complies with subsection (1) above if—

(a) it is in writing;

(b) it is a notice of intention to bring the tenancy to an end;

(c) where the notice is to take effect at the termination of the stipulated endurance of the lease, it is given not less than one year nor more than 2 years before that date;

(d) in the case of a lease continued in force by tacit relocation, it gives not less than one year nor more than 2 years' notice.

(4) The provisions of the Sheriff Courts (Scotland) Act 1907 relating to removings shall, in the case of an agricultural holding, have effect subject to this section.

(5) Notice to quit shall be given either—

(a) in the same manner as notice of removal under section 6 of the Removal Terms (Scotland) Act 1886; or

(b) in the form and manner prescribed by the Sheriff Courts (Scotland) Act 1907, and such notice shall come in place of the notice required by the said Act of 1907.

(6) Nothing in this section shall affect the right of the landlord of an agricultural holding to remove a tenant whose estate has been sequestrated under the Bankruptcy (Scotland) Act 1985 or the Bankruptcy (Scotland) Act 1913, or who by failure to pay rent or otherwise has incurred irritancy of his lease or other liability to be removed.

(7) This section shall not apply—

(a) to a notice given in pursuance of a stipulation in a lease entitling the landlord to resume land for building, planting, feuing or other purposes (not being agricultural purposes); or

(b) in relation to subjects let under a lease for any period less than a year, not being a lease which by virtue of section 2 of this Act takes effect as a lease from year to year.

22. Restrictions on operation of notices to quit

(1) Where not later than one month from the giving of a notice to quit an agricultural holding (or, in a case where section 23(3) of this Act applies, within the extended period therein mentioned) the tenant serves on the landlord a counter-notice in writing requiring that this subsection shall apply to the notice to quit, subject to subsection (2) below and to section 25 of this Act, the notice to quit shall not have effect unless the Land Court consent to the operation thereof.

(2) Subsection (1) above shall not apply where—

(a) the notice to quit relates to land being permanent pasture which the landlord has been in the habit of letting annually for seasonal grazing or of keeping in his own occupation and which has been let to the tenant for a definite and limited period for cultivation as arable land on the condition that he shall, along with the last or waygoing crop, sow permanent grass seeds;

(b) the notice to quit is given on the ground that the land is required for use, other than agriculture, for which permission has been granted on an application made under the enactments relating to town and country planning, or for which (otherwise than by virtue of any provision of those enactments) such permission is not required;

(c) the Land Court, on an application in that behalf made not more than 9 months before the giving of the notice to quit, were satisfied that the tenant was not fulfilling his responsibilities to farm the holding in accordance with the rules of good husbandry, and certified that they were so satisfied;

(d) at the date of the giving of the notice to quit the tenant had failed to comply with a demand in writing served on him by the landlord requiring him within 2 months from the service thereof to pay any rent due in respect of the holding, or within a reasonable time to remedy any breach by the tenant, which was capable of being remedied, of any term or condition of his tenancy which was not inconsistent with the fulfilment of his responsibilities to farm in accordance with the rules of good husbandry;

(e) at the date of the giving of the notice to quit the interest of the landlord in the holding had been materially prejudiced by a breach by the tenant, which was not capable of being remedied in reasonable time and at economic cost, of any term or condition of the tenancy which was not inconsistent with the fulfilment by the tenant of his responsibilities to farm in accordance with the rules of good husbandry;

(f) at the date of the giving of the notice to quit the tenant's apparent insolvency had been constituted in accordance with section 7 of the Bankruptcy (Scotland) Act 1985;

(g) section 25(1) of this Act applies, and the relevant notice complies with section 25(2)(a), (b) and (d) of this Act;

and, where any of paragraphs (a) to (f) above applies, the ground under the appropriate paragraph on which the notice to quit proceeds is stated in the notice.

23. Consent by Land Court or arbitration on notices to quit

(1) An application by a landlord for the consent of the Land Court under section 22 of this Act to the operation of a notice to quit shall be made within one month after service on the landlord by the tenant of a counter-notice requiring that subsection (1) of that section shall apply to the notice to quit.

(2) A tenant who has been given a notice to quit in connection with which any question arises under section 22(2) of this Act shall, if he requires such question to be determined by arbitration under this Act, give notice to the landlord to that effect within one month after the notice to quit has been served on him.

(3) Where the award of the arbiter in an arbitration required under subsection (2) above is such that section 22(1) of this Act would have applied to the notice to quit if a counter-notice had been served within the period provided for in that subsection, that period shall be extended up to the expiry of one month from the issue of the arbiter's award.

(4) Where such an arbitration as is referred to in subsection (2) above has been required by the tenant, or where an application has been made to the Land Court for their consent to the operation of a notice to quit, the operation of the notice to quit shall be suspended until the issue of the arbiter's award or of the decision of the Land Court, as the case may be.

(5) Where the decision of the Land Court giving their consent to the operation of a notice to quit, or the award of the arbiter in such an arbitration as is referred to in subsection (2) above, is issued at a date later than 6 months before the date on which the notice to quit is expressed to take effect, the Land Court, on application made to them in that behalf at any time not later than one month after the issue of the decision or award aforesaid, may postpone the operation of the notice to quit for a period not exceeding 12 months.

(6) If the tenant of an agricultural holding receives from the landlord notice to quit the holding or a part thereof and in consequence thereof gives to a sub-tenant notice to quit that holding or part, section 22(1) of this Act shall not apply to the notice given to the sub-tenant; but if the notice to quit given to the tenant by the landlord does not have effect, then the notice to quit given by the tenant to the sub-tenant shall not have effect.

(7) For the purposes of subsection (6) above, a notice to quit part of the holding which under section 30 of this Act is accepted by the tenant as notice to quit the entire holding shall be treated as a notice to quit the holding.

(8) Where notice is served on the tenant of an agricultural holding to quit the holding or a part thereof, being a holding or part which is subject to a sub-tenancy, and the tenant serves on the landlord a counter-notice in accordance with section 22(1) of this Act, the tenant shall also serve on the sub-tenant notice in writing that he has served such counter-notice on the landlord and the sub-tenant shall be entitled to be a party to any proceedings before the Land Court for their consent to the notice to quit.

24. Consents for purposes of section 22

(1) Subject to subsection (2) below and to section 25(3) of this Act, the Land Court shall consent under section 22 of this Act to the operation of a notice to quit an agricultural holding or part of an agricultural holding if, but only if, they are satisfied as to one or more of the following matters, being a matter or matters specified by the landlord in his application for their consent—
  (a) that the carrying out of the purpose for which the landlord proposes to terminate the tenancy is desirable in the interests of good husbandry as respects the land to which the notice relates, treated as a separate unit;
  (b) that the carrying out thereof is desirable in the interests of sound management of the estate of which that land consists or forms part;
  (c) that the carrying out thereof is desirable for the purposes of agricultural research, education, experiment or demonstration, or for the purposes of the enactments relating to allotments, smallholdings or such holdings as are referred to in section 64 of the Agriculture (Scotland) Act 1948;
  (d) that greater hardship would be caused by withholding than by giving consent to the operation of the notice;
  (e) that the landlord proposes to terminate the tenancy for the purpose of the land being used for a use, other than for agriculture, not falling within section 22(2)(b) of this Act.

(2) Notwithstanding that they are satisfied as aforesaid, the Land Court shall withhold consent to the operation of the notice to quit if in all the circumstances it appears to them that a fair and reasonable landlord would not insist on possession.

(3) Where the Land Court consent to the operation of a notice to quit they may (subject to section 25(4) of this Act) impose such conditions as appear to them requisite for securing that the land to which the notice relates will be used for the purpose for which the landlord proposes to terminate the tenancy.

(4) Where, on an application by the landlord in that behalf the Land Court are satisfied that by reason of any change of circumstances or otherwise any condition imposed under subsection (3) above ought to be varied or revoked, they shall vary or revoke the condition accordingly.

25. Termination of tenancies acquired by succession

(1) This section applies where notice to quit is duly given to the tenant of an agricultural holding who acquired right to the lease of the holding—
  (a) under section 16 of the Succession (Scotland) Act 1964; or
  (b) as a legatee, under section 11 of this Act.

(2) Notice to quit is duly given to a tenant to whom this section applies if—
  (a) it complies with section 21 of this Act; and
  (b) it specifies as its effective date—
        (i) where, when he acquired right to the lease, the unexpired period of the lease exceeded 2 years, the term of outgo stipulated in the lease;
        (ii) where, when he acquired right to the lease, the unexpired period was 2 years or less, the term of outgo stipulated in the lease or the corresponding date in any subsequent year, being a date not less than one nor more than 3 years after the said acquisition;
  (c) where he was a near relative of the deceased tenant from whom he acquired right, it specifies the Case set out in Schedule 2 to this Act under which it is given; and
  (d) where he was not a near relative of the deceased tenant from whom he acquired right, he acquired right to the lease after August 1, 1958.

(3) Section 22(1) of this Act shall apply and section 24 of this Act shall not apply where subsection (2)(c) above applies and notice to quit is duly given in accordance with subsection (2)(a) to (c) above; and in such a case the Land Court shall consent to the operation of a notice duly given—
  (a) where the holding was let before January 1, 1984, if they are satisfied that the

circumstances are as specified in any Case set out in Part I of Schedule 2 to this Act:

(b) where the holding was let on or after that date and the notice specifies any of Cases 4, 5 or 7 in that Schedule, unless the tenant satisfies them that the circumstances are not as specified in that Case (provided that, for the purposes of Case 7, the tenant shall not be required to prove that he is not the owner of any land);

(c) where the holding was let on or after that date, if they are satisfied that the circumstances are as specified in Case 6 in that Schedule;

except that where any of Cases 1, 2, 3, 6 or 7 in that Schedule is specified, the Court shall withhold consent on that ground if it appears to them that a fair and reasonable landlord would not insist on possession.

(4) Where consent is given because the circumstances are as specified in Case 2 or 6 in Schedule 2 to this Act, the Land Court shall impose such conditions as appear to them necessary to secure that the holding to which the notice relates will, within 2 years after the termination of the tenancy, be amalgamated with the land specified in the notice; and section 27 of this Act shall, with any necessary modifications, apply to a condition imposed under this subsection as that section applies to a condition imposed under section 24 of this Act.

(5) Part III of Schedule 2 to this Act shall have effect for the purposes of interpretation of this section and that Schedule.

26. Certificates of bad husbandry

(1) For the purposes of section 22(2)(c) of this Act, the landlord of an agricultural holding may apply to the Land Court for a certificate that the tenant is not fulfilling his responsibilities to farm in accordance with the rules of good husbandry, and the Land Court, if satisfied that the tenant is not fulfilling his said responsibilities, shall grant such a certificate.

(2) In determining whether to grant a certificate under this section, the Land Court shall disregard any practice adopted by the tenant in compliance with any obligation imposed on him by or accepted by him under section 31B of the Control of Pollution Act 1974.

27. Penalty for breach of condition

(1) Where, on giving consent under section 22 of this Act to the operation of a notice to quit an agricultural holding or part of an agricultural holding, the Land Court imposes a condition under section 24(3) of this Act, and it is proved, on an application to the Land Court on behalf of the Crown that the landlord—

(a) has failed to comply with the condition within the period allowed, or

(b) has acted in breach of the condition,

the Land Court may impose on the landlord a penalty of an amount not exceeding 2 years' rent of the holding at the rate at which rent was payable immediately before the termination of the tenancy, or, where the notice to quit related to a part only of the holding, of an amount not exceeding the proportion of the said 2 years' rent which it appears to the Land Court is attributable to that part.

(2) A penalty imposed under this section shall be a debt due to the Crown and shall, when recovered, be paid into the Consolidated Fund.

28. Effect on notice to quit of sale of holding

(1) This section shall apply where a contract for the sale of the landlord's interest in land which comprises or forms part of an agricultural holding is made after the giving of a notice to quit and before its expiry.

(2) Unless, within the period of 3 months ending with the date on which a contract to which this section applies is made, the landlord and the tenant have agreed in writing whether or not the notice to quit shall continue to have effect—

(a) the landlord shall—

      (i) within 14 days after the making of the contract; or

      (ii) before the expiry of the notice to quit,

whichever is the earlier, give notice to the tenant of the making of the contract; and

(b) the tenant may, before the expiry of the notice to quit and not later than one month after he has received notice under paragraph (a) above, give notice in writing to the landlord that he elects that the notice to quit shall continue to have effect.

(3) Where this section applies, unless—

(a) the landlord and tenant have agreed that the notice to quit shall continue to have effect;

(b) the tenant has so elected, under subsection (2)(b) above; or

(c) the landlord having failed to give notice of the making of the contract in accordance with subsection (2)(a) above, the tenant quits the holding in consequence of the notice to quit,

the notice to quit shall cease to have effect.

(4) Where this section applies and there is an agreement between the landlord and the tenant that the notice to quit shall continue to have effect, the notice shall not be invalid by reason only that the agreement is conditional.

29. Notice to quit part of holding to be valid in certain cases

(1) A notice to quit part of an agricultural holding held on a tenancy from year to year shall not be invalid on the ground that it relates to part only of the holding if it is given—

(a) for the purpose of adjusting the boundaries between agricultural units or of amalgamating agricultural units or parts thereof, or

(b) with a view to the use of the land to which the notice relates for any of the purposes mentioned in subsection (2) below,

and the notice states that it is given for that purpose or with a view to such use, as the case may be.

(2) The purposes referred to in subsection (1)(b) above are—

(a) the erection of farm labourers' cottages or other houses with or without gardens;

(b) the provision of gardens for farm labourers' cottages or other houses;

(c) the provision of allotments;

(d) the provision of small holdings under the Small Landholders (Scotland) Acts 1886 to 1931, or of such holdings as are referred to in section 64 of the Agriculture (Scotland) Act 1948;

(e) the planting of trees;

(f) the opening or working of coal, ironstone, limestone, brickearth, or other minerals, or of a stone quarry, clay, sand, or gravel pit, or the construction of works or buildings to be used in connection therewith;

(g) the making of a watercourse or reservoir;

(h) the making of a road, railway, tramroad, siding, canal or basin, wharf, or pier, or work connected therewith.

30. Tenant's right to treat notice to quit part as notice to quit entire holding

Where a notice to quit part of an agricultural holding is given to a tenant, being a notice which is rendered valid by section 29 of this Act, and the tenant within 28 days after—

(a) the giving of the notice, or

(b) where the operation of the notice depends on any proceedings under the foregoing provisions of this Act, the time when it is determined that the notice has effect,

whichever is later, gives to the landlord a counter-notice in writing that he accepts the notice as a notice to quit the entire holding, to take effect at the same time as the original notice, the notice to quit shall have effect accordingly.

31. Reduction of rent where tenant dispossessed of part of holding

(1) Where—
  (a) the tenancy of part of an agricultural holding terminates by reason of a notice to quit which is rendered valid by section 29 of this Act; or
  (b) the landlord of an agricultural holding resumes possession of part of the holding in pursuance of a provision in that behalf contained in the lease,

the tenant shall be entitled to a reduction of rent of an amount, to be determined by arbitration, proportionate to that part of the holding, together with an amount in respect of any depreciation of the value to him of the residue of the holding caused by the severance or by the use to be made of the part severed.

(2) Where subsection (1)(b) above applies, the arbiter, in determining the amount of the reduction, shall take into account any benefit or relief allowed to the tenant under the lease in respect of the part whose possession is being resumed.

32. Further restrictions on operation of certain notices to quit

(1) Subsections (2) to (5) below shall apply where—
  (a) notice to quit an agricultural holding or part of an agricultural holding is given to a tenant; and
  (b) the notice includes a statement in accordance with section 22(2) of this Act and paragraph (d) thereof to the effect that it is given by reason of the tenant's failure to remedy a breach of a kind referred to in section 66(1) of this Act.

(2) If not later than one month from the giving of the notice to quit the tenant serves on the landlord a counter-notice in writing requiring that this subsection shall apply to the notice to quit, subject to subsection (3) below, the notice to quit shall not have effect (whether as a notice to which section 22(1) of this Act does or does not apply) unless the Land Court consent to the operation thereof.

(3) A counter-notice under subsection (2) above shall be of no effect if within one month after the giving of the notice to quit the tenant serves on the landlord an effective notice under section 23(2) of this Act requiring the validity of the reason stated in the notice to quit to be determined by arbitration.

(4) Where—
  (a) the tenant has served on the landlord a notice of the kind referred to in subsection (3) above;
  (b) the notice to quit would, apart from this subsection, have effect in consequence of the arbitration; and
  (c) not later than one month from the date on which the arbiter's award is delivered to the tenant the tenant serves on the landlord a counter-notice in writing requiring that this subsection shall apply to the notice to quit;

the notice to quit shall not have effect (whether as a notice to which section 22(1) of this Act does or does not apply) unless the Land Court consent to the operation thereof.

(5) On an application made in that behalf by the landlord, the Land Court shall consent under subsection (2) or (4) above or (6) below to the operation of the notice to quit unless in all the circumstances it appears to them that a fair and reasonable landlord would not insist on possession.

(6) Where a notice to quit is given in accordance with section 66(3) of this Act in a case where the arbitration under that section followed an earlier notice to quit to which subsection (1) above applied, if the tenant serves on the landlord a counter-notice in writing within one month after the giving of the subsequent notice to quit (or, if the date specified in that notice for the termination of the tenancy is earlier, before that date), the notice to quit given under section 66(3) of this Act shall not have effect unless the Land Court consent to the operation thereof.

PART IV

COMPENSATION FOR IMPROVEMENTS

33. Improvements

In this Part the following are referred to as 'improvements'—

'1923 Act improvement' means an improvement carried out on an agricultural holding, being an improvement specified in Schedule 3 to this Act, and begun before July 31, 1931;

'1931 Act improvement' means an improvement so carried out, being an improvement specified in Schedule 4 to this Act and begun on or after July 31, 1931 and before November 1, 1948;

'old improvement' means a 1923 Act improvement or a 1931 Act improvement;

'new improvement' means an improvement carried out on an agricultural holding, being an improvement specified in Schedule 5 to this Act begun on or after November 1, 1948.

34. Right to compensation for improvements

(1) Subject to subsections (2) to (4), (7) and (8) below, and to sections 36 and 39 to 42 of this Act, a tenant of an agricultural holding shall be entitled, on quitting the holding at the termination of the tenancy, to compensation from the landlord in respect of improvements carried out by the tenant.

(2) A tenant whose lease was entered into before January 1, 1921 shall not be entitled to compensation under this section for an improvement which he was required to carry out by the terms of his tenancy.

(3) A tenant shall not be entitled to compensation under this section for an old improvement carried out on land which, at the time the improvement was begun, was not a holding within the meaning of the Agricultural Holdings (Scotland) Act 1923 as originally enacted, or land to which provisions of that Act relating to compensation for improvements and disturbance were applied by section 33 of that Act.

(4) Nothing in this section shall prejudice the right of a tenant to any compensation to which he is entitled—

(a) in the case of an old improvement, under custom, agreement or otherwise;

(b) in the case of a new improvement, under an agreement in writing between the landlord and the tenant;

in lieu of any compensation provided by this section.

(5) Where a tenant has remained in an agricultural holding during two or more tenancies, he shall not be deprived of his right to compensation under subsection (1) above by reason only that the improvements were not carried out during the tenancy on the termination of which he quits the holding.

(6) Subject to section 36(4) of this Act, a tenant shall be entitled to compensation under this section in respect of the 1931 Act improvement specified in paragraph 28 of Schedule 4 to this Act, or the new improvement specified in paragraph 32 of Schedule 5 to this Act (laying down of temporary pasture), notwithstanding that the laying down or the leaving at the termination of the tenancy of temporary pasture was in contravention of the terms of the lease or of any agreement made by the tenant respecting the method of cropping the arable lands; but, in ascertaining the amount of the compensation, the arbiter shall take into account any injury to or deterioration of the holding due to the contravention (except insofar as the landlord may have recovered damages therefor).

(7) Where under an agreement in writing entered into before January 1, 1921 a tenant is entitled to compensation which is fair and reasonable having regard to the circumstances existing at the time of the making of the agreement, for an old improvement specified in Part III of Schedule 3 to this Act or in Part III of Schedule 4 to this Act,

such compensation shall, as respects that improvement, be substituted for compensation under subsection (1) above.

(8) Compensation shall not be payable under this Part of this Act in respect of repairs of the kind specified in paragraph 29 of Schedule 3 to this Act or in paragraph 29 of Schedule 4 to this Act unless, before beginning to execute any such repairs, the tenant gave to the landlord notice in writing under paragraph (29) of Schedule 1 to the Agricultural Holdings (Scotland) Act 1923, or under paragraph (30) of Schedule 1 to the Small Landholders and Agricultural Holdings (Scotland) Act 1931, of his intention to execute the repairs, together with particulars thereof, and the landlord failed to exercise the right conferred on him by the said paragraph (29) or, as the case may be, the said paragraph (30) to execute the repairs himself within a reasonable time after receiving the notice.

35. Payment of compensation by incoming tenant

(1) This section applies to compensation which is payable or has been paid to an outgoing tenant of an agricultural holding by the landlord under or in pursuance of this Act or the Agricultural Holdings (Scotland) Act 1923, the Small Landholders and Agricultural Holdings (Scotland) Act 1931, the Agriculture (Scotland) Act 1948 or the 1949 Act.

(2) Subject to subsection (3) below, any agreement made after November 1, 1948 between an incoming tenant and his landlord whereby the tenant undertakes to pay to the outgoing tenant or to refund to the landlord any compensation to which this section applies shall be null and void.

(3) Subsection (2) above shall not apply in the case of an improvement of a kind referred to in Part III of Schedule 5 to this Act, where the agreement is in writing and states a maximum amount which may be payable thereunder by the incoming tenant.

(4) Where, on entering into occupation of an agricultural holding, a tenant, with the consent in writing of the landlord pays to the outgoing tenant compensation to which this section applies—

(a) in respect of an old improvement, in pursuance of an agreement in writing made before November 1, 1948; or

(b) where subsection (3) above applies,

the incoming tenant shall be entitled, on quitting the holding, to claim compensation for the improvement or part in like manner, if at all, as the outgoing tenant would have been entitled if the outgoing tenant had remained tenant of the holding and quitted it at the time at which the tenant quits it.

(5) Where, in a case not falling within subsection (2) or (3) above, a tenant, on entering into occupation of an agricultural holding, paid to his landlord any amount in respect of the whole or part of a new improvement, he shall, subject to any agreement in writing between the landlord and the tenant, be entitled on quitting the holding to claim compensation in respect of the improvement or part in like manner, if at all, as he would have been entitled if he had been tenant of the holding at the time when the improvement was carried out and the improvement or part thereof had been carried out by him.

36. Amount of compensation under this Part

(1) Subject to subsections (2) to (4) below, the amount of any compensation payable to a tenant under this Part of this Act shall be such sum as fairly represents the value of the improvement to an incoming tenant.

(2) In the ascertainment of the amount of compensation payable in respect of an old improvement, there shall be taken into account any benefit which the landlord has given or allowed to the tenant (under the lease or otherwise) in consideration of the tenant carrying out the improvement.

(3) In the ascertainment of the amount of compensation payable under this section for a new improvement, there shall be taken into account—

(a) any benefit which the landlord has agreed in writing to give the tenant in consideration of the tenant carrying out the improvement; and

(b) any grant out of moneys provided by Parliament which has been or will be made to the tenant in respect of the improvement.

(4) In ascertaining the amount of any compensation payable under section 34(6) of this Act, the arbiter shall take into account any injury to or deterioration of the holding due to the contravention of the lease or agreement referred to in that subsection, except in so far as the landlord has recovered damages in respect of such injury or deterioration.

37. Consents necessary for compensation for some improvements

(1) Compensation under this Part of this Act shall not be payable for—
  (a) a 1923 Act improvement specified in Part I of Schedule 3 to this Act;
  (b) a 1931 Act improvement specified in Part I of Schedule 4 to this Act; or
  (c) a new improvement specified in Part I of Schedule 5 to this Act;
unless, before the improvement was carried out, the landlord consented to it in writing (whether unconditionally or upon terms as to compensation or otherwise agreed on between the parties).

(2) Where such consent was given on terms agreed as to compensation, the compensation payable under the agreement shall be substituted for compensation under section 34 of this Act.

38. Notice required of certain improvements

(1) Subject to subsections (2) to (6) below, compensation under this Act shall not be payable for—
  (a) a 1923 Act improvement specified in Part II of Schedule 3 to this Act;
  (b) a 1931 Act improvement specified in Part II of Schedule 4 to this Act;
  (c) a new improvement specified in Part II of Schedule 5 to this Act;
unless the tenant gave notice to the landlord in accordance with subsection (3) below of his intention to carry it out and of the manner in which he proposed to do so.

(2) Subsection (1) above shall not apply in the case of an improvement mentioned in subsection (1)(a) or (b) above, if the parties agreed by the lease or otherwise to dispense with the requirement for notice under subsection (3).

(3) Notice shall be in accordance with this subsection if it is in writing and—
  (a) in the case of an improvement mentioned in subsection (1)(a) above, it was notice under section 3 of the Agricultural Holdings (Scotland) Act 1923, given not more than 3 nor less than 2 months,
  (b) in the case of an improvement mentioned in subsection (1)(b) above, it was notice under the said section 3, given not more than 6 nor less than 3 months,
  (c) in the case of an improvement mentioned in subsection (1)(c) above, it was given not less than 3 months,
before the tenant began to carry out the improvement.

(4) In the case of an improvement mentioned in subsection (1)(a) or (b) above, compensation shall not be payable unless—
  (a) the parties agreed on the terms as to compensation or otherwise on which the improvement was to be carried out;
  (b) where no such agreement was made and the tenant did not withdraw the notice, the landlord failed to exercise his right under the said section 3 to carry out the improvement himself within a reasonable time; or
  (c) in the case of an improvement mentioned in subsection (1)(b) above, where the landlord gave notice of objection and the matter was referred under section 28(2) of the Small Landholders and Agricultural Holdings (Scotland) Act 1931 for determination by the appropriate authority, that authority was satisfied that the improvement should be carried out and the improvement was carried out in accordance with any directions given by that authority as to the manner of so doing.

(5) If the parties agreed (either after notice was given under this section or by an agreement to dispense with it) on terms as to compensation, the compensation payable under the agreement shall be substituted for compensation under this Part of this Act.

(6) In subsection (4) above, 'the appropriate authority' means—

(a) in relation to the period before September 4, 1939, the Department of Agriculture for Scotland;

(b) in relation to the period starting on that day, the Secretary of State.

39. Compensation for Sch 5, Pt II, improvements conditional on approval of Land Court in certain cases

(1) Subject to subsections (2) to (4) below, compensation under this Part of this Act shall not be payable in respect of a new improvement specified in Part II of Schedule 5 to this Act if, within one month after receiving notice under section 38(3) of this Act from the tenant of his intention to carry out the improvement, the landlord gives notice in writing to the tenant that he objects to the carrying out of the improvement or to the manner in which the tenant proposes to carry it out.

(2) Where notice of objection has been given under subsection (1) above, the tenant may apply to the Land Court for approval of the carrying out of the improvement, and on such application the Land Court may approve the carrying out of the improvement either—

(a) unconditionally, or

(b) upon such terms, as to reduction of the compensation which would otherwise be payable or as to other matters, as appears to them to be just,

or may withhold their approval.

(3) If, on an application under subsection (2) above, the Land Court grant their approval, the landlord may, within one month after receiving notice of the decision of the Land Court, serve notice in writing on the tenant undertaking to carry out the improvement himself.

(4) Where, on an application under subsection (2) above the Land Court grant their approval, then if either—

(a) no notice is served by the landlord under subsection (3) above, or

(b) such a notice is served but, on an application made by the tenant in that behalf, the Land Court determines that the landlord has failed to carry out the improvement within a reasonable time,

the tenant may carry out the improvement and shall be entitled to compensation under this Part of this Act in respect thereof as if notice of objection had not been given by the landlord, and any terms subject to which the approval was given shall have effect as if they were contained in an agreement in writing between the landlord and the tenant.

PART V

OTHER PROVISIONS REGARDING COMPENSATION

Market gardens

40. Market gardens

(1) This section applies to any agricultural holding which, by virtue of an agreement in writing made on or after January 1, 1898, is let or is to be treated as a market garden.

(2) This section also applies where—

(a) a holding was, on January 1, 1898 under a lease then current, in use or cultivation as a market garden with the knowledge of the landlord; and

(b) an improvement of a kind specified in Schedule 6 to this Act (other than such an alteration of a building as did not constitute an enlargement thereof) has been carried out on the holding; and

(c) the landlord did not, before the improvement was carried out, serve on the tenant a written notice dissenting from the carrying out of the improvement;
in relation to improvements whether carried out before or after January 1, 1898.

(3) In the application of Part IV of this Act to an agricultural holding to which this section applies, subject to subsections (5) and (7) below, the improvements specified in Schedule 6 to this Act shall be included in the improvements specified in Part III of each of Schedules 3, 4 and 5 to this Act.

(4) In the case of an agricultural holding to which this section applies—
(a) section 18 of this Act shall apply to every fixture or building affixed or erected by the tenant to or upon the holding or acquired by him since December 31, 1900 for the purposes of his trade or business as a market gardener;
(b) it shall be lawful for the tenant to remove all fruit trees and fruit bushes planted by him on the holding and not permanently set out, but if the tenant does not remove such fruit trees and fruit bushes before the termination of his tenancy they shall remain the property of the landlord and the tenant shall not be entitled to any compensation in respect thereof; and
(c) the right of an incoming tenant to claim compensation in respect of the whole or part of an improvement which he has purchased may be exercised although the landlord has not consented in writing to the purchase.

(5) Where a tenancy of a kind described in subsection (2) above was a tenancy from year to year, the compensation payable in respect of an improvement of a kind referred to in that subsection shall be such (if any) as could have been claimed if the 1949 Act had not been passed.

(6) Where the land to which this section applies consists of part only of an agricultural holding this section shall apply as if that part were a separate holding.

(7) Nothing in this section shall confer a right to compensation for the alteration of a building (not being an alteration constituting an enlargement of the building) where the alteration was begun before 1st November 1948.

41. Direction by Land Court that holding be treated as market garden

(1) Where—
(a) the tenant of an agricultural holding intimates to the landlord in writing his desire to carry out on the holding or any part thereof an improvement specified in Schedule 6 to this Act;
(b) the landlord refuses, or within a reasonable time fails, to agree in writing that the holding, or that part thereof, shall be treated as a market garden;
(c) the tenant applies to the Land Court for a direction under this subsection; and
(d) the Land Court is satisfied that the holding or that part thereof is suitable for the purposes of market gardening;
the Land Court may direct that section 40 of this Act shall apply to the holding or, as the case may be, part of a holding either—
(i) in respect of all the improvements specified in Schedule 6 to this Act, or
(ii) in respect of some only of those improvements,
and that section shall apply accordingly as respects any improvement carried out after the date on which the direction is given.

(2) A direction under subsection (1) above may be given subject to such conditions, if any, for the protection of the landlord as the Land Court may think fit and, in particular, where the direction relates to part only of the holding, the direction may, on the application of the landlord, be given subject to the condition that the tenant shall consent to the division of the holding into two parts (one such part being the part to which the direction relates) to be held at rents agreed by the landlord and tenant or in default of agreement determined by arbitration, but otherwise on the same terms and conditions (so far as applicable) as those on which the holding is held.

(3) Where a direction is given under subsection (1) above, if the tenancy is terminated—

(a) by notice of intention to quit given by the tenant, or

(b) by reason of the tenant's apparent insolvency being constituted under section 7 of the Bankruptcy (Scotland) Act 1985,

the tenant shall not be entitled to compensation in respect of improvements specified in the direction unless he produces an offer which complies with subsection (4) below and the landlord fails to accept the offer within 3 months after the production thereof.

(4) An offer complies with this subsection if—

(a) it is in writing;

(b) it is made by a substantial and otherwise suitable person;

(c) it is produced by the tenant to the landlord not later than one month after the date of the notice of intention to quit or constitution of apparent insolvency as the case may be, or at such later date as may be agreed;

(d) it is an offer to accept a tenancy of the holding from the termination of the existing tenancy on the terms and conditions of the existing tenancy so far as applicable;

(e) it includes an offer, subject to subsection (5) below, to pay to the outgoing tenant all compensation payable under this Act or under the lease;

(f) it is open for acceptance for a period of 3 months from the date on which it is produced.

(5) If the landlord accepts an offer which complies with subsection (4) above the incoming tenant shall pay to the landlord on demand all sums payable to him by the outgoing tenant on the termination of the tenancy in respect of rent or breach of contract or otherwise in respect of the holding.

(6) Any amount paid by the incoming tenant under subsection (5) above may, subject to any agreement between the outgoing tenant and incoming tenant, be deducted by the incoming tenant from any compensation payable by him to the outgoing tenant.

(7) A tenancy created by the acceptance of an offer which complies with subsection (4) above shall be deemed for the purposes of section 13 of this Act not to be a new tenancy.

42. Agreements as to compensation relating to market gardens

(1) Where under an agreement in writing a tenant of an agricultural holding is entitled to compensation which is fair and reasonable having regard to the circumstances existing at the time of making the agreement, for an improvement for which compensation is payable by virtue of section 40 of this Act, such compensation shall, as respects that improvement, be substituted for compensation under this Act.

(2) The landlord and the tenant of an agricultural holding who have agreed that the holding shall be let or treated as a market garden may by agreement in writing substitute, for the provisions as to compensation which would otherwise be applicable to the holding, the provisions as to compensation in section 41(3) to (6) of this Act.

Miscellaneous

43. Compensation for disturbance

(1) Where the tenancy of an agricultural holding terminates by reason of—

(a) a notice to quit given by the landlord; or

(b) a counter-notice given by the tenant under section 30 of this Act,

and in consequence the tenant quits the holding, subject to subsections (2) to (8) below, compensation for the disturbance shall be payable by the landlord to the tenant.

(2) Compensation shall not be payable under this section where the application of section 22(1) of this Act to the notice to quit is excluded by any of paragraphs (a) or (c) to (f) of subsection (2) of that section.

(3) Subject to subsection (4) below, the amount of the compensation payable under

this section shall be the amount of the loss or expense directly attributable to the quitting of the holding which is unavoidably incurred by the tenant upon or in connection with the sale or removal of his household goods, implements of husbandry, fixtures, farm produce or farm stock on or used in connection with the holding, and shall include any expenses reasonably incurred by him in the preparation of his claim for compensation (not being expenses of an arbitration to determine any question arising under this section).

(4) Where compensation is payable under this section—

(a) the compensation shall be an amount equal to one year's rent of the holding at the rate at which rent was payable immediately before the termination of the tenancy without proof by the tenant of any such loss or expense as aforesaid;

(b) the tenant shall not be entitled to claim any greater amount than one year's rent of the holding unless he has given to the landlord not less than one month's notice of the sale of any such goods, implements, fixtures, produce or stock as aforesaid and has afforded him a reasonable opportunity of making a valuation thereof;

(c) the tenant shall not in any case be entitled to compensation in excess of 2 years' rent of the holding.

(5) In subsection (4) above 'rent' means the rent after deduction of such an amount as, failing agreement, the arbiter finds to be the amount payable by the landlord in respect of the holding for the year in which the tenancy was terminated by way of any public rates, taxes or assessments or other public burdens, the charging of which on the landlord would entitle him to relief in respect of tax under Part II of the Income and Corporation Taxes Act 1988.

(6) Where the tenant of an agricultural holding has lawfully sub-let the whole or part of the holding, and in consequence of a notice to quit given by his landlord becomes liable to pay compensation under this section to the sub-tenant, the tenant shall not be debarred from recovering compensation under this section by reason only that, owing to not being in occupation of the holding or part of the holding, on the termination of his tenancy he does not quit the holding or that part.

(7) Where the tenancy of an agricultural holding terminates by virtue of a counter-notice given by the tenant under section 30 of this Act and—

(a) the part of the holding affected by the notice to quit given by the landlord, together with any part of the holding affected by any previous notice to quit given by the landlord which is rendered valid by section 29 of this Act, is either less than a quarter of the area of the original holding or of a rental value less than one quarter of the rental value of the original holding, and

(b) the holding as proposed to be diminished is reasonably capable of being farmed as a separate holding,

compensation shall not be payable under this section except in respect of the part of the holding to which the notice to quit relates.

(8) Compensation under this section shall be in addition to any compensation to which the tenant may be entitled apart from this section.

44. Compensation for continuous adoption of special standard of farming

(1) Where the tenant of an agricultural holding proves that the value of the holding to an incoming tenant has been increased during the tenancy by the continuous adoption of a standard of farming or a system of farming which has been more beneficial to the holding than—

(a) the standard or system required by the lease, or

(b) in so far as no system of farming is so required, the system of farming normally practised on comparable holdings in the district,

the tenant shall be entitled, on quitting the holding, to obtain from the landlord such compensation as represents the value to an incoming tenant of the adoption of that more beneficial standard or system.

(2) Compensation shall not be recoverable under subsection (1) above unless—

(a) the tenant has, not later than one month before the termination of the tenancy, given to the landlord notice in writing of his intention to claim such compensation; and

(b) a record of the condition of the fixed equipment on, and the cultivation of, the holding has been made under section 8 of this Act;

and shall not be so recoverable in respect of any matter arising before the date of the record so made or, where more than one such record has been made during the tenancy, before the date of the first such record.

(3) In assessing the compensation to be paid under subsection (1) above, due allowance shall be made for any compensation agreed or awarded to be paid to the tenant under Part IV of this Act for any improvement which has caused or contributed to the benefit.

(4) Nothing in this section shall entitle a tenant to recover, in respect of any improvement, any compensation which he would not be entitled to recover apart from this section.

45. Compensation to landlord for deterioration etc. of holding

(1) The landlord of an agricultural holding shall be entitled to recover from the tenant, on his quitting the holding on termination of the land tenancy compensation—

(a) where the landlord shows that the value of the holding has been reduced by dilapidation, deterioration or damage caused by;

(b) where dilapidation, deterioration or damage has been caused to any part of the holding or to anything in or on the holding by;

non-fulfilment by the tenant of his responsibilities to farm in accordance with the rules of good husbandry.

(2) The amount of compensation payable under subsection (1) above shall be—

(a) where paragraph (a) of that subsection applies, (insofar as the landlord is not compensated for the dilapidation, deterioration or damage under paragraph (b) thereof) an amount equal to the reduction in the value of the holding;

(b) when paragraph (b) of that subsection applies, the cost, as at the date of the tenant's quitting the holding, of making good the dilapidation, deterioration or damage.

(3) Notwithstanding anything in this Act, the landlord may, in lieu of claiming compensation under subsection (1)(b) above, claim compensation in respect of matters specified therein, under and in accordance with a lease in writing, so however that—

(a) compensation shall be so claimed only on the tenant's quitting the holding on the termination of the tenancy;

(b) subject to section 46(4) of this Act compensation shall not be claimed in respect of any one holding both under such a lease and under subsection (1) above;

and compensation under this subsection shall be treated, for the purposes of subsection (2)(a) above and of section 46(2) of this Act as compensation under subsection (1)(b) above.

46. Compensation for failure to repair or maintain fixed equipment

(1) This section applies where, by virtue of section 4 of this Act, the liability for the maintenance or repair of an item of fixed equipment is transferred from the tenant to the landlord.

(2) Where this section applies, the landlord may within the period of one month beginning with the date on which the transfer takes effect require that there shall be determined by arbitration, and paid by the tenant, the amount of any compensation which would have been payable under section 45(1)(b) of this Act in respect of any previous failure by the tenant to discharge the said liability, if the tenant had quitted the holding on the termination of his tenancy at the date on which the transfer takes effect.

(3) Where this section applies, any claim by the tenant in respect of any previous failure by the landlord to discharge the said liability shall, if the tenant within the period of

one month referred to in subsection (2) above so requires, be determined by arbitration, and any amount directed by the award to be paid by the landlord shall be paid by him to the tenant.

(4) For the purposes of section 45(3)(b) of this Act any compensation under this section shall be disregarded.

47. Provisions supplementary to ss 45 and 46

(1) Compensation shall not be recoverable under section 45 of this Act, unless the landlord has, not later than 3 months before the termination of the tenancy, given notice in writing to the tenant of his intention to claim compensation thereunder.

(2) Subsection (3) below shall apply to compensation—

(a) under section 45 of this Act, where the lease was entered into after July 31, 1931; or

(b) where the lease was entered into on or after November 1, 1948.

(3) When this subsection applies, no compensation shall be recoverable—

(a) unless during the occupancy of the tenant a record of the condition of the fixed equipment on, and cultivation of, the holding has been made under section 8 of this Act;

(b) in respect of any matter arising before the date of the record referred to in paragraph (a) above, or

(c) where more than one such record has been made during the tenant's occupancy, in respect of any matter arising before the date of the first such record.

(4) If the landlord and the tenant so agree in writing a record of the condition of the holding shall, notwithstanding that it was made during the occupancy of a previous tenant, be deemed, for the purposes of subsection (3) above, to have been made during the occupancy of the tenant and on such date as may be specified in the agreement and shall have effect subject to such modifications (if any) as may be so specified.

(5) Where the tenant has remained in his holding during 2 or more tenancies, his landlord shall not be deprived of his right to compensation under section 45 of this Act in respect of any dilapidation, deterioration or damage by reason only that the tenancy during which the relevant act or omission occurred was a tenancy other than the tenancy at the termination of which the tenant quit the holding.

48. Landlord not to have right to penal rent or liquidated damages

Notwithstanding any provision to the contrary in a lease of an agricultural holding, the landlord shall not be entitled to recover any sum, by way of higher rent, liquidated damages or otherwise, in consequence of any breach or non-fulfilment of a term or condition of the lease, which is in excess of the damage actually suffered by him in consequence of the breach or non-fulfilment.

49. Compensation provisions to apply to parts of holdings in certain cases

(1) Where—

(a) the tenancy of part of an agricultural holding terminates by reason of a notice to quit which is rendered valid by section 29 of this Act; or

(b) the landlord of an agricultural holding resumes possession of part of the holding in pursuance of a provision in that behalf contained in the lease;

the provisions of this Act with respect to compensation shall apply as if that part of the holding were a separate holding which the tenant had quitted in consequence of a notice to quit.

(2) In a case falling within subsection (1)(b) above, the arbiter, in assessing the amount of compensation payable to the tenant, shall take into account any benefit or relief allowed to the tenant under the lease in respect of the land possession of which is resumed by the landlord.

(3) Where any land comprised in a lease is not an agricultural holding within the meaning of this Act by reason only that the land so comprised includes land to which

subsection (4) below applies, the provisions of this Act with respect to compensation for improvements and for disturbance shall, unless it is otherwise agreed in writing, apply to the part of the land exclusive of the land to which subsection (4) below applies as if that part were a separate agricultural holding.

(4) This subsection applies to land which, owing to the nature of the building thereon or the use to which it is put, would not, if it had been separately let, be an agricultural holding.

50. Determination of claims for compensation where holding is divided

Where the interest of the landlord in an agricultural holding has become vested in several parts in more than one person and the rent payable by the tenant of the holding has not been apportioned with his consent or under any statute, the tenant shall be entitled to require that any compensation payable to him under this Act shall be determined as if the holding had not been divided; and the arbiter shall, where necessary, apportion the amount awarded between the persons who for the purposes of this Act together constitute the landlord of the holding, and any additional expenses of the award caused by the apportionment shall be directed by the arbiter to be paid by those persons in such proportions as he shall determine.

51. Compensation not to be payable for things done in compliance with this Act

(1) Notwithstanding anything in the foregoing provisions of this Act or any custom or agreement—
   (a) no compensation shall be payable to the tenant of an agricultural holding in respect of anything done in pursuance of a direction under section 9(2) of this Act;
   (b) in assessing compensation to an outgoing tenant of an agricultural holding where land has been ploughed up in pursuance of a direction under section 9(2) of this Act, the value per hectare of any tenant's pasture comprised in the holding shall be taken not to exceed the average value per hectare of the whole of the tenant's pasture comprised in the holding on the termination of the tenancy.

(2) In subsection (1)(b) above 'tenant's pasture' means pasture laid down at the expense of the tenant or paid for by the tenant on entering the holding.

(3) The tenant of an agricultural holding shall not be entitled to compensation for an improvement specified in Part III of any of Schedules 3 to 5 to this Act, being an improvement carried out for the purposes of—
   (a) the proviso to section 35(1) of the Agricultural Holdings (Scotland) Act 1923;
   (b) the proviso to section 12(1) of the 1949 Act; or
   (c) section 9 of this Act.

52. Compensation for damage by game

(1) Subject to subsection (2) below, where the tenant of an agricultural holding has sustained damage to his crops from game, the right to kill and take which is vested neither in him nor in anyone claiming under him other than the landlord, and which the tenant has not permission in writing to kill, he shall be entitled to compensation from his landlord for the damage if it exceeds in amount the sum of 12 pence per hectare of the area over which it extends.

(2) Compensation shall not be recoverable under subsection (1) above, unless—
   (a) notice in writing is given to the landlord as soon as is practicable after the damage was first observed by the tenant, and a reasonable opportunity is given to the landlord to inspect the damage—
        (i) in the case of damage to a growing crop, before the crop is begun to be reaped, raised or consumed;
        (ii) in the case of damage to a crop reaped or raised, before the crop is begun to be removed from the land; and
   (b) notice in writing of the claim, together with the particulars thereof, is given to the

landlord within one month after the expiry of the calendar year, or such other period of 12 months as by agreement between the landlord and the tenant may be substituted therefor, in respect of which the claim is made.

(3) The amount of compensation payable under subsection (1) above shall, in default of agreement made after the damage has been suffered, be determined by arbitration.

(4) Where the right to kill and take the game is vested in some person other than the landlord, the landlord shall be entitled to be indemnified by that other person against all claims for compensation under this section; and any question arising under this subsection shall be determined by arbitration.

(5) In this section 'game' means deer, pheasants, partridges, grouse and black game.

53. Extent to which compensation recoverable under agreements

(1) Unless this Act makes express provision to the contrary, where provision is made in this Act for compensation to be paid to a landlord or tenant—
  (a) he shall be so entitled notwithstanding any agreement, and
  (b) he shall not be entitled to compensation except under that provision.

(2) Where the landlord and the tenant of an agricultural holding enter into an agreement in writing for such a variation of the terms of the lease as could be made by direction under section 9 of this Act, the agreement may provide for the exclusion of compensation in the same manner as under section 51(1) of this Act.

(3) A claim for compensation by a landlord or tenant of an agricultural holding in a case for which this Act does not provide for compensation shall not be enforceable except under an agreement in writing.

PART VI

ADDITIONAL PAYMENTS

54. Additional payments to tenants quitting holdings

(1) Where compensation for disturbance in respect of an agricultural holding or part of such a holding becomes payable—
  (a) to a tenant, under this Act; or
  (b) to a statutory small tenant, under section 13 of the 1931 Act;
  subject to this Part of this Act, there shall be payable by the landlord to the tenant, in addition to the compensation, a sum to assist in the reorganisation of the tenant's affairs of the amount referred to in subsection (2) below.

(2) The sum payable under subsection (1) above shall be equal to 4 times the annual rent of the holding or, in the case of part of a holding, times the appropriate portion of that rent, at the rate at which the rent was payable immediately before the termination of the tenancy.

55. Provisions supplementary to s 54

(1) Subject to subsection (2) below no sum shall be payable under section 54 of this Act in consequence of the termination of the tenancy of an agricultural holding or part of such a holding by virtue of a notice to quit where—
  (a) the notice contains a statement that the carrying out of the purpose for which the landlord proposes to terminate the tenancy is desirable on any grounds referred to in section 24(1)(a) to (c) of this Act and, if an application for consent in respect of the notice is made to the Land Court in pursuance of section 22(1) of this Act, the Court consent to its operation and state in the reasons for their decision that they are satisfied that termination of the tenancy is desirable on that ground;
  (b) the notice contains a statement that the landlord will suffer hardship unless the notice has effect and, if an application for consent in respect of the notice is made to the Land Court in pursuance of section 22(1) of this Act, the Court consent to

its operation and state in the reasons for their decision that they are satisfied that greater hardship would be caused by withholding consent than by giving it;

(c) the notice is one to which section 22(1) of this Act applies by virtue of section 25(3) of this Act and the Land Court consent to its operation and specify in the reasons for their decision the Case in Schedule 2 to this Act as regards which they are satisfied; or

(d) section 22(1) of this Act does not apply to the notice by virtue of section 29(4) of the Agriculture Act 1967 (which relates to notices to quit given by the Secretary of State or a Rural Development Board with a view to boundary adjustments or an amalgamation).

(2) Subsection (1) above shall not apply in relation to a notice to quit where—

(a) the reasons given by the Land Court for their decision to consent to the operation of the notice include the reason that they are satisfied as to the matter referred to in section 24(1)(e) of this Act; or

(b) the reasons so given include the reason that the Court are satisfied as to the matter referred to in section 24(1)(b) of this Act or, where the tenant has succeeded to the tenancy as the near relative of a deceased tenant, as to the matter referred to in any of Cases 1, 3, 5 and 7 in Schedule 2 to this Act; but the Court state in their decision that they would have been satisfied also as to the matter referred to in section 24(1)(e) of this Act if it had been specified in the application for consent.

(3) In assessing the compensation payable to the tenant of an agricultural holding in consequence of the compulsory acquisition of his interest in the holding or part of it or the compulsory taking of possession of the holding or part of it, no account shall be taken of any benefit which might accrue to the tenant by virtue of section 54 of this Act.

(4) Any sum payable in pursuance of section 54 of this Act shall be so payable notwithstanding any agreement to the contrary.

(5) The following provisions of this Act shall apply to sums claimed or payable in pursuance of section 54 of this Act as they apply to compensation claimed or payable under section 43 of this Act—

sections 43(6);

section 50;

section 74.

(6) No sum shall be payable in pursuance of section 54 of this Act in consequence of the termination of the tenancy of an agricultural holding or part of such a holding by virtue of a notice to quit where—

(a) the relevant notice is given in pursuance of section 25(2)(a), (b) and (d) of this Act;

(b) the landlord is terminating the tenancy for the purpose of using the land for agriculture only; and

(c) the notice contains a statement that the tenancy is being terminated for the said purpose.

(7) If any question arises between the landlord and the tenant as to the purpose for which a tenancy is being terminated, the tenant shall, notwithstanding section 61(1) of this Act, refer the question to the Land Court for determination.

(8) In this section—

(a) references to section 54 of this Act do not include references to it as applied by section 56 of this Act; and

(b) for the purposes of subsection (1)(a) above, the reference in section 24(1)(c) of this Act to the purposes of the enactments relating to allotments shall be ignored.

56. Additional payments in consequence of compulsory acquisition etc of agricultural holdings

(1) This section applies where, in pursuance of any enactment providing for the acquisition or taking of possession of land compulsorily, any person (referred to in this section and in sections 57 and 58 of and Schedule 8 to this Act as 'an acquiring authority')

acquires the interest of the tenant in, or takes possession of, an agricultural holding or any part of an agricultural holding or the holding of a statutory small tenant.

(2) Subject to subsection (3) below and sections 57 and 58 of this Act, where this section applies section 54 of this Act shall apply as if the acquiring authority were the landlord of the holding and compensation for disturbance in respect of the holding or part in question had become payable to the tenant on the date of the acquisition or taking of possession.

(3) No compensation shall be payable by virtue of this section in respect of an agricultural holding held under a tenancy for a term of 2 years or more unless the amount of such compensation is less than the aggregate of the amounts which would have been payable by virtue of this section if the tenancy had been from year to year: and in such a case the amount of compensation payable by virtue of this section shall (subject to section 57(4) of this Act) be equal to the difference.

57. Provisions supplementary to s 56

(1) For the purposes of section 56 of this Act, a tenant of an agricultural holding shall be deemed not to be a tenant of it in so far as, immediately before the acquiring of the interest or taking of possession referred to in that section, he was neither in possession, nor entitled to take possession, of any land comprised in the holding: and in determining, for those purposes, whether a tenant was so entitled, any lease relating to the land of a kind referred to in section 2(1) of this Act which has not taken effect as a lease of the land from year to year shall be ignored.

(2) Section 56(1) of this Act shall not apply—

(a) where the acquiring authority require the land comprised in the holding or part in question for the purposes of agricultural research or experiment or of demonstrating agricultural methods or for the purposes of the enactments relating to smallholdings;

(b) where the Secretary of State acquires the land under section 57(1)(c) or 64 of the Agricultural (Scotland) Act 1948.

(3) Where an acquiring authority exercise, in relation to any land, power to acquire or take possession of land compulsorily which is conferred on the authority by virtue of section 102 or 110 of the Town and Country Planning (Scotland) Act 1972 or section 7 of the New Towns (Scotland) Act 1968, the authority shall be deemed for the purposes of subsection (2) above not to require the land for any of the purposes mentioned in that subsection.

(4) Schedule 8 to this Act shall have effect in relation to payments under section 56 of this Act.

58. Effect of early resumption clauses on compensation

(1) Where—

(a) the landlord of an agricultural holding resumes land under a provision in the lease entitling him to resume land for building, planting, feuing or other purposes (not being agricultural purposes); or

(b) the landlord of the holding of a statutory small tenant resumes the holding or part thereof on being authorised to do so by the Land Court under section 32(15) of the 1911 Act; and

(c) in either case, the tenant has not elected that section 55(2) of the Land Compensation (Scotland) Act 1973 (right to opt for notice of entry compensation) should apply to the notice;

compensation shall be payable by the landlord to the tenant (in addition to any other compensation so payable apart from this subsection) in respect of the land.

(2) The amount of compensation payable under subsection (1) above shall be equal to the value of the additional benefit (if any) which would have accrued to the tenant if the land had, instead of being resumed at the date of resumption, been resumed at the expiry of 12 months from the end of the current year of the tenancy.

(3) Section 55(4) and (5) of this Act shall apply to compensation claimed or payable under subsection (1) above with the substitution for references to section 54 of this Act of references to this section.

(4) In the assessment of the compensation payable by an acquiring authority to a statutory small tenant in the circumstances referred to in section 56(1) of this Act, any authorisation of resumption of the holding or part thereof by the Land Court under section 32(15) of the 1911 Act for any purpose (not being an agricultural purpose) specified therein shall—

(a) in the case of an acquisition, be treated as if it became operative only on the expiry of 12 months from the end of the year of the tenancy current when notice to treat in respect of the acquisition was served or treated as served on the tenant; and

(b) in the case of a taking of possession, be disregarded;

unless compensation assessed in accordance with paragraph (a) or (b) above would be less than would be payable but for this subsection.

(5) For the purposes of subsection (1) above, the current year of a tenancy for a term of 2 years or more is the year beginning with such day in the period of 12 months ending with a date 2 months before the resumption mentioned in that subsection as corresponds to the day on which the term would expire by the effluxion of time.

59. Interpretation etc of Part VI

In sections 54 to 58 of and Schedule 8 to this Act—

'acquiring authority' has the meaning assigned to it by section 56(1) of this Act;

'statutory small tenant' and 'holding' in relation to a statutory small tenant have the meanings given in section 32(1) of the 1911 Act; and

references to the acquisition of any property are references to the vesting of the property in the person acquiring it.

PART VII

ARBITRATION AND OTHER PROCEEDINGS

60. Questions between landlord and tenant

(1) Subject to subsection (2) below and except where this Act makes express provision to the contrary, any question or difference between the landlord and the tenant of an agricultural holding arising out of the tenancy or in connection with the holding (not being a question or difference as to liability for rent) shall, whether such question or difference arises during the currency or on the termination of the tenancy, be determined by arbitration.

(2) Any question or difference between the landlord and the tenant of an agricultural holding which by or under this Act or under the lease is required to be determined by arbitration may, if the landlord and the tenant so agree, in lieu of being determined by arbitration be determined by the Land Court, and the Land Court shall, on the joint application of the landlord and the tenant, determine such question or difference accordingly.

61. Arbitrations

(1) Any matter which by or under this Act, or by regulations made thereunder, or under the lease of an agricultural holding is required to be determined by arbitration shall, whether the matter arose before or after the passing of this Act, be determined, notwithstanding any agreement under the lease or otherwise providing for a different method of arbitration, by a single arbiter in accordance with the provisions of Schedule 7 to this Act, and the Arbitration (Scotland) Act 1894 shall not apply to any such arbitration.

(2) An appeal by application to the Land Court by any party to an arbitration under section 13(1) of this Act (variation of rent) against the award of an arbiter appointed by the Secretary of State or the Land Court on any question of law or fact (including the amount of the award) shall be competent.

(3) An appeal under subsection (2) above must be brought within 2 months of the date of issue of the award.

(4) The Secretary of State may by regulations made by statutory instrument subject to annulment in pursuance of a resolution of either House of Parliament make such provision as he thinks desirable for expediting, or reducing the expenses of, proceedings on arbitrations under this Act.

(5) The Secretary of State shall not make regulations under subsection (4) above which are inconsistent with the provisions of Schedule 7 to this Act.

(6) Section 62 of this Act shall apply to the determination by arbitration of any claims which arise—

(a) under this Act or any custom or agreement, and

(b) on or out of the termination of the tenancy of an agricultural holding or part thereof.

(7) This section and section 60 of this Act shall not apply to valuations of sheep stocks, dung, fallow, straw, crops, fences and other specific things the property of an outgoing tenant, agreed under a lease to be taken over from him at the termination of a tenancy by the landlord or the incoming tenant, or to any questions which it may be necessary to determine in order to ascertain the sum to be paid in pursuance of such an agreement, whether such valuations and questions are referred to arbitration under the lease or not.

(8) Any valuation or question mentioned in subsection (7) above falling to be decided by reference to a date after May 16, 1975, which would, if it had fallen to be decided by reference to a date immediately before that day, have been decided by reference to fiars prices, shall be decided in such manner as the parties may by agreement determine or, failing such agreement, shall, notwithstanding the provisions of that subsection, be decided by arbitration under this Act.

62. Claims on termination of tenancy

(1) Without prejudice to any other provision of this Act, any claim by a tenant of an agricultural holding against his landlord or by a landlord of an agricultural holding against his tenant, being a claim which arises, under this Act or under any custom or agreement, on or out of the termination of the tenancy (or of part thereof) shall, subject to subsections (2) to (5) below, be determined by arbitration.

(2) Without prejudice to any other provision of this Act, no claim to which this section applies shall be enforceable unless before the expiry of 2 months after the termination of the tenancy the claimant has given notice in writing to his landlord or his tenant, as the case may be, of his intention to make the claim.

(3) A notice under subsection (2) above shall specify the nature of the claim, and it shall be a sufficient specification thereof if the notice refers to the statutory provision, custom, or term of an agreement under which the claim is made.

(4) The landlord and the tenant may within 4 months after the termination of the tenancy by agreement in writing settle any such claim and the Secretary of State may upon the application of the landlord or the tenant made within that period extend the said period by 2 months and, on a second such application made during these 2 months, by a further 2 months.

(5) Where before the expiry of the period referred to in subsection (4) above and any extension thereof under that subsection any such claim has not been settled, the claim shall cease to be enforceable unless before the expiry of one month after the end of the said period and any such extension, or such longer time as the Secretary of State may in special circumstances allow, an arbiter has been appointed by agreement between the landlord and the tenant under this Act or an application for the appointment of an

arbiter under those provisions has been made by the landlord or the tenant.

(6) Where a tenant lawfully remains in occupation of part of an agricultural holding after the termination of a tenancy, references in subsections (2) and (4) above to the termination of the tenancy thereof shall be construed as references to the termination of the occupation.

63. Panel of arbiters, and remuneration of arbiter

(1) Such number of persons as may be appointed by the Lord President of the Court of Session, after consultation with the Secretary of State, shall form a panel of persons from whom any arbiter appointed, otherwise than by agreement, for the purposes of this Act shall be selected.

(2) The panel of arbiters constituted under subsection (1) above shall be subject to revision by the Lord President of the Court of Session, after consultation with the Secretary of State, at such intervals not exceeding 5 years, as the Lord President and the Secretary of State may from time to time agree.

(3)—
  (a) the remuneration of an arbiter appointed by the Secretary of State under Schedule 7 to this Act shall be such amount as is fixed by the Secretary of State;
  (b) the remuneration of an arbiter appointed by the parties to an arbitration under this Act shall, in default of agreement between those parties and the arbiter, be such amount as, on the application of the arbiter or of either of the parties, is fixed by the auditor of the sheriff court, subject to appeal to the sheriff;
  (c) the remuneration of an arbiter, when agreed or fixed under this subsection, shall be recoverable by the arbiter as a debt due from either of the parties;
  (d) any amount paid in respect of the remuneration of the arbiter by either of the parties in excess of the amount (if any) directed by the award to be paid by that party in respect of the expenses of the award shall be recoverable from the other party.

64. Appointment of arbiter in cases where Secretary of State is a party

Where the Secretary of State is a party to any question or difference which under this Act is to be determined by arbitration or by an arbiter appointed in accordance with this Act, the arbiter shall, in lieu of being appointed by the Secretary of State, be appointed by the Land Court, and the remuneration of the arbiter so appointed shall be such amount as may be fixed by the Land Court.

65. Recovery of compensation and other sums due

Any award or agreement under this Act as to compensation, expenses or otherwise may, if any sum payable thereunder is not paid within one month after the date on which it becomes payable, be recorded for execution in the Books of Council and Session or in the sheriff court books, and shall be enforceable in like manner as a recorded decree arbitral.

66. Power to enable demand to remedy a breach to be modified on arbitration

(1) Where a question or difference required by section 60 of this Act to be determined by arbitration relates to a demand in writing served on a tenant by a landlord requiring the tenant to remedy a breach of any term or condition of his tenancy by the doing of any work of provision, repair, maintenance or replacement of fixed equipment, the arbiter may—
  (a) in relation to all or any of the items specified in the demand, whether or not any period is specified as the period within which the breach should be remedied, specify such period for that purpose as appears in all the circumstances to the arbiter to be reasonable;
  (b) delete from the demand any item or part of an item which, having due regard to the interests of good husbandry as respects the holding and of sound-manage-

ment of the estate of which the holding forms part or which the holding constitutes, the arbiter is satisfied is unnecessary or unjustified;

(c) substitute, in the case of any item or part of an item specified in the demand, a
different method or material for the method or material which the demand would
otherwise require to be followed or used where, having regard to the purpose
which that item or part is intended to achieve, the arbiter is satisfied that—

   (i) the latter method or material would involve undue difficulty or expense,

   (ii) the first-mentioned method or material would be substantially as
effective for the purpose, and

   (iii) in all the circumstances the substitution is justified.

(2) Where under subsection (1)(a) above an arbiter specifies a period within which a
breach should be remedied or the period for remedying a breach is extended by virtue
of subsection (4) below, the Land Court may, on the application of the arbiter or the
landlord, specify a date for the termination of the tenancy by notice to quit in the event
of the tenant's failure to remedy the breach within that period, being a date not earlier
than whichever of the two following dates is the later, that is to say—

(a) the date on which the tenancy could have been terminated by notice to quit
served on the expiry of the period originally specified in the demand, or if no such
period is so specified, on the date of the giving of the demand, or

(b) 6 months after the expiry of the period specified by the arbiter or, as the case may
be, of the extended period.

(3) A notice to quit on a date specified in accordance with subsection (2) above shall
be served on the tenant within one month after the expiry of the period specified by the
arbiter or the extended time, and shall be valid notwithstanding that it is served less
than 12 months before the date on which the tenancy is to be terminated or that that
date is not the end of a year of the tenancy.

(4) Where—

(a) notice to quit to which 22(2)(d) of this Act applies is stated to be given by reason
of the tenant's failure to remedy within the period specified in the demand a
breach of any term or condition of his tenancy by the doing of any work of provision, repair, maintenance or replacement of fixed equipment, or within that
period as extended by the landlord or the arbiter; and

(b) it appears to the arbiter on an arbitration required by notice under section 23(2)
of this Act that, notwithstanding that the period originally specified or extended
was reasonable, it would, in consequence of any happening before the expiry of
that period, have been unreasonable to require the tenant to remedy the breach
within that period;

the arbiter may treat the period as having been extended or further extended and make
his award as if the period had not expired; and where the breach has not been remedied
at the date of the award, the arbiter may extend the period as he considers reasonable,
having regard to the length of period which has elapsed since the service of the demand.

67. Prohibition of appeal to sheriff principal

Where jurisdiction is conferred by this Act on the sheriff, there shall be no appeal to the
sheriff principal.

Sheep stock valuation

68.—(1) This section and sections 69 to 72 of this Act shall apply where under a lease
of an agricultural holding, the tenant is required at the termination of the tenancy to
leave the stock of sheep on the holding to be taken over by the landlord or by the incoming tenant at a price or valuation to be fixed by arbitration, referred to in this section
and sections 69 to 72 of this Act as a 'sheep stock valuation.'

(2) In a sheep stock valuation where the lease was entered into before or on
November 6, 1946, the arbiter shall in his award show the basis of valuation of each

class of stock and state separately any amounts included in respect of acclimatisation or hefting or of any other consideration or factor for which he has made special allowance.

(3) In a sheep stock valuation where the lease was entered into after November 6, 1946, the arbiter shall fix the value of the sheep stock in accordance—

   (a) in the case of a valuation made in respect of a tenancy terminating at Whitsunday in any year, with Part I of Schedule 9 to this Act if the lease was entered into before December 1, 1986, otherwise with Part I of Schedule 10 to this Act; or

   (b) in the case of a valuation made in respect of a tenancy terminating at Martinmas in any year, with the provisions of Part II of Schedule 9 to this Act, if the lease was entered into before December 1, 1986, otherwise with Part II of Schedule 10 to this Act,

and subsection (2) above shall apply in such a case as if for the words from 'show the basis' to the end of the subsection there were substituted the words 'state separately the particulars set forth in Part III of Schedule 9 (or, as the case may be, Schedule 10) to this Act.'

(4) Where an arbiter fails to comply with any requirement of subsection (2) or (3) above, his award may be set aside by the sheriff.

(5) The Secretary of State may, by order made by statutory instrument subject to annulment in pursuance of a resolution of either House of Parliament, vary the provisions of Schedule 10 to this Act, in relation to sheep stock valuations under leases entered into on or after the date of commencement of the order.

69. Submission of questions of law for decision of sheriff

(1) In a sheep stock valuation where the lease was entered into after June 10, 1937 the arbiter may, at any stage of the proceedings, and shall, if so directed by the sheriff (which direction may be given on the application of either party) submit, in the form of a stated case for the decision of the sheriff, any question of law arising in the course of the arbitration.

(2) The decision of the sheriff on questions submitted under subsection (1) above shall be final unless, within such time and in accordance with such conditions as may be prescribed by Act of Sederunt, either party appeals to the Court of Session, from whose decision no appeal shall lie.

(3) Where a question is submitted under subsection (1) above for the decision of the sheriff, and the arbiter is satisfied that, whatever the decision on the question may be, the sum ultimately to be found due will be not less than a particular amount, it shall be lawful for the arbiter, pending the decision of such question, to make an order directing payment to the outgoing tenant of such sum, not exceeding that amount, as the arbiter may think fit, to account of the sum that may ultimately be awarded.

70. Determination by Land Court of questions as to value of sheep stock

(1) Any question which would fall to be decided by a sheep stock valuation—

   (a) where the lease was entered into before or on November 6, 1946 may, on the joint application of the parties; and

   (b) where the lease was entered into after that date shall, on the application of either party,

in lieu of being determined in the manner provided in the lease, be determined by the Land Court.

(2) The Land Court shall determine any question or difference which they are required to determine, in a case where subsection (1)(b) above applies, in accordance with the appropriate provisions—

   (a) where the lease was entered into before December 1, 1986, of Schedule 9 to this Act;

   (b) where the lease was entered into on or after that date, of Schedule 10 to this Act.

71. Statement of sales of stock

(1) Where any question as to the value of any sheep stock has been submitted for determination to the Land Court or to an arbiter, the outgoing tenant shall, not less than 28 days before the determination of the question, submit to the Court or to the arbiter, as the case may be—

(a) a statement of the sales of sheep from such stock—

(i) in the case of a valuation made in respect of a tenancy terminating at Whitsunday during the preceding three years; or

(ii) in the case of a valuation made in respect of a tenancy terminating at Martinmas during the current year and in each of the two preceding years; and

(b) such sale-notes and other evidence as may be required by the Court or the arbiter to vouch the accuracy of such statement.

(2) Any document submitted by the outgoing tenant in pursuance of this section shall be open to inspection by the other party to the valuation proceedings.

72. Interpretation of sections 68 to 71

In sections 68 to 71 of this Act—

(a) 'agricultural holding' means a piece of land held by a tenant which is wholly or in part pastoral, and which is not let to the tenant during and in connection with his continuance in any office, appointment, or employment held under the landlord;

(b) 'arbiter' includes an oversman and any person required to determine the value or price of sheep stock in pursuance of any provision in the lease of an agricultural holding, and 'arbitration' shall be construed accordingly; and

(c) 'sheep stock valuation' shall be construed in accordance with section 68(1) of this Act.

PART VIII

MISCELLANEOUS

73. Power of Secretary of State to vary Schedules 5 and 6

(1) The Secretary of State may, after consultation with persons appearing to him to represent the interests of landlords and tenants of agricultural holdings, by order vary the provisions of Schedules 5 and 6 to this Act.

(2) An order under this section may make such provision as to the operation of this Act in relation to tenancies current when the order takes effect as appears to the Secretary of State to be just having regard to the variation of the said Schedules effected by the order.

(3) Nothing in any order made under this section shall affect the right of a tenant to claim, in respect of an improvement made or begun before the date on which such order comes into force, any compensation to which, but for the making of the order, he would have been entitled.

(4) Orders under this section shall be made by statutory instrument which shall be of no effect unless approved by resolution of each House of Parliament.

74. Power of limited owners to give consents, etc

The landlord of an agricultural holding, whatever may be his estate or interest in the holding, may for the purposes of this Act give any consent, make any agreement, or do or have done to him any act which he might give or make or do or have done to him if he were the owner of the dominium utile of the holding.

75. Power of tenant and landlord to obtain charge on holding

(1) Where any sum has become payable to the tenant of an agricultural holding in respect of compensation by the landlord and the landlord has failed to discharge his liability therefor within one month after the date on which the sum became payable, the Secretary of State may, on the application of the tenant and after giving not less than 14 days' notice of his intention so to do to the landlord, create, where the landlord is the owner of the dominium utile of the holding, a charge on the holding, or where the landlord is the lessee of the holding under a lease recorded under the Registration of Leases (Scotland) Act 1857 a charge on the lease for the payment of the sum due.

(2) For the purpose of creating a charge of a kind referred to in subsection (1) above, the Secretary of State may make in favour of the tenant a charging order charging and burdening the holding or the lease, as the case may be, with an annuity to repay the sum due together with the expenses of obtaining the charging order and recording it in the General Register of Sasines or registering it in the Land Register of Scotland.

(3) Where the landlord of an agricultural holding, not being the owner of the dominium utile of the holding, has paid to the tenant of the holding the amount due to him under this Act, or under custom or agreement, or otherwise, in respect of compensation for an improvement or in respect of compensation for disturbance, or has himself defrayed the cost of an improvement proposed to be executed by the tenant, the Secretary of State may, on the application of the landlord and after giving not less than 14 days notice to the absolute owner of the holding, make in favour of the landlord a charging order charging and burdening the holding with an annuity to repay the amount of the compensation or of the cost of the improvement, as the case may be, together with the expenses of obtaining the charging order and recording it in the General Register of Sasines or registering it in the Land Register of Scotland.

(4) Section 65(2), (4) and (6) to (10) of the Water (Scotland) Act 1980 shall, with the following and any other necessary modifications, apply to any such charging order as is mentioned in subsection (2) or (3) above, that is to say—

(a) for any reference to an islands or district council there shall be substituted a reference to the Secretary of State;

(b) for any reference to the period of 30 years there shall be substituted—

    (i) where subsection (1) above applies, a reference to such period (not exceeding 30 years) as the Secretary of State may determine;

    (ii) in the case of a charging order made in respect of compensation for, or of the cost of, an improvement, a reference to the period within which the improvement will, in the opinion of the Secretary of State, have become exhausted;

(c) for references to Part V of the said Act of 1980 there shall be substituted references to this Act.

(5) Where subsection (3) above applies, an annuity constituted a charge by a charging order recorded in the General Register of Sasines or registered in the Land Register of Scotland shall be a charge on the holding specified in the order and shall rank after all prior charges heritably secured thereon.

(6) The creation of a charge on a holding under this section shall not be deemed to be a contravention of any prohibition against charging or burdening contained in the deed or instrument under which the holding is held.

76. Power of land improvement companies to advance money

Any company incorporated by Parliament or incorporated under the Companies Act 1985 or under the former Companies Acts within the meaning of that Act and having power to advance money for the improvement of land, or for the cultivation and farming of land, may make an advance of money upon a charging order duly made and recorded or registered under this Act, on such terms and conditions as may be agreed upon between the company and the person entitled to the order.

77. Appointment of guardian to landlord or tenant

Where the landlord or the tenant of an agricultural holding is a pupil or a minor or is of unsound mind, not having a tutor, curator or other guardian, the sheriff, on the application of any person interested, may appoint to him, for the purposes of this Act, a tutor or a curator, and may recall the appointment and appoint another tutor or curator if and as occasion requires.

78. Validity of consents, etc

It shall be no objection to any consent in writing or agreement in writing under this Act signed by the parties thereto or by any persons authorised by them that the consent or agreement has not been executed in accordance with the enactments regulating the execution of deeds in Scotland.

PART IX

SUPPLEMENTARY

Crown and Secretary of State

79. Application to Crown land

(1) This Act shall apply to land belonging to Her Majesty in right of the Crown, with such modifications as may be prescribed; and for the purposes of this Act the Crown Estate Commissioners or other proper officer or body having charge of the land for the time being, or if there is no such officer or body, such person as Her Majesty may appoint in writing under the Royal Sign Manual, shall represent Her Majesty and shall be deemed to be the landlord.

(2) This Act shall apply to land notwithstanding that the interest of the landlord or the tenant thereof belongs to a government department or is held on behalf of Her Majesty for the purposes of any government department with such modifications as may be prescribed.

80. Determination of matters where Secretary of State is landlord or tenant

(1) This section applies where the Secretary of State is the landlord or the tenant of an agricultural holding.

(2) Where this section applies, any provision of this Act—

(a) under which any matter relating to the holding is referred to the decision of the Secretary of State; or

(b) relating to an arbitration concerning the holding,

shall have effect with the substitution for every reference to 'the Secretary of State' of a reference to 'the Land Court,' and any provision referred to in paragraph (a) above which provides for an appeal to an arbiter from the decision of the Secretary of State shall not apply.

81. Expenses and receipts

(1) All expenses incurred by the Secretary of State under this Act shall be paid out of moneys provided by Parliament.

(2) All sums received by the Secretary of State under this Act shall be paid into the Consolidated Fund.

82. Powers of entry and inspection

(1) Any person authorised by the Secretary of State in that behalf shall have power at all reasonable times to enter on and inspect any land for the purpose of determining

whether, and if so in what manner, any of the powers conferred on the Secretary of State by this Act are to be exercised in relation to the land, or whether, and if so in what manner, any direction given under any such power has been complied with.

(2) Any person authorised by the Secretary of State who proposes to exercise any power of entry or inspection conferred by this Act shall, if so required, produce some duly authenticated document showing his authority to exercise the power.

(3) Admission to any land used for residential purposes shall not be demanded as of right in the exercise of any such power unless 24 hours notice of the intended entry has been given to the occupier of the land.

(4) Save as provided by subsection (3) above, admission to any land shall not be demanded as of right in the exercise of any such power unless notice has been given to the occupier of the land that it is proposed to enter during a period, specified in the notice, not exceeding 14 days and beginning at least 24 hours after the giving of the notice and the entry is made on the land during the period specified in the notice.

(5) Any person who obstructs a person authorised by the Secretary of State exercising any such power shall be guilty of an offence and shall be liable on summary conviction to a fine not exceeding level 2 on the standard scale.

Land Court

83. Proceedings of the Land Court

The provisions of the Small Landholders (Scotland) Acts 1886 to 1931 relating to the Land Court shall apply, with any necessary modifications, for the purposes of the determination by the Land Court of any matter referred to them under this Act, as they apply for the purposes of the determination of matters referred to them under those Acts.

Service of notices

84. Service of notices, etc

(1) Any notice or other document required or authorised by or under this Act to be given to or served on any person shall be duly given or served if it is delivered to him, or left at his proper address, or sent to him by registered post or recorded delivery.

(2) Any such document required or authorised to be given to or served on an incorporated company or body shall be duly given or served if it is delivered to or sent by registered post or recorded delivery to the registered office of the company or body.

(3) For the purposes of this section and of section 7 of the Interpretation Act 1978, the proper address of any person to or on whom any such document as aforesaid is to be given or served shall, in the case of the secretary or clerk of any incorporated company or body, be that of the registered or principal office of the company or body, and in any other case be the last known address of the person in question.

(4) Unless or until the tenant of an agricultural holding shall have received notice that the person previously entitled to receive the rents and profits of the holding (hereinafter referred to as 'the original landlord') has ceased to be so entitled, and also notice of the name and address of the person who has become so entitled, any notice or other document served on or delivered to the original landlord by the tenant shall be deemed to have been served on or delivered to the landlord of the holding.

Interpretation

85. Interpretation

(1) In this Act, unless the context otherwise requires—
'the 1911 Act' means the Small Landholders (Scotland) Act 1911;
'the 1949 Act' means the Agricultural Holdings (Scotland) Act 1949;
'agricultural holding' (except in sections 68 to 72 of this Act) and 'agricultural land' have the meanings assigned to them by section 1 of this Act;

'agricultural unit' means land which is an agricultural unit for the purposes of the Agriculture (Scotland) Act 1948;

'agriculture' includes horticulture, fruit growing; seed growing; dairy farming; livestock breeding and keeping; the use of land as grazing land, meadow land, osier land, market gardens and nursery grounds; and the use of land for woodlands where that use is ancillary to the farming of land for other agricultural purposes: and 'agricultural' shall be construed accordingly;

'building' includes any part of a building;

'fixed equipment' includes any building or structure affixed to land and any works on, in, over or under land, and also includes anything grown on land for a purpose other than use after severance from the land, consumption of the thing grown or of produce thereof, or amenity, and, without prejudice to the foregoing generality, includes the following things, that is to say—

(a) all permanent buildings, including farm houses and farm cottages, necessary for the proper conduct of the agricultural holding;

(b) all permanent fences, including hedges, stone dykes, gate posts and gates;

(c) all ditches, open drains and tile drains, conduits and culverts, ponds, sluices, flood banks and main water courses;

(d) stells, fanks, folds, dippers, pens and bughts necessary for the proper conduct of the holding;

(e) farm access or service roads, bridges and fords;

(f) water and sewerage systems;

(g) electrical installations including generating plant, fixed motors, wiring systems, switches and plug sockets;

(h) shelter belts,

and references to fixed equipment on land shall be construed accordingly;

'improvement' shall be construed in accordance with section 33 of this Act, and 'new improvement,' 'old improvement,' '1923 Act improvement' and '1931 Act improvement' have the meanings there assigned to them;

'Land Court' means the Scottish Land Court;

'Lands Tribunal' means the Lands Tribunal for Scotland;

'landlord' means any person for the time being entitled to receive the rents and profits or to take possession of an agricultural holding, and includes the executor, assignee, legatee, disponee, guardian, curator bonis, tutor, or permanent or interim trustee (within the meaning of the Bankruptcy (Scotland) Act 1985), of a landlord;

'lease' means a letting of land for a term of years, or for lives, or for lives and years, or from year to year;

'livestock' includes any creature kept for the production of food, wool, skins or fur, or for the purpose of its use in the farming of land;

'market garden' means a holding, cultivated, wholly or mainly, for the purpose of the trade or business of market gardening;

'prescribed' means prescribed by the Secretary of State by regulations made by statutory instrument which shall be subject to annulment in pursuance of a resolution of either House of Parliament;

'produce' includes anything (whether live or dead) produced in the course of agriculture;

'tenant' means the holder of land under a lease of an agricultural holding and includes the executor, assignee, legatee, disponee, guardian, tutor, curator bonis, or permanent or interim trustee (within the meaning of the Bankruptcy (Scotland) Act 1985) of a tenant;

'termination,' in relation to a tenancy, means the termination of the lease by reason of effluxion of time or from any other cause;

(2) Schedules 5 and 6 to the Agriculture (Scotland) Act 1948, (which have effect respectively for the purpose of determining for the purposes of that Act whether the owner of agricultural land is fulfilling his responsibilities to manage it in accordance with the rules of good estate management and whether the occupier of such land is fulfilling his responsibilities to farm it in accordance with the rules of good husbandry) shall have effect for the purposes of this Act as they have effect for the purposes of that Act.

(3) References in this Act to the farming of land include references to the carrying on in relation to the land of any agricultural activity.

(4) References to the terms, conditions, or requirements of a lease of or of an agreement relating to, an agricultural holding shall be construed as including references to any obligations, conditions or liabilities implied by the custom of the country in respect of the holding.

(5) Anything which by or under this Act is required or authorised to be done by, to or in respect of the landlord or the tenant of an agricultural holding may be done by, to or in respect of any agent of the landlord or of the tenant.

86. Construction of references in other Acts to holdings as defined by earlier Acts

References, in whatever terms, in any enactment, other than an enactment contained in—

this Act,
the Agricultural Holdings (Scotland) Acts 1923 and 1931, or,
Part I of the Agriculture (Scotland) Act 1948

to a holding within the meaning of the Agricultural Holdings (Scotland) Act 1923 or of the Agricultural Holdings (Scotland) Acts 1923 to 1948 shall be construed as references to an agricultural holding within the meaning of this Act.

87. Savings

Schedule 12 to this Act, which exempts from the operation of this Act certain cases current at the commencement of this Act and contains other transitional provisions and savings shall have effect.

Consequential amendments and repeals

88. Consequential amendments and repeals

(1) The enactments specified in Schedule 11 to this Act shall be amended in accordance with that Schedule.

(2) The enactments specified in Schedule 13 to this Act are repealed to the extent there specified.

Citation, commencement and extent

89. Citation, commencement and extent

(1) This Act may be cited as the Agricultural Holdings (Scotland) Act 1991.

(2) This Act shall come into force at the end of the period of 2 months beginning with the date on which it is passed.

(3) This Act shall extend to Scotland only, except for those provisions in Schedule 11 which amend enactments which extend to England and Wales or to Northern Ireland.

SCHEDULES

Section 4   # SCHEDULE 1

PROVISIONS REQUIRED IN LEASES

1. The names of the parties.
2. Particulars of the holding with sufficient description, by reference to a map or plan, of the fields and other parcels of land comprised therein to identify the extent of the holding.

3. The term or terms for which the holding or different parts thereof is or are agreed to be let.

4. The rent and the dates on which it is payable.

5. An undertaking by the landlord in the event of damage by fire to any building comprised in the holding to reinstate or replace the building if its reinstatement or replacement is required for the fulfilment of his responsibilities to manage the holding in accordance with the rules of good estate management, and (except where the interest of the landlord is held for the purposes of a government department or a person representing Her Majesty under section 79 of this Act is deemed to be the landlord, or where the landlord has made provision approved by the Secretary of State for defraying the cost of any such reinstatement or replacement) an undertaking by the landlord to insure to their full value all such buildings against damage by fire.

6. An undertaking by the tenant, in the event of the destruction by fire of harvested crops grown on the holding for consumption thereon, to return to the holding the full equivalent manurial value of the crops destroyed, in so far as the return thereof is required for the fulfilment of his responsibilities to farm in accordance with the rules of good husbandry, and (except where the interest of the tenant is held for the purposes of a government department or where the tenant has made provision approved by the Secretary of State in lieu of such insurance) an undertaking by the tenant to insure to their full value all dead stock on the holding and all such harvested crops against damage by fire.

Section 25   SCHEDULE 2

GROUNDS FOR CONSENT TO OPERATION OF NOTICES TO QUIT A TENANCY WHERE SECTION 25(3) APPLIES

PART I

GROUNDS FOR CONSENT TO OPERATION OF NOTICE TO QUIT A TENANCY LET BEFORE JANUARY 1, 1984

Case 1

The tenant has neither sufficient training in agriculture nor sufficient experience in the farming of land to enable him to farm the holding with reasonable efficiency.

Case 2

(a) The holding or any agricultural unit of which it forms part is not a two-man unit;

(b) the landlord intends to use the holding for the purpose of effecting an amalgamation within 2 years after the termination of the tenancy; and

(c) the notice specifies the land with which the holding is to be amalgamated.

Case 3

The tenant is the occupier (either as owner or tenant) of agricultural land which—

(a) is a two-man unit;

(b) is distinct from the holding and from any agricultural unit of which the holding forms part; and

(c) has been occupied by him since before the death of the person from whom he acquired right to the lease of the holding;

and the notice specifies the agricultural land.

PART II

GROUNDS FOR CONSENT TO OPERATION OF NOTICE TO QUIT A TENANCY LET ON OR AFTER JANUARY 1, 1984

Case 4

The tenant does not have sufficient financial resources to enable him to farm the holding with reasonable efficiency.

Case 5

The tenant has neither sufficient training in agriculture nor sufficient experience in the farming of land to enable him to farm the holding with reasonable efficiency:

Provided that this Case shall not apply where the tenant has been engaged, throughout the period from the date of death of the person from whom he acquired right to the lease, in a course of relevant training in agriculture which he is expected to complete satisfactorily within 4 years from the said date, and has made arrangements to secure that the holding will be farmed with reasonable efficiency until he completes that course.

Case 6

(a) The holding or any agricultural unit of which it forms part is not a two-man unit;
(b) the landlord intends to use the holding for the purpose of effecting an amalgamation within 2 years after the termination of the tenancy; and
(c) the notice specifies the land with which the holding is to be amalgamated.

Case 7

The tenant is the occupier (either as owner or tenant) of agricultural land which—
(a) is a two-man unit;
(b) is distinct from the holding; and
(c) has been occupied by him throughout the period from the date of giving of the notice;
and the notice specifies the land.

PART III

SUPPLEMENTARY

1. For the purposes of section 25 of this Act and this Schedule—
'amalgamation' means a transaction for securing that agricultural land which is comprised in a holding to which a notice to quit relates and which together with other agricultural land could form an agricultural unit, shall be owned and occupied in conjunction with that other land (and cognate expressions shall be construed accordingly);
'near relative' in relation to a deceased tenant of an agricultural holding means a surviving spouse or child of that tenant, including a child adopted by him in pursuance of an adoption order (as defined in section 23(5) of the Succession (Scotland) Act 1964); and
'two-man unit' means an agricultural unit which in the opinion of the Land Court is capable of providing full-time employment for an individual occupying it and at least one other man.

2. For the purposes of determining whether land is a two-man unit, in assessing the capability of the unit of providing employment it shall be assumed that the unit is farmed under reasonably skilled management, that a system of husbandry suitable for the district is followed and that the greater part of the feeding stuffs required by any livestock kept on the unit is grown there.

3. For the purposes of Case 7 of this Schedule, occupation of agricultural land—

(a) by a company which is controlled by the tenant shall be treated as occupation by the tenant; and

(b) by a Scottish partnership shall, notwithstanding section 4(2) of the Partnership Act 1890, be treated as occupation by each of its partners.

Section 33

SCHEDULE 3

1923 ACT IMPROVEMENTS FOR WHICH COMPENSATION MAY BE PAYABLE

PART I

IMPROVEMENTS FOR WHICH CONSENTS REQUIRED

1. Erection, alteration, or enlargement of buildings.
2. Formation of silos.
3. Laying down of permanent pasture.
4. Making and planting of osier beds.
5. Making of water meadows or works of irrigation.
6. Making of gardens.
7. Making or improvement of roads or bridges.
8. Making or improvement of watercourses, ponds, wells, or reservoirs, or of works for the application of water power or for supply of water for agricultural or domestic purposes.
9. Making or removal of permanent fences.
10. Planting of hops.
11. Planting of orchards or fruit bushes.
12. Protecting young fruit trees.
13. Reclaiming of waste land.
14. Warping or weiring of land.
15. Embankments and sluices against floods.
16. Erection of wirework in hop gardens.
17. Provision of permanent sheep dipping accommodation.
18. In the case of arable land, the removal of bracken, gorse, tree roots, boulders, or other like obstructions to cultivation.

PART II

IMPROVEMENTS FOR WHICH NOTICE REQUIRED

19. Drainage.

PART III

IMPROVEMENTS FOR WHICH NO CONSENTS OR NOTICE REQUIRED

20. Chalking of land.
21. Clay-burning.
22. Claying of land or spreading blaes upon land.
23. Liming of land.
24. Marling of land.
25. Application to land of purchased artificial or other manure.
26. Consumption on the holding by cattle, sheep, or pigs, or by horses other than those regularly employed on the holding, of corn, cake, or other feeding stuff not produced on the holding.
27. Consumption on the holding by cattle, sheep, or pigs, or by horses other than

those regularly employed on the holding, of corn proved by satisfactory evidence to have been produced and consumed on the holding.

28. Laying down temporary pasture with clover, grass, lucerne, sainfoin, or other seeds, sown more than 2 years prior to the termination of the tenancy, in so far as the value of the temporary pasture on the holding at the time of quitting exceeds the value of the temporary pasture on the holding at the commencement of the tenancy for which the tenant did not pay compensation.

29. Repairs to buildings, being buildings necessary for the proper cultivation or working of the holding, other than repairs which the tenant is himself under an obligation to execute.

Section 33   SCHEDULE 4

1931 ACT IMPROVEMENTS FOR WHICH COMPENSATION MAY BE PAYABLE

PART I

IMPROVEMENTS FOR WHICH CONSENT REQUIRED

1. Erection, alteration, or enlargement of buildings.
2. Laying down of permanent pasture.
3. Making and planting of osier beds.
4. Making of water meadows or works of irrigation.
5. Making of gardens.
6. Planting of orchards or fruit bushes.
7. Protecting young fruit trees.
8. Warping or weiring of land.
9. Making of embankments and sluices against floods.

PART II

IMPROVEMENTS OF WHICH NOTICE REQUIRED

10. Drainage.
11. Formation of silos.
12. Making or improvement of roads or bridges.
13. Making or improvement of watercourses, ponds or wells, or of works for the application of water power or for the supply of water for agricultural or domestic purposes.
14. Making or removal of permanent fences.
15. Reclaiming of waste land.
16. Repairing or renewal of embankments and sluices against floods.
17. Provision of sheep dipping accommodation.
18. Provision of electrical equipment other than moveable fittings and appliances.

PART III

IMPROVEMENTS FOR WHICH NO CONSENT OR NOTICE REQUIRED

19. Chalking of land.
20. Clay-burning.
21. Claying of land or spreading blaes upon land.
22. Liming of land.
23. Marling of land.
24. Eradication of bracken, whins, or gorse growing on the holding at the com-

mencement of a tenancy and in the case of arable land the removal of tree roots, boulders, stones or other like obstacles to cultivation.

25. Application to land of purchased artificial or other manure.

26. Consumption on the holding by cattle, sheep, or pigs, or by horses other than those regularly employed on the holding, of corn, cake, or other feeding stuff not produced on the holding.

27. Consumption on the holding by cattle, sheep, or pigs, or by horses other than those regularly employed on the holding, of corn proved by satisfactory evidence to have been produced and consumed on the holding.

28. Laying down temporary pasture with clover, grass, lucerne, sainfoin, or other seeds, sown more than 2 years prior to the termination of the tenancy, in so far as the value of the temporary pasture on the holding at the time of quitting exceeds the value of the temporary pasture on the holding at the commencement of the tenancy for which the tenant did not pay compensation.

29. Repairs to buildings, being buildings necessary for the proper cultivation or working of the holding, other than repairs which the tenant is himself under an obligation to execute.

Section 33   SCHEDULE 5

NEW IMPROVEMENTS FOR WHICH COMPENSATION MAY BE PAYABLE

PART I

IMPROVEMENTS FOR WHICH CONSENT IS REQUIRED

1. Laying down of permanent pasture.
2. Making of water-meadows or works of irrigation.
3. Making of gardens.
4. Planting of orchards or fruit bushes.
5. Warping or weiring of land.
6. Making of embankments and sluices against floods.
7. Making or planting of osier beds.
8. Haulage or other work done by the tenant in aid of the carrying out of any improvement made by the landlord for which the tenant is liable to pay increased rent.

PART II

IMPROVEMENTS FOR WHICH NOTICE IS REQUIRED

9. Land drainage.
10. Construction of silos.
11. Making or improvement of farm access or service roads, bridges and fords.
12. Making or improvement of watercourses, ponds or wells, or of works for the application of water power for agricultural or domestic purposes or for the supply of water for such purposes.
13. Making or removal of permanent fences, including hedges, stone dykes and gates.
14. Reclaiming of waste land.
15. Renewal of embankments and sluices against floods.
16. Provision of stells, fanks, folds, dippers, pens and bughts necessary for the proper conduct of the holding.
17. Provision or laying on of electric light or power, including the provision of generating plant, fixed motors, wiring systems, switches and plug sockets.
18. Erection, alteration or enlargement of buildings, making or improvement of permanent yards, loading banks and stocks and works of a kind referred to in paragraph

13(2) of Schedule 8 to the Housing (Scotland) Act 1987 (subject to the restrictions mentioned in that subsection).

19. Erection of hay or sheaf sheds, sheaf or grain drying racks, and implement sheds.

20. Provision of fixed threshing mills, barn machinery and fixed dairying plant.

21. Improvement of permanent pasture by cultivation and re-seeding.

22. Provision of means of sewage disposal.

23. Repairs to fixed equipment, being equipment reasonably required for the efficient farming of the holding, other than repairs which the tenant is under an obligation to carry out.

PART III

IMPROVEMENTS FOR WHICH NO CONSENT OR NOTICE REQUIRED

24. Protecting fruit trees against animals.

25. Clay burning.

26. Claying of land.

27. Liming (including chalking) of land.

28. Marling of land.

29. Eradication of bracken, whins or broom growing on the holding at the commencement of the tenancy and, in the case of arable land, removal of tree roots, boulders, stones or other like obstacles to cultivation.

30. Application to land of purchased manure and fertiliser, whether organic or inorganic.

31. Consumption on the holding of corn (whether produced on the holding or not) or of cake or other feeding stuff not produced on the holding by horses, cattle, sheep, pigs or poultry.

32. Laying down temporary pasture with clover, grass, lucerne, sainfoin, or other seeds, sown more than 2 years prior to the termination of the tenancy, in so far as the value of the temporary pasture on the holding at the time of quitting exceeds the value of the temporary pasture on the holding at the commencement of the tenancy for which the tenant did not pay compensation.

Section 40   # SCHEDULE 6

MARKET GARDEN IMPROVEMENTS

1. Planting of fruit trees or bushes permanently set out.

2. Planting of strawberry plants.

3. Planting of asparagus, rhubarb, and other vegetable crops which continue productive for 2 or more years.

4. Erection, alteration or enlargement of buildings for the purpose of the trade or business of a market gardener.

Section 61   # SCHEDULE 7

ARBITRATIONS

APPOINTMENT OF ARBITERS

1. A person agreed upon between the parties or, in default of agreement, appointed on the application in writing of either of the parties by the Secretary of State from among the members of the panel constituted under this Act for the purpose, shall be appointed arbiter.

2. If a person appointed arbiter dies, or is incapable of acting, or for 7 days after notice from either party requiring him to act fails to act, a new arbiter may be appointed as if no arbiter had been appointed.

3. Neither party shall have the power to revoke the appointment of the arbiter without the consent of the other party.

4. An appointment, notice, revocation and consent of a kind referred to in any of paragraphs 1 to 3 of this Schedule must be in writing.

PARTICULARS OF CLAIM

5. Each of the parties to the arbitration shall, within 28 days from the appointment of the arbiter, deliver to him a statement of that party's case with all necessary particulars; and—

(a) no amendment or addition to the statement or particulars delivered shall be allowed after the expiration of the said 28 days except with the consent of the arbiter;

(b) a party to the arbitration shall be confined at the hearing to the matters alleged in the statement and particulars so delivered and any amendment thereof or addition thereto duly made.

EVIDENCE

6. The parties to the arbitration, and all persons claiming through them respectively, shall, subject to any legal objection—

(a) submit to be examined by the arbiter on oath or affirmation in relation to the matters in dispute; and

(b) produce before the arbiter;

all samples, books, deeds, papers, accounts, writings, and documents, within their possession or power respectively which may be required or called for, and do all other things which during the proceedings the arbiter may require.

7. The arbiter shall have power to administer oaths, and to take the affirmation of parties and witnesses appearing, and witnesses shall, if the arbiter thinks fit, be examined on oath or affirmation.

AWARD

8. The arbiter shall make and sign his award within 3 months of his appointment or within such longer period as may, either before or after the expiry of the aforesaid period be agreed to in writing by the parties, or be fixed by the Secretary of State.

9. The arbiter may, if he thinks fit, make an interim award for the payment of any sum on account of the sum to be finally awarded.

10. An arbiter appointed by the Secretary of State or the Land Court in an arbitration under section 13(1) of this Act shall, in making his award, state in writing his findings of fact and the reasons for his decision and shall make that statement available to the Secretary of State and to the parties.

11. The award and any statement made under paragraph 10 of this Schedule shall be in such form as may be specified by statutory instrument made by the Secretary of State.

12. The arbiter shall—

(a) state separately in his award the amounts awarded in respect of the several claims referred to him; and

(b) on the application of either party, specify the amount awarded in respect of any particular improvement or any particular matter which is the subject of the award.

13. Where by virtue of this Act compensation under an agreement is to be substituted for compensation under this Act for improvements, the arbiter shall award

compensation in accordance with the agreement instead of in accordance with this Act.

14. The award shall fix a day not later than one month after delivery of the award for the payment of the money awarded as compensation, expenses or otherwise.

15. Subject to section 61(2) of this Act, the award shall be final and binding on the parties and the persons claiming under them respectively.

16. The arbiter may correct in an award any clerical mistake or error arising from any accidental slip or omission.

EXPENSES

17. The expenses of and incidental to the arbitration and award shall be in the discretion of the arbiter, who may direct to and by whom and in what manner those expenses or any part thereof are to be paid, and the expenses shall be subject to taxation by the auditor of the sheriff court on the application of either party, but that taxation shall be subject to review by the sheriff.

18. The arbiter shall, in awarding expenses, take into consideration the reasonableness or unreasonableness of the claim of either party whether in respect of amount or otherwise, and any unreasonable demand for particulars or refusal to supply particulars, and generally all the circumstances of the case, and may disallow the expenses of any witness whom he considers to have been called unnecessarily and any other expenses which he considers to have been incurred unnecessarily.

19. It shall not be lawful to include in the expenses of and incidental to the arbitration and award, or to charge against any of the parties, any sum payable in respect of remuneration or expenses to any person appointed by the arbiter to act as clerk or otherwise to assist him in the arbitration unless such appointment was made after submission of the claim and answers to the arbiter and with either the consent of the parties to the arbitration or the sanction of the sheriff.

STATEMENT OF CASE

20. Subject to paragraph 22 of this Schedule, the arbiter may at any stage of the proceedings, and shall, if so directed by the sheriff (which direction may be given on the application of either party), state a case for the opinion of the sheriff on any question of law arising in the course of the arbitration.

21. Subject to paragraph 22 of this Schedule, the opinion of the sheriff on any case stated under the last foregoing paragraph shall be final unless, within such time and in accordance with such conditions as may be specified by act of sederunt, either party appeals to the Court of Session, from whose decision no appeal shall lie.

22. Where the arbiter in any arbitration under section 13(1) of this Act has been appointed by the Secretary of State or by the Land Court, paragraphs 20 and 21 of this Schedule shall not apply, and instead the arbiter may at any stage of the proceedings state a case (whether at the request of either party or on his own initiative) on any question of law arising in the course of the arbitration, for the opinion of the Land Court, whose decision shall be final.

REMOVAL OF ARBITER AND SETTING ASIDE OF AWARD

23. Where an arbiter has misconducted himself the sheriff may remove him.

24. When an arbiter has misconducted himself, or an arbitration or award has been improperly procured, the sheriff may set the award aside.

FORMS

25. Any forms for proceedings in arbitrations under this Act which may be specified by statutory instrument made by the Secretary of State shall, if used, be sufficient.

Section 57   SCHEDULE 8

SUPPLEMENTARY PROVISIONS WITH RESPECT TO PAYMENTS UNDER
SECTION 56

1. Subject to paragraph 4 of this Schedule, any dispute with respect to any sum which may be or become payable by virtue of section 56(1) of this Act shall be referred to and determined by the Lands Tribunal for Scotland.

2. If in any case the sum to be paid by virtue of the said section 56(1) to the tenant of an agricultural holding or to a statutory small tenant by an acquiring authority would, apart from this paragraph and paragraph 3 of this Schedule, fall to be ascertained in pursuance of section 54(2) of this Act by reference to the rent of the holding at a rate which was not—

(a) determined by arbitration under section 13 or 15 of this Act;

(b) determined by the Land Court in pursuance of section 61(2) of this Act; or

(c) in the case of a statutory small tenant, fixed by the Scottish Land Court in pursuance of section 32(7) and (8) of the 1911 Act;

and which the authority consider is unduly high, the authority may make an application to the Lands Tribunal for Scotland for the rent to be considered by the tribunal.

3. Where, on an application under paragraph 2 above, the tribunal are satisfied that—

(a) the rent to which the application relates is not substantially higher than the rent which in their opinion would be determined for the holding in question on a reference to arbitration duly made in pursuance of—

(i) section 13 of this Act; or

(ii) in the case of a statutory small tenancy, the equitable rent which in their opinion would be fixed by the Land Court under section 32(7) and (8) of the 1911 Act; (hereafter in this paragraph referred to as 'the appropriate rent'); or

(b) the rent to which the application relates is substantially higher than the appropriate rent but was not fixed by the parties to the relevant lease with a view to increasing the amount of any compensation payable, or of any sum to be paid by virtue of section 56(1) of this Act, in consequence of the compulsory acquisition or taking of possession of any land included in the holding,

they shall dismiss the application; and if the tribunal do not dismiss the application in pursuance of the foregoing provisions of this paragraph they shall determine that, in the case to which the application relates, the sum to be paid by virtue of section 56(1) of this Act shall be ascertained in pursuance of the said section 13 by reference to the appropriate rent instead of by reference to the rent to which the application relates.

4. For the purposes of paragraph 3(a) above, section 13(1) of this Act shall have effect as if for the reference therein to the next ensuing day there were substituted a reference to the date of the application referred to in paragraph 3(a) above.

5. The enactments mentioned in paragraph 6 of this Schedule shall, subject to any necessary modifications, have effect in their application to such an acquiring of an interest or taking of possession as is referred in section 56(1) of this Act (hereafter in this paragraph referred to as 'the relevant event')—

(a) in so far as those enactments make provision for the doing, before the relevant event, of any thing connected with compensation (including in particular provision for determining the amount of the liability to pay compensation or for the deposit of it in a Scottish bank or otherwise), as if references to compensation, except compensation for damage or injurious affection, included references to any sum which will become payable by virtue of section 56 of this Act in consequence of the relevant event; and

(b) subject to sub-paragraph (a) above, as if references to compensation (except compensation for damage or injurious affection) included references to sums payable or, as the context may require, to sums paid by virtue of section 56 of this Act in the consequence of the relevant event.

6. The enactments aforesaid are—

(a) sections 56 to 60, 62, 63 to 65, 67 to 70, 72, 74 to 79, 83 to 87, 114, 115 and 117 of the Lands Clauses (Scotland) Act 1845;

(b) paragraph 3 of Schedule 2 to the Acquisition of Land (Authorisation Procedure) (Scotland) Act 1947;

(c) Parts I and II and section 40 of the Land Compensation (Scotland) Act 1963;

(d) paragraph 4 of Schedule 6 to the New Towns (Scotland) Act 1968;

(e) any provision in any local or private Act, in any instrument having effect by virtue of an enactment, or in any order or scheme confirmed by Parliament or brought into operation in accordance with special parliamentary procedure, corresponding to a provision mentioned in sub-paragraph (a), (b) or (d) above.

Section 70

SCHEDULE 9

VALUATION OF SHEEP STOCK IN SCOTLAND IN RESPECT OF OLD LEASES

PART I

VALUATION MADE IN RESPECT OF A TENANCY TERMINATING AT WHITSUNDAY

1. The Land Court or the arbiter (in Part I and Part II of this Schedule referred to as 'the valuer') shall ascertain the number of, and the prices realised for, the ewes and the lambs sold off the hill from the stock under valuation at the autumn sales in each of the 3 preceding years, and shall determine by inspection the number of shotts present in the stock at that time of the valuation.

2. The valuer shall calculate an average price per ewe, and an average price per lamb, for the ewes and lambs sold as aforesaid for each of the 3 preceding years. In calculating the average price for any year the valuer shall disregard such number of ewes and lambs so sold in that year, being the ewes or lambs sold at the lowest prices, as bears the same proportion to the total number of ewes or lambs so sold in that year as the number of shotts as determined bears to the total number of ewes or lambs in the stock under valuation.

3. The valuer shall then ascertain the mean of the average prices so calculated for the 3 preceding years for ewes and for lambs, respectively. The figures so ascertained or ascertained, in a case to which paragraph 4 below applies, in accordance with that paragraph, are in this Part of this Schedule referred to as the '3-year average price for ewes' and the '3-year average price for lambs.'

4. In the case of any sheep stock in which the number of ewes or the number of lambs sold off the hill at the autumn sales during the preceding 3 years has been less than half the total number of ewes or of lambs sold, the 3-year average price for ewes or the 3-year average price for lambs, as the case may be, shall, in lieu of being ascertained by the valuer as aforesaid, be determined by the Land Court on the application of the parties; and the Land Court shall determine such prices by reference to the prices realised at such sales for ewes and for lambs respectively from similar stocks kept in the same district and under similar conditions.

5. The 3-year average price for ewes shall be subject to adjustment by the valuer within the limits of 20 per cent. (in the case of leases entered into before May 15, 1963, 50 pence) upwards or downwards as he may think proper having regard to the general condition of the stock under valuation and to the profit which the purchaser may reasonably expect it to earn. The resultant figure shall be the basis of the valuation of the ewes, and is in this Part of this Schedule referred to as the 'basic ewe value.'

The valuer shall similarly adjust the 3 year average price for lambs, and the resultant figure shall be the basis for the valuation of the lambs and is in this Part of this Schedule referred to as the 'basic lamb value.'

6. In making his award the valuer shall value the respective classes of stock in accordance with the following rules, that is to say—

(a) ewes of all ages (including gimmers) shall be valued at the basic ewe value with the addition of 30 per cent. (in the case of leases entered into before May 15, 1963, 75 pence) of such value per head;

(b) lambs shall be valued at the basic lamb value; so however that twin lambs shall be valued at such price as the valuer thinks proper;

(c) ewe hoggs shall be valued at two-thirds of the combined basic values of a ewe and a lamb subject to adjustment by the valuer within the limits of 10 per cent. (in the case of leases entered into before May 15, 1963, 25 pence) per head upwards or downwards as he may think proper, having regard to their quality and condition;

(d) tups shall be valued at such price as in the opinion of the valuer represents their value on the farm having regard to acclimatisation or any other factor for which he thinks it proper to make allowance:

(e) eild sheep shall be valued at the value put upon the ewes subject to such adjustment as the valuer may think proper having regard to their quality and condition; and

(f) shotts shall be valued at such value not exceeding two-thirds of the value put upon good sheep of the like age and class on the farm as the valuer may think proper.

PART II

VALUATION MADE IN RESPECT OF A TENANCY TERMINATING AT MARTINMAS

7. The valuer shall ascertain the number of, and the prices realised for, the ewes sold off the hill from the stock under valuation at the autumn sales in the current year and in each of the 2 preceding years, and shall calculate an average price per ewe so sold for each of the said years. In calculating the average price for any year the valuer shall disregard one-tenth of the total number of ewes so sold in that year being the ewes sold at the lowest price.

8. The mean of the average prices so calculated shall be subject to adjustment by the valuer within the limits of 10 per cent. (in the case of leases entered into before May 15, 1963, 25 pence) upward or downwards as he may think proper having regard to the general condition of the stock under valuation and to the profit which the purchaser may reasonably expect it to earn. The resultant figure shall be the basis of the valuation of the ewes and is in this Part of this Schedule referred to as the 'basic ewe value.'

9. In making his award the valuer shall assess the respective classes of stock in accordance with the following rules, that is to say—

(a) ewes of all ages (including gimmers) shall be valued at the basic ewe value with the addition of 30 per cent. (in the case of leases entered into before May 15, 1963, 75 pence) of such value per head;

(b) ewe lambs shall be valued at the basic ewe value subject to adjustment by the valuer within the limits of 10 per cent. (in the case of leases entered into before May 15, 1963, 25 pence) per head upwards or downwards as he may think proper having regard to their quality and condition; and

(c) tups shall be valued at such price as in the opinion of the valuer represents their value on the farm having regard to acclimatisation or any other factor for which he thinks it proper to make allowance.

PART III

PARTICULARS TO BE SHOWN IN AN ARBITER'S AWARD

10. The 3-year average price for ewes and the 3-year average price for lambs ascertained under Part I, or the mean of the average prices calculated under Part II, of this Schedule, as the case may be.

11. Any amount added or taken away by way of adjustment for the purpose of fixing the basic ewe value or the basic lamb value, and the grounds on which such adjustment was made.

12. The number of each class of stock valued (ewes and gimmers of all ages with lambs being taken as one class, and eild ewes and eild gimmers being taken as separate classes at a Whitsunday valuation, and ewes and gimmers of all ages being taken as one class at a Martinmas valuation) and the value placed on each class.

13. Any amount added to or taken away by way of adjustment in fixing the value of ewe hoggs at a Whitsunday valuation, or the value of ewe lambs at a Martinmas valuation, and the grounds on which such adjustment was made.

Part IV

Interpretation

14. In this Schedule the expressions 'ewe,' 'gimmer,' 'eild ewe,' 'eild gimmer,' 'lamb,' 'ewe hogg,' 'eild sheep' and 'tup' shall be construed as meaning respectively sheep of the classes customarily known by those designations in the locality in which the flock under valuation is maintained.

Section 70   # Schedule 10

Valuation of Sheep Stock in Scotland in Respect of Leases Entered into after December 1, 1986

Part I

Valuation Made in Respect of a Tenancy Terminating at Whitsunday

1. The Land Court or the arbiter (in Part I and Part II of this Schedule referred to as 'the valuer') shall ascertain the number of, and the prices realised for, the regular cast ewes and the lambs sold off the hill from the stock under valuation at the autumn sales in each of the 3 preceding years, and shall determine by inspection the number of shotts present in the stock at that time of the valuation.

2. The valuer shall calculate an average price per ewe, and an average price per lamb, for the regular cast ewes and lambs sold as aforesaid for each of the 3 preceding years. In calculating the average price for any year the valuer shall disregard such number of regular cast ewes and lambs so sold in that year, being the ewes or lambs sold at the lowest prices, as bears the same proportion to the total number of regular cast ewes or lambs so sold in that year as the number of shotts as determined bears to the total number of ewes or lambs in the stock under valuation.

3. The valuer shall then ascertain the mean of the average prices so calculated for the 3 preceding years for regular cast ewes and for lambs, respectively. The figures so ascertained or ascertained, in a case to which paragraph 4 below applies, in accordance with that paragraph, are in this Part of this Schedule referred to as the '3-year average price for regular cast ewes' and the '3-year average price for lambs.'

4. In the case of any sheep stock in which the number of regular cast ewes or the number of lambs sold off the hill at the autumn sales during the preceding 3 years has been less than half the total number of regular cast ewes or of lambs sold, the 3-year average price for regular cast ewes or the 3-year average price for lambs, as the case may be shall, in lieu of being ascertained by the valuer as aforesaid, be determined by the Land Court on the application of the parties; and the Land Court shall determine such prices by reference to the prices realised at such sales for regular cast ewes and for lambs respectively from similar stocks kept in the same district and under similar conditions.

5. The 3-year average price for regular cast ewes shall be subject to adjustment by the valuer within the limits of 30 per cent. upwards or downwards as he may think proper having regard to the general condition of the stock under valuation and to the profit which the purchaser may reasonably expect it to earn. The resultant figure shall be the basis of the valuation of the ewes, and is in this Part of this Schedule referred to as the 'basic ewe value.'

The valuer shall adjust the 3 year average price for lambs within the limits of 20 per cent, upwards or downwards as he may think proper having regard to their quality and condition. The resultant figure shall be the basis for the valuation of the lambs and is in this Part of this Schedule referred to as the 'basic lamb value.'

6. In making his award the valuer shall value the respective classes of stock in accordance with the following rules, that is to say—

(a) ewes of all ages (including gimmers) shall be valued at the basic ewe value with the addition of 30 per cent. of such value per head;

(b) lambs shall be valued at the basic lamb value but twin lambs shall be valued at such price as the valuer thinks proper;

(c) ewe hoggs shall be valued at three quarters of the combined basic values of a ewe and a lamb subject to adjustment by the valuer within the limits of 25 per cent. per head upwards or downwards as he may think proper, having regard to their quality and condition;

(d) tups shall be valued at such price as in the opinion of the valuer represents their value on the farm having regard to acclimatisation or any other factor for which he thinks it proper to make allowance;

(e) eild sheep shall be valued at the value put upon the ewes subject to such adjustment as the valuer may think proper having regard to their quality and condition; and

(f) shotts shall be valued at such value not exceeding two-thirds of the value put upon good sheep of the like age and class on the farm as the valuer may think proper.

PART II

VALUATION MADE IN RESPECT OF A TENANCY TERMINATING AT MARTINMAS

7. The valuer shall ascertain the number of, and the prices realised for, the regular cast ewes sold off the hill from the stock under valuation at the autumn sales in the current year and in each of the 2 preceding years, and shall calculate an average price per ewe so sold for each of the said years. In calculating the average price for any year the valuer shall disregard one-fifth of the total number of regular cast ewes so sold in that year being the ewes sold at the lowest price.

8. The mean of the average prices so calculated shall be subject to adjustment by the valuer within the limits of 30 per cent. upward or downwards as he may think proper having regard to the general condition of the stock under valuation and to the profit which the purchaser may reasonably expect it to earn. The resultant figure shall be the basis of the valuation of the ewes and is in this Part of this Schedule referred to as the 'basic ewe value.'

9. In making his award the valuer shall assess the respective classes of stock in accordance with the following rules, that is to say—

(a) ewes of all ages (including gimmers) shall be valued at the basic ewe value with the addition of 30 per cent. of such value per head;

(b) ewe lambs shall be valued at the basic ewe value subject to adjustment by the valuer within the limits of 20 per cent. per head upwards or downwards as he may think proper having regard to their quality and condition; and

(c) tups shall be valued at such price as in the opinion of the valuer represents their value on the farm having regard to acclimatisation or any other factor for which he thinks it proper to make allowance.

PART III

PARTICULARS TO BE SHOWN IN AN ARBITER'S AWARD

10. The 3-year average price for regular cast ewes and the 3-year average price for lambs ascertained under Part I, or the mean of the average prices calculated under Part II, of this Schedule, as the case may be.

11. Any amount added or taken away by way of adjustment for the purpose of fixing the basic ewe value or the basic lamb value, and the grounds on which such adjustment was made.

12. The number of each class of stock valued (ewes and gimmers of all ages with lambs being taken as one class, and eild ewes and eild gimmers being taken as separate classes at a Whitsunday valuation, and ewes and gimmers of all ages being taken as one class at a Martinmas valuation) and the value placed on each class.

13. Any amount added to or taken away by way of adjustment in fixing the value of ewe hoggs at a Whitsunday valuation, or the value of ewe lambs at a Martinmas valuation, and the grounds on which such adjustment was made.

PART IV

INTERPRETATION

14. In this Schedule the expressions 'regular cast ewes,' 'ewe,' 'gimmer,' 'eild ewe', 'eild gimmer,' 'lamb', 'ewe hogg,' 'eild sheep' and 'tup' shall be construed as meaning respectively sheep of the classes customarily known by those designations in the locality in which the flock under valuation is maintained.

Section 88

SCHEDULE 11

CONSEQUENTIAL AMENDMENTS OF ENACTMENTS

Hill Farming Act 1946 (c 73)

1. In section 9, as substituted by the Seventh Schedule to the 1949 Act,—
(a) in subsection (1), for 'Agricultural Holdings (Scotland) Act 1949' substitute 'Agricultural Holdings (Scotland) Act 1991, referred to in subsections (2) and (4) below as 'the 1991 Act';
(b) in subsections (2) and (4), for 'the said Act of 1949' substitute 'the 1991 Act';
(c) in subsection (2)—
    (i) for 'Part I or Part II of the First Schedule' substitute 'Part I or II of Schedule 5,';
    (ii) in paragraph (a), for 'section fifty of that Act' substitute 'section 37 of the 1991 Act';
    (iii) in paragraph (b), for 'section fifty-one of that Act' substitute 'section 38 of the 1991 Act';
    (iv) in paragraph (b), for 'section fifty-two of that Act' substitute 'section 39 of the 1991 Act,';
    (v) for 'the said section fifty or the said fifty-one' substitute 'section 37 or 38 of the 1991 Act';
(d) in subsection (3), for 'section eight of the Agricultural Holdings (Scotland) Act 1949' substitute 'section 15 of the 1991 Act.'

Reserve and Auxiliary Forces (Protection of Civil Interests) Act 1951 (c 65)

2. In section 21—
(a) in subsection (2) for 'Subsection (1) of section twenty-five of the Agricultural Holdings (Scotland) Act 1949' substitute 'section 22 of the Agricultural

Holdings (Scotland) Act 1991,' and for 'section twenty-six of that Act' substitute 'section 24 of that Act,';

(b) in subsection (3) for 'section twenty-five' in both places where it occurs substitute 'section 22,' and for 'section twenty-six' substitute 'section 24';

(c) in subsection (8) for 'the said Act of 1949' substitute 'the Agricultural Holdings (Scotland) Act 1991.'

3. In section 22(4)(a), for 'subsection (1) of section twenty five of the Agricultural Holdings (Scotland) Act 1949' substitute 'section 22(1) of the Agricultural Holdings (Scotland) Act 1991.'

4. In section 38(6)(a)(i), for 'Agricultural Holdings (Scotland) Act 1949' substitute 'Agricultural Holdings (Scotland) Act 1991.'

Crofters (Scotland) Act 1955 (c 21)

5. In section 14(10), for 'Agricultural Holdings (Scotland) Act 1949' substitute 'Agricultural Holdings (Scotland) Act 1991.'

6. In section 37(1), in the definition of 'fixed equipment,' for 'Agricultural Holdings (Scotland) Act 1949' substitute 'Agricultural Holdings (Scotland) Act 1991.'

7. In Schedule 2, paragraph 10, for 'section 15 of the Agricultural Holdings (Scotland) Act 1949' substitute 'section 52 of the Agricultural Holdings (Scotland) Act 1991.'

Agriculture (Safety, Health and Welfare Provisions) Act 1956 (c 49)

8. In section 25(4), for the words from 'the provisions' to 'section eighteen' substitute 'section 5(2), (3) and (5) of the Agricultural Holdings (Scotland) Act 1991 (liabilities of landlord and tenant of agricultural holding regarding fixed equipment) and section 10.'

9. In section 25(5), for 'section eight of the Agricultural Holdings (Scotland) Act 1949' substitute 'section 15 of the Agricultural Holdings (Scotland) Act 1991.'

10. In section 25(10), in the definition of 'agricultural holding,' 'fixed equipment' and 'landlord,' for 'the Agricultural Holdings (Scotland) Act, 1949' substitute 'the Agricultural Holdings (Scotland) Act 1991.'

Coal Mining (Subsidence) Act 1957 (c 59)

11. In section 10(1)(a), for 'Agricultural Holdings (Scotland) Act 1949' substitute 'Agricultural Holdings (Scotland) Act 1991.'

Opencast Coal Act 1958 (c 69)

12. In section 14A—

(a) in subsection (3), for the words 'Agricultural Holdings (Scotland) Act 1949 in this Act referred to as the Scottish Act of 1949' substitute 'the Scottish Act of 1991,';

(b) in subsection (4), for 'the Scottish Act of 1949' substitute 'the Scottish Act of 1991';

(c) in subsection (5), for 'the Scottish Act of 1949' substitute 'the Scottish Act of 1991';

(d) in subsection (6)—

(i) for 'section 25(2) of the Scottish Act of 1949' substitute 'section 22(2) of the Scottish Act of 1991'; and

(ii) for '(c)' substitute '(b)';

(e) in subsection (7), for the words from 'For the purposes' to 'paragraph (e) of subsection (1)' substitute 'The condition specified in section 24(1)(e) of the Scottish Act of 1991 (consent of Land Court to notice to quit where land to be used for purposes other than agriculture)';

(f) in subsection (8), for 'section 7 of the Scottish Act of 1949' substitute 'section 13 of the Scottish Act of 1991';

(g) in subsection (9), for 'section 8 of the Scottish Act of 1949' substitute 'section 15 of the Scottish Act of 1991.'

13. For section 24(10) substitute—

'(10) In the application of this section to Scotland, for references—

(a) to the Act of 1986 and to sections 70 and 83(4) of that Act there shall be substituted respectively references to the Scottish Act of 1991 and to sections 44 and 62(3) of that Act;

(b) to subsections (1), (2) and (3) of section 69 of the Act of 1986 there shall be substituted respectively references to sections 34(5) and 35(4) and (5) of the Scottish Act of 1991 (as they apply to new improvements);

(c) to Parts I and II of Schedule 7 to the Act of 1986 and to the first day of March 1948 there shall be substituted respectively references to Parts I and II of Schedule 5 to the Scottish Act of 1991 and to the first day of November 1948; and

(d) to sub-paragraphs (1) and (2) of paragraph 5 of Part I of Schedule 9 to the 1986 Act there shall be substituted respectively references to sections 34(5) and 35(4) of the Scottish Act of 1991 (as they apply to old improvements).'.

14. For section 25(3) substitute—

'(3) In the application of this section to Scotland, for paragraphs (a) and (b) of subsection (1) above there shall be substituted the words 'under section 45 of the Scottish Act of 1991 (which relates to compensation for deterioration of a holding or part thereof for which a tenant is responsible).'.

15. In section 26(6) after 'Scotland' insert '(a)' and for the words from 'in subsection (3)' to the end substitute—

'(b) in subsection (3) of this section for the reference to the Act of 1986 there shall be substituted a reference to the Scottish Act of 1991; and

(c) in subsection (5) of this section there shall be substituted—

(i)   for the reference to section 91 of the Act of 1986 a reference to section 73 of the Scottish Act of 1991;

(ii) for the reference to Schedule 8 to the Act of 1986 a reference to Part III of Schedule 5 to the Scottish Act of 1991;

(iii) for the reference to Parts I, II and III of the Fourth Schedule to this Act a reference to Parts IV and V of that Schedule.'.

16. In section 27(4), for 'section fourteen of the Scottish Act of 1949' substitute 'section 18 of the Scottish Act of 1991.'

17. In section 28(6)—

(a) for 'to section sixty-five of the Scottish Act of 1949 and to paragraph (b) of subsection (1) of that section' substitute 'section 40 of the Scottish Act of 1991 and to subsection (4)(a) of that section';

(b) for 'to subsection (1) of section sixty-six of the Scottish Act of 1949 and to section 14 of that Act' substitute 'to section 41(1) and to section 18 of the Scottish Act of 1991';

(c) for 'to section seventy-nine of the Scottish Act of 1949 and to the Fourth Schedule to that Act' substitute 'to section 73 of the Scottish Act of 1991 and to Schedule 6 thereto.'

18. In section 52(2)—

(a) in the definition of 'agricultural holding,' for '1949' substitute '1991';

(b) for the definition of 'the Scottish Act of 1949' substitute ' "the Scottish Act of 1991" means the Agricultural Holdings (Scotland) Act 1991;'.

19. In section 52(5)(a)—

(a) for 'the Scottish Act of 1949' where it first occurs substitute 'the Scottish Act of 1991'; and

(b) for sections fifty-seven and fifty-eight of the Scottish Act of 1949' substitute 'section 45 of the Scottish Act of 1991.'

20. In Schedule 6, paragraph 31, for 'section 2(1) of the Scottish Act of 1949' substitute 'section 2 of the Scottish Act of 1991.'

21. For Schedule 7, paragraph 25(a) substitute—
'(a) for references—

      (i)   to the Act of 1986 and to sections 12, 13, 23 and 84 of that Act there shall be substituted respectively references to the Scottish Act of 1991 and to sections 13, 15, 10 and 61 of that Act;

      (ii) to section 10 of the Act of 1986 and to subsections (3) and (4) of that section there shall be substituted respectively references to section 18 of the Scottish Act of 1991 and to subsections (2) and (3) of that section; and

      (iii) to subsection (3) of section 79 of the Act of 1986 there shall be substituted references to section 40(4)(a) of the Scottish Act of 1991.'.

Horticulture Act 1960 (c 22)

22. In section 1(1)(b), for 'Agricultural Holdings (Scotland) Act 1949' substitute 'Agricultural Holdings (Scotland) Act 1991.'

Crofters (Scotland) Act 1961 (c 58)

23. In section 13(1), for 'the Agricultural Holdings (Scotland) Act 1949' substitute 'the Agricultural Holdings (Scotland) Act 1991.'

Succession (Scotland) Act 1964 (c 41)

24. In section 16—

(a) in subsections (2)(c) and (3)(b)(i), for 'section 20 of the Act of 1949' substitute 'section 11 of the 1991 Act';

(b) in subsection (6)(b), for 'section 27(2) of the Act of 1949' substitute 'section 23(2) and (3) of the 1991 Act' and for 'section 25(2)(f)' substitute 'section 22(2)(e)';

(c) in subsection (8), for 'subsections (2) to (7) of section 20 of the Act of 1949' substitute 'section 11(2) to (8) of the 1991 Act';

(d) in subsection 9—

      (i) in the definition of 'agricultural lease,' for 'the Act of 1949' substitute 'the 1991 Act';

      (ii) for the definition of 'the Act of 1949' substitute ' "the 1991 Act" means the Agricultural Holdings (Scotland) Act 1991;'.

25. In section 29(2), for 'section 20 of the Agricultural Holdings (Scotland) Act 1949' substitute 'section 11 of the Agricultural Holdings (Scotland) Act 1991.'

Agriculture Act 1967 (c 22)

26. In section 26(1), for 'the Agricultural Holdings (Scotland) Act 1949' substitute 'the Agricultural Holdings (Scotland) Act 1991.'

27. In section 27(5B), for 'the Agricultural Holdings (Scotland) Act 1949' substitute 'the Agricultural Holdings (Scotland) Act 1991.'

28. In section 28(1)(a), for 'section 35 of the Agricultural Holdings (Scotland) Act 1949' substitute 'section 43 of the Agricultural Holdings (Scotland) Act 1991.'

29. In section 29—

(a) in subsection (3)(a), for 'section 35 of the Agricultural Holdings (Scotland) Act 1949' substitute 'section 43 of the Agricultural Holdings (Scotland) Act 1991'; and

(b) in subsection (4), for 'section 25(1) of the Agricultural Holdings (Scotland) Act 1949' substitute 'section 22(1) of the Agricultural Holdings (Scotland) Act 1991.'

30. In section 48(2)(a), for 'section 35 of the Agricultural Holdings (Scotland) Act 1949' substitute 'section 43 of the Agricultural Holdings (Scotland) Act 1991.'

31. In Schedule 3, paragraph 7(5)—

(a) for 'sections 75 and 77 of the Agricultural Holdings (Scotland) Act 1949' substitute 'sections 61 and 64 of the Agricultural Holdings (Scotland) Act 1991'; and

(b) for 'sections 78 and 87(2)' substitute 'sections 60(2) and 80(2).'

Conveyancing and Feudal Reform (Scotland) Act 1970 (c 35)

32. In Schedule 1 in paragraph 5(a), for 'Agricultural Holdings (Scotland) Act 1949' substitute 'Agricultural Holdings (Scotland) Act 1991.'

Land Compensation (Scotland) Act 1973 (c 56)

33. In section 31(3)(c) for 'Agricultural Holdings (Scotland) Act 1949' substitute 'Agricultural Holdings (Scotland) Act 1991.'

34. In section 44—

(a) in subsection (2)(a)(i) for 'section 25(2)(c) of the Agricultural Holdings (Scotland) Act 1949' substitute 'section 22(2)(b) of the Agricultural Holdings (Scotland) Act 1991';

(b) in subsection (2)(a)(ii)—

    (i) for 'section 26(1)(e)' substitute 'section 24(1)(e)'; and

    (ii) for 'section 25(2)(c)' substitute 'section 22(2)(b)';

(c) in subsection (3)(a) for 'sections 25(2)(c) and 26(1)(e)' substitute 'sections 22(2)(b) and 24(1)(e)';

(d) in subsection (4), for 'section 12 of the Agriculture (Miscellaneous Provisions) Act 1968' substitute 'section 56 of the Agricultural Holdings (Scotland) Act 1991.'

35. In section 52—

(a) in subsection (3)(d) for 'Agricultural Holdings (Scotland) Act 1949' substitute 'Agricultural Holdings (Scotland) Act 1991'; and

(b) in subsection (4) for 'section 59(1) of the Agricultural Holdings (Scotland) Act 1949' substitute 'section 47(1) of the Agricultural Holdings (Scotland) Act 1991' and for 'the said section 59(1)' substitute 'the said section 47(1).'

36. In section 55—

(a) for subsection (1)(b) substitute—

    '(b) either—

        (i) section 22(1) of the Agricultural Holdings (Scotland) Act 1991 does not apply by virtue of subsection (2)(b) of that section; or

        (ii) the Scottish Land Court have consented to the notice on the ground set out in section 24(1)(e) of that Act.';

(b) in subsection (2)(a), for 'section 12 of the Agriculture (Miscellaneous Provisions) Act 1968' substitute 'section 56 of the Agricultural Holdings (Scotland) Act 1991';

(c) in subsection (2)(b) for 'Agricultural Holdings (Scotland) Act 1949' substitute 'Agricultural Holdings (Scotland) Act 1991,' and for 'sections 9 and 15(3) of the Agriculture (Miscellaneous Provisions) Act 1968' substitute 'sections 54 and 58(1) and (2) of that Act';

(d) in subsection (6) for 'section 33 of the Agricultural Holdings (Scotland) Act 1949' substitute 'section 30 of the Agricultural Holdings (Scotland) Act 1991.'

37. In section 80(1), in the definitions of 'agricultural holding' and 'holding' for 'Agricultural Holdings (Scotland) Act 1949' substitute 'Agricultural Holdings (Scotland) Act 1991.'

Land Tenure Reform (Scotland) Act 1974 (c 38)

38. In section 8(5)(a), for 'Agricultural Holdings (Scotland) Act 1949' substitute 'Agricultural Holdings (Scotland) Act 1991.'

Control of Pollution Act 1974 (c 40)

39. In section 31(B)(2)(a), for the words 'an absolute owner (within the meaning of section 93 of the Agricultural Holdings (Scotland) Act 1949)' substitute 'the owner of the dominium utile.'

Matrimonial Homes (Family Protection) (Scotland) Act 1981 (c 59)

40. In section 13(8), in the definition of 'agricultural holding,' for 'Agricultural Holdings (Scotland) Act 1949' substitute 'Agricultural Holdings (Scotland) Act 1991.'

Rent (Scotland) Act 1984 (c 58)

41. For section 25(1)(iii) substitute—
'(iii) the Agricultural Holdings (Scotland) Act 1991.'

Law Reform (Miscellaneous Provisions) (Scotland) Act 1985 (c 73)

42. In section 7(2), in the definition of 'agricultural holding,' for 'section 1 of the Agricultural Holdings (Scotland) Act 1949' substitute 'the Agricultural Holdings (Scotland) Act 1991.'

Agriculture Act 1986 (c 49)

43. In section 14(a) for 'the Agricultural Holdings (Scotland) Act 1949' substitute 'the 1991 Act.'

44. In section 16—
(a) in subsection (2), for 'section 7 of the 1949 Act' substitute 'section 13 of the 1991 Act'; and
(b) in subsection (7), for 'section 86 of the 1949 Act' substitute 'section 79 of the 1991 Act.'

45. In section 18(6) for the words from 'the absolute owner' to '1949' substitute 'the owner of the dominium utile.'

46. In section 19(4) for 'the Crofters (Scotland) Act 1955' substitute 'the 1955 Act.'

47. After section 23 insert—
'23A. In this Act—
'the 1886 Act' means the Crofters Holdings (Scotland) Act 1886;
'the 1911 Act' means the Small Landholders (Scotland) Act 1911;
'the 1955 Act' means the Crofters (Scotland) Act 1955; and
'the 1991 Act' means the Agricultural Holdings (Scotland) Act 1991.'

48. In Schedule 2, paragraph 1(1)—
(a) in the definition of 'landlord'—
(i) in sub-paragraph (a), for 'the 1949 Act' substitute 'the 1991 Act' and for 'section 93(1)' substitute 'section 85(1)'; and
(ii) in sub-paragraph (c), for 'the 1949 Act' substitute 'the 1991 Act';
(b) in the definition of 'tenancy,' for 'the 1949 Act' substitute 'the 1991 Act'; and
(c) in the definition of 'tenant'—
(i) in sub-paragraph (a), for 'the 1949 Act' substitute 'the 1991 Act' and for 'section 93(1)' substitute 'section 85(1)'; and
(ii) in sub-paragraph (c), for 'the 1949 Act' substitute 'the 1991 Act.'

49. In Schedule 2, paragraph 3(1)(b), for 'section 20 of the 1949 Act' substitute 'section 11 of the 1991 Act.'

50. In Schedule 2, paragraph 7—
(a) in sub-paragraph (2), for 'the 1949 Act' where it first occurs substitute 'the 1991 Act' and for 'section 7 of the 1949 Act' substitute 'section 13 of the 1991 Act'; and
(b) in sub-paragraph (4)—
(i) in sub-paragraph (a)(i), for 'section 93 of the 1949 Act' substitute 'section 85 of the 1991 Act';

(ii) in sub-paragraph (a)(iii), for 'the 1949 Act' substitute 'the 1991 Act' and

(iii) in sub-paragraph (b), for 'section 93 of the 1949 Act' substitute 'section 85 of the 1991 Act.'

51. In Schedule 2, paragraph 10(1)—

(a) sub-paragrah (a), for 'the 1949 Act' substitute 'the 1991 Act' and for 'section 78' substitute 'section 60(2)'; and

(b) for 'section 75 (or, where the circumstances require, sections 77 and 87) of the 1949 Act' substitute 'section 60(1) (or, where the circumstances require, sections 64 and 80) of the 1991 Act.'

52. In Schedule 2, paragraph 11—

(a) in sub-paragraph (1)(a), for 'the 1949 Act' substitute 'the 1991 Act' and for 'section 78' substitute 'section 60(2)';

(b) in sub-paragraph (4)—

(i) for 'section 75 (or, where the circumstances require, sections 77 and 87) of the 1949 Act' substitute 'section 60(1) (or, where the circumstances require, sections 64 and 80) of the 1991 Act'; and

(ii) for 'paragraph 13 of the Sixth Schedule' substitute 'paragraph 14 of Schedule 7'; and

(c) in sub-paragraph (5), for 'section 61 of the 1949 Act' substitute 'section 50 of the 1991 Act.'

53. In Schedule 2, for paragraph 12 substitute—

'Sections 65 and 75(1), (2), (4) and (6) of the 1991 Act (recovery of sums due and power of tenant to obtain charge on holding) shall apply in relation to any sum payable to the tenant under this Schedule as they apply to sums payable under that section.'.

Housing (Scotland) Act 1987 (c 26)

54. In section 256(1) and (3) for 'Agricultural Holdings (Scotland) Act 1949' substitute 'Agricultural Holdings (Scotland) Act 1991.'

55. In section 338(1), in the definition of 'agricultural holding,' for 'Agricultural Holdings (Scotland) Act 1949' substitute 'Agricultural Holdings (Scotland) Act 1991.'

56. In Schedule 8, Part IV, paragraph 13—

(a) in sub-paragraph (1)—

(i) for 'Section 8 of the Agricultural Holdings (Scotland) Act 1949' substitute 'Section 15 of the Agricultural Holdings (Scotland) Act 1991';

(ii) for 'the said section 8' substitute 'the said section 15';

(b) in sub-paragraph (2)—

(i) for 'paragraph 18 of Schedule 1 to the said Act of 1949' substitute 'paragraph 18 of Schedule 5 to the Agricultural Holdings (Scotland) Act 1991';

(ii) for 'section 79' substitute 'section 73';

(iii) for 'the said Schedule 1' substitute 'the said Schedule 5';

(iv) for 'sections 51 and 52' substitute 'sections 38 and 39';

(v) for 'section 49 of the said Act of 1949' substitute 'section 36 of that Act.'

Housing (Scotland) Act 1988 (c 43)

57. In Schedule 4 in paragraph 6(a), for 'Agricultural Holdings (Scotland) Act 1949' substitute 'Agricultural Holdings (Scotland) Act 1991.'

Section 87

SCHEDULE 12

TRANSITIONALS AND SAVINGS

Continuation of savings

1. The repeal by this Act of an enactment which repealed a previous enactment subject to a saving shall not affect the continued operation of that saving.

Construction of references to old and new law

2.—(1) Where an enactment contained in this Act repeals and re-enacts an earlier enactment—

(a) for the purpose of giving effect to any instrument or other document it shall be competent, so far as the context permits, to construe a reference to either enactment as a reference to the other;

(b) anything done or required to be done for the purposes of either enactment may, so far as the context permits, be treated as having been done or as something required to be done for the purposes of the other.

(2) In this paragraph, a reference to an enactment reenacted in this Act includes a reference to any such enactment repealed by the Agricultural Holdings Act 1923, the 1949 Act or the Agricultural Holdings (Amendment) (Scotland) Act 1983.

Savings for specific enactments

3. Nothing in this Act shall affect any provision of the Allotments (Scotland) Act 1922.

4. Section 21 of the Reserve and Auxiliary Forces (Protection of Civil Interests) Act 1951 (as read with section 24 of that Act) shall continue to have effect—

(a) in subsections (2) and (3) with the substitution for references to the Secretary of State of references to the Land Court; and

(b) with the reference in subsection (6) to section 27 of the 1949 Act being construed as a reference to that section as originally enacted.

Compensation

5. Notwithstanding section 16 of the Interpretation Act 1978, rights to compensation conferred by this Act shall be in lieu of rights to compensation conferred by any enactment repealed by this Act.

Section 88

SCHEDULE 13

REPEALS AND REVOCATIONS

PART I

REPEALS

Chapter	Short title	Extent of repeal
1 Edw 8 & 1 Geo 6 c 34.	Sheep Stocks Valuation (Scotland) Act 1937.	The whole Act.
9 & 10 Geo 6 c 73.	Hill Farming Act 1946.	Sections 28 to 31. Second Schedule.
11 & 12 Geo 6 c 45.	Agriculture (Scotland) Act 1948.	Section 52. In section 54, the definitions of 'deer', 'occupier of an agricultural holding' and 'woodlands'.

Chapter	Short title	Extent of repeal
12, 13 and 14 Geo 6 c 75.	Agricultural Holdings (Scotland) Act 1949.	The whole Act.
14 & 15 Geo 6 c 18.	Livestock Rearing Act 1951.	In section 1(2)(b) the words 'in paragraph (d) of subsection (1) of section 8 of the Agricultural Holdings (Scotland) Act 1949'.
14 & 15 Geo 6 c 65.	Reserve and Auxiliary Forces (Protection of Civil Interests) Act 1951.	In section 24(b), the words from 'for references' to 'twenty-seven thereof'.
6 & 7 Eliz 2 c 71.	Agriculture Act 1958.	Section 3. Schedule 1.
1963 c 11.	Agriculture (Miscellaneous Provisions) Act 1963.	Section 21
1964 c 41.	Succession (Scotland) Act 1964.	In Schedule 2, paragraphs 19 to 23.
1968 c 34.	Agriculture (Miscellaneous Provisions) Act 1968.	Part II. Schedules 4 and 5.
1973 c 65.	Local Government (Scotland) Act 1973.	Section 228(5).
1976 c 21.	Crofting Reform (Scotland) Act 1976.	Schedule 2, para. 25.
1976 c 55.	Agriculture (Miscellaneous Provisions) Act 1976.	Section 13 and 14.
1980 c 45.	Water (Scotland) Act 1980.	In Schedule 10, Part II, the entry relating to the 1949 Act.
1983 c 46.	Agricultural Holdings (Amendment) (Scotland) Act 1983.	The whole Act.
1985 c 73.	Law Reform (Miscellaneous Provisions) (Scotland) Act 1985.	Section 32.
1986 c 5.	Agricultural Holdings Act 1986.	In Schedule 14, paras. 25(8), 26(11) and 33(8).
1986 c 49.	Agriculture Act 1986.	In Schedule 2, para. 1, the definitions of 'the 1986 Act', 'the 1911 Act', 'the 1949 Act' and 'the 1955 Act'.

PART II

REVOCATIONS OF SUBORDINATE LEGISLATION

Number	Citation	Extent of revocation
SI 1950/1553.	The Agricultural Holdings (Scotland) Regulations 1950.	The whole Instrument.
S I 1978/798	The Agricultural Holdings (Scotland) Act 1949 (Variation of First Schedule) Order 1978.	The whole Order.

Number	Citation	Extent of revocation
S I 1986/1823.	The Hill Farming Act 1946 (Variation of Second Schedule) (Scotland) Order 1986.	The whole Order.

TABLE OF DERIVATIONS

Notes: The following abbreviations are used in this Table—

1937	=	The Sheep Stocks Valuation (Scotland) Act 1937 (1 Edw 8 & 1 Geo 6 c 34).
1946	=	The Hill Farming Act 1946 (9 & 10 Geo 6 c 73).
1948	=	The Agriculture (Scotland) Act 1948 (11 & 12 Geo 6 c 45).
1949	=	The Agricultural Holdings (Scotland) Act 1949 (12, 13 & 14 Geo 6 c 75).
1958	=	The Agriculture Act 1958 (c 71).
1963	=	The Agriculture (Miscellaneous Provisions) Act 1963 (c 11).
1964	=	The Succession (Scotland) Act 1964 (c 41).
1968	=	The Agriculture (Miscellaneous Provisions) Act 1968 (c 34).
1973	=	The Local Government (Scotland) Act 1973 (c 65).
1976	=	The Agriculture (Miscellaneous Provisions) Act 1976 (c 55).
1983	=	The Agriculture Holdings (Amendment) (Scotland) Act 1983 (c 46).
1986	=	The Agriculture Holdings Act 1986 (c 5).
SI 1950/1553	=	The Agriculture Holdings (Scotland) Regulations 1950 (SI 1950/1553).
SI 1978/798	=	The Agricultural Holdings (Scotland) Act 1949 (Variation of First Schedule) Order 1978 (SI 1978/798).
SI 1986/1823	=	The Hill Farming Act 1946 (Variation of Second Schedule) (Scotland) Order 1986.

Provision of Act	Derivation
1	1949 s 1; 1958 s 9(1).
2	1949 s 2.
3	1949 s 3; 1949 s 24(1).
4	1949 s 4, s 6(4).
5	1949 s 5.
6	1949 s 23.
7	1949 s 12; 1958 Sch 1, Pt II, para 33.
8	1949 s 17.
9	1949 s 9; 1958 Sch 1, Pt II, para 32.
10	1949 s 18.
11	1949 s 20; 1964 s 34(1), Sch 2, paras 19, 20 and 21.
12	1949 s 21; 1964 s 34(1), Sch 2, para 22.
13	1949 s 7; 1983 s 2.
14	1949 s 6(3).
15	1949 s 8.
16	1949 s 10.
17	1949 s 13.
18	1949 s 14.
19	1949 s 22.
20	1949 s 19.
21	1949 s 24; 1958 Sch 1, Pt II, para 34.
22	1949 s 25; 1958 s 3(1), (3), Sch 1, Pt II, para 35.

Provision of Act	Derivation
23	1949 s 27; 1958 Sch 1, Pt II, para 37.
24	1949 s 26; 1958 s 3(2), (3), Sch 1, Pt II, para 36; 1983 s 4(1).
25	1949 s 26A; 1983 s 3, s 4(2).
26	1949 s 28; 1958 Sch 1, Pt II, para 38; 1989 (c 15) Sch 25, para 12.
27	1949 s 30; 1958 Sch 1, Pt II, para 40.
28	1949 s 31.
29	1949 s 32
30	1949 s 33
31	1949 s 34.
32	1976 s 14.
33	1949 s 36; s 47.
34	1949 s 37, s 41, s 42, s 43, s 44(4), s 45, s 48, s 53, s 54.
35	1949 s 11, s 46, s 55
36	1949 s 38, s 43, s 44(1), s 49, s 53.
37	1949 s 39, s 50
38	1949 s 40, s 51.
39	1949 s 52; 1958 Sch 1, Pt II, para 41.
40	1949 s 65
41	1949 s 66; 1958 Sch 1, Pt II, para 43.
42	1949 s 67
43	1949 s 35.
44	1949 s 56.
45	1949 s 57; s 58
46	1949 s 6(1), (2); s 57(3); SI 1950/1553
47	1949 s 59.
48	1949 s 16.
49	1949 s 60.
50	1949 s 61.
51	1949 s 63; 1958 Sch 1, Pt II, para 42; SI 1977/2007.
52	1949 s 15; SI 1977/2007.
53	1949 s 64.
54	1968 s 9, s 16, Sch 5, para 1.
55	1968 s 11.
56	1968 s 12, s 16.
57	1968 s 14; 1972 (c 52) Sch 21, Pt II.
58	1968 s 15, s 16, Sch 5, para 5.
59	1968 s 16, s 17.
60	1949 s 74, s 78
61	1949 s 68, s 75; 1973 s 228(5); 1983 s 5(1).
62	1949 s 68.
63	1949 s 76; 1971 (c 58) s 4.
64	1949 s 77, s 87(2); 1986 s 17(3).
65	1949 s 69.
66	1976 s 13.
67	1949 s 91; 1971 (c 58) s 4.
68	1937 s 1; 1946 s 28; 1985 (c 73) s 32; SI 1986/1823.
69	1937 s 2.
70	1937 s 3; 1946 s 29; SI 1986/1823.
71	1946 s 30.
72	1937 s 4.
73	1949 s 79.
74	1949 s 80.
75	1949 s 70, s 82; 1980 (c 45) Sch 10.
76	1949 s 83.

Provision of Act	Derivation
77	1949 s 84.
78	1949 s 85.
79	1949 s 86; 1968 s 17(3)
80	1949 s 87(1).
81	1949 s 88.
82	1949 s 89; 1975 (c 21) s 289; 1977 (c 45) s 31
83	1949 s 73; 1976 s 14(6); 1976 (c 21) Sch 2, para 25.
84	1949 s 90; 1985 (c 6) s 725 (1).
85	1949 s 93.
86	1949 s 95.
87	1949 s 99(2).
88	1949 s 97.
89	1949 s 101.
Schedule 1	1949 Sch 5.
Schedule 2	1983 Sch 1.
Schedule 3	1949 Sch 2.
Schedule 4	1949 Sch 3.
Schedule 5	1949 Sch 1; SI 1978/798.
Schedule 6	1949 Sch 4.
Schedule 7	1949 Sch 6; 1983 s 5(2).
Schedule 8	1968 Sch 4; Sch 5, para 6, para 7.
Schedule 9	1946 Sch 2; 1963 s 21.
Schedule 10	1946 Sch 2; 1963 s 21.

The Agricultural Holdings (Specification of Forms) (Scotland) Order 1991
(SI 1991/2154)

The Secretary of State, in exercise of the powers conferred on him by paragraphs 11 and 25 of Schedule 7 to the Agricultural Holdings (Scotland) Act 1991, and of all other powers enabling him in that behalf, and after consultation with the Council on Tribunals as provided in section 10 of the Tribunals and Inquiries Act 1971, hereby makes the following Order:

1.—(1) This Order may be cited as the Agricultural Holdings (Specification of Forms) (Scotland) Order 1991, and shall come into force on 25th September 1991.

(2) In this Order 'the Act' means the Agricultural Holdings (Scotland) Order 1991.

2. The form specified in Schedule 1 to this Order shall, modified as circumstances may require, be the form of an award in an arbitration under the Act.

3. The forms specified in Schedule 2 to this Order, or forms as near thereto as circumstances may require, may be used for proceedings in arbitrations under the Act as follows:—

(a) for the making of an application for appointment by the Secretary of State of an arbiter to determine claims, questions of differences (except as to determination of rent) arising between the landlord and tenant of an agricultural holding – Form A;

(b) for the making of an application for appointment by the Secretary of State of an arbiter to determine the rent of an agricultural holding – Form B;

(c) for the making of an application to the Secretary of State by an arbiter for extension of time for making his award in an arbitration – Form C.

4.—(1) The Agricultural Holdings (Specification of Forms) (Scotland) Order 1983 is hereby revoked.

(2) Anything whatsoever done under or by virtue of the instrument revoked by this Order shall be deemed to have been done under or by virtue of the corresponding provision of this Order and anything whatsoever begun under any article of the said instrument may be continued under this Order as if begun under this Order.

SCHEDULE 1

FORM OF AWARD

Agricultural Holdings (Scotland) Act 1991

Award in Arbitration between A. B. (name and address), the [outgoing] tenant, and C. D. (name and address), the landlord, with regard to the holding known as   (insert name of holding, district and region), [lately] in the occupation of the said tenant.

Whereas under the Agricultural Holdings (Scotland) Act 1991, the claims, questions or differences set forth in the Schedule to this Award are referred to arbitration in accordance with the provisions set out in Schedule 7 to the said Act:

And whereas by appointment dated the   day of   19 , signed by (on behalf of) the said tenant and landlord [or, as the case may be – given under the seal of the Secretary of State], I, (insert name and address), was duly appointed under the said Act to be the arbiter for the purpose of

¹ ⎰ settling the said claims
⎱ settling the said questions or differences
  determining the rent to be paid in respect of the said
  holding as from²

in accordance with the provisions set out in Schedule 7 to the said Act:

[And whereas the time for making my Award has been extended by

¹ ⎰ the written agreement of the said tenant and landlord, dated the   day of
⎱   19 , order of the Secretary of State, dated the   day of
        19 , to the   day of   19 .]

And whereas I, the said (insert name)   , having accepted the appointment as arbiter, and having heard the parties (agents for the parties) and examined the documents and other productions lodged and the evidence led and having fully considered the whole matters referred to me, do hereby make my final Award as follows:—³

I award and determine that the said landlord shall pay to the said tenant the sum of   pounds and   pence, as compensation in respect of the claims set forth in the [first part of the] Schedule to this Award, and amount awarded in respect of each claim being as there stated.

I award and determine that the said tenant shall pay to the said landlord the sum of   pounds and   pence, in respect of the claims set forth in the [second part of the] Schedule to this Award, the amount awarded in respect of each claim being as there stated.

I determine the questions or differences set forth in the [third part of the] Schedule to this Award, as follows, namely:—

¹I fix and determine the rent to be paid by the said tenant to the said landlord, as from   to be the sum of   per annum. [My findings in fact and the reasons for my decision are set forth in the [fourth part of the] Schedule to this award.]

I award and direct that each party shall bear his own expenses and one half of the other expenses of and incidental to the arbitration and Award, including my remuneration [and that of the clerk].

(or otherwise as the arbiter may see fit to direct in light of the provisions of section 63(3) of, and paragraphs 17 to 19 of Schedule 7 to, the said Act) and that, subject to the provisions of the said Act, all sums including any expenses, payable under or by virtue of this Award shall be so paid not later than⁴

In witness whereof I have signed this Award this   day of   19   , in the presence of the following witnesses.

Signature . . . .
Designation . . . .
Address . . . .

. . . .

(Arbiter)

Signature . . . .
Designation . . . .
Address . . . .

Schedule to the above Award

In the case of appointment by the Secretary of State or by the Scottish Land Court of an arbiter to determine claims questions or differences (except as to rent) the arbiter must, if either party so requests, state the reasons for any determination arrived at (section 12 of the Tribunals and Inquiries Act 1971).

In the case of appointment by the Secretary of State or by the Scottish Land Court of an arbiter to determine the rent of an agricultural holding under section 13 of the 1991 Act, the arbiter shall, in every case, and regardless of whether or not he is requested to do so, state in writing his findings of fact and the reasons for his decision under the headings set forth in Part IV of this Schedule (paragraph 10 of Schedule 7 to the 1991 Act).

Claims,[5] questions or differences to be determined.

Part I—
Claims made by the tenant.

Part II—
Claims made by the landlord.

Part III—
Questions or differences (including questions of rent in cases where the arbiter is not appointed by the Secretary of State or Scottish Land Court).

Part IV—
Variation of rent cases under section 13 of the 1991 Act in which the arbiter is appointed by the Secretary of State or Scottish Land Court.

In such cases a statement under the following headings must be provided as a Schedule to the Award and made available to the parties to the case and the Secretary of State—
  (i)   a summary of the statement of case submitted by or on behalf of the landlord;
  (ii) a summary of the statement of case submitted by or on behalf of the tenant;
  (iii) details of any evidence of the condition of the holding, including the state of the landlord's and tenant's fixed equipment, which emerged at the inspection of the holding and were taken into account;
  (iv) a summary of the relevant evidence considered at any hearing;
  (v)   an appraisal of the evidence submitted under (i) to (iv);
  (vi) details of any other evidence of open market rents for comparable subjects introduced by the arbiter on which the parties had an opportunity to comment and which the arbiter took into account;
  (vii) the reasons for seeking evidence (in terms of the factors specifically listed in section 13(4) of the 1991 Act) other than evidence of open market rents for comparable subjects in the surrounding area;

(viii) details of the factors specified in section 13(4) of the 1991 Act which the arbiter considers it desirable to take into account;

(ix) an indication of the weight attached by the arbiter to the various criteria taken into account;

(x) an explanation of any adjustment made by the arbiter to take account of differences in holdings used for comparative purpose;

(xi) any other explanation necessary to clarify the arbiter's decision.

1 Adapt to meet the circumstances.

2 Insert date from which revised rent is to run. (Where variation of rent under section 13 of the 1991 Act is concerned, the date will be the next ensuing day on which the tenancy could have been terminated by notice to quit given at the date of demanding the reference of the rent question to arbitration – usually a term of Whitsunday or Martinmas.)

3 Such parts of the following four paragraphs as may be appropriate should be incorporated in the award, adaptations to meet the particular circumstances being made as necessary.

4 The date of payment specified must not be later than one calendar month after the delivery of the Award.

5 Where claims are made under Schedules 3, 4, 5 or 6 to the 1991 Act, the amounts awarded must, if either party so requires, be shown separately against each numbered item as set out in those Schedules. Where claims are made by either party under agreement or custom and not under statute, the amounts awarded must be separately stated.

SCHEDULE 2   Article 3

FORM A

(Application for appointment by the Secretary of State of an arbiter to determine claims, questions or differences (except as to determination of rent) arising between the landlord and tenant of an agricultural holding.)

AGRICULTURAL HOLDINGS (SCOTLAND) ACT 1991

To the Secretary of State,

In default of agreement between the landlord and the tenant of the holding specified in the Schedule to this application as to the person to act as arbiter and in the absence of any provision in any lease or agreement between them relating to the appointment of an arbiter, I/we hereby apply to the Secretary of State to appoint an arbiter for the purpose of settling the claims questions or differences set out in the Schedule to this application.

Signature(s) ....
                    1
                    ....

Date ....

SCHEDULE

(Applicants seeking determination of a claim for compensation associated with questions and differences should answer questions 1 to 11 inclusive.)

Particulars required	Replies

SECTION A – *To be completed by all applicants*

1. Name and address of holding.	Holding: District: Region:
2. Name and address of landlord.	
3. Name and address of landlord's agent.²	
4. Name and address of tenant.	
5. Name and address of tenant's agent.²	
6. If the tenancy has terminated state date of termination.	
7. Approximate area in hectares of holding.	
8. Description of holding.³	

SECTION B – *To be completed ONLY by applicants seeking determination of a claim for compensation*

9. If an extension of time has been granted under section 62(4) of the Agricultural Holdings (Scotland) Act 1991 for the settlement of claims, state date on which extension expires.

10. Nature of claim to be referred to arbitration.
(a) State claim for compensation for improvements by the tenant, and give short particulars of any further claims by the tenant.
(b) Give short particulars of any claims by the landlord.

SECTION C – *To be completed ONLY by applicants seeking determination of questions or differences*

11. State questions or differences to be referred to arbitration.

1 State whether landlord or tenant. If an agent signs state on whose behalf he is signing. The appointment will be expedited if the application is made by both parties.
2 If no agent, insert 'None'.
3 Describe holding briefly, eg mixed, arable, dairying, market garden.

FORM B

(Application for appointment by the Secretary of State of an arbiter to determine rent of an agricultural holding.)

AGRICULTURAL HOLDINGS (SCOTLAND) ACT 1991

To the Secretary of State,

In default of agreement between the landlord and the tenant of the holding specified in the Schedule to this application as to the person to act as arbiter and in the absence of any provision in any lease or agreement between them relating to the appointment of an arbiter, I/we hereby apply to the Secretary of State to appoint an arbiter to determine the rent to be paid for the said holding as from   19   : *(enter appropriate date in accordance with note 1 below)*

Signature(s) . . . .
² . . . .
Date . . . .

SCHEDULE

Particulars required	Replies
1. Name and address of holding.	Holding: District: Region:
2. Name and address of landlord.	
3. Name and address of landlord's agent.³	
4. Name and address of tenant.	
5. Name and address of tenant's agent.³	
6. Approximate area in hectares of holding.	
7. Description of holding.⁴	
8. Date of demand in writing for reference to arbitration.	
9. Date at which tenancy of holding could be terminated by notice to quit.	

10. (a) Date of commencement of tenancy.
    (b) Effective date of any previous increase or
        reduction of rent.
    (c) Effective date of any previous direction of
        an arbiter that the rent continue
        unchanged.

1 Where variation of rent under section 13 of the 1991 Act is concerned, the date will be the next
  ensuing day on which the tenancy could have been terminated by notice to quit given at the date
  of demanding the reference of the rent question to arbitration – usually a term of Whitsunday
  or Martinmas.
2 State whether landlord or tenant. If an agent signs, state on whose behalf he is signing. The
  appointment will be expedited if the application is made by both parties.
3 If no agent, insert 'None' in second column.
4 Describe holding briefly, eg mixed, arable, dairying, market garden.

FORM C

(Application to the Secretary of State by an arbiter for extension of time for making his award in an arbitration.)

AGRICULTURAL HOLDINGS (SCOTLAND) ACT 1991

To the Secretary of State,

As the time for making the Award in the arbitration detailed below will expire/expired on the   day of   19   , I hereby apply for an extension of the time for making the said Award to the   day of   19

(Signature of arbiter
or arbiter's clerk) . . . .
Date . . . .

Details to be supplied—
1. Name of holding and district and region in which situated.
2. Name and address of landlord (and agent, if any).
3. Name and address of tenant (and agent, if any).
4. Name and address of arbiter (and clerk, if any).
5. (a) Date on which arbiter appointed.
   (b) Whether appointed by agreement of parties or by the Secretary of State or Scottish Land Court.

The Scottish Land Court Rules 1992
(SI 1992/2656)

The Scottish Land Court, in exercise of the powers conferred upon them by section 29 of the Crofters Holdings (Scotland) Act 1886, and now vested in them and by section 3(12) of the Small Landholders (Scotland) Act 1911 and all other powers enabling them in that behalf, and with the approval of the Treasury, hereby make the following Rules:

1. These rules may be cited as the Scottish Land Court Rules 1992 and shall come into force on 1st November 1992.

2. The rules of the Scottish Land Court shall be as set forth in the Schedule to these rules.

3. The rules of the Scottish Land Court 1979 are revoked.

DEFINITIONS

1. In the construction of these Rules (unless the context otherwise requires)—
(a) The word 'Court' shall mean the Scottish Land Court and shall include the Full Court and any Divisional Court; the expression 'Full Court' shall mean the court constituted for hearing appeals under section 25(5) of the Small Landholders (Scotland) Act 1911 and the expression 'Divisional Court' shall mean any member or any two members sitting or acting with any legal assessor by virtue of any powers delegated under the said subsection either by a quorum of the whole Court or by these Rules.
(b) The word 'Chairman' shall mean the Chairman of the Court.
(c) The expression 'Principal Clerk' shall mean the Principal Clerk and Legal Secretary to the Court and shall include (except for the purposes of Rule 102) every person who for the time being is authorised or deputed in the absence of the Principal Clerk to discharge the duties of the Principal Clerk to the Court.
(d) The word 'Auditor' shall mean the Auditor of the Land Court, who is the Principal Clerk.
(e) The word 'Order' shall include decree, award and determination in any proceeding before the Court.
(f) The word 'month' shall, in the computation of time for the purposes of these Rules and of any Order made by the Court, mean calendar month.
(g) The word 'revaluation' shall mean fixing of a second and every subsequent Fair Rent or Equitable Rent for a holding.
(h) The word 'hearing' shall include trial, proof and debate in any Application or any proceeding accessory or incidental thereto.
(i) The word 'landlord' shall mean any person for the time being entitled to receive the rents and profits or to take possession of any holding and shall include the trustees, executors, administrators, assignees, legatee, disponee or next-of-kin, spouse, guardian, curator bonis, trustee in bankruptcy or judicial factor of a landlord.
(j) The word 'person' shall include any body or association of persons, incorporated or unincorporated.
(k) The expression 'Final Order' shall mean an Order of the Court which, either by

itself or taken along with a previous Order or Orders, disposes of the subject-matter of the Application, though all the questions of law or of fact arising in the Application shall not have been decided and though expenses, if found due, shall not have been modified, taxed or decerned for.

OFFICE AND SITTINGS OF THE COURT

2. The office of the Court in Edinburgh shall be open to the public on every day of the year from 9 o'clock am until 4 o'clock pm except Saturdays, Sundays, public holidays and any other days on which the office may be closed by Order of the Court.

3. The Court shall hold sittings for the purpose of hearing Applications, including Appeals and Motions for Rehearing, at such places as they shall from time to time intimate to parties.

4. Any sitting and any hearing in any Application or proceeding may be postponed or adjourned either to a fixed day or to a day to be afterwards fixed by the Court.

PROCEDURE

Applications

5. All applications to the Court shall be framed as nearly as reasonably may be in accordance with the forms provided by the Court. These forms, with relative copy forms for service, may be procured by intending Applicants from the Court free of charge. They may be varied and the initial conclusions altered, supplemented or combined, so far as necessary to adapt the forms to any special case or to Applications for which no special form has been issued.

6. No application shall be incompetent solely on the ground that a declaratory Order only is applied for.

7. Except as otherwise provided, Applications shall be signed by the Applicant or by a solicitor or counsel or, where an Applicant is furth of Scotland, by any person duly authorised in writing, on his behalf. Applications by a landlord may be signed by his factor.

8. Where an Applicant cannot sign his name and is not represented by a solicitor, he may instead adhibit his X or mark in the presence of at least one witness above eighteen years of age who shall certify in writing on the Application that it was read over and explained to the Applicant before his mark was adhibited.

9. All Applications shall be addressed to the Land Court at their office at 1 Grosvenor Crescent, Edinburgh, EH12 5ER and shall be posted, or delivered to, the said office, together with

(1) (except in the cases provided for under Rules 17, 18 and 22 and in Applications where no service is necessary, e.g. joint Applications by landlord and tenant) a copy, or as many copies as are required, duly completed, for service on the Respondent or Respondents and

(2) the appropriate fee, as specified in the current Table of Court Fees a copy of which is available from the Principal Clerk.

10. Tenants who hold pasture, grazing or other rights in common or whose holdings are situated in the same township and on the estate of the same landlord may join as Applicants, or be called by their landlord as Respondents, in one Application to fix fair rents for their holdings.

Service, intimation, etc

11. If an Application, posted or delivered as aforesaid, appears to be in proper form and if the appropriate fee in accordance with the Table of Court Fees has been paid (unless the Application is one in which the Court fees fall to be assessed by the Court), the Principal Clerk shall, after satisfying himself of the accuracy of the service copy or copies lodged along with the Application, effect service of the Application by transmitting such copy or copies, duly certified, by first class recorded delivery service or registered post letter to the Respondent, or each Respondent, at the address, or addresses, stated in the Application.

12. Any notice, order, summons or proceeding in any Application shall similarly be served or intimated by first class recorded delivery service or registered post letter containing a certified copy of such notice, order, summons or proceeding, directed by the Principal Clerk (or by an Applicant or Respondent if so ordered) to the person, or persons, on or to whom such service or intimation is required.

13. Any period which begins to run from service or intimation shall be reckoned from the expiry of twenty-four hours after the time of posting such recorded delivery service or registered letter.

14. Service on, or intimation to, a landlord may be effected by first class recorded delivery service or registered post letter containing a certified copy of the Application, order, notice, summons or proceeding, of which service or intimation is required, directed to him at the address of his factor or the solicitor to whom the tenant, or other Applicant or Applicants, has usually paid rent.

15. Service on, or intimation to, any association, board, firm, company or corporation may be effected in like manner by first class recorded delivery service or registered post letter containing a certified copy as aforesaid, directed to such association, board, firm, company or corporation under the name or description which they ordinarily use, at the principal office or place of business or (if the principal office or place of business be situated outwith Scotland) at any office or place within Scotland (including the office of a clerk, secretary or representative) where they carry on business.

16. In every case where a party to an Application is represented by a solicitor, any order, summons, notice or other proceeding may be served on, or intimated to, such party in like manner by first class recorded delivery service or registered post letter directed to such solicitor at his office or place of business, unless and until the other parties and the Principal Clerk are notified that such solicitor no longer acts for such party.

17. In any Application where a Respondent's address in unknown to the Applicant the Court may allow or direct the Applicant to give notice of intimation of the Application, or any proceeding therein, by the publication in a newspaper circulating in the area of the holding of the Respondent, or of the last known address of the Respondent or elsewhere of an advertisement in such form as the Court may order.

Prior to allowing or directing intimation by way of an advertisement in a newspaper, the Applicant will satisfy the Court that he has taken all reasonable steps to trace the whereabouts of the Respondent.

Where intimation by way of an advertisement, the Applicant will lodge a copy of the newspaper containing said advertisement with the Principal Clerk.

If, after intimation by advertisement as aforesaid, the address of the Respondent becomes known, the Court may allow the Application to be amended subject to such conditions as to re-service, intimation and expenses as seems just.

18. In any application for resumption of, or otherwise relating to, common grazings, when the number of persons called as Respondents, or to whom intimation is ordered, exceeds twenty, the Court may allow the Applicant, or Applicants, or other parties, as the case may be, to give notice or intimation of the Application or any proceeding therein to all such persons by advertisement in each of two successive weeks in any newspaper circulating in the district or by service of such notice or intimation on the clerk to the grazings committee or in such other manner as the Court may think sufficient, in substitution for intimation or service made by recorded delivery service or registered post letter by the Principal Clerk to or on each Respondent so called or each person to whom intimation is so ordered.

19. If any person who is named as a Respondent or who has an interest to intervene in, or who it is proposed should be made a party to, an Application has no known factor or solicitor and no known residence or place of business within Scotland, but has a known residence or place of business outwith Scotland, notice or intimation of such

Application, or of any order or proceeding therein, shall be given to him by first class recorded delivery service or registered post letter containing a certified copy thereof directed to such residence or place of business.

20. The receipt of the Post Office for a first class recorded delivery or registered post letter duly directed, which is certified by the Principal Clerk or is otherwise proved to have contained a true copy of the Application, order, notice, summons or proceeding intended to be served on, or intimated to, the person to whom such recorded delivery service or registered post letter was directed, shall be sufficient prima facie proof of due service on, or intimation to, such person of such Application, order, summons, notice or proceeding having been effected at the time at which said recorded delivery service or registered post letter would have been delivered in ordinary course of post. A first class recorded delivery service or registered post letter shall be deemed, until the contrary is proved, to have been duly directed to the person on or to whom service or intimation was intended to be so made, when it has been directed to him either

(1) at the address stated by him in any Application or pleading or proceeding in the Application or

(2) at the address of his factor or solicitor or

(3) at his last known residence or place of business.

21. When intimation is made under Rule 17 or 18 copies of the newspaper containing the advertisement shall be deemed prima facie proof of such intimation.

22. Any person named as a Respondent in an Application or made a party thereto, may by a signed endorsement on the Application or by statement in open Court or by letter to the Applicant or other party moving, or entitled to move, for service or intimation or to the Principal Clerk, agree to dispense with service or intimation of such Application or order, summons, notice or proceeding therein.

23. No party who appears in Court or lodges objections or answers or other pleading shall be entitled to state any objection to the regularity of the service on, or of the intimation to, himself.

24. If there has been any insufficiency of, or irregularity in, any service on, or intimation to, a person who has not appeared in Court or lodged objections or answers to other pleading or if it seems expedient that service or intimation of any Application, order, notice, summons or proceeding on or to any person should be made of new or in any other or further manner than by first class recorded delivery service or registered post letter as aforesaid, the Court may authorise or direct new, or further, service or intimation accordingly, on such conditions as the Court may think proper, in any manner allowed by the law and practice of Scotland.

25. As soon as an Application has been received by the Principal Clerk, it shall be deemed to be in dependence before the Court and shall not be abandoned or withdrawn without leave of the Court on such conditions as to expenses or otherwise as the Court may think just.

Process

26. Any party to an Application or his solicitor or other authorised representative may

(1) require the Principal Clerk to exhibit the Application, or any part of the process therein, in his custody at the office of the Court during office hours, free of charge;

(2) make a handwritten copy of the Application or any order pronounced therein or any answers, minutes, writings, plans or other documents in process in such custody at the said office, during office hours, and under supervision of the Principal Clerk, free of charge.

Where a party to an Application or his solicitor or other authorised representative obtains from the Principal Clerk a photocopy of the Application, or any Order pronounced therein, or any answers, minutes, writings, plans or other documents in process, the Principal Clerk will be entitled to charge a fee therefor in terms of the Court's current Table of Fees.

27. No person shall be allowed, without leave of the Court, to borrow the principal Application or any original deed, writing, plan, document or other production forming part of the process therein: but the Principal Clerk may, when duly requested, issue a certified copy, or copies, thereof to any party to an Application or his solicitor or other authorised representative at the charge specified in the Table of Court Fees.

28. Any solicitor acting for a party to an Application may borrow

(1) any part or parts of the process therein, other than those specified in the preceding Rule and

(2) also the part or parts so specified, by leave of the Court, or by permission of the Principal Clerk,

in each case upon granting a borrowing receipt and undertaking to return the productions borrowed to the office of the Court within 48 hours after demand by the Principal Clerk.

29. After the issue of a Final Order in any Application the Principal Clerk shall, if he considers it appropriate, return any productions lodged by any party to that party or his agents upon the granting of an appropriate receipt.

CONSIGNATION

30. In any Application which raises questions regarding any claims for payment of money which the Court has power to decide, any party may consign a sum of money in Court to be dealt with according as the rights of parties may be determined in course of the proceedings.

31. Any sum of money which a party desires, or has been ordered, to consign shall be consigned in the hands of the Principal Clerk in the same manner as in an ordinary action in the Sheriff Court and shall be held by the Principal Clerk subject to the directions of the Court. No consigned money shall be paid or uplifted without leave of the Court or the consent in writing of all parties interested.

TIME LIMITS

32. Any period limited in these Rules, or in any Order, for any act or proceeding, which expires on a Saturday, Sunday, public holiday or any other day on which the office of the Court is closed by Order of the Court, shall be extended to the next lawful day.

33. Any period limited by an Order for any act or proceeding may be extended by the Court, on cause shown, either before or after the expiry of such period.

PLEADINGS

34. A respondent is not required to lodge Answers unless, and until, Answers are ordered by the Court. Answers, Replies, Objections or other written pleadings shall be lodged by hand, by post or by FAX with the Principal Clerk, unless otherwise directed by the Court, and the Principal Clerk shall note receipt of the same on the Application. If Answers ordered by the Court are not lodged within the time specified in the Order, the Principal Clerk shall certify to that effect on the Application.

35. The Court, at any stage of an Application, order any party therein to lodge a statement or pleadings, where this has not been done, or to revise his statement or pleadings, and also to make specific any statement, answer or reply contained in his Application, Objections, Answers or other pleadings, relating to material facts disputed; and either to admit or deny definitely any statement, made by any opposing party in that party's Application or Answers or Objections or other pleadings, relating to disputed material facts, when the Court is of opinion that such statement or pleadings, specification, admission or denial is necessary to define, or determine, the real matter or matters in dispute; or to withdraw or expunge any irrelevant and improper matter contained in his Application, Answers, Replies, Objections or other pleadings.

36. All Answers, Replies, Objections or other pleadings shall be subscribed as provided for under Rules 7 and 8.

37. Where a party is represented by a solicitor or factor who lodges any Answers, Replies, Objections or other pleadings by hand or by post he shall be obliged to lodge, at the same time, a copy thereof for each of the other parties to the Application.

Amendment, conjunction, etc

38. The Court may, either of their own accord or on the motion of any person interested, at any time before a Final Order has been pronounced in an Application and upon such terms or conditions as to notice, intimation or service and expenses or otherwise, as the Court shall think proper.

(a) Amend any error, omission or defect in the Applications or any pleadings or proceeding therein;

(b) Qualify, restrict, enlarge, or add new conclusions to, the conclusions of the Application, notwithstanding that, by such amendment or addition, additional or alternative remedies may be sought or a larger sum of money or an additional area of land or other interests in land may thereby be subjected to the adjudication of the Court;

(c) Strike out the names of any persons who have, improperly or unnecessarily, been made parties to the Application;

(d) Substitute or add names or proper characters of any persons as Applicants or Respondents who ought to have been, but were not, made parties, Applicant or Respondent as the case may be, or not made parties in their proper character, representative, individual or otherwise;

(e) Substitute or add the names of any persons as parties, Applicant or Respondent as the case may be, who by reason of any assignation or renunciation by, or the marriage, sequestration or death of, any of the parties to the Application, or of any other event occurring during its dependence, have acquired any right or interest, or become subject to any liability, in respect of the matters to which the Application relates;

and, when necessary to enable the Court effectually to determine, or adjudicate on, the real matters in dispute, such amendments or additions shall be made or allowed.

39. Where the same, or similar, questions of law or fact arise in, or where there is a relation between, two or more depending Applications, the Court

(1) may sist one or more of such Applications and appoint the other Application or Applications to proceed or

(2) where it appears more convenient that they should be heard together, may conjoin such Applications and dispose of them, either together or separately, as may be found expedient.

40. The Court may appoint a curator ad litem to any party in an Application who is under the age of 16 years or is of defective capacity and has no known curator or other guardian.

ADMISSIONS, WITNESSES, PRODUCTION OF DOCUMENTS, ETC

41. As soon as an Order appointing a time and place for hearing has been pronounced in any Application, any party thereto shall be entitled, shall be entitled, unless the Order limits the hearing to matters of law or procedure,

(1) to call upon any opposing party by written notice, delivered or transmitted by first class recorded delivery service or registered post letter, not less than seven days before the time so appointed to admit, but only as between the parties giving and receiving such notice and solely for the purposes of the particular Application, any specific fact or facts stated in such notice and relating to the subject-matter of the Application. If the party so called upon unnecessarily refuses or delays to admit such specific fact or facts, he may be found liable in the expenses incurred in proving any specific fact which he so refused or delayed to admit; and

(2) to move the Court to grant a summons requiring the persons therein named and designed to attend at such appointed time and place, and any adjourned hearing, for the purpose of (a) giving evidence and/or (b) producing the writings, documents, business books, plans or articles therein specified or described.

42. Further, on special cause shown or of their own accord the Court may, at any stage of the proceedings in an Application—

(1) Order any party thereto (a) to lodge in process all writings, documents, business books, plans or articles in his possession or under his control, whether founded on by such party or not, relating to any matter in dispute therein, which are specified or described in such Order and (b) to state whether all or any writings, documents, business books, plans or articles specified or described in such Order are, or have at any time been, in his possession or under his control and whether he has parted with the same and what has become of them or any of them; and

(2) Order any person to attend the Court at a fixed time and place, or a time and place to be afterwards notified to such person, and to produce all writings, documents, business books, plans or articles in his possession, or under his control, which are specified or described in such Order, the production of which may be deemed by the Court material and proper for the determination of matters in dispute.

43. All writings, plans, books, or excerpts from books, or other documents or productions which are founded on in any Application or in any Answers, Objections, Minutes or other pleadings, or certified copies thereof, shall be lodged in process along with such Application or Answers, Objections, Minutes or other pleadings.

44. All writings, plans, books, or excerpts from books, or other documents or productions which any party intends to refer to or to use or put in evidence at any hearing shall, if in his possession or under his control, be lodged in process by such a party at least seven clear days before the time appointed for such hearing and the party lodging the same shall forthwith intimate to the other parties to the Application that he has done so. Any party to an Application lodging any such production shall be obliged to lodge, with the principal production, as many copies thereof for use by the Court as may be directed by or agreed with the Principal Clerk.

45. The Court may allow any Answers, Objections, Minutes or other pleadings and any writing, plan, book, or excerpt from a book, or other document or productions, which ought to have been, but were not, timeously lodged, to be received or to be referred to or to be used or put in evidence at the hearing, if satisfied that such omission was in the circumstances excusable, upon such conditions as to expenses, adjournment, further allowance of proof or otherwise as the Court may think proper.

46. When an Application or pleading or other original document has been lost or destroyed, a copy thereof proved and authenticated to the satisfaction of the Court may be substituted for the original to all effects and purposes.

47. If any Applicant or Respondent fails to lodge any statement or pleading or to produce any writing, plan, book, or excerpt from a book, or other document or article, which the Court have ordered to be lodged or produced, or to obey any Order of Court

(a) where the Applicant is the party in default, the Application may be in respect thereof dismissed, with expenses

(b) where the Respondent is the party in default, any pleas, objections or claims stated by him may be in respect thereof repelled, with expenses or

(c) the party in default, whether Applicant or Respondent, may be merely found liable in expenses occasioned by such default.

48. A copy of any Order summoning persons therein named to attend the Court for the purpose of giving evidence and/or producing documents, which has been certified by the Principal Clerk or by the solicitor of the party who obtained it as correct, in so far as concerns the person on or to whom the said Order is to be served, intimated or directed, shall be held as equivalent to the original Order to all effects and purposes.

49. Parties may, orally in open Court or by letters or Minutes, renounce proof or dispense with a hearing, either generally or as regards particular matters or questions.

HEARING

50. At the time and place appointed for the hearing of an Application the parties shall lead or tender such oral and documentary evidence as they desire to lead or tender on any matters of fact in dispute, unless the hearing has been, by Order, limited to matters of law or procedure.

51. When the Application is called in Court at the appointed time and place,

(a) if no appearance is made by or on behalf of an Applicant, but appearance is made by or on behalf of a Respondent, (1) the Application may be dismissed in respect of such failure to appear, with or without expenses or (3) the Respondent may proceed, on any matters of fact in dispute, or lead evidence, so far as consistent with the terms of the Order appointing a hearing, or as allowed by the Court, and may thereafter move for an Order disposing of the subject-matter of the Application;

(b) if no appearance is made by or on behalf of a Respondent, but appearance is made by or on behalf of an Applicant, (1) the Application may be continued (2) any defence, objection or claim pleaded by such Respondent may be repelled in respect of such failure to appear, with or without expenses or (3) the Applicant may proceed to lead evidence, so far as consistent with the terms of the Order appointing a hearing, or as allowed by the Court, either on any matters of fact in dispute or only on matters in regard to which the burden of proof rests upon him and may thereafter move for an Order disposing of the subject-matter of the Application;

(c) if no appearance is made by any party, the Application may be continued indefinitely or dismissed, as the Court may think proper.

52. Any person who, after being warned by the Court,

(a) wilfully disobeys any Summons or Order of the Court to attend in open Court for the purpose of producing documents and/or of giving evidence or

(b) having attended, wilfully refuses to be sworn or to affirm or to answer any proper question or to produce any book, or excerpt from a book, writing or plan or other document or article, which he has been lawfully required to produce,

may be found liable in payment of expenses occasioned by any adjournment which such disobedience or refusal renders necessary and may also be dealt with by the Court for contempt of court.

53. The Court may call and examine, or grant commission to examine, as a witness in the cause any person whose evidence appears to them to be necessary for the purpose of determining the matters in dispute, though such person has not been called or adduced by any of the parties, and may direct his fees and expenses as a witness to be paid by the parties, or any of them, in such proportion as the Court may determine, as part of the expenses of the Application.

EVIDENCE

54. Evidence shall be taken, unless otherwise agreed by parties, upon oath or affirmation.

55. Any consent or undertaking in an Application may be given by or on behalf of any party or parties thereto in letters or Minutes or verbally, either in open Court or during an inspection made by a member or members of the Court, of the lands or other subjects to which the Application relates, provided that, where such verbal consent or undertaking forms the basis of, or a material element in, any Order of Court, its tenor shall be sent out in such Order or in a note appended thereto.

56. All relevant objections to any deed or writing which is founded on in any Application may be stated and maintained by way of exception and shall for the purposes of the Application be disposed of by the Court without the necessity of proceedings being sisted in order that a reduction may be brought in the appropriate Court, unless the Court think it necessary or more convenient in the circumstances that the party challenging such deed or writing should proceed by reduction.

57. Notes of evidence may be taken down by the Court or, where all parties so desire, by a shorthand writer appointed by the Court, whose fee shall be fixed by the Court and may be ordered to be paid by the parties equally or in such proportions as the Court may think fit. Such notes of evidence may be used by the Court in any Appeal in, or Rehearing of, the Application.

Evidence on commission

58. The Court may, at any stage of the proceedings in an Application, order that the evidence of any witness whose evidence is in danger of being lost or who is resident furth of Scotland or who by reason of age or infirmity or remoteness of place of residence or other reasonable cause is unable to attend at the time and place fixed for hearing the Application or at any adjourned hearing shall be taken on commission, with or without interrogatories, by any member of the Court and/or the Principal Clerk, or any Deputy-Clerk of Court or other qualified person or persons delegated or appointed for this purpose, and shall be reported to the Court.

59. The Court may, of consent of parties, or where satisfied that such course is expedient in the interests of all parties, remit to one of their number or to the Principal Clerk or any Depute-Clerk of Court or other qualified person to take the whole evidence in the cause and report it to the Court.

APPOINTMENT OF REPORTERS, ASSESSORS, ETC

60. The Court may, either of their own accord or on the motion of any party, at any time before a Final Order has been pronounced in an Application

(1) if they consider that all, or any, of the material facts in dispute may be appropriately so ascertained, remit to any person specially qualified by skill and experience to enquire into such matters of fact and to report and may, upon such report or a further report, after affording parties an opportunity of being heard or of lodging written pleadings, if they so desire, proceed, without further inquiry or evidence, to determine the said matters of fact and any questions arising thereon or to make such other Order as the Court think just

(2) if they consider that the assistance of one or more persons specially qualified by skill and experience is desirable for the better disposal of the matters in dispute, may appoint one or more such persons to act as assessors and sit with the Court at any hearing or inspect the lands or buildings or other subjects to which the Application relates.

Such reporters, valuers or assessors shall be appointed by the Court and shall receive remuneration for their services out of funds provided by Parliament at such rates as the Treasury may sanction.

INSPECTION

61. The Court, and any assessor, valuer, surveyor or other official authorised in writing by the Court, may at any time and from time to time, during reasonable hours on any lawful day, enter upon and inspect all or any lands or buildings after notice to the parties, either in writing or verbally in open Court, to enable them to have an opportunity of attending, or being represented, at the inspection.

ABANDONMENT

62. Any Applicant or Respondent may, at any time before final decision upon, or dismissal of, the Application, abandon or withdraw his Application or Answers or Objections or pleadings by leave of the Court on such conditions as to expenses or otherwise as the Court may consider just.

FALLING ASLEEP AND WAKENING

63. If no Order has been pronounced in an Application for a year and a day it shall be held to have fallen asleep.

64. The Court may either of consent of all the parties or on the motion of one of the parties duly intimated to the other parties pronounce an Order wakening the Application and thereafter proceed with it.

REPONING

65. Where, by reason of the failure of an Applicant or Respondent to lodge any statement or pleading or to produce any writing, plan, book or excerpt from a book, or other document or article or to appear at a hearing or to obey any Order of Court or by reason of any other default, an Order has been pronounced

(1) dismissing the Application or

(2) repelling any pleas or objections or claims,

the party in such default may within thirty days from the date of intimation of such Order move the Court to be reponed against such Order; and the Court, if satisfied that such default occurred through mistake or inadvertence or was in the circumstances excusable, may, upon such terms and conditions as to expenses or further hearing or otherwise as they shall think just, recall such Order and appoint the Application to proceed as if such default had not occurred.

66. When a party, who has obtained any Order in his favour upon terms or conditions therein expressed, has failed within the time limited by such Order or by any subsequent Order (or, if no time has been so limited, then within such time as the Court think reasonable) to perform or comply with such terms or conditions, it shall be competent for any party in whose interest such terms or conditions were imposed to move the Court in the Application before it has been disposed of by a Final Order, or afterwards in a Rehearing, to recall or vary such Order.

APPEAL AGAINST DIVISIONAL COURT ORDER

67. Any competent Appeal to the Full Court against an Order by a Divisional Court shall be taken by a note dated and signed by the appellant or his solicitor or counsel or factor or by any person duly authorised in writing on behalf of the appellant.

68. Such note shall be delivered, or transmitted by first class recorded delivery service or registered post letter, to the Principal Clerk and shall be in the following or similar terms:

'The Applicant ...., (or Respondent or other Party) appeals to the Full Court in the Application Record No. ...., ..., Region (or .... District) (or .... Islands Area) on the following grounds, namely, ...'

Dated   *Signature*

The Appellant shall at the same time

(1) lodge a statement setting forth the grounds on which the Appeal is to be maintained

(2) lodge with the Principal Clerk a copy of the said note and statement for service, by the Principal Clerk, on each of the other parties and

(3) pay to the Principal Clerk the fee specified in the Table of Court Fees.

69. At least two weeks prior to the hearing of an Appeal the Appellant shall lodge a Note with the Principal Clerk stating whether or not he desires that proof, or additional proof should be led or allowed at the hearing of the appeal or that the subjects should be inspected or re-inspected.

70. Any party who has lodged an Appeal in terms of Rule No. 67 may, on cause shown, apply for leave to amend his grounds of Appeal at any time.

71. It shall not be competent to take any Appeal after the expiry of one month from the date of intimation to parties of the Order complained of.

72. It shall not be competent to take any Appeal except to the Full Court or against any Order other than

(a) A Final Order, or

(b) An Order against which the Court which pronounced it has granted leave to appeal.

73. The Full Court may give judgment in any Appeal—

(a) Without ordering either written pleadings or a hearing, where all parties so agree

(b) Upon written pleadings only, where all parties so agree.

Except as above provided, the Court shall appoint a time and place at which parties shall be heard on the Appeal.

74. Every competent Appeal shall submit to review at the instance not only of the Appellant but of every other party appearing in the Appeal the whole Orders pronounced in the Application, to the effect of enabling the Full Court to do justice between the parties without hindrance from the terms of any previous Order.

75. In the event of an Appellant obtaining leave to withdraw or abandon his Appeal, any other party appearing in the Appeal may insist in such Appeal (if otherwise competent) in the same manner and to the same effect as if it had originally been taken by himself.

76. When the Order appealed against is a Final Order, the Application shall not be remitted to the Court which pronounced it, unless special circumstances render a remit expedient, but shall be completely decided by the Full Court.

77. When the Order appealed against is an Order appealed by leave of the Court, the taking of the Appeal shall not stay procedure before the said Court in the Application. The said Court may make such interim Order, or Orders, concerning the preservation of evidence, consignation or payment of money, custody or production of documents or other like matters as just regard to the manner in which the final decision is likely to affect the parties' interests may require. Such interim Order or Orders shall not be subject to review except by the Full Court when the Appeal is heard or determined.

RE-HEARING

78. Any party to an Application whose interests are directly affected by a Final Order pronounced in an Application may move the Court, on one or more of the grounds enumerated in Rule 82, to order that the Application shall be reheard, in whole or in part, upon such terms or conditions as to expenses, or otherwise, as the Court shall think right.

79. Such motion shall be made by a note dated and signed by the party moving or his solicitor or counsel or factor or by any person duly authorised in writing on behalf of such party.

80. Such note shall be delivered, or transmitted by first class recorded delivery service or registered post letter, to the Principal Clerk and shall be in the following or similar terms:

'The Applicant ...., (or Respondent or other party) moves for a Rehearing of the Application Record No. .... Region (or .... District) (or .... Islands Area), in which a Final Order was pronounced on .... (insert date) on the following grounds, namely ....'

Dated   *Signature*

The party moving shall at the same time

(1) lodge with the Principal Clerk a statement specifying

(a) whether the whole, or a part (and, if so, what part), of the Final Order is craved to be varied, amended or recalled,

(b) whether it is desired that proof should be led or allowed at the Rehearing and, if so,

(c) to what points proof is to be directed and

(d) whether it is desired that the land, or other subjects, to which the Application relates, should be inspected or re-inspected

(2) lodge with the Principal Clerk a copy of the said note and statement for service, by the Principal Clerk, on each of the other parties and

(3) pay to the Principal Clerk the fee specified in the Table of Court Fees.

81. It shall not be competent to move for Rehearing after the expiry of three months from the date of intimation to parties of the Final Order in the Application except

(1) where all the parties whose interests may be directly affected concur in the motion or

(2) where leave to move is granted on special cause shown.

82. A motion for Rehearing may be made upon one or more of the following grounds:

(1) that the Order or Orders sought to be varied, recalled or annulled

(a) proceeded upon essential error, either shared by all the parties, or induced by one or more of the opposing parties, or

(b) were obtained or procured by fraud or fabrication of documents or subornation of perjury or other like misconduct on the part of one or more of the opposing parties in course of the Application;

(2) that pertinent and important evidence as to disputed matters of fact was tendered and erroneously rejected or disallowed;

(3) that the party moving is prepared to adduce pertinent and important evidence, of the tenor set forth in his statement, which was unknown to, and could not reasonably have been discovered by, him before a Final Order was pronounced;

(4) that the opposing party or parties has or have, without reasonable excuse, failed substantially to fulfil or comply with conditions imposed in the interest of the party moving by the Order or Orders sought to be varied or recalled;

(5) that owing to a change of circumstances the Final Order sought to be recalled has not been given effect to or is no longer appropriate.

83. Every motion for Rehearing or for leave to move for Rehearing and every Rehearing ordered shall be determined, or adjudicated on, by the Full Court.

84. The Court may dispose of the motion for Rehearing or for leave to move for Rehearing, either after hearing parties or, if parties agree to dispense with a hearing, on their written pleadings only.

85. Where the Court, having regard to any of the grounds mentioned in Rule 82, are satisfied that if the Order or Orders complained of are allowed to stand, a substantial wrong, or miscarriage of justice, which cannot by any other process be so conveniently remedied or set right, is likely to be thereby occasioned, they may order a rehearing of the Application, in whole or in part, in such manner and on such terms and conditions as they shall think just.

86. Neither a motion for Rehearing nor an Order granting a Rehearing nor any subsequent procedure therein shall have the effect of staying proceedings under, or implement of, the Order or Orders complained of, unless it be so ordered.

87. In any Appeal or Rehearing the Court may vary, recall or annul any Orders appealed or complained against either

(a) in whole or

(b) in so far only as affecting any separate and distinct part of the matters in dispute or

(c) as between some of the parties only,

and may make any Order or Orders which should have been made and also such other or further Order or Orders as they may think necessary to deal with any change of circumstances occurring after the date of the Order or Orders appealed or complained against, or to set right any substantial error, omission, defect, wrong or miscarriage of justice, and that upon such terms and conditions as they shall think just.

SPECIAL CASE

88. Any party to an Application who intends to require that a special case shall be stated on any question, or questions, of law for the opinion of a Division of the Court of Session shall, within one month after the date of intimation to parties of the decision complained of, lodge with the Principal Clerk a requisition to that effect, and also a draft statement of the case specifying

(a) the facts out of which such question, or questions, of law are alleged to have arisen

(b) the decision complained of

(c) in what respect and to what extent such decision is maintained to be erroneous in point of law and

(d) the question, or questions, of law proposed to be submitted to the Court of Session.

89. The said party shall at the same time lodge with the Principal Clerk a copy of the said requisition and draft statement of the case for service, by the Principal Clerk, on each of the parties in the Application. Any of these parties may, within three weeks after intimation of such copy or copies, lodge with the Principal Clerk a note of any proposed alterations, or observations, on the said draft statement and question or questions which they may deem necessary.

90. After adjustment by parties or the Court, the draft case shall be settled by the Court or the Chairman and returned by the Principal Clerk to the party making the requisition, in order that a fair copy of the same may be made for lodging. The fair copy special case and the settled draft shall, within fourteen days after the settled draft has been posted to such party, be lodged with the Principal Clerk in order that the special case may be authenticated by the Court.

91. On the special case being authenticated in terms of Rule 105, the Principal Clerk shall transmit the same, with relative productions, if any, which have been made part of the case, to the Deputy Principal Clerk of the Court of Session and shall notify such transmission to the parties thereto.

Within fourteen days after the receipt of the special case by the said Deputy Principal Clerk of Session, or, where such fourteen days expire during a vacation or recess of the Court of Session, then on or before the first sederunt day of the Court of Session thereafter ensuing, the party on whose requisition the special case has been stated shall lodge the same, together with any productions made part of the case, in the General Department of the Court of Session, along with a process and copies of productions for the use of the Court in terms of Rules of the Court of Session Nos. 20 and 26(b) and shall at the same time intimate the lodging of the special case to the opposite party or his solicitor and deliver to him at least ten copies of the said case; and shall also deliver three copies of the said case to the Principal Clerk of the Land Court.

In the event of such party failing to lodge the special case within the time above prescribed, any other party thereto may within the like period of time from such failure, lodge the special case and also lodge copies with the Principal Clerk of the Land Court all in like manner.

In the event of the special case not being lodged as above prescribed, the special case shall, unless the Court of Session otherwise orders, be deemed to have been withdrawn or abandoned by all the parties thereto and shall be re-transmitted by the Deputy Principal Clerk of Session to the Principal Clerk of the Land Court; and the Land Court may thereafter determine any questions of expenses relating to the preparation and settling of the special case and shall otherwise proceed where any further procedure is necessary or expedient, as if no special case had been required.

92. Neither the requisition for a special case nor any subsequent proceeding therein shall have the effect of staying procedure in the Application, or Applications, in course of which the said question, or questions, of law are alleged to have arisen, unless it be so ordered.

93. The party on whose requisition the special case has been stated, whom failing the other party or parties thereto, shall, as soon as reasonably may be after the Court of Session has pronounced opinion upon the question, or questions, of law therein set forth, take the proper steps to cause a certified copy of the said opinion together with the relative productions, if any, to be transmitted to the Principal Clerk of the Land Court.

94. When the opinion of the Court of Session has been received by the Principal Clerk, the Land Court shall, if and in so far as necessary, bring their decision on the matters in regard to which the said question or questions of law have arisen into conformity with the said opinion.

EXPENSES

95. In all proceedings before the Court the fees stated in the current Table of Fees payable to solicitors in ordinary Sheriff Court actions shall be the fees and emoluments ordinarily chargeable by, and payable to, solicitors for all professional services rendered in connection with an Application, but subject to the power of the Court, which is hereby reserved, to deal in such manner with expenses as shall in each case seem just. No higher fees or remuneration shall (unless specially sanctioned by the Court) be recoverable or be allowed between party and party or (except on such sanction or under special written agreement) between solicitor and client.

96. The Court may sanction the employment of counsel in Applications of difficulty and general importance and fix the fees payable or chargeable in such cases.

97. Accounts of expenses, charged by solicitors against clients, or awarded by the Court as between party and party, in relation to any proceeding before the Court may be remitted for taxation and report by the Auditor. Expenses may be modified at a fixed sum by the Court as between party and party.

98. When any person other than a solicitor or counsel appears by leave of the Court on behalf of any party or parties to an Application, the Court may allow, or direct the Auditor to allow, him reasonable outlays and also a remuneration for time and trouble proportionate to his services and the value of the cause.

99. In any Application the Court may make an award of expenses (and outlays) in favour of a party litigant and where expenses of a party litigant are ordered to be paid by any other party to the Application the Auditor of Court may allow as expenses such sums as appear to him to be reasonable, having regard to all the circumstances in respect of:—

(a) work done which was reasonably required in connection with the case, up to a maximum of two-thirds of the sum allowable to a solicitor for that work under the current Table of Fees for solicitors, and

(b) outlays reasonably incurred for the proper conduct of the case.

Without prejudice to the generality of the above, the circumstances to which the Auditor of Court will have regard in determining what sum, if any, to allow in respect of any work done may include

(1) the nature of the work;

(2) the time taken and the time reasonably required to do the work;

(3) the amount of time spent in respect of which there was no loss of earnings;

(4) the amount of any earnings lost during the time required to do the work;

(5) the importance of the case to the party litigant, and

(6) the complexity of the issues involved in the case.

100. When two or more Applications involving similar questions and arising with respect to holdings which are held under the same landlord are heard at the same sitting, whether formally conjoined or not, the Court may award to the solicitor, or the person appearing by leave of the Court, who conducts the same, an inclusive fee or remuneration in respect of all the Applications in which he so appears and may settle the proportions in which it shall be paid by the respective parties.

ORDERS OF THE COURT, ETC

101. Every Order shall be in writing and shall be signed by at least one member of the Court and initialled by the Principal Clerk or one of the Clerks of Court.

102. Any one member of the Court or the Principal Clerk may sign, and power is hereby delegated to each member or the Principal Clerk, to sign for the Court any Order which merely

(a) appoints answers, replies, objections, minutes, statements or other pleadings or documents or articles founded on by a party to be lodged with the Principal Clerk or

(b) directs any service, notice or intimation to be made or given on or to any party or parties or

(c) grants a summons to attend a sitting of the Court for the purpose of giving evidence and/or producing documents or articles or

(d) fixes or alters the date of any sitting of the Court or hearing or

(e) requires borrowed productions to be returned.

Any one member of the Court may sign and power is hereby delegated to each member to sign for the Court any Final Order in an unopposed Application and any Order which

(a) interpones authority or effect to a joint minute for parties or

(b) allows any Application or Appeal or motion for Rehearing or other proceeding therein to be, by consent of all parties, amended or abandoned or withdrawn as the case may be, or the like. Such signature by any one member or the Principal Clerk shall in these cases be sufficient.

103. Any verbal, clerical or casual error or omission or informality in an Order may be corrected or supplied de recenti, or of consent of parties, by the member or members of Court or the Principal Clerk who signed it.

104. Every Order which disposes of the subject-matter of an Application or of any separate controverted part thereof, otherwise than by consent of parties, shall be signed by the member, or members, of the Court by whom the case, or such separate part thereof, has been considered and determined; provided that in the event of a difference of opinion among such members it shall be competent for the majority to omit signing the order and to record dissent from any of the findings in the Order by note appended thereto; and the signatures of the majority shall in that case be sufficient.

105. All extracts or copies of Orders by the Court, required for the purpose of being used in any proceeding before any court of law, arbiter, public department or any public authority whatever, and all special cases stated shall be authenticated by the signature of the Chairman or of the Principal Clerk and sealed with the seal of the Court before being issued from the office.

106. The Principal Clerk shall at the request of any interested party issue, free of charge, an extract of any Order pronounced by the Court provided that, where applicable, the time limit for lodging an appeal in terms of Rule No. 71 has passed. Every such extract shall include a warrant for execution in the following terms:— 'and the Court grant warrant for all lawful execution hereon'.

GENERAL

107. In matters of procedure or evidence which are not provided for by statute or by these Rules the Court shall have regard to the general practice of courts of law so far as applicable and appropriate to the conduct of its business.

APPENDIX 2

Styles

Lease for arable/stock farm

LEASE[1]

between

[   ]
(hereinafter referred to as 'the Landlord')
<u>ON THE ONE PART</u>

and

[   ]
(hereinafter referred to as 'the Tenant')[2]
<u>ON THE OTHER PART</u>

IT IS CONTRACTED AND AGREED BETWEEN the landlord and the Tenant as follows:

(**ONE**)   The Landlord lets to the Tenant (but expressly excluding legatees, successors by virtue of any statute, assignees, whether legal or voluntary, trustees, sub-tenants, creditors and trustees or managers for creditors) [and declaring for the avoidance of doubt that this Lease shall *ipso facto* terminate on the dissolution of the Tenant][2] ALL and WHOLE the Farm and Lands of [   ] extending to [   ] hundred hectares and [   ] decimal or one hundredth parts of an hectare ([ ] ha) or thereby all as specified in the schedule of acreages annexed and signed as relative hereto and [ ] shown outlined in red on the demonstrative Plan annexed and signed as relative hereto[3], which areas and Plan, although believed to be correct, are not warranted (which Farm, is hereinafter referred to as 'the Farm').

---

1 As a 'style' this lease will require modification to take account of the particular details relating to the farm to be let.

It is a 'landlord's' style. A tenant might want to negotiate for an exclusion or restriction of some of the clauses.

2 If the lease is to a partnership and the intention is that the lease should continue, even where there are changes in the partners, then the lease must be to the 'House'; see *IRC v Graham's Trs* 1971 SC(HL) 1.

3 If part (or all) of the farm is hill ground or permanent pasture, then the hill ground and permanent pasture should be specified here or in the schedule.

(TWO)   The Lease shall run from [   ] until [   ][4].

(THREE)
(a) The rent payable by the Tenant shall be [   ] THOUSAND POUNDS
(£[   ]) STERLING per annum (of which, for the purposes of calculating Value Added Tax (VAT) the sum of [   ] Pounds (£[   ]) is allocated to [   ] Farmhouse) exclusive of Value Added Tax payable
thereon (whether or not at the Landlord's option) payable half-yearly in
arrears at the term days of Martinmas (Twenty eighth November) and
Whitsunday (Twenty eighth May) in each year commencing with the first
payment of [   ] pounds (£[   ]) exclusive of VAT as at Twenty eighth
[   ], [   ] for the period from Twenty eighth [   ] to Twenty eighth
[   ] and the next half year's rent of [   ] THOUSAND POUNDS
(£[   ]) as at Twenty eighth [   ] for the half year preceding and so on
half-yearly thereafter with interest at a rate per annum [   ] on each half
year's payment until paid, and no claim for damages or other claim whatever shall entitle the Tenant to withhold or delay payment of the rent or
any part thereof unless the Landlord has in writing admitted the claim in
full or has been found liable thereunder by the Court or an arbiter.
(b) No stock or crop shall be removed until all rent and any other sums due to
the Landlord have been paid.
(c) The Landlord shall not be barred from irritating the Lease or claiming
damages at any time from the Tenant for failure to implement any obligation under this Lease or any subsequent agreement, formal or otherwise,
relating thereto by reason of the fact that the Landlord may have accepted
rent after the cause of action arose.

(FOUR)   There are reserved to the Landlord:
(A)[5] generally the whole strata beneath the tillable surface together with all
oils, petroleum, natural gases (and their hydrocarbons), stone, sand,
gravel, mines, pits, quarries, metals and minerals of every description on,
in or under the Farm with power to the Landlord and those authorised
by him to do everything necessary to search for, win, work and carry
away the same and any other material found on, in or under the Farm or
any other lands belonging to the Landlord or any other proprietor, and to
erect buildings, plant, factories and works, and to carry out any other
operation that the Landlord may consider necessary for these purposes
or other purposes related thereto declaring that (i) in the event of loss or
damage to crop, stock, implements, fixtures or fixed equipment by surface damage, or subsidence, in connection with the exercise of the rights
reserved by the Landlord in terms of this paragraph, compensation shall
be paid to the Tenant as may be determined failing agreement, by arbitration (ii) the Tenant shall have no claim against the Landlord for injury
or damage resulting from mining operations carried out prior to the commencement of this Lease (iii) the Landlord shall have full power and liberty to work stone, sand, gravel, limestone, and others for use on the

---

4 The most appropriate duration is for one year from entry to the anniversary of that entry term. The
lease then continues by tacit relocation under the 1991 Act, s 3. The landlord then has the option of
serving a notice to quit, which he cannot do except at a term, when the lease may be terminated.
5 If there is coal under the farm the landlord might be wise to reserve the right to permit opencast
coal extraction without substantial payments to the tenant; see Opencast Coal Act 1958.

Farm without abatement of rent or compensation for surface damage and (iv) compensation for ground resumed for any of the purposes specified in this sub-clause shall be determined in accordance with the provisions of sub-Clause (D) hereof.

(B) all woods, plantations, trees, brushwood, underwood and coppice planted or to be planted on the Farm with free access for all necessary purposes to and from the same and to and from neighbouring plantations through the Farm, with power to the Landlord on payment for surface damage to the Tenant (1) to remove trees and others whether from the Farm or neighbouring plantations through lands of the Farm not in crop; (2) to plant and fence trees in place of trees removed and hedges; (3) to plant and fence trees or hedges along the boundaries of the Farm and (4) to pasture or cut and remove the grass growing within woods and plantations.

(C) all existing rights of way, wayleaves and servitudes, with power to grant further wayleaves and servitudes for all necessary estate and other purposes, subject to payment to the Tenant for any surface damage occasioned by the granting of any such further rights.

(D) power to resume (a) any part or parts of the Farm at any time and from time to time on giving three months' written notice for any purpose whatever other than for agriculture or pasture[6]; the Landlord where necessary, being bound to erect and thereafter to maintain stock-proof fences round all ground so resumed, declaring that (i) the Tenant shall be allowed deductions from the rent for ground so resumed from the respective dates of resumption such deductions being *pro rata* on the basis of the area resumed in relation to the total area of the Farm immediately prior to the resumption, which deductions shall represent the annual value of the ground resumed and (ii) the Tenant shall be entitled to the statutory claim for disturbance but the said deductions and claim shall be held to cover the Tenant's whole claim for loss of profit and all other claims of whatever nature arising out of or caused by such resumptions and (b) any dwellinghouse and garden on the Farm which has not been occupied for a continuous period of six months by someone engaged in agricultural activities on the Farm forthwith without alteration in the rent payable.

(E) power to alter marches and to excamb land, and later the direction of the roads on the Farm and to make new roads, but not in such a way as to have a materially adverse effect on the farming activities of the Tenant, the annual value of any land that may be added to the Farm by excambion or alteration of marches being added to the rent of the Farm, and the annual value of any land that may be taken away from the Farm being deducted from the said rent, in either case the alteration in rent to be made according to the method of calculating compensation for ground resumed as stipulated in sub-clause (D) of this Clause and, failing agreement, to be determined by arbitration.

---

6 If the landlord wants to reserve a right to resume for agricultural purposes this must be clearly specified. Resumptions for agricultural purposes are subject to the notice to quit procedures; see 1991 Act, s 21(7).

(F)[7]   all shootings on the Farm and all fishings in the waters on, adjoining or flowing through the Farm, with the exclusive right of sporting, hunting and coursing on the Farm and of taking and killing birds and wild animals of every description and fish, subject to Clause Twenty three hereof and the tenant's rights under the Ground Game (Scotland) Acts: and the Tenant, so far as in his power, shall protect the game, deer, wildfowl and fish on the Farm and shall immediately give notice to the Landlord or his factor of any poaching or suspected poaching.

(G)   right to use for all necessary purposes all common roads and means of access to other lands belonging to the Landlord.

(H)   all water whether in rivers, burns, springs, lochs, ponds, wells, reservoirs, dams, drains, conduits, aqueducts, canals or underground channels with right of access thereto and with power to the Landlord to regulate the flow and to alter the route of any watercourse or drain, and to conduct springs and streams of water from or through the Farm, subject however to the overriding right of the Tenant to the use of water by the Tenant for domestic purposes and for watering and dipping stock only, the quality and quantity of the water source, being neither guaranteed nor warranted by the Landlord, declaring that (i) any extension of the use of water by the Tenant shall require the prior written consent of the Landlord and be carried out at the Tenant's sole cost, and (ii) the Tenant is expressly prohibited from causing any pollution of water on, adjoining or flowing through the Farm particularly, and without prejudice to the generality, from the washing of sheep, sheep dip, silage and sewerage effluents.

(**FIVE**)   No straw shall be burned on the Farm without prior notice being given to the Landlord and then only in accordance with the Straw and Stubble Burning Code issued by the National Farmers Union of Scotland. The Tenant shall be responsible for notifying adjoining occupiers of any proposed burning and for employing sufficient people to control the fire or fires. The Tenant shall be liable for any resultant damage caused to property, whether belonging to the Landlord or other parties, including damage to crops, buildings, plantations, fences and others. No heather, whin or broom may be burned on the Farm except between Thirtieth September and Sixteenth April and that only with prior permission in writing from the Landlord who shall be entitled to direct the burning and the Tenant shall be responsible for notifying adjoining proprietors and for employing sufficient people to control the fires and reimbursing the proprietors (including the Landlord) of adjoining property for any loss sustained by them as a result of muirburning carried out by the Tenant.

(**SIX**)   The Landlord undertakes (1) in the event of damage by fire to any building belonging to the Landlord comprised in the Farm to reinstate or replace the building if its reinstatement or replacement is required for the fulfilment of his responsibilities to manage the Farm in accordance with the rules of good land

---

7 If the farm is a hill farm with moorland on which grouse shooting or stalking takes place this clause should make provision for the tenant either to ingather his stock before shooting or stalking days or not to gather or otherwise disturb the grouse or deer in the days leading up to a shooting or stalking day.

management and (2) to insure to their full value all such buildings against damage by fire. The Tenant undertakes to notify the Landlord or his factor immediately of any outbreak of fire. Further the Tenant undertakes (1) in the event of the destruction by fire of harvested crops grown on the Farm for consumption thereon, to return to the Farm the full equivalent manurial value of the crops destroyed in so far as the return thereof is required for the fulfilment of its responsibilities to farm in accordance with the rules of good husbandry; (2) to insure to their full value against damage by fire (a) all live and dead stock on the Farm and (b) all such harvested crops as aforesaid and (3) to insure to their full value all livestock against electrocution risks; and (4) to produce to the Landlord on request, the policies and renewal receipts of said insurances.

(**SEVEN**)   All rates, charges for water and sewerage services, charges for the inspection of private water supplies serving the Farm and other burdens levied or to be levied by the Local Authority or others in respect of the Farm except ground burdens, shall be paid by the Tenant.

(**EIGHT**)   The Tenant shall always keep the Farm fully stocked and equipped with his own *bona fide* stock, crop, fixtures, implements, machinery and equipment[8]. Without prejudice to the foregoing generality the Tenant shall not let or sell grazing on the Farm or permit the stock of any other person to be grazed or fed on the Farm[9]. For the purpose of fulfilling the requirements of this clause, any implements, machinery or equipment held by the Tenant on hire-purchase or leasing agreement shall for the purpose of this Clause be regarded as the Tenant's own *bona fide* implements, machinery or equipment and, at any time, if called upon to do so, the Tenant shall produce proof of ownership to the Landlord.

(**NINE**)   The Tenant shall manage, manure, cultivate, clean, crop, and stock the Farm according to the most approved system of good husbandry and in particular shall (a) use its best endeavours to prevent the introduction onto the Farm of any notifiable disease, (b) notify the Landlord forthwith of any outbreak or suspected outbreak of any notifiable disease on the Farm and (c) take all reasonable precautions to prevent the introduction or spread of eelworm on the Farm. The Tenant shall leave the Farm in such rotation as may be agreed on in writing and that not in patches, but in whole fields. The whole hay, straw and other fodder shall not be sold or removed from the Farm unless provision has been made to return the full equivalent manurial value of any such fodder as may be sold off or removed from the Farm.

(**TEN**)   In the last year of the Lease if so required by the Landlord the Tenant shall sow out, harrow and roll free of charge on one-sixth of the arable land not already in grass a good and suitable grass and clover seed mixture to be approved by the Landlord or provided by the incoming tenant, the cost of which seeds (excluding cultivations) on production of vouchers, shall except insofar as provided by the incoming tenant or the Landlord, be repaid by the

---

8 This clause is more often honoured in the breach; eg where the lease is to an individual, but the farm is farmed through the medium of a (eg family) partnership. If the tenant plans to farm through a partnership this clause will need to be modified.

9 This clause is necessary, because a prohibition against sub-letting may not exclude the letting of seasonal grazing; see p 45 note 1. Such a clause would put a transferee of milk quota in bad faith if he took a grazing lease from the tenant for the purpose of transferring milk quota of the holding; see p 301.

incoming tenant or the Landlord provided that the young grass is carefully hained from stock and otherwise preserved from injury.

(ELEVEN)   All dung and slurry made on the Farm shall be applied to the land timeously each year. No dung or slurry shall be sold or removed from the Farm. All dung and slurry made on the Farm after the First June immediately prior to the date of waygoing and not applied to the land before waygoing shall be taken over at valuation by the Landlord or, in the Landlord's option, the incoming tenant provided that the dung is properly heaped in the midden and the slurry stored in slurry tanks. All older dung shall be left free of charge.

(TWELVE)   In addition to the items specified in the two immediately preceding clauses the Tenant shall be entitled to receive payment at waygoing from the Landlord or, in his option, the incoming tenant for the items specified in Part III of the Fifth Schedule to the Agricultural Holdings (Scotland) Act 1991, the silage in the pit, the straw provided it has been baled and protected from the weather and the first, second and third year grasses on the Farm. At its ingoing the Tenant shall be obliged to pay the outgoing tenant for the items specified in this and the two preceding clauses.

(THIRTEEN)   The Tenant shall not carry out any such improvements as are specified in Part II of the Fifth Schedule to the Agricultural Holdings (Scotland) Act 1991 unless at least three months' written notice has been given to the Landlord sent by Recorded Delivery Postal Service and the plans, specifications, proposed method of construction and siting have first received the approval in writing of the Landlord, without prejudice to the Landlord's right of objection to the Tenant's proposal to carry out any such works under section 39 of the Agricultural Holdings (Scotland) Act 1991.

(FOURTEEN)[10]   The Farm is let as an arable and stock rearing farm only, and is specifically not let as a dairy farm or for market gardening, fruit, flower and vegetable growing, or other intensive forms of agriculture. If the Tenant shall carry on a dairy or dairying on the Farm it shall be at its own risk and on its own authority only and the Landlord shall not be bound to carry out any alterations or improvements to the Farm or to any buildings or other fixed equipment thereon or to the water supply for the purposes of a dairy or dairying, nor shall the Tenant be entitled to require the Landlord to do any act or

---

10 If the farm is let as a hill sheep farm, then this clause may require to specify that 'a full and sufficient sheep stock' is carried or the maximum numbers of the sheep flock that are to be carried on the farm. It may provide for a bound sheep stock to be taken over (with quota) on the termination of the tenancy; eg:

'The tenant shall be bound to take over the existing sheep from the landlord or outgoing at a valuation to be fixed by arbitration in terms of sections 68 to 71 and [Schedule 10] of the Agricultural Holdings (Scotland) Act 1991 [and the sheep annual premium quota pertaining to said sheep stock at a valuation to be fixed by said arbiter] and thereafter the sheep stock on the holding shall be limited to [number] breeding ewes [with lambs at foot] and [number] hoggs with no more than [number] tups, for the duration of the lease, which numbers of sheep stock the tenant shall be bound to keep and maintain on the holding in accordance with the rules of good husbandry, in regular ages, all of good quality. On the termination of the tenancy, the tenant shall be bound to deliver to the landlord or the incoming tenant a like number of breeding ewes [but not lambs at foot], hoggs and tups in regular ages, all of good quality, at a valuation to be fixed by arbitration under said sections and schedule of the 1991 Act or any amending legislation [together with the sheep annual premium quota and/or any other quota subsequently allocated and pertaining to said sheep stock at a valuation to be fixed by said arbiter].'

to carry out any works in terms of legislation or regulations in regard to milk and dairies and in the event of the Tenant executing such works at its own cost, it shall not be entitled to compensation therefor in terms of any statute or otherwise. The Farm may not be used for any purpose other than agriculture and livestock breeding for the duration of these presents.

(**FIFTEEN**) The Tenant accepts the fixed equipment (as defined in the Agricultural Holdings (Scotland) Act 1991[11]) on the Farm as in a thorough state of repair and in every way suitable and sufficient to fulfil the Landlord's obligations relating to the provision of fixed equipment under section 5 of the Agricultural Holdings (Scotland) Act 1991 and the Tenant shall be bound to maintain the same in a good and sufficient state of repair and further undertakes in the case of fixed equipment, provided, improved or renewed by the Landlord during the tenancy to maintain such in as good a state as it was in immediately after it had been so provided, improved, replaced or renewed by the Landlord and shall leave the whole fixed equipment in accordance with the foregoing provisions of this clause at its removal or pay for any deterioration that may have taken place thereon, fair wear and tear excepted, as the same failing agreement shall be settled by arbitration and in particular but without prejudice to the foregoing generality the Tenant shall

(a) when necessary repair or renew rhones, gutters, downpipes, internal plumbing and sanitary pipes and fittings serving the Farm and also repair or renew all taps, water ballcocks, troughs and cisterns,

(b) when necessary repair, renew or replace all internal and boundary fences, hedges, dykes, gates, gateposts and all roadside and riverside fences whether internal or boundary and in particular maintain all internal and boundary dykes 'slap free',

(c) keep clean and free-running all internal and boundary ditches on the Farm scouring them as often as may be required and at least every two years, except that where a ditch forms the boundary between the Farm and other lands (including woodlands) belonging to the Landlord or between the Farm and lands belonging to other proprietors, the Tenant shall be liable for only half the cost of such work,

(d) maintain to their present standard at its own expense the roads and bridges on or serving the Farm but declaring that where damage is caused to roads or bridges by the Landlord or those authorised by him, the same shall be made good by the Landlord,

(e) once in every five years from the beginning of the tenancy and in the last year thereof thoroughly clean, prepare and paint the outside woodwork and ironwork of all buildings and all rhones, rhone water, soil and other cast iron pipes with at least two coats of good oil paint of approved colours,

(f) once in every five years, or otherwise as may be expressly agreed following consultation with the Landlords, clean, colour, whiten, paper and paint with materials of suitable quality the inside of the farmhouse and all other buildings on the Farm which have been previously so treated, also prepare and paint metalwork of roof trusses and tubular trevisses and in each year limewash the inside of all farm buildings which have been previously so treated,

---

11 If the parties are not to have a Record made [1991 Act, s 5(1)], then the fixed equipment should be listed and described in a Schedule.

(g) renew all broken or cracked slates, tiles, fibre cement sheets, pointing or harling from time to time as damage occurs and replace all slipped slates and tiles; renew, when damaged or replace, after slipping, any corrugated iron sheets or zinc or asbestos ridging and capping and keep clean and in good working order all roof valleys, rhones and downpipes, and

(h) repair all damage to doors, windows, trevisses or hecks done or caused by the Tenant or its employees or stock on the Farm and replace all broken glass in windows, roof lights and deadlights.

In the event of the Tenant failing to carry out its obligations under this Clause within three months of its being notified by or on behalf of the Landlord in writing sent to the Tenant at the Farm of such failure, the Landlord, without prejudice to his rights under section 22(2)(e) of the Agricultural Holdings (Scotland) Act 1991, shall in his sole option be entitled to carry out the necessary repairs at the expense of the Tenant and to recover the cost thereof from it as the same may be certified by the Landlord on demand.

(**SIXTEEN**)   The Landlord and any persons authorised by him shall have, at all reasonable times, the right to enter all parts of the Farm for viewing the state thereof or for fulfilling his obligations or exercising his rights thereover or to take soil samples and the right to walk, ride and drive on the Farm subject to making good any material damage arising therefrom.

(**SEVENTEEN**)   The Tenant shall not allow any dwellinghouse or any part thereof to be occupied by anyone not employed on the Farm, other than the families of farmworkers.

(**EIGHTEEN**)   The Tenant shall provide to the Landlord as and when submitted copies of all agricultural returns, Integrated Administration and Control System forms and the like which he is required to make to the Scottish Office Agriculture Environment and Fisheries Department. All such returns and others shall be timeously and accurately made and the Tenant shall free, relieve and indemnify the Landlord from and against all losses and costs which the Landlord may suffer or incur, whether during or after the termination of the tenancy, or of any failure to meet the requirements for the obtaining of Arable Area Payments including set-aside. Insofar as the Tenant is not prevented by law from contracting so to do the Tenant agrees not to enter into any voluntary, guaranteed or other 'set-aside' agreement or similar provision which restricts agricultural activity on any part of the Farm for more than one season without the prior written consent of the Landlord.

(**NINETEEN**)   The Tenant shall produce to the Landlord annually a true record of the cropping on the Farm for the preceding crop year. The Tenant is expressly prohibited from:

(a) [ploughing[12]]

(b) entering into, without the express prior consent of the Landlord, any Farm Woodland Premium or other forms of schemes or arrangements in respect of tree planting on any part of the Farm.

---

12 It is usual to provide that the tenant is prohibited from ploughing hill ground or permanent pasture except for the purposes of reseeding, perhaps only with the landlord's consent (see note 3 above and the 1991 Act, s 9).

(TWENTY)   The Tenant is prohibited from nailing, stapling or otherwise fixing netting, fencing wire or any other metal or iron work or other object to growing trees.

(TWENTY-ONE)   The Tenant shall not breed or keep and shall not permit others to breed or keep dogs on the Farm except dogs used solely for the working of sheep and cattle on the Farm or as domestic pets.

(TWENTY-TWO)   Camping, caravans, huts, posters, signs, advertisements, chalets, car parking, picnic parties, hikers, pony trekkers, squatters, tinkers and vagrants are prohibited on the Farm and the Tenant shall take all reasonable steps to prevent the same being present on the Farm. No tipping or commercial riding establishments are permitted on the Farm.

(TWENTY-THREE)   The Tenant shall attempt to destroy rabbits, moles and other ground vermin on the Farm. The Tenant shall also cut down or effectively spray before flowering all thistles, docks, ragwort, nettles and other weeds and, in the event of failure of the Tenant so to do, the Landlord reserves the right, but shall not be bound, to have the work done at the Tenant's expense and in such event he shall be entitled to charge the cost thereof to the Tenant who shall pay the same on demand with interest at a rate per annum from the date of demand until paid. The Tenant is prohibited from (a) clearing any whins and brooms and from tearing out or removing hedges and dykes on the Farm and (b) applying any chemicals or other preparations to the Farm in such a way as to cause damage to hedges, game birds and other creatures (whether on the Farm or adjoining land) or crops on adjoining land and shall forthwith make good or pay compensation for any damage caused by the foregoing prohibited activities.

(TWENTY-FOUR)   Fixtures introduced on to the Farm by the Tenant shall be kept in proper repair and working order by the Tenant at all times and at waygoing the Landlord or the incoming tenant shall have the option of taking over any or all of the said fixtures at valuation. If at waygoing the Tenant removes any such as are not taken over as aforesaid, the Tenant shall be obliged to restore any damage caused by the removal of such fixtures.

(TWENTY-FIVE)   Any valuation relating to dung, straw, crops or other specific things which require to be valued at the commencement or termination of this Lease for taking over by or from the Tenant shall, unless otherwise agreed, be settled by a single valuer mutually chosen and if the parties shall fail to agree on the identity of a valuer within one month after either shall first have proposed the name of a competent person, either party may apply by summary Petition or otherwise to the Sheriff of [   ] for the nomination of a Valuer.

(TWENTY-SIX)[13]   The Tenant shall not without the prior consent in writing of the Landlord, which consent shall not be unreasonably withheld or

---

13 As this lease is for an arable and stock rearing farm no particular provision is made in regard to milk quota. If the farm is let as a dairy farm with milk quota, then a clause should be included requiring the tenant to transfer the same volume of the milk quota (less official reductions) to the landlord at the termination of the tenancy. There should be a stipulation that the tenant should do nothing to transfer quota off the holding or to prejudice the quantity of milk quota attached to the farm. There should be a requirement that the tenant exhibit to the landlord all relevant milk quota returns and an extract of the milk quota register. In theory the landlord's position is protected as an interested party, but this is better fortified by contractual provisions.

delayed, sell, assign or dispose of to any third party its rights to any quotas, subsidies, premia, allowances or other payments or production rights relating to crops, livestock and other farm produce whether pertaining to the Farm or not or which the Tenant may acquire in any manner or way from third parties and which the Tenant uses specifically in connection with the Tenant's occupation of the Farm and including in particular the rights to suckler cow and sheep annual premia which the Tenant acquires from the outgoing tenant; declaring that (ONE) in the event of the Tenant selling or otherwise disposing of all or any of his entitlement to any such quotas and others the Landlord without prejudice to his right of irritancy in terms of this Lease will be entitled to recover from the Tenant either (a) the full price received or estimated value receivable (as shall be determined, failing agreement, by arbitration) by the Tenant for such sale or other disposal as the case may be or (b) the estimated loss of value of the Farm as a consequence of such sale or other disposal whichever is the greater, and the Tenant shall be bound and obliged to pay the whole costs of the Landlord incurred in connection therewith; (TWO) the Tenant shall be bound and obliged to take all steps open to it to maintain and retain unreduced all such quotas and others and in particular (but without prejudice to the generality) to make regularly and timeously all applications required to maintain such quotas and others; (THREE) at the termination of this Lease howsoever caused (a) the Landlord, or at the Landlord's option, the incoming tenant shall be entitled to acquire any entitlement of the Tenant to any such quotas and others as the Tenant would, apart from the provisions of this clause, be entitled to dispose of separately from occupation of the Farm or any part of it at the current market value thereof and (b) that a Tenant shall take all such steps as may be necessary to have the benefit of all other such quotas and others in so far as relating to the Farm whether or not the Landlord or incoming tenant is required to pay for them in terms of this clause or otherwise transferred to the Landlord or in his option the incoming tenant.

(**TWENTY-SEVEN**)   If the Tenant [   ] shall (a) become bankrupt or apparently insolvent, (b) have its, his or her estate sequestrated or (c) grant a Trust Deed for behoof of its, his or her creditors or if the Tenant shall (a) possess the Farm directly or indirectly for the benefit of its creditors though nominally for itself (b) allow one half year's rent to remain unpaid for three months after it shall have become due, (c) assign or sublet the Farm or any part thereof without the Landlord's written consent, or (d) commit any other breach which is not remediable or which is not remedied by the Tenant within a reasonable time after being required so to do by notice in writing from the Landlord of any material term or condition of this Lease or any subsequent Minute to this Lease or any agreement in writing regulating any matter relative to the tenancy, then in any of these events the Landlord shall, in his option, be at liberty forthwith to terminate this Lease and to resume possession of the Farm without any declarator or process of law and without prejudice to any claims which he may have against the Tenant under this Lease or otherwise with exclusive right to all sown crops and to all possessions of the Tenant on the Farm in so far as required to meet all obligations due by the Tenant declaring that (a) the Landlord shall not be prevented from exercising his right of irritancy contained in this clause by reason that he is or may be in breach of any obligations incumbent on him under this Lease or otherwise relating to the Farm and (b) such irritancies are hereby declared to be pactional and not purgeable at the Bar.

(TWENTY-EIGHT)   Each party will be responsible for their own expenses incurred in the preparation and negotiation of this Lease and any agreement relating thereto. The stamp duty will be paid by the Tenant.

(TWENTY-NINE)   The Tenant binds and obliges itself to comply with all obligations incumbent on any party in occupation of and/or in control of land imposed whether under the Public Health (Scotland) Act 1897, the Environmental Protection Act 1990, the Environment Act 1995, the Control of Pollution Act 1974 (as amended by Schedule 23 of the Water Act 1989), Natural Heritage (Scotland) Act 1991 and any subsequent amending, re-enacting, consolidating or other environmental protection or control of pollution legislation from time to time in force and all rules, regulations and orders issued pursuant thereto and to indemnify the Landlord against all actions, proceedings, damages, penalties, costs, charges, claims and demands in respect of any failure to comply as aforesaid and in respect of proceedings raised under the common law relating to any such failure. The terms of this clause will be a continuing enforceable obligation on the Tenant notwithstanding the termination of the tenancy howsoever caused.

(THIRTY)   The Landlord and the Tenant consent to the registration of this Lease and of all decrees arbitral, interim or final issued hereunder for preservation and execution.

(THIRTY-ONE)   It is hereby certified that this Lease is not a lease which gives effect to an agreement for lease as interpreted by the Inland Revenue in terms of guidance note dated Thirtieth June Nineteen hundred and ninety four referring to section 240 of the Finance Act 1994.

IN WITNESS WHEREOF

Post-lease agreement

MINUTE OF AGREEMENT[1]

between

[   ]
(hereinafter referred to as 'the Landlord')
ON THE ONE PART

and

[   ]
(hereinafter referred to as 'the Tenant')
ON THE OTHER PART

Whereas the parties have entered into a Lease[2] dated [   ] of the farm and Lands of [   ] extending in total to [   ] hectares and [   ] decimal or one hundredth parts of an hectare or thereby, which subjects are hereinafter referred to as 'the farm', with entry at [   ], prior to the execution hereof and in terms of section Five, sub-section Three of the Agricultural Holdings (Scotland) Act 1991 they have agreed to vary the obligations expressed or implied by statute in the said Lease incumbent on the Landlord and the Tenant respectively for the provision and maintenance of fixed equipment on the Farm during the tenancy NOW THEREFORE it is hereby agreed as follows:-

FIRST The tenant accepts the buildings and all fixtures and fittings therein and the roads, bridges, dykes, all boundary fences, hedges, gates, ditches (whether within or outside the Farm), drains, streams and the whole other fixed equipment of the Farm in their state at the date of entry as being in a thorough state of repair and in every way suitable and sufficient to fulfil the Landlord's obligations relating to the condition and provision of fixed equipment under section 5 of the said Act and the Tenant shall be bound at his own expense to (a) maintain the same and any fixed equipment provided, improved

---

1 As a 'style' this minute will require modification to take account of the particular details relating to the fixed equipment on the farm to be let.

It is a 'landlord's' style. A tenant might want to negotiate for an exclusion or restriction of some of the obligations transferred to him or some of the clauses.

2 The agreement must be 'made ... after the lease has been entered into'; 1991 Act, s 5(3). In *Murray v Fane* (22 Apr 1996, unreported) Perth Sh Ct (under appeal), the sheriff, in a stated case, held that a 'post-lease' agreement was not valid and enforceable in circumstances (1) where the missives settling the terms of the lease provided that the tenant was to enter into a post-lease agreement and, alternatively (2) where the parties had negotiated the terms of the lease and post-lease agreement at the same time and the documents were then sent to the tenant who signed the lease and post-lease agreement before sending the documents to the landlord for signature. The sheriff said that 'the whole act of making the agreement (including execution by the tenant) requires to take place after the lease has been entered into, that is to say after the lease has been executed by both parties.'

or renewed by the Landlord during the tenancy in as good a state as it was in at the date of entry or immediately after it had been so provided, improved, replaced or renewed by the Landlord, (b) carry out on behalf of the Landlord all repairs, replacement and renewals of the fixed equipment on the Farm, which may be required whether as a result of natural decay, the passage of time, fair wear and tear or otherwise during the tenancy and to maintain the same in at least as good condition as that in which it stood at the date of entry or when first introduced on to the Farm and (c) in particular but without prejudice to the foregoing generality the Tenant shall

(a) when necessary repair or renew rhones, gutters, downpipes, internal plumbing and sanitary pipes and fittings serving the Farm and also repair or renew all taps, water ballcocks, troughs and cisterns;

(b) when necessary, repair, renew or replace all internal and boundary fences, hedges, gates, gateposts and all roadside and riverside fences, whether internal or boundary and in particular maintain all internal and boundary dykes 'slap free';

(c) keep clean and free-running all internal and boundary ditches on the Farm scouring them as often as may be required, except that where a ditch forms the boundary between the Farm and other lands (including woodlands) belonging to the Landlord or between the Farm and lands belonging to other Proprietors, the Tenant shall be liable for only half the cost of such works;

(d) maintain to their present standard the roads and bridges on or serving the Farm but declaring that where damage is caused to roads or bridges by the Landlord or those authorised by him, the same shall be made good by the Landlord;

(e) once in every five years from the beginning of the tenancy and in the last year thereof thoroughly clean, prepare and paint the outside woodwork and ironwork of all buildings and all rhones, water, soil and other cast iron pipes with at least two coats of good oil paint of approved colours;

(f) renew all broken or cracked slates, tiles, fibre cement sheets, pointing or harling from time to time as damage occurs and replace all slipped slates and tiles; renew, when damaged or replace, after slipping, any corrugated iron sheets or zinc or asbestos ridging and capping and keep clean and in good working order all roof valleys, rhones and downpipes;

(g) repair all damage to doors, windows, trevisses or hecks done or caused by the Tenant or his employees or stock on the Farm and replace all broken glass in windows, roof lights and deadlights; and

(h) once in every five years and in any event in the last year of the tenancy clean, colour, whiten, paper and paint with materials of suitable quality the inside of the farmhouse and all other buildings on the Farm which have been previously so treated, and also prepare and paint metalwork of roof trusses and tubular trevisses and in each second year limewash the inside of all farm buildings which have been previously so treated.

SECOND In the event of the Tenant during the currency of the tenancy not implementing its obligations under Clause *FIRST* hereof the Landlord shall be entitled to call upon the Tenant to implement such obligations within such time as may be reasonably specified by the Landlord and failing compliance by the Tenant shall be entitled if he so elects, to perform such repairs or renewal as may be required by himself or by contractors employed by him and in such event the Tenant shall be bound to pay the expense of such operations on demand as the same shall be ascertained by the accounts of the Landlord or the contractor employed or be sufficiently vouched by a certificate under the hand of the Landlord; Declaring that it is entirely in the discretion of the Landlord whether or not he shall exercise the foregoing provision and if he elects not to exercise this option this shall in no way prejudice any claims competent to him against the Tenant for dilapidations under Statute or Common Law or any rights and pleas open to him following upon a Notice to Quit.

THIRD   The Tenant shall be obliged at its own expense to (a) carry out any works may which may be required at any time during the currency of the Tenancy to comply with the requirements of any local or public authority in relation to the provision, repair, replacement, improvement or removal of any item of fixed equipment including the water source and supply to the Farm and (b) pay all annual rates or other charges arising in respect thereof.

FOURTH   The rent payable under the said Lease shall be reduced from [   ] THOUSAND POUNDS (£[   ]) to [   ] THOUSAND POUNDS (£[   ]) with effect from the term day immediately preceding this Agreement or in the event of this Agreement being signed prior to the date of entry under the said Lease, with effect from the date of entry.

FIFTH   Except in so far as modified by these presents, the said Lease shall continue in full force and effect.

SIXTH   The parties hereto consent to the registration hereof and of any certificates issued hereunder for preservation and execution: IN WITNESS WHEREOF

CONTRACT OF LIMITED PARTNERSHIP[1]

between

(who and his assignees, executors or successors in his interest in the Partnership are hereinafter referred to as 'the Limited Partner')

and

(hereinafter together referred to as 'the General Partners' which expression shall include where the context so requires the survivor of the said

)

WHEREAS the Parties hereto have agreed and do hereby agree to enter into these presents to farm the farm and lands of
in the County of   extending in total to   acres/hectares or thereby shown delineated in red and coloured pink on the plan annexed and signed as relative hereto (hereinafter referred to as 'the Farm') on the terms and conditions hereinafter set forth:

NOW THEREFORE THE PARTIES HEREBY AGREE as follows:-

FIRST   The Limited Partner and the General Partners shall carry on the business of farming at the Farm under the firm or partnership name of   .

SECOND[2]   The Partnership shall commence notwithstanding the date or dates hereof on the   day of   Nineteen hundred and ninety   and shall endure subject as aftermentioned until the   day of   Two Thousand and   and year to year thereafter unless and until terminated (a) on the last mentioned date or (b) on the Twenty-eighth day of November (Martinmas) in any year thereafter by any party giving to the others not less than twelve months' previous notice in writing to that effect.

THIRD   Notwithstanding the terms of Section 6(5)(e) of the Limited Partnerships Act 1907 ('the Act') the Limited Partner shall be entitled to dissolve the Partnership at any time during its

---

1 See 'Sham transactions' (p 33) and 'The limited partnership' (p 37).
2 It is sometimes better to provide that the partnership will terminate on [date], unless the parties agree in writing that it is to continue after that date from year to year until terminated by notice.

currency by notice forthwith in the event of any breach by the General Partners or either of them of any condition of this Contract not being remedied within a period of one month after the date on which the Limited Partner shall have given written notice requiring that the breach be remedied. The death or bankruptcy (meaning apparent insolvency or the signing of a Trust Deed for behoof of creditors or entering into any Agreement with them to pay them less than in full) of either one of the General Partners shall not dissolve the Partnership which shall continue between the Limited Partner and the survivor or solvent of the General Partners.

FOURTH The Partnership shall be registered under and in accordance with the Act.

FIFTH The General Partners shall be solely responsible for the management of the partnership business and shall effect compliance by the Partnership at all times with the obligations of the Partnership as the tenant of the Farm. The Limited Partner shall not take any part in the management of the Partnership business and shall not have power to bind the Partnership but may by himself or his agent consult with the General Partners thereon and for that purpose shall be entitled to require the General Partners to attend not more than two Partners' meetings in each calendar year.

SIXTH The Limited Partner shall contribute to the Partnership the sum of (£) POUNDS and the General Partners shall contribute all further capital required for the purposes of the said business in such proportions as may be agreed between them. The Limited Partner shall not during the continuance of the Partnership either directly or indirectly withdraw all or any part of his capital but shall be entitled on an annual basis to a share of profits of (£) POUNDS. The said share of profit without deducting any tax payable by the Partnership in respect thereof shall be payable to the Limited Partner in full once the accounts have been finalised by the Partnership Accountants at the end of each financial year.

SEVENTH The Limited Partner may assign his share in the Partnership to any person, firm or body and the assignee shall thereupon become the Limited Partner for the purposes of these presents. The General Partners shall forthwith upon intimation of any such assignation execute all documents necessary to have the assignee registered as Limited Partner in the firm in terms of the said Act. For the avoidance of doubt it is specifically provided that (One) in the event of the death of the Limited Partner his share, interest, rights and obligations under the Partnership shall vest in his executors and (Two) if the Limited Partner shall become apparently insolvent then his share, interest, rights and obligations under the Partnership shall vest in his trustee in bankruptcy, so that in either of these events the Partnership hereby constituted shall continue in existence notwithstanding such death or apparent insolvency, subject to the whole other terms hereto.

Neither of the General Partners shall be entitled to assign his or her share in the Partnership nor shall the General Partners be entitled to assume any other Partner into the firm.

EIGHTH Full records shall be kept of all correspondence and transactions of the Partnership which shall be conducted on partnership stationery and in accordance with the Business Names Act 1985 and regular separate books of account shall be kept and brought to a balance annually at   in each year. The annual accounts and balance sheet shall be circulated for approval to all the Partners and if not challenged by any Partner within one month of circulation shall be final and binding except in the case of manifest error. The Limited Partner by himself alone may at any time inspect the books of the Partnership business and examine the state and prospects thereof and consult with the General Partners in respect of any aspects thereof. A bank account in the name of the Limited Partnership shall be opened and maintained during the currency of the Partnership and shall be operated solely by the General Partners. An application to HM Customs and Excise for registration of the firm for Value Added Tax shall be made by the General Partners in the firm name.

NINTH The whole profits of the Partnership in each year after payment of the share of profits referred to in Clause SIXTH hereof to the Limited Partner as hereinbefore specified shall be paid to the General Partners in such proportions as they may from time to time agree. The total liability of the Limited Partner for the duration of these presents for the losses of the Partnership shall be limited to the capital of   Pounds to be contributed by him at the commencement of the Partnership. The losses of the firm shall not be debited to the account of the Limited Partner until they have exhausted the capital contributed by the General Partners.

TENTH In the event of the termination of the Partnership the Limited Partner shall be paid (One) his said capital contribution of   Pounds and (Two) any balance in the aforementioned share of profits due to him under deduction of any tax payable by the Partnership in respect thereof, all within fourteen days of the termination with interest thereon at the rate per annum of Four *per centum* above the unsecured lending rate of the Bank of Scotland for the time being in force calculated on a day to day basis from the date of the same becoming due to the date of receipt of payment or such lesser sum as may be realised from the Partnership assets after payment of the Partnership debts.

ELEVENTH In the event of any rights to production quotas, premia, government subsidies or the like which are allocated and bound to the Farm being issued in respect of the production of crops or livestock such rights shall be registered in the name of the Partnership where possible and where not possible in the name of the General Partners who shall hold the same as trustees for the firm. On dissolution of the Partnership all such rights, unless otherwise agreed with the Limited Partner, shall be transferred to such person as may be nominated by the Limited Partner at the current market value thereof, if any. All returns and other communications with the Scottish Office Agricultural and Fisheries Department and other official bodies shall disclose the firm as occupier of the Farm.

TWELFTH   All questions, difference or disputes which may arise between or among the parties or their respective representatives as to the true intent and meaning of these presents or the implement thereof in any way, failing agreement, shall be referred on the application of either party to the amicable decision of an Arbiter to be mutually agreed, failing which to be appointed by the Sheriff of   at   on the application of any of the parties or the representatives of any of them and all awards of such Arbiter shall, notwithstanding the provisions of Section 3(1) of the Administration of Justice (Scotland) Act 1972, which are hereby expressed excluded, be final and binding upon all concerned.

LASTLY   All parties hereto consent to registration for preservation and execution:

IN WITNESS WHEREOF

Index